Fullerenes
Principles and Applications

# RSC Nanoscience & Nanotechnology

Series Editors

Paul O' Brien, *University of Manchester, UK*
Sir Harry Kroto FRS, *University of Sussex, UK*
Harold Craighead, *Cornell University, USA*

This series will cover the wide-ranging areas of Nanoscience and Nanotechnology. In particular, the series will provide a comprehensive source of information on research associated with nanostructured materials and miniaturised lab on chip technologies.

Topics covered will include the characterisation, performance and properties of materials and technologies associated with miniaturised lab on chip systems. The books will also focus on potential applications and future developments of the materials and devices discussed.

Ideal as an accessible reference and guide to investigations at the interface of chemistry with subjects such as materials science, engineering, biology, physics and electronics for professionals and researchers in academia and industry.

Titles in the Series:

**Nanotubes and Nanowires**
C.N.R. Rao FRS and A. Govindaraj, *Jawaharlal Nehru Centre for Advanced Scientific Research, Bangalore, India*

**Fullerenes: Principles and Applications**
F. Langa, *Faculty of Environmental Science, University of Castilla La Mancha, Toledo, Spain*, Jean-Francois Nierengarten, *Laboratory of Coordination Chemistry, CNRS, Toulouse, France*

Visit our website at www.rsc.org/nanoscience

For further information please contact:
Sales and Customer Care, Royal Society of Chemistry, Thomas Graham House, Science Park, Milton Road, Cambridge, CB4 0WF, UK
Telephone: +44 (0)1223 432360, Fax: +44 (0)1223 426017, Email: sales@rsc.org

# Fullerenes
# Principles and Applications

**Edited by**

**Fernando Langa**
*Faculty of Environmental Science, University of Castilla La Mancha, Toledo, Spain*

**Jean-Francois Nierengarten**
*Laboratory of Coordination Chemistry, CNRS, Toulouse, France*

RSCPublishing

ISBN-13: 978-0-85404-551-8

A catalogue record for this book is available from the British Library

© The Royal Society of Chemistry 2007

*All rights reserved*

*Apart from fair dealing for the purposes of research for non-commercial purposes or for private study, criticism or review, as permitted under the Copyright, Designs and Patents Act 1988 and the Copyright and Related Rights Regulations 2003, this publication may not be reproduced, stored or transmitted, in any form or by any means, without the prior permission in writing of The Royal Society of Chemistry, or in the case of reproduction in accordance with the terms of licences issued by the Copyright Licensing Agency in the UK, or in accordance with the terms of the licences issued by the appropriate Reproduction Rights Organization outside the UK. Enquiries concerning reproduction outside the terms stated here should be sent to The Royal Society of Chemistry at the address printed on this page.*

Published by The Royal Society of Chemistry,
Thomas Graham House, Science Park, Milton Road,
Cambridge CB4 0WF, UK

Registered Charity Number 207890

For further information see our web site at www.rsc.org

Typeset by Macmillan India Ltd, Bangalore, India
Printed by Henry Lings Ltd, Dorchester, Dorset, UK

# *Preface*

Observations of the interstellar space by radioastronomers suggested the existence of chains of carbon atoms in some particular stars, the red giant ones. With the aim of mimicking the conditions existing in these stars to produce such carbon chains in a laboratory, Sir Harold W. Kroto, Robert F. Curl and the late Richard E. Smalley were certainly not suspected that they were on the way of a discovery, which would be awarded by the Nobel price in Chemistry a few years later. The analysis of the carbon clusters produced by directing an intense pulse of laser light at a carbon surface revealed the existence of caged molecules exclusively constituted by carbon atoms: the *fullerenes*. After graphite and diamond, it was a new form of carbon that was thus discovered. Whereas the two former ones are infinite-atom arrays, fullerenes are well defined molecules.

Following the first evidence for fullerenes in 1985, other important breakthroughs in fullerene science occurred. It was Roger Taylor's expertise in chromatography that enabled him to obtain the first pure samples of $C_{60}$ and $C_{70}$, and hence to make a decisive contribution to the Sussex group's "one-line proof" of the structure of $C_{60}$ in the now-classic 1990 paper on "*Isolation, Separation, and Characterisation of the Fullerenes $C_{60}$ and $C_{70}$: The Third Form of Carbon*" (R. Taylor, J.P. Hare, A.K. Abdul-Sada, H.W. Kroto, *J. Chem. Soc., Chem. Commun.*, 1990, 1423). With the availability of initially tiny amounts of the two fullerenes, it was possible to begin the exploration of the chemistry, spectroscopy and properties of the fullerenes that has been such a major theme in chemistry in the last 15 years. The development of the macroscopic scale fullerenes synthesis by Wolfgang Krätschmer and Donald Huffman was certainly the decisive step allowing intensive research in this new branch of chemistry, with consequences in such diverse areas as superconductivity, biology and materials science. Finally, Sumio Iijima made another major development in 1991 with the discovery of carbon nanotubes, the missing link between graphite and *fullerenes*, as well as a perfectly suited material to be used in the emerging field of nanotechnology.

The purpose of this book is to summarize the basic principle of *fullerene* chemistry, but also to highlight some of the most remarkable advances that occurred in the field during the last recent years. Indeed, the rapid advances in fullerene synthetic chemistry have moved towards the creation of functional systems with increased attention to potential applications. *Fullerene* research is now a truly interdisciplinary branch of science. *Fullerene*-based derivatives have shown a wide range of physical and chemical properties that make them attractive for the preparation of supramolecular assemblies, nanostructures and new advanced materials for optoelectronic devices. On the other hand, recent studies have shown that *fullerenes* exhibit interesting biological activities. All these aspects of *fullerene* science are summarized in the different chapters collected in this book. An additional chapter is devoted to a related field, namely carbon nanotube chemistry. The latter is still at an early stage; however many of its developments are directly related to the knowledge accumulated in *fullerene* chemistry over the last 20 years.

The different chapters of the present book have been prepared by prominent colleagues and we would like to warmly thank all of them for their contributions and their enthusiasm to participate to this adventure. We have a special thought for Roger Taylor, who prepared his chapter while being seriously affected by his disease. Roger is deceased on February 1, 2006. He was one of the pioneers of *fullerene* chemistry and a research collaborator and good friend of many working in our field.

Our gratitude is extended to the Royal Society of Chemistry for its support in this enterprise and to all the RSC team for the efficient handling of all the chapters. We further thank Juan Luis Delgado de la Cruz for the design of the cover picture.

# Contents

| Chapter 1 | Production, Isolation and Purification of Fullerenes<br>*Roger Taylor, Glenn A. Burley* | 1 |
|---|---|---|
| | 1.1 Introduction | 1 |
| | 1.2 Production | 3 |
| |     1.2.1 The Hufmann–Krätschmer Method | 3 |
| |     1.2.2 The Combustion Process | 4 |
| |     1.2.3 Condensation of Polycyclic Aromatic Hydrocarbons through Pyrolytic Dehydrogenation or Dehydrohalogenation | 4 |
| | 1.3 Isolation and Purification of Fullerenes | 4 |
| | 1.4 Formation and Stability | 5 |
| | 1.5 Fullerenes with Incarcerated Atoms: *incar*Fullerenes (Endohedrals) | 6 |
| |     1.5.1 Nitrogen | 6 |
| |     1.5.2 Noble Gases | 6 |
| |     1.5.3 Hydrogen | 7 |
| |     1.5.4 Other Atoms | 7 |
| |     1.5.5 The *incar*trimetalnitridofullerenes, $i\text{-}NR_3C_n$ | 9 |
| | 1.6 Concluding Remarks | 9 |
| | References | 9 |
| Chapter 2 | Basic Principles of the Chemical Reactivity of Fullerenes<br>*Fernando Langa, Pilar de la Cruz* | 15 |
| | 2.1 [60]Fullerene Reactivity | 15 |
| |     2.1.1 Introduction | 15 |
| |     2.1.2 [4 + 2] Cycloadditions: Diels–Alder Reaction | 21 |
| |     2.1.3 [3 + 2] Cycloaddition Reactions | 26 |
| | 2.2 Multiaddition Reactions | 35 |
| | 2.3 Reactivity of Higher Fullerenes ($C_{70}$, $C_{76}$, $C_{84}$) | 40 |
| | References | 43 |

| Chapter 3 | Three Electrodes and a Cage: An Account of Electrochemical Research on $C_{60}$, $C_{70}$ and their Derivatives | 51 |

*Maurizio Carano, Massimo Marcaccio, Francesco Paolucci*

| | | |
|---|---|---|
| 3.1 | Introduction | 51 |
| 3.2 | The Electrochemical Properties of Fullerenes | 53 |
| | 3.2.1 Increasing the Electronegativity of Fullerenes | 53 |
| | 3.2.2 Fulleropyrrolidines: Mono and Bis-Adducts | 54 |
| | 3.2.3 Polyphenylated Derivatives | 55 |
| | 3.2.4 $C_{60}Ph_5Cl$, a Very Special Polyphenylated Fullerene Derivative | 57 |
| 3.3 | Electrochemically Induced Reactivity in Fullerene Derivatives | 58 |
| | 3.3.1 The Electrochemically Induced Retro-cyclopropanation Reaction | 60 |
| | 3.3.2 The Role of Digital Simulation of the CV Experiments in Mechanistic Studies | 61 |
| 3.4 | Fullerenes as Active Components of Molecular Devices | 63 |
| | 3.4.1 Photoactive Dyads | 63 |
| | 3.4.2 Fullerene-Based Photoactive Liquid Crystals | 65 |
| | 3.4.3 Fullerene-$[Ru(bpy)_3]^{2+}$ Systems | 66 |
| | 3.4.4 Fullerenes Containing Crown Ethers | 69 |
| | 3.4.5 Trannulenes | 69 |
| | 3.4.6 Electropolymerizable $C_{60}$ Derivatives | 71 |
| | 3.4.7 $C_{60}$ as a Mediator in Bio Electrochemical Sensors | 71 |
| 3.5 | Back to the Basics: The Oxidative Electrochemistry of $C_{60}$ | 72 |
| 3.6 | Final Remarks | 74 |
| References | | 75 |

| Chapter 4 | Light-Induced Processes in Fullerene Multicomponent Systems | 79 |

*Nicola Armaroli, Gianluca Accorsi*

| | | |
|---|---|---|
| 4.1 | Introduction | 79 |
| 4.2 | Photophysical Properties of $C_{60}$ and its Derivatives | 80 |
| 4.3 | Dyads with Oligophenylenevinylenes | 81 |
| 4.4 | Fullerodendrimers | 88 |
| | 4.4.1 Fullerene Inside | 89 |
| | 4.4.2 Fullerene Outside | 99 |
| 4.5 | $C_{60}$: Metal Complex Arrays | 105 |
| 4.6 | $C_{60}$-Porphyrin Assemblies | 109 |

Contents   ix

|  |  | 4.7 Conclusions | 113 |
|---|---|---|---|
|  |  | Acknowledgements | 115 |
|  |  | References | 115 |

**Chapter 5** **Encapsulation of [60]Fullerene into Dendritic Materials to Facilitate their Nanoscopic Organization**    127
*Jean-François Nierengarten, Nathalie Solladie, Robert Deschenaux*

|  | 5.1 | Introduction | 127 |
|---|---|---|---|
|  | 5.2 | Langmuir-Blodgett Films with Fullerene-Containing Dendrimers | 128 |
|  | 5.3 | Fullerene-Containing Thermotropic Liquid Crystals | 140 |
|  |  | 5.3.1 Non-Covalent Fullerene-Containing Liquid Crystals | 140 |
|  |  | 5.3.2 Covalent Fullerene-Containing Thermotropic Liquid Crystals | 142 |
|  | 5.4 | Conclusions | 149 |
|  |  | References | 149 |

**Chapter 6** **Hydrogen Bonding Donor–Acceptor Carbon Nanostructures**    152
*M. Ángeles Herranz, Francesco Giacalone, Luis Sánchez, Nazario Martín*

|  | 6.1 | Introduction | 152 |
|---|---|---|---|
|  | 6.2 | Hydrogen Bonded $C_{60}$•Donor ($C_{60}$•D) Ensembles | 154 |
|  |  | 6.2.1 H-bonding interfaced metallomacrocycles•$C_{60}$ dyads | 154 |
|  |  | 6.2.2 H-Bonding Tethered π-Conjugated Oligomer•$C_{60}$ Dyads | 158 |
|  | 6.3 | Other Electron Donor Moieties H-Bonding Interfaced with [60]fullerene | 163 |
|  | 6.4 | H-Bonded Supramolecular $C_{60}$-Based Polymers | 166 |
|  | 6.5 | Non-Covalent Functionalization of Carbon Nanotubes (CNTs) | 172 |
|  |  | 6.5.1 Polymer Wrapping | 174 |
|  |  | 6.5.2 Electrostatic Interactions | 178 |
|  |  | 6.5.3 van der Waals and Complementary Electrostatic Interactions | 179 |
|  | 6.6 | Conclusions and Outlook | 181 |
|  |  | Acknowledgments | 182 |
|  |  | References | 182 |

**Chapter 7** **Fullerenes for Material Science**    191
*Stéphane Campidelli, Aurelio Mateo-Alonso, Maurizio Prato*

|  | 7.1 | Introduction | 191 |
|---|---|---|---|
|  | 7.2 | Donor–Acceptor Systems | 191 |

|  |  | 7.2.1 | Covalently Linked Donor–Acceptor Systems | 192 |
|---|---|---|---|---|
|  |  | 7.2.2 | Donor–Acceptor Systems Assembled by Supramolecular Interactions | 193 |
|  |  | 7.2.3 | Polyads | 198 |
|  | 7.3 | Fullerenes for Nonlinear Optical Applications | | 200 |
|  |  | 7.3.1 | Functionalized Fullerenes for Nonlinear Optics | 201 |
|  |  | 7.3.2 | Donor–Acceptor Derivatives | 202 |
|  | 7.4 | Amphiphilic Fullerenes | | 205 |
|  |  | 7.4.1 | Langmuir Films | 205 |
|  |  | 7.4.2 | Fullerene in Smectite Clays | 208 |
|  |  | 7.4.3 | Self Organization | 208 |
|  | 7.5 | Conclusion | | 211 |
|  | References | | | 211 |

## Chapter 8 Plastic Solar Cells Using Fullerene Derivatives in the Photoactive Layer 221
*Piétrick Hudhomme, Jack Cousseau*

|  | 8.1 | Introduction | | 221 |
|---|---|---|---|---|
|  | 8.2 | From Inorganic to Fullerene-Based Organic Solar Cells | | 222 |
|  | 8.3 | Principle of Fullerene-Based Organic Solar Cells | | 224 |
|  | 8.4 | Polymer – $C_{60}$ Derivatives Heterojunctions | | 226 |
|  |  | 8.4.1 | p/n Heterojunction Devices | 226 |
|  |  | 8.4.2 | "bulk-heterojunction" Devices | 230 |
|  |  | 8.4.3 | "Double-cable" Polymer-$C_{60}$ Derivatives | 242 |
|  | 8.5 | Molecular Scale $C_{60}$-Based Heterojunctions | | 244 |
|  |  | 8.5.1 | Low-Molecular-Weight Materials | 244 |
|  |  | 8.5.2 | Molecular $\pi$-Donor – $C_{60}$ Derivatives Solar Cells | 245 |
|  | 8.6 | Supramolecular Nanostructured $C_{60}$–Based Devices | | 249 |
|  |  | 8.6.1 | Self-Assembled Monolayers | 249 |
|  |  | 8.6.2 | Langmuir and Layer by Layer (LBL) Films | 252 |
|  |  | 8.6.3 | Hydrogen-Bonding Supramolecular Devices | 254 |
|  | 8.7 | Conclusion and Outlook | | 254 |
|  | References | | | 259 |

## Chapter 9 Fullerene Modified Electrodes and Solar Cells 266
*Hiroshi Imahori, Tomokazu Umeyama*

|  | 9.1 | Introduction | 266 |
|---|---|---|---|
|  | 9.2 | Langmuir–Blodgett Films | 267 |
|  | 9.3 | Self-Assembled Monolayers | 269 |

|  |  |  |
|---|---|---|
| | 9.3.1 Self-Assembled Monolayers of Fullerene-Containing Systems on Gold Electrodes | 269 |
| | 9.3.2 Self-Assembled Monolayers of Fullerene-Containing Systems on ITO Electrodes | 275 |
| 9.4 | Layer-by-Layer Deposition | 280 |
| 9.5 | Vacuum Deposition | 281 |
| 9.6 | Electrochemical Deposition | 283 |
| | 9.6.1 Fullerenes and Their Derivatives | 283 |
| | 9.6.2 Donor–Acceptor Linked Systems Involving Fullerene | 284 |
| | 9.6.3 Donor and Fullerene Composite Systems | 285 |
| | 9.6.4 Pre-Organized Multi-Donor Systems | 287 |
| 9.7 | Chemical Adsorption and Spin Coating Deposition | 292 |
| 9.8 | Summary | 295 |
| References | | 295 |

**Chapter 10 Biological Applications of Fullerenes** — 301
*Alberto Bianco, Tatiana Da Ros*

| | | |
|---|---|---|
| 10.1 | Methodologies for Fullerene Solubilisation | 301 |
| 10.2 | Health and Environment Impact of Fullerenes | 304 |
| 10.3 | Fullerenes for Drug Delivery | 307 |
| 10.4 | Neuroprotection and Antioxidant Activity | 310 |
| 10.5 | DNA Photocleavage and Photodynamic Approach Using Fullerenes | 312 |
| 10.6 | Antibacterial, Antiviral Activity and Enzymatic Inhibition | 317 |
| 10.7 | Immunological Properties of Fullerenes | 319 |
| 10.8 | Biological Applications of Radio Labelled Fullerenes | 322 |
| 10.9 | General Conclusions | 324 |
| References | | 324 |

**Chapter 11 Covalent and Non-Covalent Approaches Toward Multifunctional Carbon Nanotube Materials** — 329
*Vito Sgobba, G.M. Aminur Rahman, Christian Ehli, Dirk M. Guldi*

| | | |
|---|---|---|
| 11.1 | Introduction | 329 |
| 11.2 | End Tips and Defect Functionalization | 331 |
| | 11.2.1 Oxidation | 332 |

|  |  |  |  |
|---|---|---|---|
|  | 11.2.2 | Derivatization of the Carboxylic Functionalities | 333 |
|  | 11.2.3 | Solvent-Free Amination and Thiolation | 338 |
| 11.3 | Sidewall Chemistry | | 338 |
|  | 11.3.1 | Fluorination | 338 |
|  | 11.3.2 | Derivatization of Fluorinated Carbon Nanotubes | 339 |
|  | 11.3.3 | 1,3 Dipolar Cycloadditions – Ozonolysis, Cycloaddition of Azomethine Ylide, and Cycloaddition of Nitrile Ymine | 340 |
|  | 11.3.4 | Diels–Alder Cycloaddition | 344 |
|  | 11.3.5 | Osmylation | 344 |
|  | 11.3.6 | [2 + 1] Cycloadditions – Carbenes or Nitrenes | 344 |
|  | 11.3.7 | [2 + 2] Cycloaddition of Singlet $O_2$ and Sidewall Oxidation | 346 |
|  | 11.3.8 | Reductive Hydrogenation/Alkylation/Arylation of CNT Sidewall | 346 |
|  | 11.3.9 | Addition of Radicals | 348 |
|  | 11.3.10 | Replacements of Carbon Atoms – Chemical Doping | 349 |
|  | 11.3.11 | Mechanochemical Functionalization | 350 |
| 11.4 | Non-Covalent Functionalization of Carbon Nanotubes | | 350 |
|  | 11.4.1 | π-π Interactions with π-Electron Rich Polymeric Blocks – Polymer Wrapping | 351 |
|  | 11.4.2 | π-π Interactions with π-Electron Rich Molecular Building Blocks | 357 |
| 11.5 | Separation of Metallic and Semiconducting CNT | | 361 |
| References | | | 363 |

**Subject Index**     **380**

CHAPTER 1
# Production, Isolation and Purification of Fullerenes

ROGER TAYLOR[†] AND GLENN A. BURLEY

Faculty of Chemistry and Pharmacy, Ludwig-Maximilians University, Munich, Germany

## 1.1 Introduction

Eight fullerenes have been obtained in significant quantities. These are [60-$I_h$]-, [70-$D_{5h}$]-, [76-$D_2$]-, [78-$D_3$]-, [78-$C_{2v}$(I) ]-, [78-$C_{2v}$(II)]-, [84-$D_2$(IV)]-, [84-$D_{2d}$(II)]- fullerenes and are depicted in Figure 1.[1,2] Of the fullerene family, [60]fullerene and [70]fullerene are the major isomers obtained in 75 and 24%, respectively, *via* the arc-discharge method of Hufmann and Krätschmer. The remaining 1% constitutes a variety of higher order fullerenes ranging from $C_{74}$ to beyond $C_{100}$. The colours of the fullerene family vary according to their molecular weight and symmetry. Their colours in solution are magenta ($C_{60}$), port-wine red ($C_{70}$), brown ($C_{76}$ and $C_{78}$), and yellow-green ($C_{84}$). A major obstacle in higher order fullerene research is confronted when investigating these fullerenes [Figure 1(c)–(h)]. A gradual increase in the size of fullerenes is accompanied by an increase in the number of isomers of the same symmetry, therefore making definitive assignment of the fullerene structure difficult. Coupled with the difficulty of separation and decreasing solubility of higher order fullerenes with increasing size, makes further studies of these larger fullerenes unattractive.

The numbering system adopted for fullerene assignment in this review is the IUPAC system that has been in place for over a decade.[3] The Roman numerals used for subdividing fullerenes exhibiting the same symmetry are those given by Fowler and Manolopoulos.[4]

---

[†] Deceased, February 1, 2006.

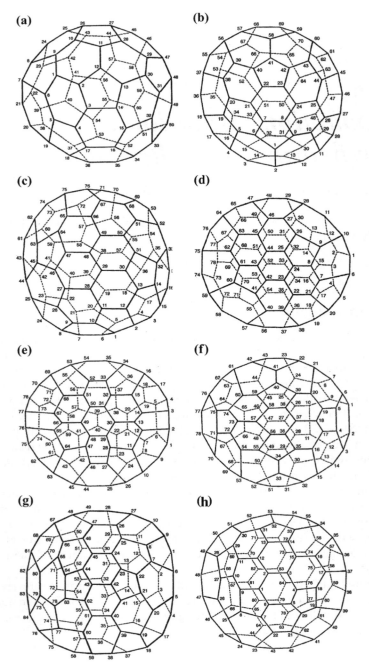

**Figure 1** Structures of (a) [60-$I_h$]fullerene, (b) [70-$D_{5h}$]fullerene, (c) [76-$D_2$]fullerene, (d) [78-$D_3$]fullerene, (e) [78-$C_{2v}(I)$]fullerene, (f) [78-$C_{2v}(II)$]fullerene, (g) [84-$D_2(IV)$]fullerene, (h) [84-$D_{2d}(II)$]fullerene

## 1.2 Production

Three methods have been used to make fullerenes. These are

(i) The Hufmann–Krätschmer procedure involving arc-discharge between graphite rods in an atmosphere of helium.[5]
(ii) Combustion of benzene in a deficiency of oxygen.[6]
(iii) Condensation of polycyclic aromatic hydrocarbons through pyrolytic dehydrogenation or dehydrohalogenation.[7]

### 1.2.1 The Hufmann–Krätschmer Method

Hitherto this has been the most important method for fullerene production, and its introduction marked the real beginning of fullerene science. It is the preferred method because the only by-product is graphite, which if necessary, can be reformed into rods and recycled. The Hufmann–Krätschmer method involves arc-discharge between high-purity carbons rods of *ca.* 6 mm diameter in an atmosphere of 100–200 torr helium.[1,5] Argon may also be used but is less effective.[8] The temperature required for fullerene formation is *ca.* 2000°C, and obviously a small gap between the rods is necessary to prevent a fall in temperature. The need for this gap was shown at an early stage[8] and was confirmed subsequently.[9,10] The yield of fullerenes in the soot produced is approximately 5% and is higher when taken from the reactor at greater distances from the arc source, which implies that the initially formed fullerenes are subsequently degraded by UV irradiation.

Numerous ingenious variations in reactor design were introduced at an early stage, including, for example, a carousel[11] and one with an autoloading device (Figure 2).[12] The latter method, designed by Bezmelnitsyn and Eletskii is especially notable, using carbon strips 7×3.5×400 mm cut from a reactor moderator block. A stack of 24 of these (anode) are housed in a 450 mm long by 280 mm diameter water-cooled chamber equipped with a hinged door fitted with a O-ring seal. The cathode consists of slowly rotating 70 mm-diameter carbon wheel which passes a scraper to remove accumulated slag. The strips are gravity fed and the lowest strip is slowly wheel-driven into the cathode. When consumed it drops away exposing the next strip, and the process continues

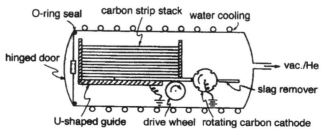

**Figure 2** *Schematic depiction of an autoloading version of an arc-discharge apparatus used for fullerene production*[12]

during 24 h to yield 100–200 g of fullerene-containing soot, accessed by opening the end door of the reactor.

### 1.2.2 The Combustion Process

Howard and co-workers[6] discovered that combustion of benzene in a deficiency of oxygen resulted in the formation of both [60]- and [70]fullerenes. This continuous method has been developed to the extent that a purpose-built factory has been erected in Japan, capable of producing 5000 ton of fullerenes per year, but currently running at about one-tenth of that capacity. One envisages that this investment must be driven by the expectation or knowledge that large-scale applications of fullerene lie ahead.

### 1.2.3 Condensation of Polycyclic Aromatic Hydrocarbons through Pyrolytic Dehydrogenation or Dehydrohalogenation

These methods produce fullerenes but not in sufficient quantities for practical applications. Rather, they provide a means of deducing the mechanisms of fullerene synthesis. For example, $C_{60}$ consists of six dehydronaphthalene moieties located at the octahedral sites, and pyrolysis of naphthalene does indeed produce $C_{60}$,[7] as does corannulene, (which has been detected as a precursor in the combustion process), 7,10-*bis*(2,2′-dibromovinyl)fluoranthene, and 11,12-benzofluoranthene. The dehydrogenation involved is a high-energy process and dehydrohalogenation of precursors is more successful,[13] a feature made use of in formation of $C_{60}$ from a chloroaromatic precursor (Scheme 1).[14] No other fullerene has yet been made in this way.

## 1.3 Isolation and Purification of Fullerenes

The fullerene-containing soot is extracted using a Soxhlet apparatus with either chloroform (slowest), toluene, or 1,2-dichlorobenzene (fastest, but removal of traces of solvent requires high vacuum). If carbon disulfide is used at any stage during the extraction/concentration, then it must be vigorously removed under vacuum, otherwise the fullerene will be contaminated with sulfur.[15]

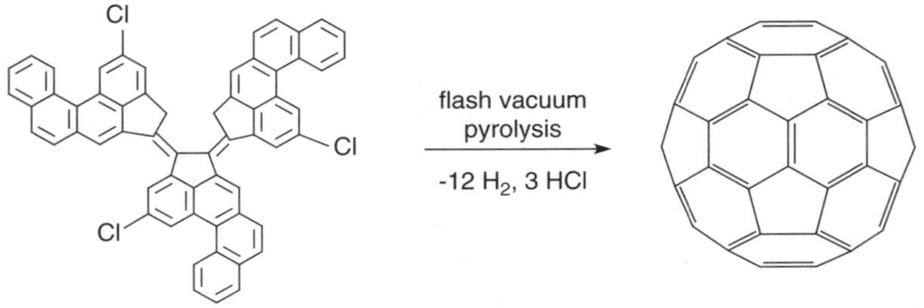

**Scheme 1**  *Formation of [60]fullerene through dehydrogenation/dehydrochlorination*

Purification of fullerenes was first carried by column chromatography using neutral alumina as the stationary phase with elution by hexane.[1] This procedure was improved through the use of mixtures of alumina and other stationary phases and, *e.g.* toluene/hexane mixtures as eluents. However, [60]fullerene produced in this way is degraded slightly, and must be repeatedly washed with acetone to remove impurities. A method which renders all others obsolete involves passing a toluene solution of the fullerene mixture through a column packed with Elorit carbon (a grade used for making car battery cases). Pure $C_{60}$ is rapidly eluted and 10 g h$^{-1}$ can easily be purified in this way.[16] $C_{70}$ can be removed subsequently by elution with 1,2-dichlorobenzene, but the higher fullerenes are retained firmly by the carbon. Chromatographic separation of fullerenes is too costly for large-scale production, and is the reason that the price has not fallen in line with expectation. It is likely that other methods of purification will be needed, possibly through the use of multiple recrystallisations.

The purity of $C_{60}$ that is described in many papers is wholly unrealistic, values of 99, 99.5, and even 99.9% being claimed without any justification.[17] To begin with, there is no known way of determining the purity of a fullerene, since combustion is never complete. Also all $C_{60}$ contains a significant amount of $C_{120}O$ to the extent that it can be recovered through the use of HPLC.[18-19] Moreover, the presence of hydrocarbons can readily be detected by running an $^1$H NMR spectrum of a solution in $CS_2$. It seems improbable that $C_{60}$ of purity greater than 98% has been used in any chemical synthesis. Very pure $C_{60}$ can of course be obtained through vacuum sublimation, but here there are substantial losses due to the reaction of the fullerene with the impurities at the high temperatures involved.

## 1.4 Formation and Stability[20]

Fullerenes, especially the higher order members, degrade to an insoluble material on storage in air. The $C_{2v}(II)$ isomer of [78]fullerene is completely degraded after storage for 5 months. Strong heating of KBr discs of the products yield $CO_2$ showing that they are oxygen containing. The instability of [$C_{2v}(II)$]fullerene may account for the variable relative yield obtained by various research groups.[2] Given the instability of fullerenes, the question arises as to why they are formed. Since fullerenes are thermodynamically less stable than graphite[21] and oxidatively degrade on standing in air, why are they formed in preference to graphite in the first place? The assumption that elimination of dangling bonds is all-important maybe misleading, since graphite is able to accommodate huge numbers of dangling bonds yet remain stable because of the resonance stability arising from the sp$^2$-hybridised carbon atoms in hexagonal arrays. It is the proportion of dangling bonds in the carbon fragment that is critical. The ratio of the number of dangling bonds to the number of carbon atoms $n$ in a single-graphene sheet can be shown to be approximately $1/n$. For 1 g of graphite ($5 \times 10^{22}$ atoms), the number of dangling bonds can be calculated to lie between $5.4 \times 10^{11}$ and an upper limit of $5 \times 10^{22}$ this latter being the hypothetical case of all-separated benzenoid rings. Notable therefore, freshly prepared pyrolytic carbon has spin density of $10^{18}$ g$^{-1}$, even higher values being reported for pitch derived from pyrolysis of either anthracene or naphthalene, and these values are within the theoretical range.[22,23]

The arc-discharge process of fullerene formation must in the first instance, produce small carbon fragments possessing dangling bonds. The preferred formation of fullerenes arises from the much higher Arrhenius $A$-factors (higher collision frequency) as a result of intramolecular processes. The intermolecular reaction will be favoured by high carbon vapour concentration, which may account for the variable yields of fullerenes with changes of the inert gas pressure in arc-discharge reactors.[24] Ring closure in organic chemistry occurs faster the smaller the chains due to the more favourable $A$-factors. Similar factors may account for the roughly decreasing yields of fullerenes with increasing size, even though the thermodynamic stability increases in this direction due to the reduction in strain.

## 1.5 Fullerenes with Incarcerated Atoms: *incar*Fullerenes (Endohedrals)[25]

### 1.5.1 Nitrogen

*i*-NC$_{60}$ had been prepared by heating [60]fullerene to *ca.* 450°C in a glow discharge reactor containing nitrogen. In the recovered fullerene fractions, the ratio of nitrogen-containing to empty molecules is approximately $10^{-5}$–$10^{-6}$.[26] The nitrogen does not bond to the inner surface of the cage because the orbital coefficients on the inside of the cage are small, and also because such bonding would result in increased strain: any carbon involved in bonding would have to move towards the cage centre, increasing the strain on the three other carbon atoms to which it is attached. The absence either of bonding or of charge transfer is shown by the three-fold degenerate EPR spectrum which is however altered slightly by the presence of addends on the cage. This arises because the distortion of the cage causes the three *p*-orbitals of the nitrogen to be no longer degenerate. The nitrogen-containing and empty molecules show no differences in reactivity, which is also consistent with the lack of bonding between the nitrogen and the cage (which has been described as a 'chemical Faraday cage').[26]

### 1.5.2 Noble Gases

Helium, neon, argon, krypton, and xenon can be incarcerated in fullerenes through the application of high temperature (620°C) and pressure (*ca.* 40,000 psi). The incorporation fractions for [60]fullerene and [70]fullerene are approximately: He, 0.1%; Ne, 0.2%; Ar, 0.3%; Kr, 0.3%; Xe, 0.008%.[27] The noble gases are released from the fullerenes by heating to 1000°C.[28] Since, the energy required even for helium to pass through a hexagon is calculated to be *ca.* 200

kcal mol$^{-1}$, noble gas insertion requires rupturing and reformation of the fullerene cage. Evidently noble gas insertion is not an entirely successful procedure since approximately 50% of the fullerene is lost during the incorporation procedure.

The presence of $^3$He can be monitored by $^3$He NMR, and the spectrum is a probe for the magnetic-shielding environment inside the fullerene cavity, in turn reflecting ring currents and hence the aromaticity of the fullerene. Thus, the more aromatic a fullerene, the more upfield should be the signal, hence [70]fullerene (−28.8 ppm) is indicated to be more aromatic than [60]fullerene (−6.3 ppm).[29]

### 1.5.3 Hydrogen

Hydrogen has been incarcerated into $C_{60}$ by purely chemical means, involving the creation of an eight-membered orifice-containing oxygen, nitrogen, sulfur, and carbon atoms. This was large enough to permit ingress of a hydrogen molecule, which was followed by thermal loss of the addends to give pure $i$-H$_2$C$_{60}$.[30] This was successfully converted into the $C_{120}$ dimer, by means of high-speed mechanical milling.

### 1.5.4 Other Atoms

A wide variety of metallic atoms have been incarcerated in a range of fullerenes, with up to four atoms accommodated in some cages. Notably, the cages that are present in low yields when empty are those which produce the most stable derivatives, $C_{74}$ and $C_{82}$ in particular. The very low yield of formation of these compounds (many have been detected sprectroscopically only), and the sensitivity of many of them to air means that their isolation, where achieved, is very time-consuming. Nevertheless, preparation of substantial quantities of $i$-LaC$_{82}$ (27.4 mg) and $i$-GdC$_{82}$ (12.0 mg) have now been described.[31] The interest in the *incar*fullerenes has been focussed on the location of the atoms within the cages, their molecular motions, and electronic consequences of their presence.

The incarcerated metal transfers electrons to the fullerene cage thereby altering the properties of the latter. *Incar*fullerenes can be regarded as a 'superatom' in having a positively and a negatively charged cage.[32] For example, in the case of $i$-LaC$_{82}$, three electrons are transferred, and the electronic structure can thus be represented as $i$-La$^{3+}$C$_{82}$$^{3-}$. The *incar*fullerenes have longer HPLC retention times than their empty-cage analogues (due probably to stronger coordination of the more electron-rich cage with the stationary phase), and the retention time increases with size of the incarcerated atom.[31] If an incarcerated element can exhibit two different oxidation states (*e.g.* Sm, Eu, Tm and Yb) then the *incar*fullerenes in which the element is in the 2+ oxidation state elute with shorter retention times than those in which the element is in the 3+ oxidation state, which follows from the greater charge transfer to the cage in the latter compounds, making them more polar.

Due to the cages being more electron-rich than those of empty fullerenes, they are less soluble in non-polar solvents, but conversely can be selectively separated by polar solvents such as aniline, pyridine, and dimethylformamide.[33,34] The band gaps are also of the order of 0.2 eV compared to 1.6 eV for [60]fullerene, and the substantially reduced stabilities rules out characterisation of structures by EI mass spectrosopy since fragmentation occurs, and limits studies to laboratories that are equipped with soft-ionisation techniques. The range of fullerenes that have incarcerated atoms is shown in Table 1.

Isomers of *incar*fullerenes have been isolated by HPLC. Examples are the four isomers of $i\text{-}CaC_{82}$[72] and three isomers of $i\text{-}TmC_{82}$ (each is very air stable, but they have quite different electronic properties).[70] Three isomers of $i\text{-}Sc_2C_{84}$ have been separated similarly, and $^{13}C$ NMR shows the main one to be the

**Table 1** *Fullerenes having incarcerated elements*

| Fullerene | Metal | References |
|---|---|---|
| | | *Fullerenes incarcerating one atom* |
| $C_{28}$ | Hf, Ti, U, Zr | 35 |
| $C_{36}$ | U | 35 |
| $C_{44}$ | K, La, U | 35–38 |
| $C_{48}$ | Cs | 38 |
| $C_{50}$ | U | 35 |
| $C_{60}$ | Li, K, Ca, Co, Y, Cs, Ba, Rb, La, Ce, Pr, Nd, Sm, Eu, Gd, Tb, Dy, Ho, Er, Lu, U | 34–48 |
| $C_{70}$ | Li, Ca, Y, Ba, La, Ce, Gd, Lu, U | 35,37–48 |
| $C_{72}$ | U | 35 |
| $C_{74}$ | Sc, La, Gd, Lu | 38,47,49 |
| $C_{76}$ | La | 35,49 |
| $C_{80}$ | Ca, Sr, Ba | 50 |
| $C_{82}$ | Ca, Sc, Sr, Ba, Y, La, Ce, Pr, Nd, Sm, Eu, Gd, Er, Tm, Lu | 35,47–49,51–75 |
| $C_{84}$ | Ca, Sc, Sr, Ba, La | 49,72,74 |
| | | *Fullerenes incarcerating two atoms* |
| $C_{28}$ | $U_2$ | 35 |
| $C_{56}$ | $U_2$ | 35 |
| $C_{60}$ | $Y_2$, $La_2$, $U_2$ | 35,36,41 |
| $C_{74}$ | $Sc_2$ | 49 |
| $C_{76}$ | $La_2$ | 49 |
| $C_{80}$ | $La_2$, $Ce_2$, $Pr_2$ | 41–44,71 |
| $C_{82}$ | $Sc_2$, $Y_2$, $La_2$ | 41,49,55–57,59,76 |
| $C_{84}$ | $Sc_2$, $La_2$ | 47–49,53,57,59 |
| | | *Fullerenes incarcerating three atoms* |
| $C_{82}$ | $Sc_3$ | 49,58,59 |
| $C_{84}$ | $Sc_3$ | 46,53,55,57,59 |
| | | *Fullerenes incarcerating four atoms* |
| $C_{82}$ | $Sc_4$ | 50 |

**Table 2** *Fullerenes containing incarcerated derivatives of ammonia $NR_3$*

| Fullerene $C_n$ | $NR_3$, $NR_2R'$, or $NRR'R''$ | References |
| --- | --- | --- |
| $C_{80}$ | Sc$_3$N, ErSc$_2$N, Er$_2$ScN, Lu$_3$N, Lu(Gd/Ho)$_2$, Lu$_2$(Gd/Ho) Y$_3$N, Ho$_3$N, Tb$_3$N, Dy$_3$N | 81,83–88,90 |
| $C_{78}$ | Sc$_3$N | 89 |
| $C_{76}$ | Dy$_3$N | 90 |
| $C_{68}$ | Sc$_3$N, Sc$_2$(Tm/Er/Gd/Ho/La)N, Sc(Tm/Er/Gd/Ho/La)$_2$N | 84 |
| $C_{82-98}$ | Dy$_3$N | 90 |

[$D_{2d}$(II)] isomer, the symmetry of the spectrum showing that the positions of the scandium atoms are equivalent within the NMR timescale; the other isomers are $C_s$(II) and $C_{2v}$(III).[78] The equivalence of the scandium atoms is confirmed by $^{45}$Sc NMR which shows only one line.[79] In both *i*-LaC$_{82}$ and *i*-YC$_{82}$ the metal atoms lie off-centre, so these molecules should be dipolar.[80]

Initial electrochemistry studies show, for example, that *i*-LaC$_{82}$ has five reduction waves and two oxidation waves while *i*-YC$_{82}$ has four reduction and two oxidation waves.[62–63] Thus these compounds appear to be more readily oxidised than open fullerenes, and this is consistent with the higher electron density on the cage, thus making removal of electron easier.

### 1.5.5 The *incar*trimetalnitridofullerenes, *i*-NR$_3$C$_n$

Interest in the other *incar*fullerenes has reached a plateau in recent years due to the discovery by Dorn *et al.*[81] of the trimetallonitridofullerenes which are obtained when the arc-discharge process for *incar*fullerene formation takes place in the presence of ammonia in the buffer gas. Not only are higher yields obtained, but the products being polar are easy to separate from the empty fullerenes. These compounds possess a range of unusual properties including enhanced non-linear optical response, luminescence, paramagnetism, and anomalous redox/photochemical behaviours.[82] The range of compounds obtained to date is compiled in Table 2.

## 1.6 Concluding Remarks

The production and isolation of various fullerenes requires considerable effort in order to obtain samples of adequate purity. As a consequence, the time and effort required for the production of fullerenes is the limiting factor for commercial applications.[91] Therefore further research and development into new methods for the large-scale production of fullerenes is at present an unmet challenge.

## References

1. R. Taylor, J.P. Hare, A.K. Abdul-Sada and H.W. Kroto, *J. Chem. Soc., Chem. Commun.*, 1990, 1423.

2. (*a*) R. Ettl, I. Chao, F. Diederich and R.L. Whetten, *Nature*, 1991, **353**, 333; (*b*) F. Diederich, R. Ettl, Y. Rubin, R.L. Whetten, R. Beck, M. Alvarez, S. Anz, D. Sensharma, F. Wudl, K.C. Khemani and A. Koch, *Science*, 1991, **252**, 548; (*c*) K. Kikuchi, N. Nakahara, T. Wakabayashi, M. Honda, H. Matsumiya, T. Moriwaki, S. Suzuki, H. Shiromaru, K. Saito, K. Yamayuchi, I. Ikemoto and Y. Achiba, *Chem. Phys. Lett.*, 1992, **188**, 177; (*d*) R. Taylor, G.J. Langley, T.J.S. Dennis, H.W. Kroto and D.R.M. Walton, *J. Chem. Soc., Chem. Commun.*, 1992, 1043.
3. (*a*) E.W. Godly and R. Taylor, *Pure Appl. Chem.*, 1997, **69**, 1411; (*b*) E.W. Godly and R. Taylor, *Fullerene Science and Technology*, 1997, **5**, 1667.
4. P.W. Fowler and D.E. Manolopoulos, *Atlas of Fullerenes*, Clarendon Press, Oxford, 1995.
5. W. Krätschmer, L.D. Lamb, K. Fostiropoulos and D.R. Huffmann, *Nature*, 1990, **347**, 354.
6. (*a*) J.B. Howard, J.T. McKinnon, Y. Makarovsky, Y. Lafleur and M.E. Johnson, *Nature*, 1991, **352**, 139; (*b*) J.B. Howard, J.T. McKinnon, M.E. Johnson, Y. Makarovsky and Y. Lafleur, *J. Phys. Chem.*, 1992, **96**, 6657.
7. (*a*) R. Taylor, G.J. Langley and H.W. Kroto, *Nature*, 1993, **366**, 728; (*b*) R. Taylor, G.J. Langley and H.W. Kroto, *Mol. Mat.*, 1994, **4**, 7; (*c*) R. Taylor and G.J. Langley, *Recent Adv. Chem. Phys. Fullerenes and Rel. Mater.*, 1994, **94-24**, 68; (*d*) C.J. Crowley, H.W. Kroto, R. Taylor, D.R.M. Walton, M.S. Bratcher, P.-C. Cheng and L.T. Scott, *Tetrahedron Lett.*, 1995, 9215; (*e*) L.T. Scott, *Angew. Chem. Int. Ed.*, 2004, **43**, 4995.
8. J.P. Hare, H.W. Kroto and R. Taylor, *Chem. Phys. Lett.*, 1991, **177**, 394.
9. (*a*) D.H. Parker, K. Chatterjee, P. Wurz, K.R. Lykke, M.J. Pellin, L.M. Stock and J.C. Hemminger, *Carbon*, 1992, **30**, 1167; (*b*) W.A. Scrivens and J.M. Tour, *J. Org. Chem.*, 1992, **57**, 6932.
10. R.E. Haufler, Y. Chai, L.P.F. Chibante, J. Conceicao, C. Jin, L.-S. Wang, S. Muruyama and R.E. Smalley, *Mat. Res. Soc. Symp. Proc.*, 1991, **206**, 627.
11. G.A. Olah, I. Bucsi, C. Lambert, R. Aniszfeld, N.J. Trivedi, D.K. Sensharma and G.K.S. Prakash, *J. Am. Chem. Soc.*, 191, **113**, 9385.
12. V.N. Bezmelnitsin, A.V. Eletskii, N.G. Schepetov, A.G. Avent and R. Taylor, *J. Chem. Soc., Perkin Trans. 2*, 1997, 683.
13. M.S. Bratcher, C.C. McComas, M.D. Best and L.T. Scott, *Chem. Eur. J.*, 1998, **4**, 234.
14. L.T. Scott, M.M. Boorum, B.J. McMahon, S. Hagen, J. Mack, J. Blank, H. Wegner and A. de Meijere, *Science*, 2002, **295**, 1500.
15. A.D. Darwish, H.W. Kroto, R. Taylor and D.R.M. Walton, *Fullerene Sci. Technol.*, 1993, **1**, 571.
16. A.D. Darwish, H.W. Kroto, R. Taylor and D.R.M. Walton, *J. Chem. Soc., Chem. Commun.*, 1994, 15.
17. I.V. Kuvychko, A.V. Streletskii, A.A. Popov, S.G. Kotsiris, T. Drewello, S.H. Strauss and O.V. Boltalina, *Chem. Eur. J.*, 2005, **11**, 5426.
18. R. Taylor, M.P. Barrow and T. Drewello, *Chem. Commun.*, 1998, 2497.

19. M. Wohlers, A. Bauer, Th. Rühle, F. Neitzel, H. Werner and R. Schlögle, *Fullerene Sci. Technol.*, 1996, **4**, 781.
20. (*a*) R. Taylor, *Fullerene Sci. Technol.*, 1999, **7**, 305; (*b*) R.E. Smalley, *Acc. Chem. Res.*, 1992, **25**, 38.
21. (*a*) H.-D. Beckhaus, S. Verevkin, C. Rüchardt, F. Diederich, C. Thilgan, H.U. ter Meer, H. Mohn and W. Müller, *Angew. Chem. Int. Edn. Engl.*, 1994, **33**, 996; (*b*) H. Dieogo, M.E. Minas de Piadade, T.J.S. Dennis, J.P. Hare, H.W. Kroto, R. Taylor and D.R.M. Walton, *J. Chem. Soc., Faraday Trans.*, 1993, **89**, 3541.
22. K.A. Holbrook, *Proc. Chem. Soc.*, 1964, 418.
23. I.C. Lewis and L.S. Singer, *Adv. Chem.*, 1988, **217**, 269.
24. L.D. Lamb, in *The Chemistry of Fullerenes*, R. Taylor (ed), World Scientific, Singapore, 1995, 20–30.
25. D.S. Bethune, C.H. Kiang, M.S. de Vries, G. Gorman, R. Savoy, J. Vazquez and R. Beyers, *Nature*, 1993, **363**, 605.
26. B. Pietzak, M. Waiblinger, T.A. Murphy, A. Weidinger, M. Hoehne, E. Dietel and A. Hirsch, *A. Chem. Phys. Lett.*, 1997, **279**, 259.
27. (*a*) M. Saunders, H.A. Jimenez-Vazquez, R.J. Cross, S. Mroczowski, M.L. Gross, D.E. Giblin and R.J. Poreda, *J. Am. Chem. Soc.*, 1994, **116**, 2193; (*b*) M. Saunders, H.A. Jiménez-Vázquez, R.J. Cross and R.J. Poreda, *Science*, 1993, **259**, 1428.
28. R. Shimshi, A. Khong, H.A. Jiménez-Vázquez, R.J. Cross and M. Saunders, *Tetrahedron*, 1996, **52**, 5143.
29. M. Saunders, H.A. Jiménez-Vázquez, R.J. Cross, S. Mroczkowski, D.I. Freedberg and F.A.L. Anet, *Nature*, 1994, **367**, 256.
30. K. Komatsu, M. Murata and Y. Murata, *Science*, 2005, **307**, 238.
31. H. Funasaka, K. Sugiyama, K. Yamamoto and T. Takahashi, *J. Phys. Chem.*, 1995, **99**, 1826.
32. S. Nagase and K. Kobayashi, *J. Chem. Soc., Chem. Commun.*, 1994, 1837.
33. L.S. Wang, J.M. Alford, Y. Chai and M. Diener, *Chem. Phys. Lett.*, 1993, **207**, 354.
34. J. Ding, N. Liu, L. Weng, N. Cue and S. Yang, *Chem. Phys. Lett.*, 1996, **261**, 92; J. Ding, L. Weng and S. Yang, *J. Phys. Chem.*, 1996, **100**, 11120.
35. T. Guo, M.D. Diener, Y. Chai, M.J. Alford, R.E. Haufler, S.M. McClure, T. Ohno, J.H. Weaver, G.E. Scuseria and R.E. Smalley, *Science*, 1992, **257**, 1661.
36. J.R. Heath, S.C. O'Brien, Q. Zhang, Y. Liu, R.F. Curl, F.K. Tittel and R.E. Smalley, *J. Am. Chem. Soc.*, 1985, **107**, 7779.
37. Y. Chai, T. Guo, C.M. Jin, R.E. Haufler, L.P.F. Chibante, J. Fure, L.H. Wang, J.M. Alford and R.E. Smalley, *J. Phys. Chem.*, 1991, **95**, 7564.
38. F.D. Weiss, S.C. O'Brien, J.L. Eklund, R.F. Curl and R.E. Smalley, *J. Am. Chem. Soc.*, 1988, **110**, 4464.
39. L.S. Wang, J.M. Alford, Y. Chai, M. Diener and R.E. Smalley, *Z. Phys. D.*, 1993, **26**, 297.
40. D.S. Bethune, C.H. Kiang, M.S. de Vries, G. Gorman, R. Savoy, J. Vazquez and R. Beyers, *Nature*, 1993, **363**, 605.

41. R.E. Smalley, *Acc. Chem. Res.*, 1992, **25**, 38.
42. R.F. Curl, *Carbon*, 1992, **30**, 1149.
43. R. Huang, H. Li, W. Lu and S. Yang, *Chem. Phys. Lett.*, 1994, **228**, 111.
44. E.G. Gillan, C. Yeretzian, K.S. Min, M.M. Alvarez, R.L. Whetten and R.B. Kaner, *J. Phys. Chem.*, 1992, **96**, 6869.
45. Y. Kubozono, T. Ohta, T. Hayashibara, H. Maeda, H. Ishida, S. Kashino, K. Oshima, H. Yamazaki, S. Ukita and T. Sogabe, *Chem. Lett.*, 1995, 457.
46. Y. Kubozono, H. Maeda, Y. Takabashi, K. Hiroaka, K. Nakai, S. Kashino, S. Emura, S. Ukita and T. Sogabe, *J. Am. Chem. Soc.*, 1996, **118**, 6998.
47. L. Moro, R.S. Ruoff, C.H. Becker, D.C. Lorents and R. Malhotra, *J. Phys. Chem.*, 1993, **97**, 6801.
48. A. Gromov, W. Krätschmer, N. Krawez, R. Tellgmann and E.E.B. Campbell, *Chem. Commun.*, 1997, 2003.
49. H. Shinohara, H. Yamaguchi, N. Hayashi, H. Sato, H. Okhohchi and Y. Sato, *Mat. Sci. Eng.*, 1993, **B19**, 25; *J. Phys. Chem.*, 1993, **97**, 4259.
50. T.J.S. Dennis and H. Shinohara, *Chem. Commun.*, 1998, 883.
51. M.E.J. Boonman, P.H.M. van Loosdrecht, D.S. Bethune, I. Holleman, G.D.M. Meder and P.J.M. VanBentum, *Physica B.*, 1995, **211**, 323.
52. K. Kikuchi, S. Suzuki, Y. Nakao, M. Nakahara, T. Wakabayshi, T. Shiromaru, K. Saito and I. Ikemoto, *Chem. Phys. Lett.*, 1993, **216**, 67.
53. R.D. Johnson, M.S. de Vries, J. Salem, D.S. Bethune and C.S. Yannoni, *Nature*, 1992, **355**, 239.
54. Y. Achiba, T. Wakabayashi, T. Moriwaki, S. Suzuki and H. Shiromaru, *Mater. Sci. Eng.*, 1993, **B19**, 14.
55. A. Bartl, L. Dunsch, J. Froehner and U. Kirbach, *Chem. Phys. Lett.*, 1994, **229**, 115.
56. J.H. Weaver, Y. Chai, G.H. Kroll, C. Jin, T.R. Ohno, R.E. Haufler, T. Guo, J.M. Alford, J. Conceicao, L.P.F. Chibante, A. Jain, G. Palmer and R.E. Smalley, *Chem. Phys. Lett.*, 1992, **190**, 460.
57. H. Shinohara, H. Sato, Y. Saito, M. Ohkohchi and Y. Ando, *J. Phys. Chem.*, 1992, **96**, 3571.
58. H. Shinohara, H. Sato, M. Ohkohchi, Y. Ando, T. Komada, T. Shida, T. Kato and Y. Saito, *Nature*, 1992, **357**, 52.
59. C.S. Yannoni, M. Hoinkis, M.S. de Vries, D.S. Bethune, J.R. Salem, M.S. Crowther and R.D. Johnson, *Science*, 1992, **256**, 1191.
60. T. Suzuki, Y. Maruyama, T. Kato, K. Kikuchi and Y. Achiba, *J. Am. Chem. Soc.*, 1992, **115**, 11006.
61. T. Kato, S. Suzuki, K. Kikuchi and Y. Achiba, *J. Phys. Chem.*, 1993, **97**, 13425.
62. K. Kikuchi, Y. Nakao, S. Suzuki, Y. Achiba, T. Suzuki and Y. Maruyama, *J. Am. Chem. Soc.*, 1994, **116**, 9367.
63. K. Yamamoto, H. Funasaka, T. Takahashi, T. Akasaka, T. Suzuki and Y. Maruyama, *J. Phys. Chem.*, 1994, **98**, 12831.
64. T. Suzuki, *Nature*, 1995, **374**, 9606.

65. H. Shinohara, M. Inakuma, M. Kishida, S. Yamazki and T. Sakurai, *J. Phys. Chem.*, 1995, **99**, 13769.
66. T. Akasaka, T. Kato, K. Kobayashi, S. Nagase, K. Yamamoto, H. Funasaka and T. Takahashi, *Nature*, 1995, **374**, 600.
67. T. Akasaka, T. Kato, K. Kobayashi, S. Nagase and K. Yamamoto, *J. Chem. Soc., Chem. Commun.*, 1995, 1343.
68. T. Akasaka, T. Kato, S. Nagase, K. Kobayashi, K. Yamamoto, H. Funasaka and T. Takahashi, *Tetrahedron*, 1996, **52**, 5015.
69. Y. Saito, S. Yokoyama, M. Inakuma and H. Shinohara, *Chem. Phys. Lett.*, 1996, **250**, 80.
70. U. Kirbach and L. Dunsch, *Angew. Chem. Intl. Edn. Engl.*, 1996, **35**, 2380.
71. J. Ding and S. Yang, *J. Am. Chem. Soc.*, 1996, **118**, 11254.
72. Z. Xu, T. Nakane and H. Shinohara, *J. Am. Chem. Soc.*, 1996, **118**, 11309.
73. H. Funasaka, K. Sugiyama, K. Yamamoto and T. Takahashi, *J. Phys. Chem.*, 1995, **99**, 1826.
74. T.J.S. Dennis and H. Shinohara, *Chem. Phys. Lett.*, 1997, **278**, 107.
75. J. Ding, N. Liu, L. Weng, N. Cue and S. Yang, *Chem. Phys. Lett.*, 1996, **261**, 92; J. Ding, L. Weng and S. Yang, *J. Phys. Chem.*, 1996, **100**, 11120.
76. H. Shinohara, H. Sata, Y. Saito, M. Ohkohchi and Y. Ando, *Rapid Commun. Mass Spectrom.*, 1992, **6**, 413.
77. H. Shinohara, Reported at the Electrochemical Society Meeting, Montreal, May 1997.
78. E. Yamamoto, M. Tansho, T. Tomiyama, H. Shinohara, H. Kawahara and Y. Kobayashi, *J. Am. Chem. Soc.*, 1996, **118**, 2293.
79. Y. Miyaki, S. Suzuki, Y. Kojima, K. Kikuchi, K. Kobayashi, S. Nagase, M. Kainsho, Y. Achiba, Y. Maniwa and K. Fisher, *J. Phys. Chem B.*, 1996, **100**, 9579.
80. S. Nagase, K. Kobashi and T. Akasaka, *Bull. Chem. Soc. Jpn.*, 1996, **69**, 2131.
81. S. Stevenson, G. Rice, T. Glass, K. Harich, F. Cromer, M.R. Jordan, J. Craft, E. Hadju, R. Bible, M.M. Olmstead, K. Maitra, A.J. Fischer, A.L. Balch and H.C. Dorn, *Nature*, 1999, **401**, 55.
82. H. Shinohara, *Rep. Prog. Phys.*, 2000, **63**, 843; S.H. Yang, *Trends Chem. Phys.*, 2001, **9**, 31; L. Dunsch, M. Krause, J. Noack and P. Georgi, *Phys. Chem. Solids*, 2004, **63**, 309; L.J. Wilson, D.W. Cagle, T.P. Thrash, S.J. Kennel, S. Mirzhadeh, M. Alford and G.J. Ehrhardt, *Coord. Chem. Rev.*, 1999, **192**, 199.
83. M.M. Olmstead, A. Bettencourt-Dias, J.C. Duchamp, S. Stevenson, H.C. Dorn and A.L. Balch, *J. Am. Chem. Soc.*, 2000, **122**, 12220.
84. S. Stevenson, P.W. Fowler, T. Heine, J.C. Duchamp, G. Rice, T. Glass, K. Harich, E. Hadju, R. Bible and H.C. Dorn, *Nature*, 2000, **408**, 427.
85. E.B. Iezzi, J.C. Duchamp, K.R. Fletcher, T.E. Glas and H.C. Dorn, *Nano Lett.*, 2002, **23**, 996.
86. L. Dunsch, P. Georgi, M. Krause and C.R. Wang, *Synth. Met.*, 2003, **135**, 761.

87. M. Krause, H. Kuzmane, P. Georgi, L. Dunsch, K. Vietze and G. Siefert, *J. Chem. Phys.*, 2001, **115**, 6596.
88. L. Feng, J.X. Xu, S.J. Shi, X.R. He and Z.N. Gu, *J. Chin. Univ.*, 2002, **23**, 996.
89. M.M. Olmstead, A. Bettancourt-Dias, J.C. Duchamp, S. Stevenson, D. Marciu, H.C. Dorn and A.L. Balch, *Angew. Chem. Int. Ed.*, 2001, **40**, 1223.
90. S. Yang and L. Dunsch, *J. Phys. Chem. B.*, 2005, **109**, 12320.
91. A.A. Bogdanov, D. Deininger and G.A. Dyuzhev, *Tech. Phys.*, 2000 **45**, 1.

CHAPTER 2
# Basic Principles of the Chemical Reactivity of Fullerenes

FERNANDO LANGA AND PILAR DE LA CRUZ

Departamento de Química Inorgánica, Orgánica y Bioquímica, Facultad de Ciencias del Medio Ambiente, E-45071, Toledo, Spain

## 2.1 [60]Fullerene Reactivity

### 2.1.1 Introduction

Since Krätschmer and Huffmann discovered a procedure for the preparation of bulk quantities of $C_{60}$[1] in the early 1990s, the physical and chemical properties of fullerenes have been intensively investigated. However, members of this new family of compounds are insoluble or sparingly soluble in most solvents[2] and, consequently, $C_{60}$ is difficult to handle. Chemical functionalization allows the preparation of soluble $C_{60}$ derivatives that maintain the electronic properties of fullerenes and, after several years of extensive studies into the scope and diversity of methods for functionalization, effort is now focussed on the application of the most versatile and general methods to prepare derivatives with interesting physical properties and biological activities.

The chemistry of [60]fullerene is characteristic of electron-deficient alkenes and, as a consequence, it reacts with nucleophiles and its [6,6] bonds are good dienophiles. Cycloaddition reactions play an important role in the preparation of $C_{60}$ derivatives and a wide variety of cycloadducts have been reported. Excellent reviews have been written cover to this area.[3,4]

This chapter concerns the most widely used cycloaddition reactions for the efficient functionalization of fullerenes: cyclopropanation, Diels–Alder and [3 + 2] cycloaddition reactions.

#### 2.1.1.1 Cyclopropanation: Bingel Reaction

Since the first reported synthesis of a methanofullerene by Wudl and co-workers[5] in the infancy of fullerene chemistry, extensive work has been undertaken to develop methods for the synthesis of this family of compounds.[6] In this respect, cyclopropanation reactions have proven to be an efficient method for the preparation of fullerene derivatives with potential applications in materials science and biological applications.[7] Cyclopropanated [60]fullerenes can be obtained by three different methods: (a) thermal addition of diazo compounds followed by thermolysis or photolysis, (b) addition of free carbenes and (c) the Bingel reaction (addition–elimination).

In this section we focus on the Bingel procedure as this is probably the reaction that enables the synthesis of cyclopropanated [60]fullerenes in the highest yields under relatively mild conditions. The Bingel reaction can be carried out on α-halomalonates in the presence of a base and occurs by an addition/elimination mechanism.[8] The α-halomalonate anion is added to $C_{60}$ and this is followed by intramolecular displacement of the halide by the anionic centre generated on the fullerene core to give the corresponding methanofullerene (Scheme 1).

This nucleophilic cyclopropanation of $C_{60}$ is also possible from malonates – from which the corresponding α-halomalonate is generated *in situ* – in the presence of iodine[9] or $CBr_4$[10] and a base (DBU: 1,8-diazabicyclo[5.4.0]undec-7-ene) by direct treatment of $C_{60}$ (Scheme 2).

The preparation of $C_{60}$-donor dyads and triads has been widely performed by the Bingel procedure, as both ester bonds can be used to link donors to the fullerene cage. Polycyclic aromatic hydrocarbons (PAHs) have the required electrochemical and photophysical properties for photoinduced electron

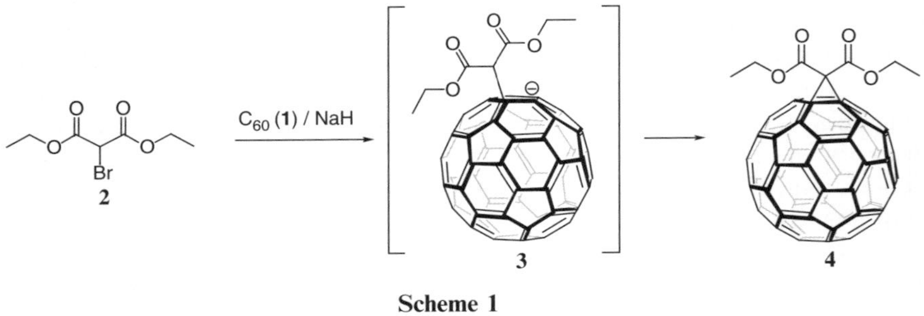

**Scheme 1**

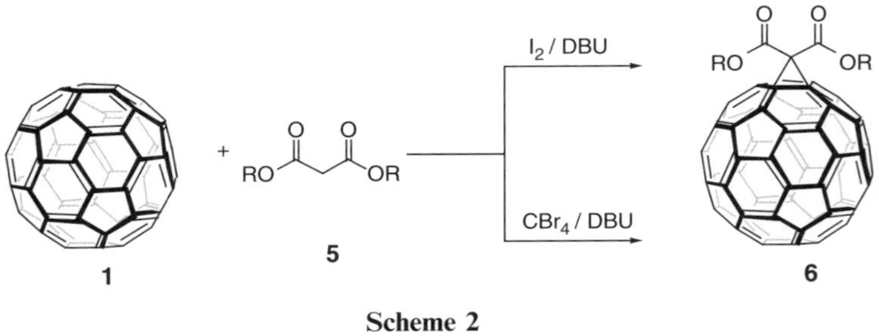

**Scheme 2**

Basic Principles of the Chemical Reactivity of Fullerenes

transfer. A classical Bingel cycloaddition reaction was used to prepare the first tetracene-[60]fullerene dyad **10** (Scheme 3).[11] This strategy prevents the usual Diels–Alder reaction, which is used to link PAHs to $C_{60}$ and would result in the loss of the tetracene aromaticity. Compound **10** does not undergo a Diels–Alder reaction, probably due to the adoption of a π-stacking conformation.

In another example, Martín and co-workers[12] described the synthesis of methanofullerene-π-extended-TTF dyads **(13)** and triads **(14)** by a Bingel cyclopropanation (Scheme 4). The bisester function allowed two donor fragments to be linked to the $C_{60}$ core. Other TTF-fullerene derivatives prepared by either the 1,3-dipolar cycloaddition of azomethine ylides or Diels–Alder cycloaddition for the study of photoinduced electron-transfer processes have only one TTF fragment.

The Bingel-type reaction allows the synthesis of a wide variety of fullerene derivatives with different properties and a diverse range of applications in the field of organic materials. A liquid-crystalline-[60]fullerene-TTF dyad **(15)** synthesized by the Bingel cyclopropanation between [60]fullerene and a bis-mesogenic fragment is shown in Chart 1.[13] Both the malonate and dyad showed smectic B and A phases. Other liquid-crystalline methanofullerenes have been prepared.[14]

The synthesis of oligoporphyrin arrays conjugated to one **(16)** or two [60]fullerene moieties **(17, 18)** has recently been described (Chart 2).[15] The authors showed that these multicomponent arrays prefer a different conformation depending on the intermolecular fullerene–porphyrin interactions. The dimer **17** exhibited an unprecedented photoinduced energy-transfer process from the fullerene to the Zn-porphyrin. Electrochemical studies clearly demonstrated the presence of an electronic interaction between the porphyrin and fullerene moieties.

An intramolecular Bingel reaction of the corresponding methano-fullerene in the presence of 1,5-dinaphtho-[38]crown-10 ether afforded a neutral [60]fullerene-based [2]catenane **(19)** and the similar [2]rotaxane **(20)** was synthesized by an intermolecular Bingel reaction, both of these systems had an

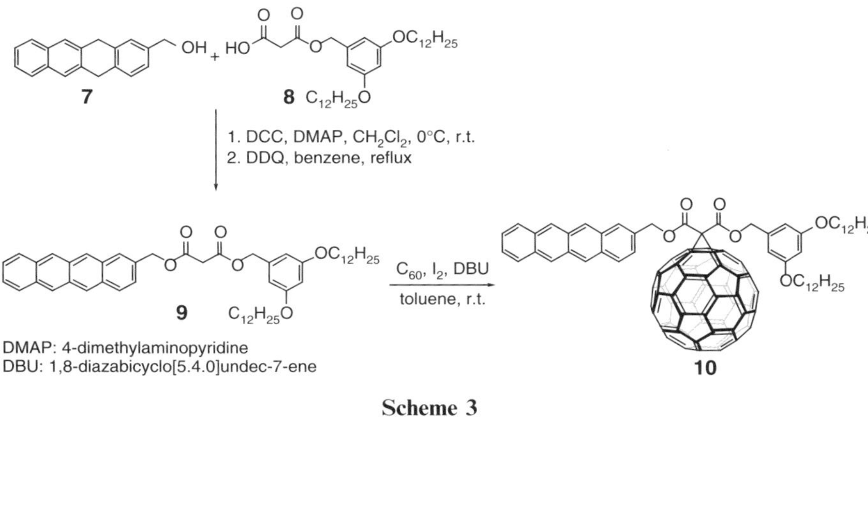

**Scheme 3**

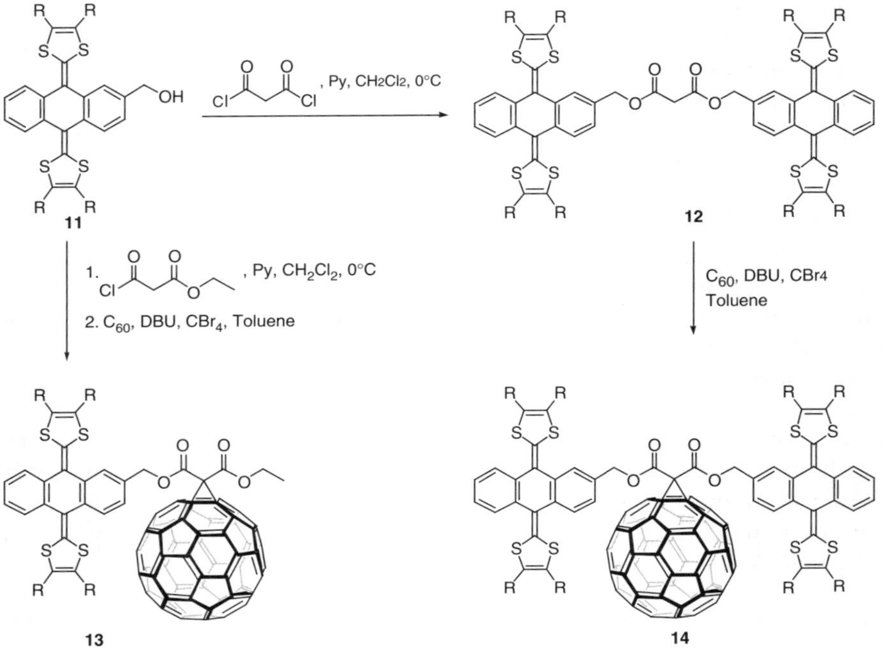

**Scheme 4**

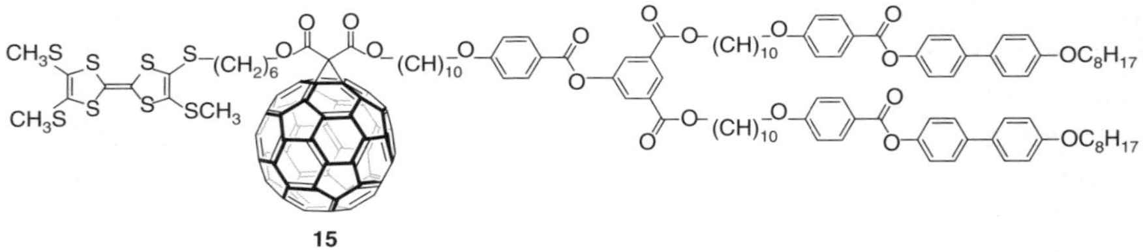

**Chart 1**

electron-deficient aromatic diimide moiety (Chart 3).[16] These molecular assemblies have attracted a great deal of interest because of their fascinating potential applications in molecular devices.

As described later in this chapter, this methodology has been extended to the regioselective formation of multiple adducts by tether-directed remote functionalization.

*2.1.1.1.1 Retro-Bingel Reaction.* The efficient removal of methano addends from fullerenes using reductive methods is called the retro-Bingel reaction.[17] Bis(alkoxycarbonyl)methano-fullerenes (**4**) formed by Bingel reactions are generally quite stable. Nevertheless, Diederich and Echegoyen[18] discovered that a bis(alkoxycarbonyl)methano addend can be electrochemically removed

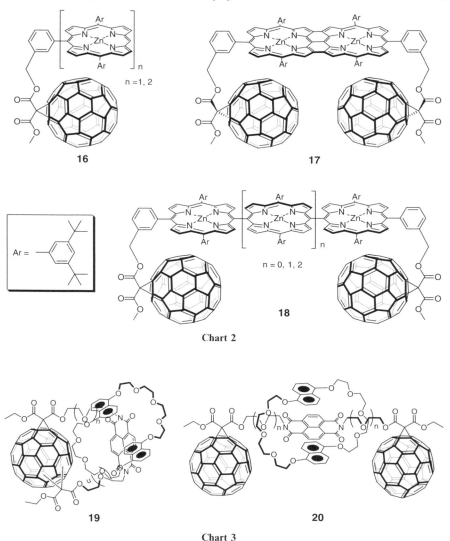

Chart 2

Chart 3

in high yield by exhaustive electrolytic reduction at constant potential (CPE) (Scheme 5).

When controlled-potential electrolysis is applied to a pure bis-methanofullerene isomer, two electrons per molecule are discharged and a competition between retro-Bingel and isomerization reactions takes place. The isomerization occurred intramolecularly (Scheme 6) and it was observed that *cis* isomers were not formed; the major isomer formed was *trans* although the *e*-isomer was also formed in a high percentage. This result could be a consequence of the higher stabilities of these isomers.[19]

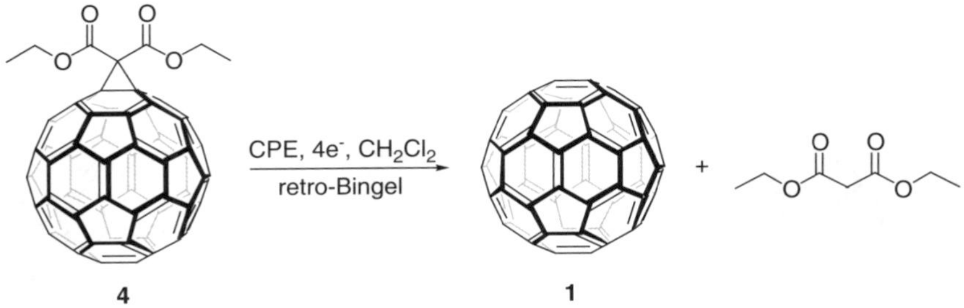

**Scheme 5**

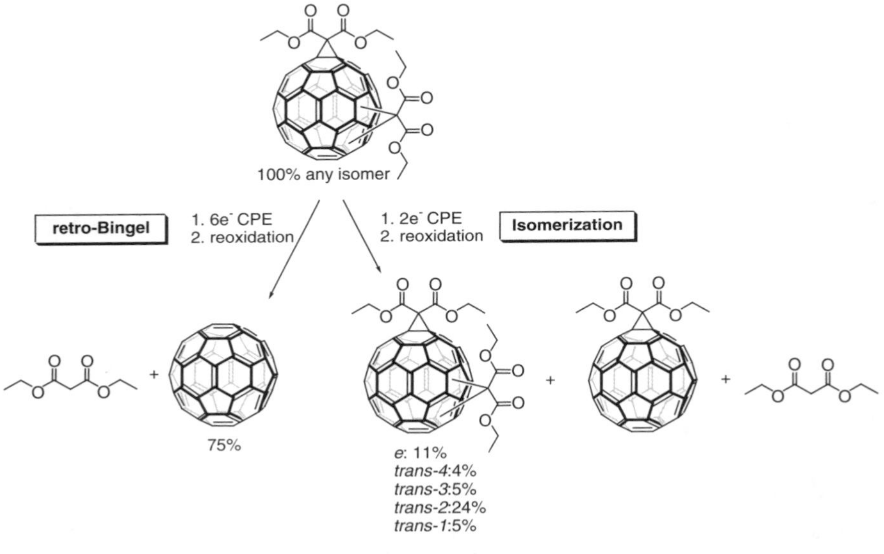

**Scheme 6**

Alternatively, a retro-cyclopropanation reaction can be carried out chemically by treatment of bis(alkoxycarbonyl)methane-fullerene **4** with amalgamated magnesium powder (10% mercuric bromide) in THF under reflux for 3 days (Scheme 7).[20] More recently, these chemical retro-cyclopropanation reactions were carried out in the presence of 18-crown-6 to avoid the use of $HgBr_2$ (Scheme 7).[21]

The retro-cyclopropanation reaction allows the selective removal of the Bingel addend from fullerenes in the presence of other addends (Scheme 8) either electrochemically (63% yield)[22] or chemically (64% yield).[20]

In general, the retro-Bingel reaction can be considered as a versatile and useful reaction. The efficient removal of methano addends from fullerenes using this method has been successfully employed in the separation of enantiomers and constitutional isomers of fullerenes, which in some cases were otherwise not accessible.

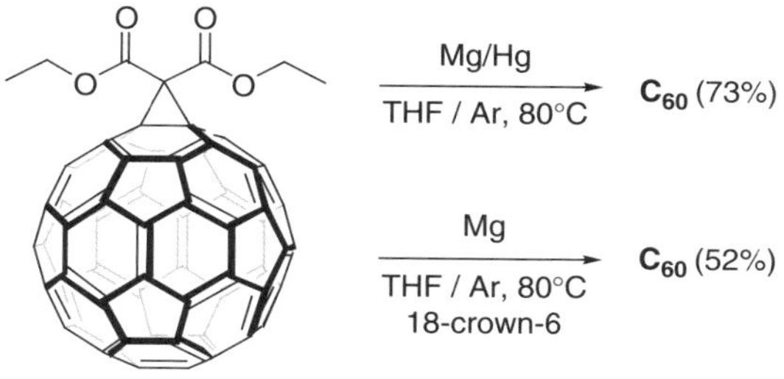

Basic Principles of the Chemical Reactivity of Fullerenes

**Scheme 7**

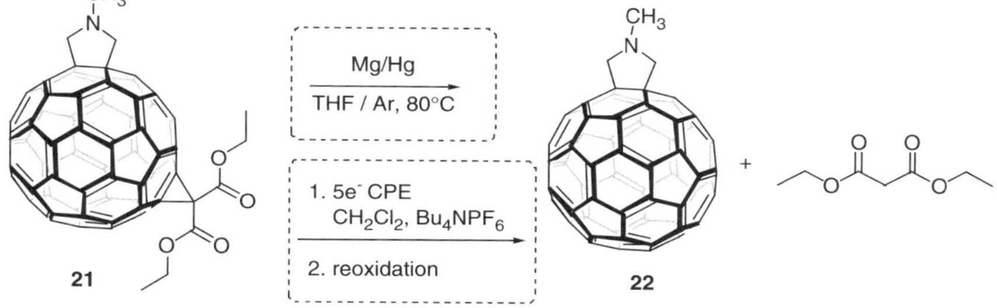

**Scheme 8**

Mono- and bis-methano[60]fullerenyl amino acid derivatives (**23**) can undergo a reductive ring-opening retro-Bingel reaction to afford the corresponding 1,2-dihydro[60]fulleneryl amino acid derivatives (**24**) upon treatment with sodium cyanoborohydride in the presence of a protic solvent or a Lewis acid (Scheme 9).[23]

## 2.1.2 [4+2] Cycloadditions: Diels–Alder Reaction

The [6,6] double bonds of [60]fullerene are excellent dienophiles (comparable to maleic anhydride). Thus, $C_{60}$ can react with different dienes by a Diels–Alder cycloaddition reaction. Depending on the reactivity of the diene, heating the reaction under reflux in a high boiling solvent may be required. This aspect of the reaction led to microwave energy being applied with excellent results.[24]

The first example of Diels–Alder reaction of $C_{60}$ was carried out using cyclopentadiene[25] as the 1,4-diene system. The addition of cyclopentadiene (**26**) to a solution of fullerene in toluene at room temperature afforded the Diels–Alder monoadduct **27** in 74% yield (Scheme 10).

Cyclopentadiene derivatives,[26] such as methylcyclopentadiene and cyclopentadienone, afford the corresponding monoadducts, which are usually more stable than monoadduct **27**. The dienophilic behaviour of $C_{60}$ in [4+2] cycloaddition reactions was also demonstrated in the reaction with anthracene (**29**) (Scheme 11)[25b,27] and other anthracene derivatives.[25b,28]

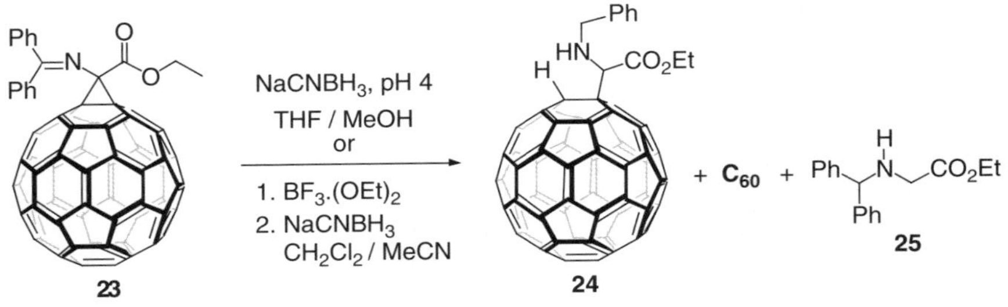

**Scheme 9**

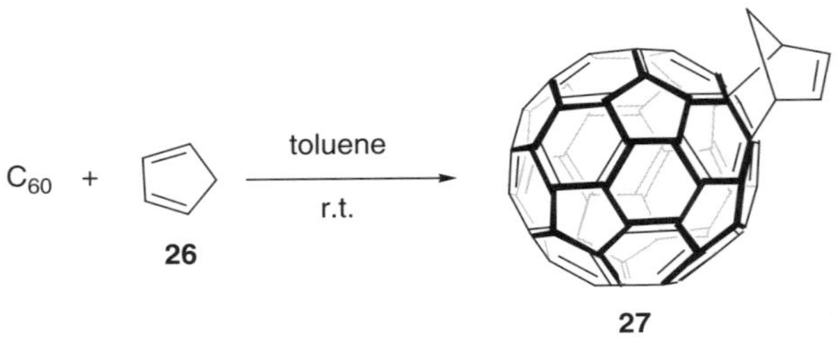

**Scheme 10**

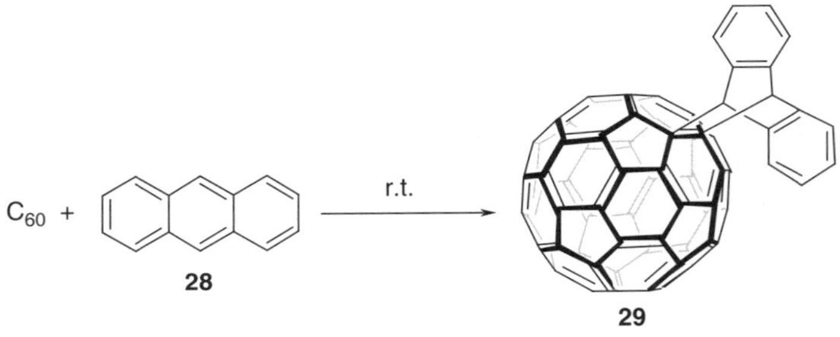

**Scheme 11**

Diels–Alder cycloadditions using either cyclopentadiene or anthracene as dienic systems do suffer from a limitation concerning the scope of the retro-Diels–Alder reaction. It was shown that there is an equilibrium between addition and elimination reactions and this could explain the low yields obtained in some cases.[27] Hydrogenation of the double bond of the resulting cycloadduct prevents the occurrence of the reverse reaction.[29]

The formation of the water-soluble poly(amidoamine) fullerodendrimer **31**[30] has been achieved by reaction of [60]fullerene with anthracenyl dendron **30** (Scheme 12). The fullerodendrimer is soluble in polar solvents and is stable for several months when stored in the dark at room temperature.

Martín et al.[31] described the Diels–Alder cycloaddition between anthracene derivatives bearing fused π-extended TTFs (**32**) and $C_{60}$, a reaction that yielded

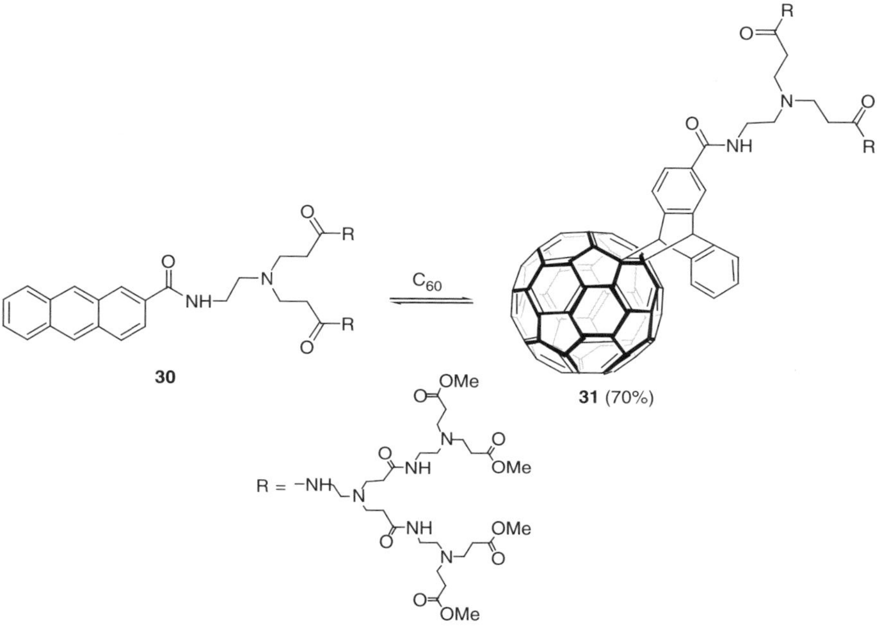

**Scheme 12**

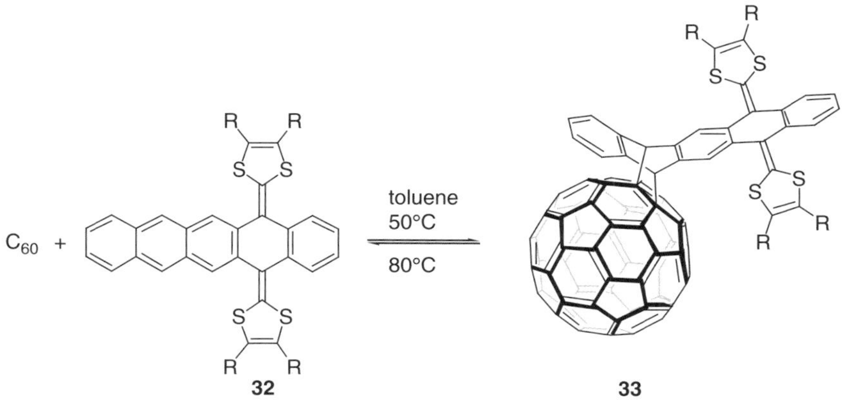

**Scheme 13**

thermally reversible donor–acceptor materials that function as fluorescence switches (Scheme 13).

As mentioned above, microwave irradiation has been applied to Diels–Alder cycloaddition reactions and it was shown that this type of reaction is accelerated and yields are higher than those obtained by classical heating. This advantage is a result of the shorter reaction times, which avoids the retro-Diels–Alder reaction and the formation of bisadducts.[32]

The efficiency of the cycloaddition reactions of anthracene systems can also be improved by performing the reaction under conditions involving high-speed vibration grinding of solid reagents.[33,34]

Other dienes have also been employed in this type of reaction.[3a,29b,35] Hirsch described the synthesis of several $C_{60}$-acceptor dyads (**35**) by irreversible [4+2] cycloadditions of anthraquinone-dienes (**34**) with $C_{60}$ (Scheme 14).[36]

One of the most important tools for the functionalization of fullerene by a Diels–Alder cycloaddition is based on the use of *o*-quinodimethanes (generated *in situ*) as dienic systems. Several methods for the generation of *o*-quinodimethanes[37] are available: iodide-induced 1,4-dehalogenation of *o*-bis(bromomethyl) benzenes,[38] thermolysis of benzocyclobutenes,[39] thermolysis of sulfolenes (**36**) (Scheme 15)[40] and thermolysis of sultines.[41]

The intramolecular 1,4-dehalogenation of α,α′-dibromo-*o*-xylenes requires the presence of an ion such as iodide to induce the 1,4-elimination and 18-crown-6 ether is needed as the phase transfer catalyst. Fullerenoquinoxalines (**41**) have been synthesized by heating the corresponding *o*-bis(bromomethyl)quinoxaline (**39**) with NaI in 1,2-dichlorobenzene (Scheme 16).[42]

This type of reaction has been used to produce photoactive compounds containing the electron-acceptor fullerene moiety and electron-donor fragments such as TTF or its analogues.[43]

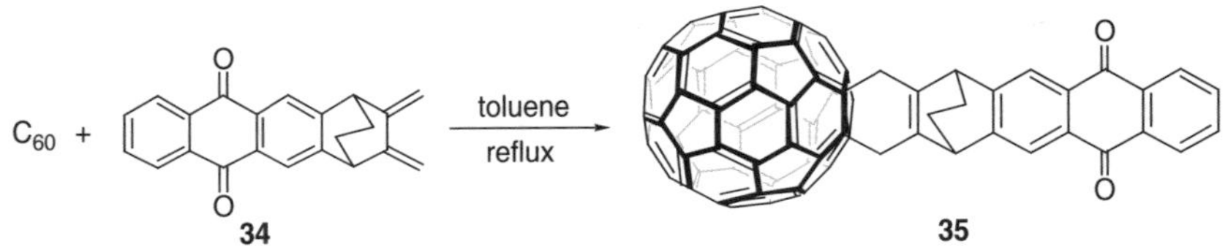

**Scheme 14**

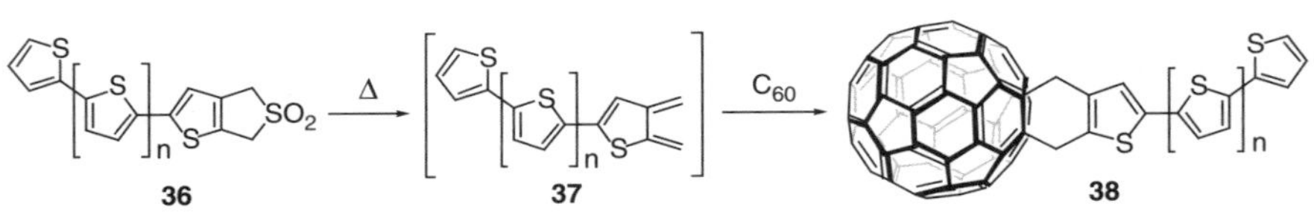

**Scheme 15**

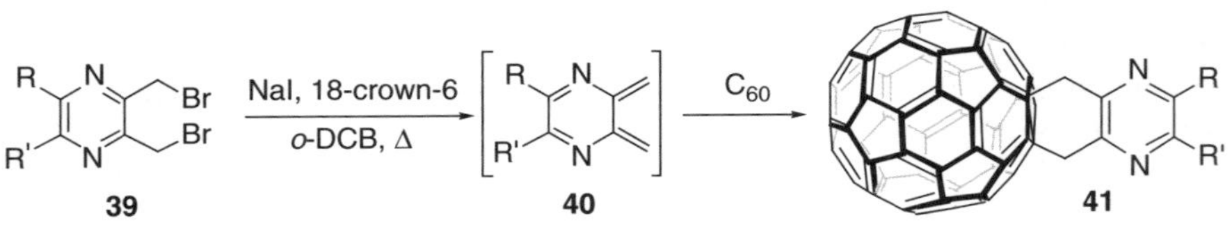

**Scheme 16**

The thermolysis of benzocyclobutenes (**42**) requires high temperatures for the ring opening reaction even when methoxy activating groups, which decrease the activation energy, are present (Scheme 17).[39]

Among the precursors for *o*-quinodimethanes, sulfolenes and sultines are good choices because they undergo pyrolysis at reasonably low temperatures and they are usually stable at room temperature. The synthesis of porphyrin-[60]fullerene-porphyrin triad **46** has been described by Diels–Alder cycloaddition between a sulfolene moiety containing a porphyrin (**45**) and $C_{60}$ (Scheme 18).[44] The porphyrin fragments are bound with a relatively flexible long chain and the symmetric structure of the two-porphyrin moieties in the triad featuring an antenna unit is able to modulate the energy transfer processes. The authors claim that this system is a promising candidate to incorporate into artificial photosynthetic units.

Dienes are formed upon elimination of sulfur dioxide not only from sulfolene derivatives but also from sultines, which are readily accessible from the commercially available rongalite and smoothly generate *o*-quinodimethanes by extrusion of sulfur dioxide. The [4+2] cycloaddition reaction of

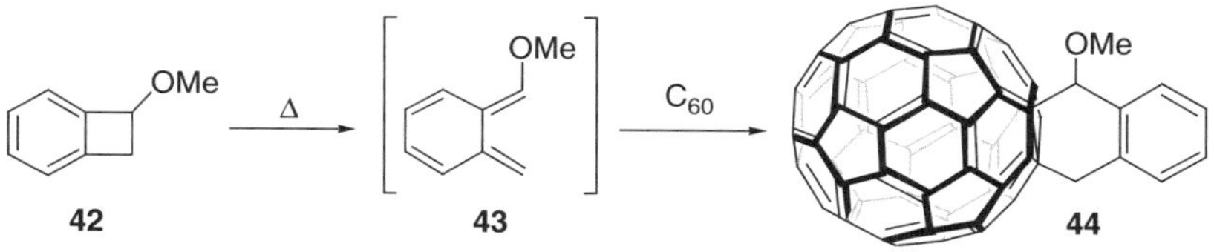

**Scheme 17**

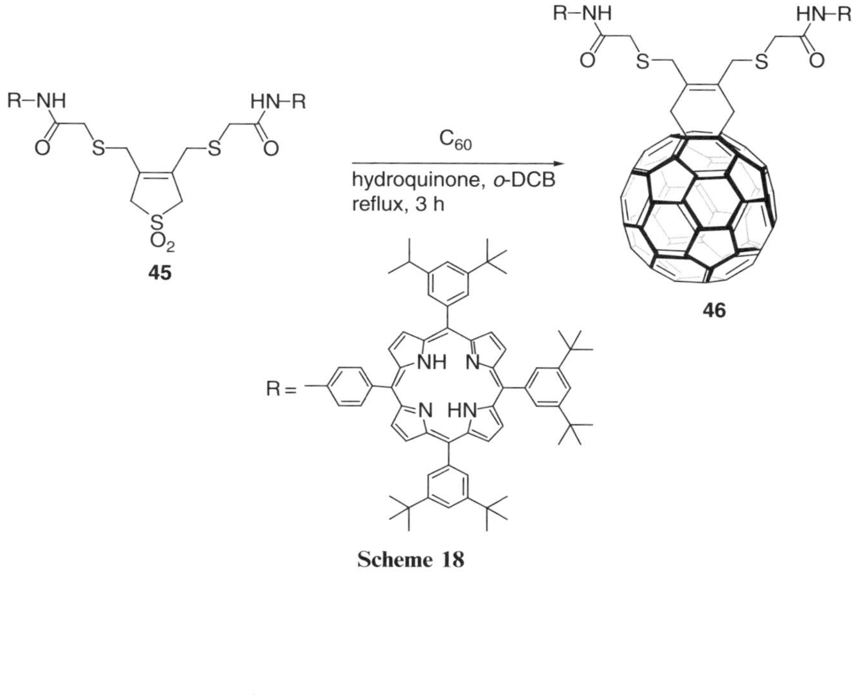

**Scheme 18**

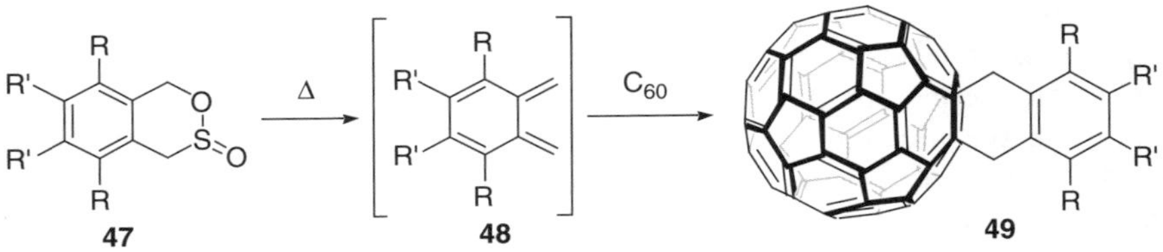

**Scheme 19**

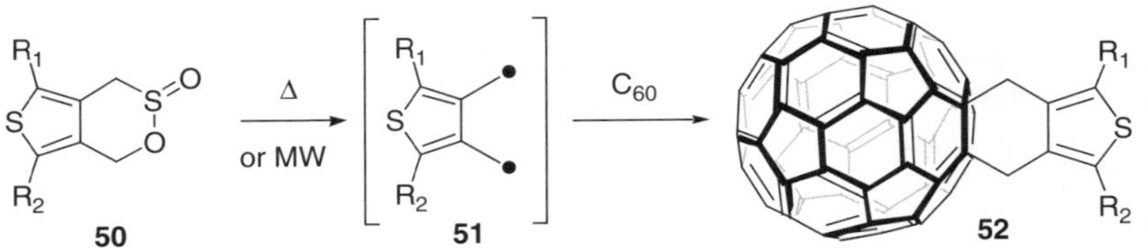

**Scheme 20**

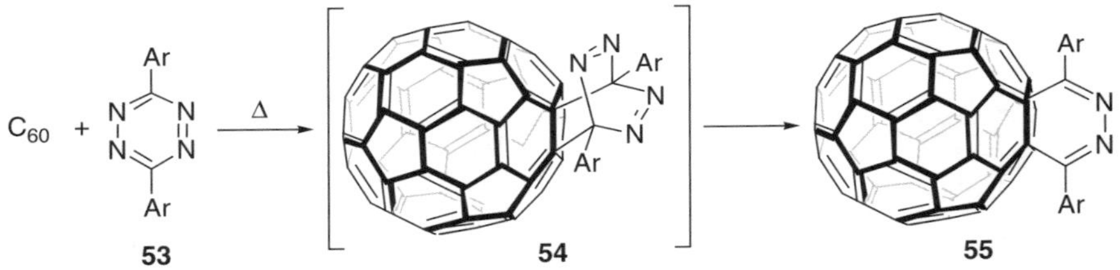

**Scheme 21**

*o*-quinodimethanes (**48**), generated *in situ* from 4,5-benzo-3,6-dihydro-1,2-oxathiin 2-oxides (**47**) (sultines), to [60]fullerene has been described (Scheme 19).[41a]

It has been demonstrated that microwave irradiation can accelerate the Diels–Alder reaction between $C_{60}$ and the corresponding *o*-quinodimethane generated *in situ* from sultine derivatives.[45] In the synthesis of the thiophene-[60]fullerene derivative **52**, microwave irradiation improves the yield and significantly reduces the reaction time (Scheme 20).[46]

Although the Diels–Alder reaction is favoured by the presence of electron-rich substituents in the diene, $C_{60}$ can also react with electron-deficient dienes – as evidenced by the inverse electron demand Diels–Alder reaction to prepare fullerenopyridazine **55** (Scheme 21).[47]

## 2.1.3 [3+2] Cycloaddition Reactions

### 2.1.3.1 Azomethine Ylides

The [3+2] cycloaddition reaction of azomethine ylides to [60]fullerene was introduced by Prato and Maggini.[48] Indeed, this is the most frequently used

Basic Principles of the Chemical Reactivity of Fullerenes 27

tool to functionalize fullerenes as a consequence of its good selectivity (the cycloaddition occurs always on the [6,6] bonds of the fullerene sphere), the fact that a wide variety of addends and functional groups are tolerated and because it is possible to introduce different substituents into three different positions of the pyrrolidine ring depending on the carbonyl compound (aldehyde or ketone) and substituted amino acid used.

In this reaction, which is already known as Prato's reaction, the azomethine ylide is generated by decarboxylation of immonium salts derived from the condensation of α-amino acids with aldehydes or ketones. The first example, described in 1993, involved the synthesis of N-methylfulleropyrrolidine **59** by reaction of sarcosine (**56**), formaldehyde (**57**) and $C_{60}$ (Scheme 22).[48a]

The availability of a wide range of aldehydes allows the synthesis of a broad range of dyad and triad donor–acceptor dyes according to this procedure, although it should be noted that the substituted carbon of the pyrrolidine ring is chiral and the corresponding stereoisomers are formed.

Azomethine ylides can also be generated by thermal ring opening of aziridines (**61**), as shown in Scheme 23.[49]

Other methods[48b] have been employed to prepare fulleropyrrolidines: acid-catalysed or thermal desilylation of trimethylsilyl amino derivatives, tautomerization of aminoester immonium salts and imines, reaction with aldehydes in the presence of aqueous ammonia, reaction with oxazolidinone or photochemical reaction with some amino derivatives.

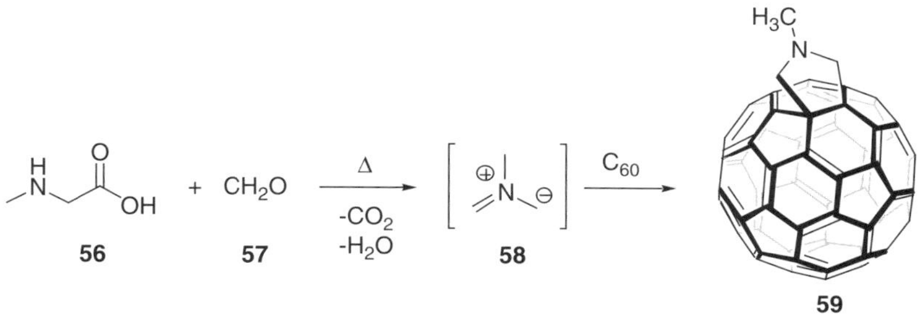

**Scheme 22**

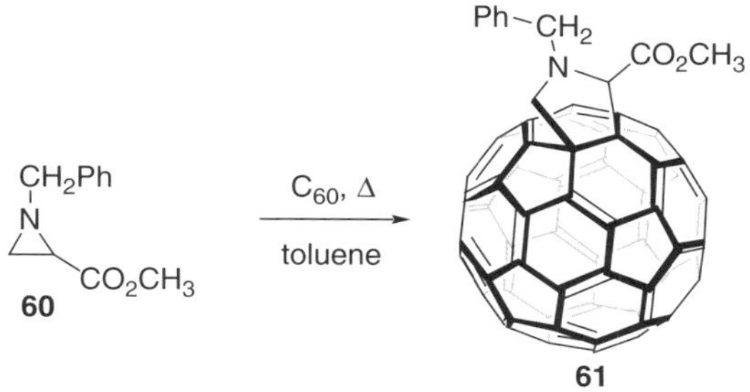

**Scheme 23**

The cycloaddition reaction of azomethine ylides can also be carried out under microwave irradiation,[50] which enables the use of milder reaction conditions, or by applying the high-speed vibration milling technique.[51]

Among the different donors used, ferrocene derivatives have been exploited intensively in donor–acceptor (D–A) systems based on fullerene with the aim of design new materials for molecular-based artificial photosynthesis.[52] Other metallocenes such as ruthenocene can be incorporated as the electron-donor fragment (**64**) with efficient photoinduced charge separation (Scheme 24).[53]

A large number of pyrrolidine[60]fullerene systems have been synthesized using Prato's reaction and different fragments with important electronic properties have been attached to the fullerene system. For example, molecules such as porphyrins,[54] subphthalocyanines,[55] TTF,[56] dendrimers[57] and conjugated oligomers[58] have all been incorporated.

[3+2] Dipolar cycloadditions of azomethine ylides using bis-aldehydes (**65**) as starting materials afford dumb-bell-like molecules containing two fullerene fragments, as shown in Scheme 25.[59] This bis-(ferrocenyl-fulleropyrrolidine) system (**66**) incorporates ferrocene as the electron-donor fragment.

Although *N*-methylglycine is the most widely used amino acid in the preparation of the azomethine ylide, other *N*-substituted glycines have also been

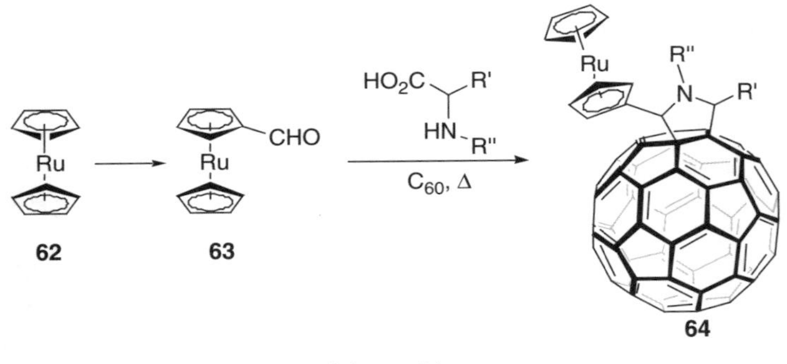

**Scheme 24**

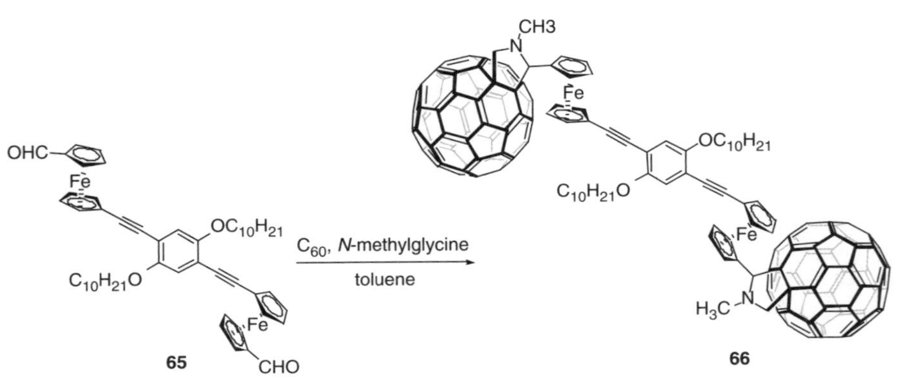

**Scheme 25**

used.[60] The synthesis of the *N*-substituted fulleropyrrolidine **68** allows esterification with anthracenyl carboxylic acid derivative **71** to yield a $C_{60}$-anthrylphenylacetylene hybrid (**70**, Scheme 26) able to form self-assembled monolayers (SAMs) by taking advantage of the excellent physical and chemical properties of $C_{60}$ and the nanoscale ordering of anthryl-based SAMs.[61]

Water-soluble fullerene derivative **72**,[62] prepared from triethylene glycol **71**, is the precursor to introduce the fullerene cage covalently into a heptapeptide (H-PPGMRPP-OH) to afford compound **73**, which has antigenic properties (Scheme 27).[63]

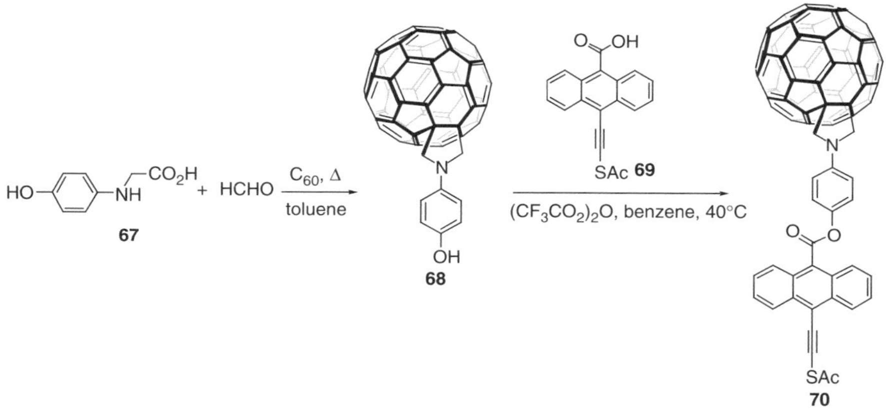

Scheme 26

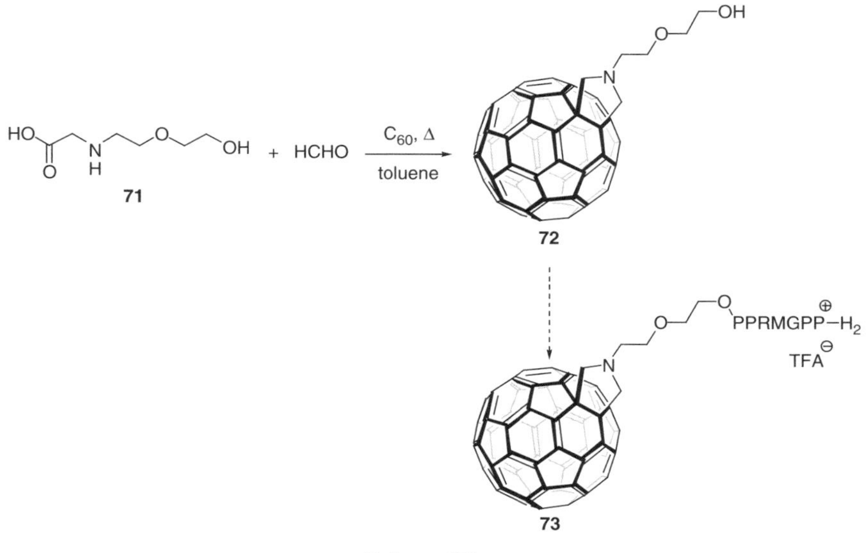

Scheme 27

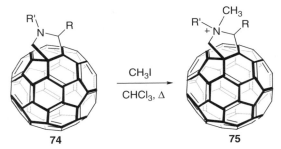

**Scheme 28**

The solubility in water of pyrrolidino[60]fullerene systems can be increased by quaternization of the nitrogen. The fulleropyrrolidinium salts **75** were prepared by methylation of pyrrolidines **74** with methyl iodide (Scheme 28).[64] These salts (**75**) are moderately soluble in polar organic solvents, tetrahydrofuran, methylene chloride, dimethyl sulfoxide and dimethyl sulfoxide/water, and these species show enhanced electron-accepting properties with respect to both the parent fulleropyrrolidines and $C_{60}$.

The nitrogen atom in fulleropyrrolidines is about six orders of magnitude less basic and three orders of magnitude less reactive than that in pyrrolidines.[65] Nevertheless, *N*-unsubstituted pyrrolidino[60]fullerene **76** can be functionalized on the nitrogen atom using different methods (Scheme 29). One of these approaches is based on the treatment of compound **76** with different acid chlorides to form an amide bond (Scheme 29, compound **77**).[66] It is also possible to carry out a nucleophilic aromatic substitution by reaction of *NH*-pyrrolidino[60]fullerene **76** with deactivated aromatic compounds (Scheme 29, compound **78**).[67] The combination of phase-transfer catalysis without solvent and microwave irradiation techniques enables different alkyl bromides to react with N–H fulleropyrrolidines to give good yields in short reaction times (Scheme 29, compound **79**).[50a]

Pyrrolidino[60]fullerenes are considered to be very stable cycloadducts under a wide variety of reaction conditions. Nevertheless, Martín *et al.*[68] recently described the efficient retrocycloaddition reaction of different pyrrolidinofullerenes (**80**) to afford quantitatively unsubstituted $C_{60}$ (Scheme 30). The reaction is carried out by heating the fullerene derivative **80** under reflux in *o*-dichlorobenzene in the presence of an excess of a highly efficient dipolarophile and a metal Lewis acid such as copper(II) trifluoromethylsulfonate ($CuTf_2$) as a catalyst.

### 2.1.3.2 Nitrile Imines

The low value of the first reduction potential of $C_{60}$ makes it an interesting partner in the preparation of donor–acceptor (D–A) systems due to the interesting electronic and optical properties that these materials can display. As a result, several groups have focussed their attention on the preparation of

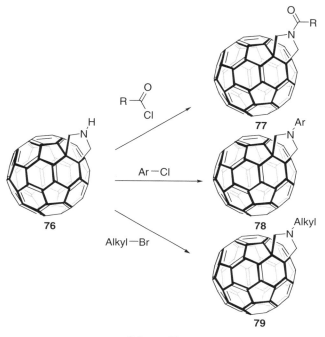

**Scheme 29**

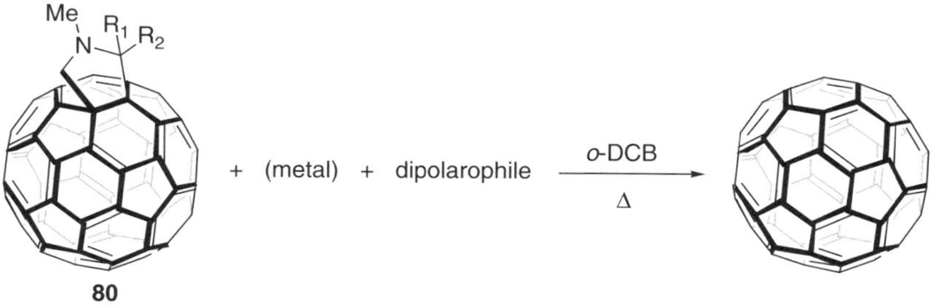

**Scheme 30**

$C_{60}$-donor dyads and the study of the existence of inter- or intramolecular electronic interactions between donor and $C_{60}$ in both the ground and excited states.

However, the reduction potentials of the monoadducts are about 100 mV more negative than those of $C_{60}$ as a consequence of the saturation of one double bond. There are very few examples in the literature that show better electron affinity than the parent $C_{60}$[69] and one example is represented by the pyrrolidinium salts[64a] mentioned above, which have a higher affinity due to the $-I$ effect of the alkylated nitrogen on the fullerene sphere.

The first 1,3-dipolar cycloadditions to $C_{60}$ were described in the early days of fullerene chemistry but the first example of a reaction between a nitrile imine

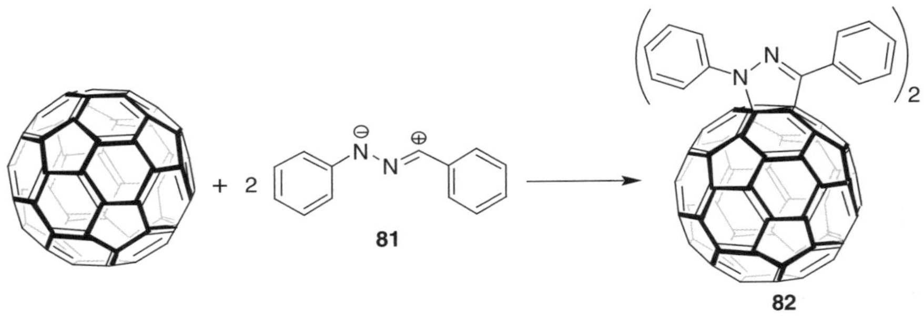

**Scheme 31**

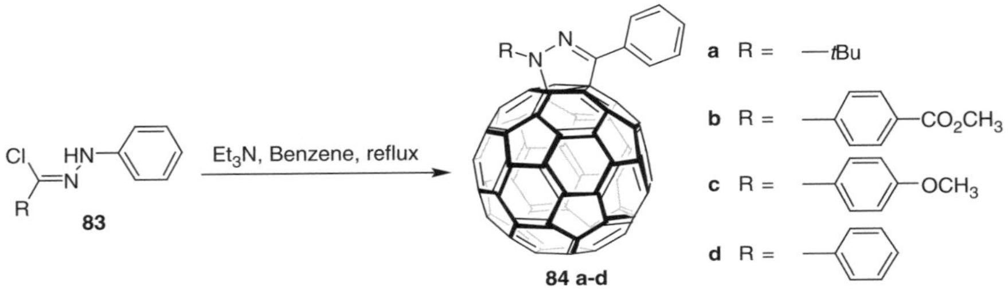

**Scheme 32**

and C$_{60}$ was not described until 1994, when Mathews et al.[70] prepared a mixture of bisadducts **82** by reaction of C$_{60}$ and 1,3-diphenylnitrile imine (**81**), generated *in situ* from benzhydrazidoyl chloride, in the presence of triethylamine. The monoadduct was not isolated (Scheme 31).

The preparation of 1,3-diphenyl-2-pyrazolinofullerenes (**84a–d**) has been described.[71] These adducts were prepared by cycloaddition of C$_{60}$ to nitrile imines generated *in situ* from the corresponding *N*-chlorobenzylidene derivatives (**83**) in the presence of triethylamine in refluxing benzene (Scheme 32).

A more useful alternative route to the pyrazolino[60]fullerenes **88a–c** (Scheme 33) involved the use of hydrazones as starting materials. The 1,3-dipoles (**87**) were generated *in situ* by treating the hydrazones (**85**) and *N*-bromosuccinimide in the presence of triethylamine and were then reacted with C$_{60}$ under microwave irradiation to yield, after isolation, the corresponding adducts **88a–c** in 20–38%.[72]

A modification of this procedure was used for the preparation of various pyrazolinofullerenes (**92a–f**) bearing different functional groups as substituents.[73] The starting hydrazones **91a–f** were treated with *N*-chlorosuccinimide in the presence of pyridine; the resulting chloro-derivatives were reacted with triethylamine and C$_{60}$ at room temperature to give the corresponding cycloadducts **92a–f** in good yields (Scheme 34).

This synthetic strategy was used in conjunction with a wide variety of hydrazones to prepare numerous different pyrazolino[60]fullerenes incorporating different organic fragments with interesting electronic properties:

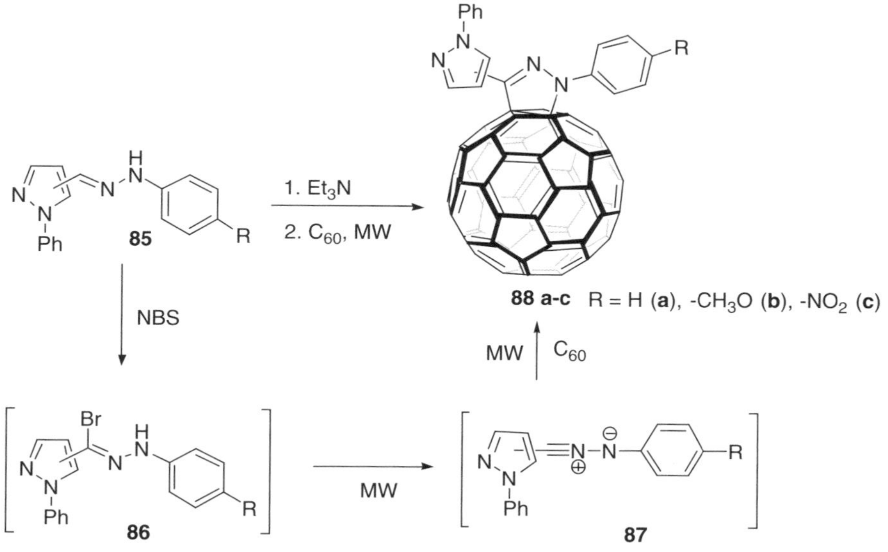

**Scheme 33**

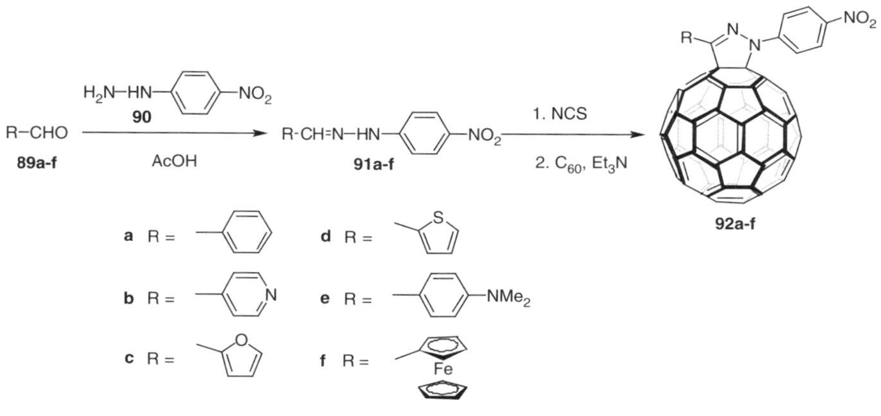

**Scheme 34**

oligophenylenevinylene (OPV),[74] phenylenevinylene dendrimers,[75] isoindazole derivatives[76] and ruthenocene.[53a]

Symmetrically substituted oligophenylenevinylene derivatives bearing terminal *p*-nitrophenyl-hydrazone groups were prepared and used for the synthesis of dumb-bell-shaped bis-(pyrazolino[60]fullerene)-oligophenylenevinylene systems (**93**)[77] (Chart 4).

This method is a general and versatile procedure for the functionalization of fullerenes in a one-pot reaction starting from the corresponding hydrazone. The hydrazones are stable compounds and are readily available from aldehydes in almost quantitative yield.[78]

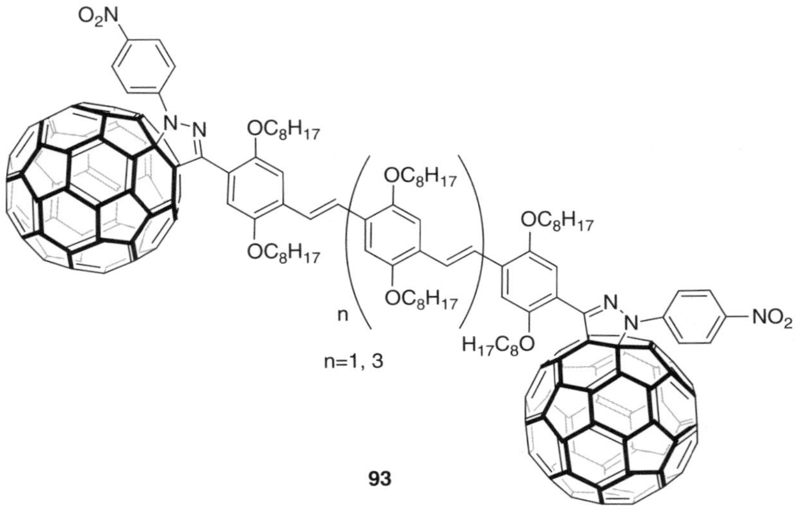

**93**

**Chart 4**

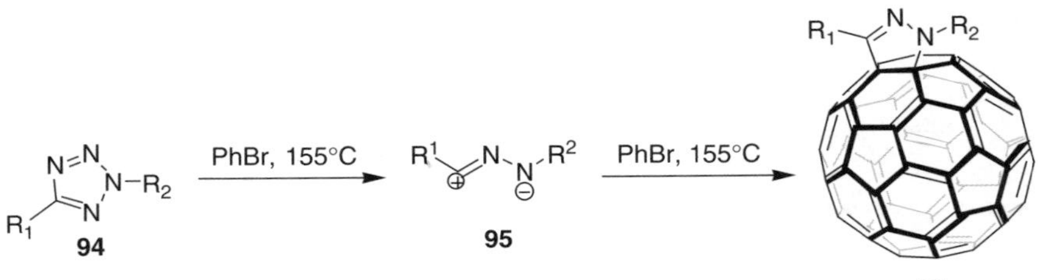

**Scheme 35**

A new synthetic route has recently been investigated.[79] Nitrile imines were generated by the thermal decomposition of 2,5-diaryltetrazoles (**94**). Cycloadducts containing aryl- and hetaryl-substituents, such as thienyl and pyridyl, in the pyrazoline fragment (**96**) were synthesized using this method (Scheme 35). The availability of 2,5-diaryltetrazoles with various structures generates a greater possibility for functionalization of the fullerene spheroid than other procedures, which are limited by the availability of the appropriate hydrazonoyl halides.

[60]Fullerene-fused pyrazolines (**99**) were also prepared by the reaction of $C_{60}$ with alkyl diazoacetates under solid-state high-speed vibration milling (HSVM) conditions as well as in toluene solution.[80] The 2-pyrazoline **99** was obtained through isomerization of the 1-pyrazoline (**98**), which were formed directly by the 1,3-dipolar cycloaddition of ethyl diazoacetate, generated *in situ* from glycine ester hydrochloride (**97**) and sodium nitrite with $C_{60}$ (Scheme 36).

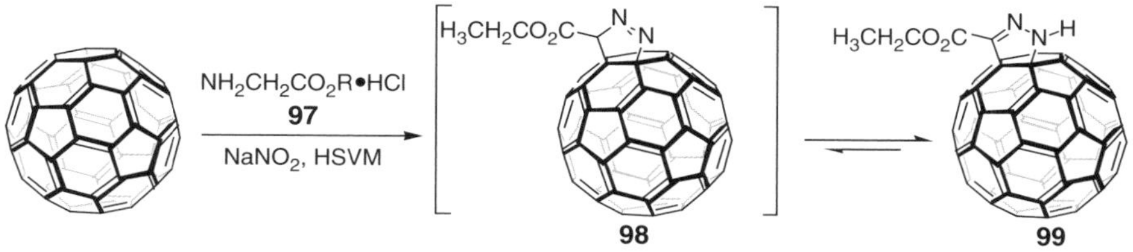

Basic Principles of the Chemical Reactivity of Fullerenes

**Scheme 36**

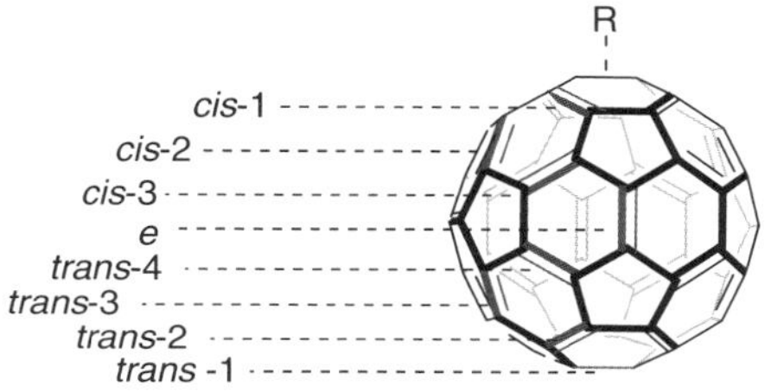

**Chart 5**

2-Pyrazolino[60]fullerenes show some advantages in comparison with other fullerene derivatives such as pyrrolidino[60]fullerene: in contrast to the formation of pyrrolidinofullerenes, in which C-2 of the pyrrolidino ring is formed as a racemic centre, the formation of stereoisomers does not occur in the pyrazolino case because this C atom is $sp^2$ and, most importantly, the systems present better reduction potentials than the parent $C_{60}$ as a consequence of the –I effect of the nitrogen atom directly linked to the $C_{60}$ cage; moreover, these systems have shown photoinduced electron transfer processes from the pyrazoline ring, a type of behaviour not observed in other derivatives such as fulleropyrrolidines.

## 2.2 Multiaddition Reactions

Bis-functionalized fullerenes have been widely used in biological studies, molecular devices and advanced materials due to their defined three-dimensional structures that contain a variety of functional groups. After functionalization, the remaining 6–6 bonds of the $C_{60}$ cage are not identical and, in principle, eight different regioisomers, which are difficult to separate, can be formed by addition of a second addend.

An unambiguous nomenclature according to IUPAC[81] can be used to name the different regioadducts but it is more intuitive to indicate the relative arrangement of the addends as proposed by Hirsch et al.[82] The relative positional relationships between the 6,6-bonds are labelled as *cis-n* ($n = 1–3$), *equatorial (e)* and *trans-n* ($n = 1–4$) (Chart 5).

Different procedures for functionalization drive certain modifications in the regioadduct distributions[48b,83] but *e* and *trans-3* positions are preferred in this order. As shown in Chart 6, the double cyclopropanation of [60]fullerene by the Bingel reaction with diethyl 2-bromomalonate affords seven isomeric bisadducts, which were isolated by chromatography.[84] When the reaction is carried out with symmetric azomethine ylides, eight isomeric bisadducts are obtained with slightly different regioselectivity to that found with bismalonates.[85]

The regioselective preparation of multiadducts of fullerenes is of great importance for the preparation of functional advanced materials as the geometry plays a crucial role in the properties of fullerne derivatives. In order to gain regiocontrol over bisadditions to fullerenes, an elegant procedure was introduced by Diederich[86] in 1994 that involved tethering directed remote functionalization (Scheme 37). This subject has been extensively reviewed.[87] Although other bisfunctional tethers have been used, the tether-directed remote functionalization is usually based on Bingel-cycloaddition reactions where the reactive malonate is connected through more or less rigid organic fragments using different spacers. The rigidity of the fragment and the length of the spacers can modify the regioselectivity of the cycloadditions reactions.[88]

The use of this functionalization has enabled the synthesis of polyadducts bearing different organic fragments. The Bingel macrocyclization of $C_{60}$ with a bismalonate containing a dibenzo[18]-crown-6 tether (**103**) yields up to 30% of the planar-chiral *trans*-1 bisadduct **104** (Scheme 38).[89]

Dendrimeric fragments were bound to the [60]fullerene sphere by this regioselective reaction. These macrocyclic bisadducts of $C_{60}$ (**105**) were synthesized

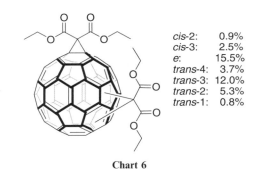

| | |
|---|---|
| *cis*-2: | 0.9% |
| *cis*-3: | 2.5% |
| *e*: | 15.5% |
| *trans*-4: | 3.7% |
| *trans*-3: | 12.0% |
| *trans*-2: | 5.3% |
| *trans*-1: | 0.8% |

**Chart 6**

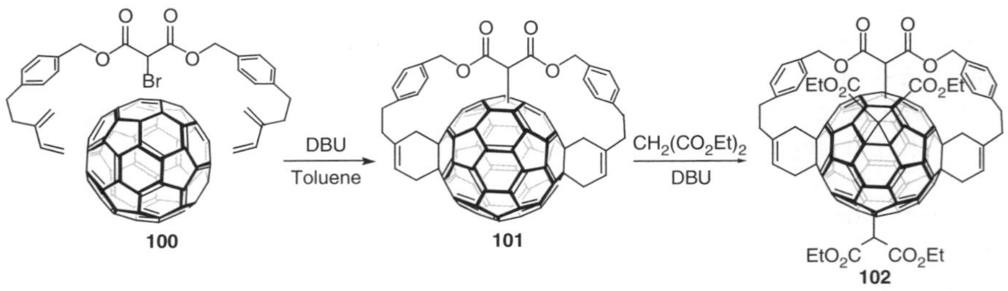

**Scheme 37**

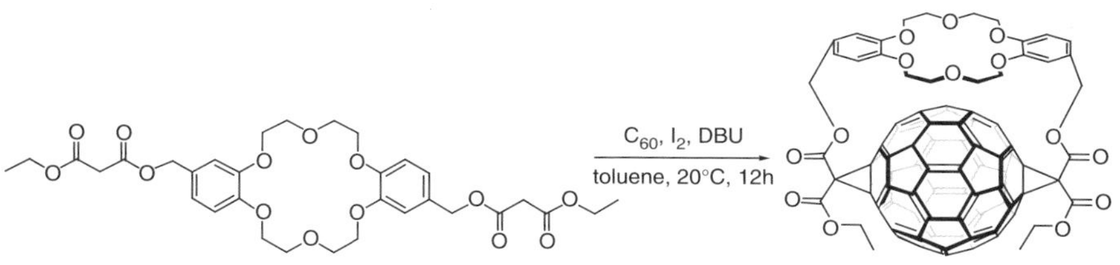

Basic Principles of the Chemical Reactivity of Fullerenes 37

**Scheme 38**

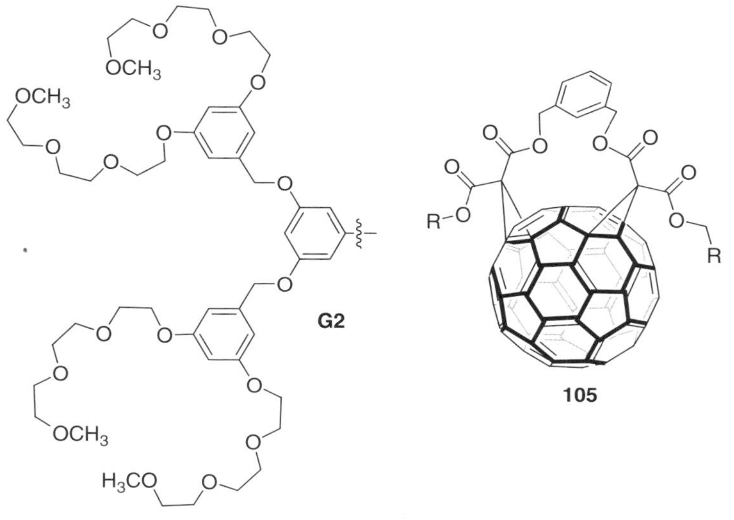

**Chart 7**

by a cyclization reaction between the $C_{60}$ and the corresponding bismalonate derivatives to introduce dendrimers of different generations (Chart 7).[90]

The first example of a regio- and stereoselective synthesis of chiral *trans*-2 bisadducts of $C_{60}$ using chiral tethers has recently been described and was based on the use of Tröger base[91] as a chiral auxiliary (Scheme 39).[92] A modified Bingel reaction of compound **106** with $C_{60}$ afforded the *trans*-2 bisadduct **107** as the major product (27%); a small amount (6%) of *e*-bisadduct was also formed.

The malonate groups are usually connected with rigid organic moieties *via* spacers in such a way that the tether directs the reactive groups to specific positions. Nevertheless, Hirsch and co-workers[93] have studied the formation of bis- (**109**) or tris-fullerene (**111**) adducts using macrocycles containing alkyl spacers between two (**108**) and three (**110**) reactive malonate groups (Scheme 40). The rigid spacers were replaced by flexible alkyl chains and the malonate groups were incorporated into a macrocyclic ring – this type of structure makes it possible to synthesize regioisomeric fullerene bis (**109**) and trisadducts (**111**) with excellent selectivity.

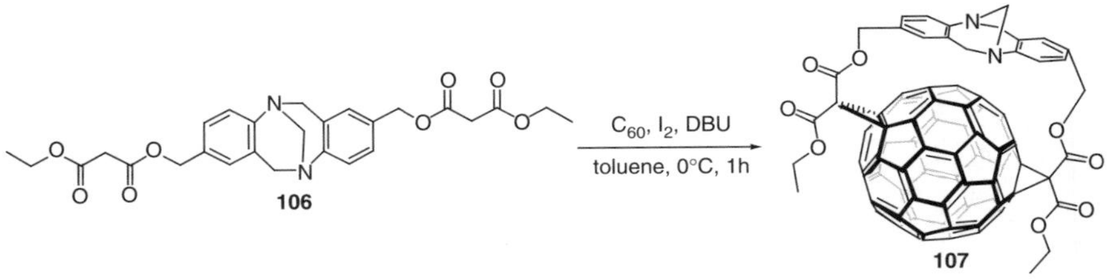

**Scheme 39**

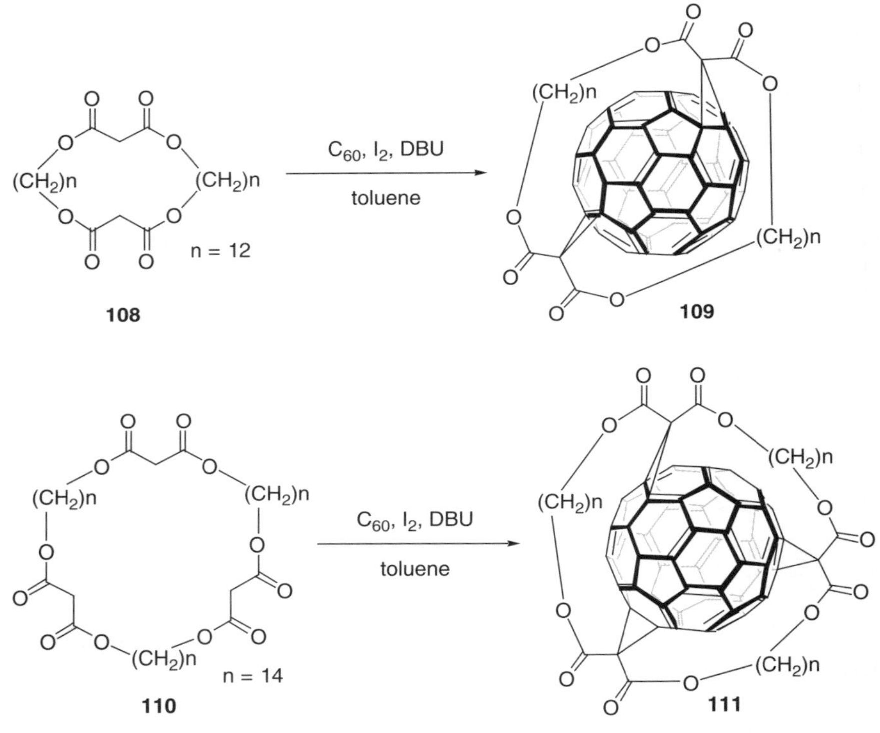

**Scheme 40**

The formation of regioselective tris-adducts is possible by appropriate tuning of the structure. The use of a tripodal tether[94] (**112**) has led to a $C_3$-symmetrical [60]fullerene tris-adducts (**113**) with an $e,e,e$ addition pattern (Scheme 41). A polar zone and a non-polar zone can be formed around the $C_{60}$ cage and the selective deprotection of both zones is possible, an appealing possibility to obtain building blocks for further selective functionalization. A similar approach was used for the regio- and stereoselective synthesis of enantiomerically pure tris-adducts with an inherently chiral $e,e,e$ addition pattern.[95]

Tether-controlled selective synthesis of fulleropyrrolidine bisadducts by the Prato reaction had not been studied until the recent report by D'Souza and

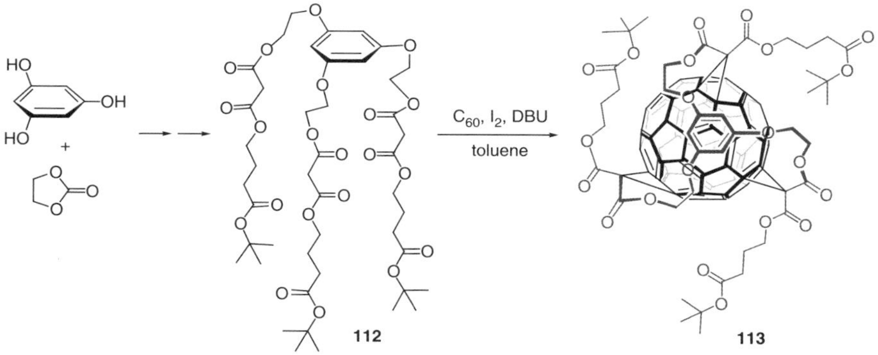

**Scheme 41**

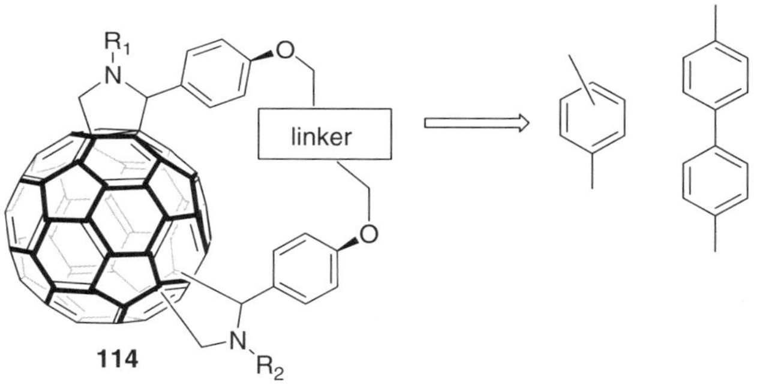

**Chart 8**

co-workers concerning the synthesis of Prato bisadducts with a dibenzo-18-crown-6 bisaldehyde as the tether. The D'Souza approach led to a bisadduct mixture containing several isomers that was directly used for further studies without separation.[96]

The first example of a systematic study on tether-directed selective synthesis of $C_{60}$ fulleropyrrolidine bisadducts was recently described and new bis(benzaldehydes) tethered by a rigid linker were prepared and used to direct the second cycloaddition of an azomethine ylide to $C_{60}$. Equatorial, *trans-4*, *trans-3*, *trans-2* and *trans-1* bisadducts **114** have been selectively prepared using this approach (Chart 8).[97] It should be noted that the nature of the tether linkages between the pyrrolidine rings is complicated by the unavoidable introduction of two new stereogenic centres.

Formation of polyadducts using tethered organic fragments has also been carried out by double Diels–Alder cycloaddition by Rubin's group. A mixture of a tethered bis(buta-1,3-diene) (**115**) and $C_{60}$ was heated under reflux to yield the *trans*-1 bisadduct **116** with a high regioselectivity and reasonable yield (30%) (Scheme 42).[98] The bisadduct **116** is an intermediate for the highly regioselective synthesis of hexakisadducts.

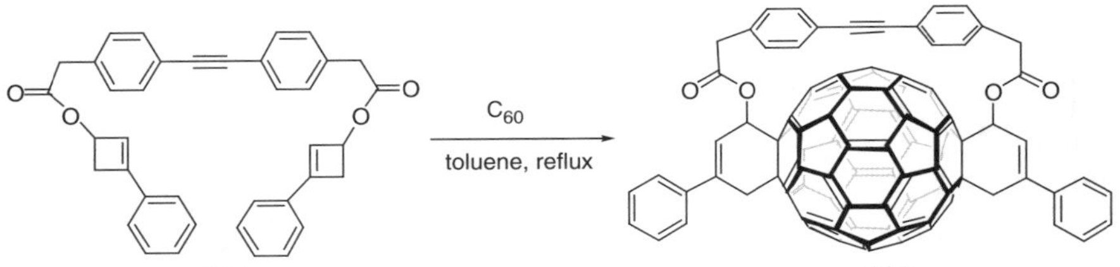

**Scheme 42**

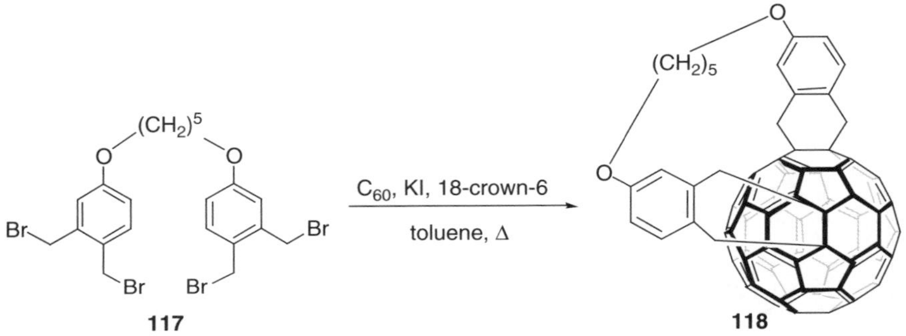

**Scheme 43**

In a different approach to the bis-(*o*-quinodimethane) system, Nishimura and co-workers found that the reaction of [60]fullerene with two α,α'-dibromo-*o*-xylene moieties (**117**), connected by an oligomethylene chain ($n = 2–5$), yields two isomers (*cis*-2 and *cis*-3) for $n = 2$ and 3 but only one isomer (**118**) (*e*) for $n = 5$ (Scheme 43).[99]

The regioselectivity of the addition processes can be controlled not only by using tethered addends but also by binding temporary, but easily removed, substituents onto the $C_{60}$ sphere. The way in which the thermal lability of anthracene adducts of fullerenes enables the synthesis of an equatorial $D_{2h}$-symmetric tetraadduct (**121**) is represented in Scheme 44.[100] The reaction of the cycloadduct **119**, which has two anthracene groups at the poles of the sphere, with an excess of diethyl bromomalonate and DBU yields the hexaadduct **120** in 95% yield. Thus, the symmetric $D_{2h}$-tetraadduct **121** is obtained in 88% yield by heating intermediate **120**. Compound **119** is formed by a solid-state and regioselective anthracene transfer synthesis that involves heating the monoadduct at 180°C for 10 min.[27a]

## 2.3 Reactivity of Higher Fullerenes ($C_{70}$, $C_{76}$, $C_{84}$)

Although the chemistry of [60]fullerene has been the most widely studied in this field since 1990, the synthesis of higher fullerene ($C_{70}$, $C_{76}$, $C_{84}$) derivatives and

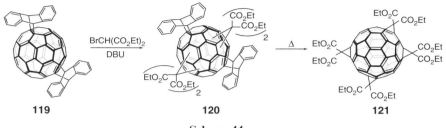

**Scheme 44**

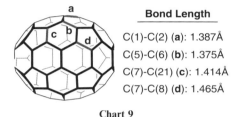

**Bond Length**

C(1)-C(2) (a): 1.387Å
C(5)-C(6) (b): 1.375Å
C(7)-C(21) (c): 1.414Å
C(7)-C(8) (d): 1.465Å

**Chart 9**

their properties represents an important part of fullerene research.[101] Nevertheless, the chemistry of the higher fullerenes is less well developed than the chemistry of $C_{60}$ because the preparation of isomerically pure higher fullerenes is limited – with the exception of $C_{70}$. In 1990, $C_{60}$ and $C_{70}$ were isolated in macroscopic quantities and at the same time evidence for the presence of higher fullerenes $C_{76}/C_{78}$ and $C_{84}$ was observed by mass spectrometry. Many higher fullerenes are chiral and the number of isomers increases rapidly with size and the symmetry simultaneously decreases, meaning that their purification is not easy.[101] Optimal chromatography conditions allowed the isolation and structural characterization of three new fullerenes, chiral $D_2$-$C_{76}$, $C_{2v}$-$C_{78}$ and $D_3$-$C_{78}$. The higher fullerenes act as electron-deficient polyenes and most of the higher fullerene derivatives are obtained by nucleophilic additions and cycloaddition reactions involving the [6,6] bonds. The most important difference with respect to $C_{60}$ is the less symmetrical cages of the higher fullerenes, which have different [6,6] bonds. This situation means that even monoadduct formation can gives rise to many isomers.

In general, the reactivity of $C_{70}$ is similar to that of $C_{60}$ and oxidation, transition-metal complex formation, hydrogenation, halogenation, radical and nucleophilic additions and cycloadditions have been described.[1,101]

Most of the additions in $C_{70}$ take place exclusively on [6,6] bonds but $C_{70}$ contains four different [6,6]bonds. The C(1)–C(2) and C(5)–C(6) bonds are the most reactive, in this order (Chart 9), followed by the C(7)–C(21) bond due to its higher degree of pyramidalization.[102] This relative reactivity was corroborated by theoretical calculations at different levels.[103] The covalent chemistry of higher fullerenes has been excellently reviewed by Diederich and co-workers.[104]

This pattern of reactivity has been experimentally confirmed. Diederich[105] described the first [70]fullerene derivative, which was synthesized by a Diels–Alder reaction between 4,5-dimethoxy-*o*-quinodimethane and $C_{70}$. The main adducts obtained correspond to C(1)–C(2) and C(5)–C(6) bonds. The same reactivity was observed by Wilson[106] in the formation of the pyrrolidino[70]fullerene system by 1,3-dipolar cycloaddition of *N*-methylazomethine ylide.

Nevertheless, the regioselectivity of the cycloaddition between *N*-methylazomethine ylide and $C_{70}$ can be modified by using microwave irradiation as the source of energy. Under microwave irradiation, and by choosing the appropriate solvent and irradiation power, the C(5)–C(6) adduct is the major product, a situation in contrast to that achieved by conventional heating, where the C(1)–C(2) isomer predominates. Moreover, the C(7)–C(21) isomer, which represents 13% of monoadducts under classical heating, is not formed at all under microwave irradiation on using ODCB as solvent.[107]

The use of higher fullerenes as building blocks for molecular materials is a promising and almost unexplored area. A photovoltaic cell built with a methano[70]fullerene (70-PCBM) shows better efficiency than one built with 60-PCBM.[108] Similarly, another photovoltaic solar cell built with the polyfluorene APFO-Green 1 as a donor (Chart 10) and the pyrazolino[70]fullerene **122**[109] as the acceptor unit has recently been described. In this example, the efficiency obtained is double that found for the analogous $C_{60}$ derivative.

The chemistry of fullerenes beyond $C_{70}$ is much less explored, but it seems that relatively few monoadducts are formed despite the large number of 6-6 bonds present. The Diels–Alder cycloaddition reaction of 4,5-dimethoxy-*o*-quinodimethane with $D_2$-$C_{76}$ affords at least six isomers of the monocycloadduct.[110] The Bingel reaction between $D_2$-$C_{76}$ and an (*S,S*)-configured 2-bromomalonate yielded three constitutionally isomeric pairs of diastereoisomeric mono-adducts and these were isolated by HPLC.[111] The addends are bound to [6,6]-type bonds of the fullerene framework. A retro-Bingel reaction allowed optically active $C_{76}$ enantiomers to be obtained.[17]

Methylation of $C_{76}$ by reaction with Al–Ni alloy/NaOH followed by quenching of the intermediate anions with methyl iodide gave a mixture of methylated and methylenated products together with oxide derivatives.[112]

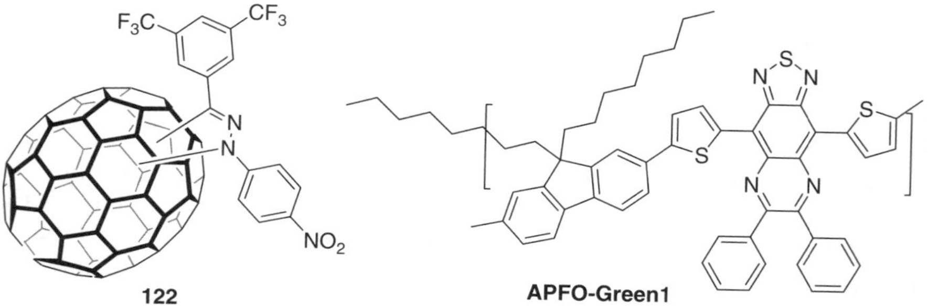

**Chart 10**

In the same way, the cyclopropanation of a 3:1 isomeric mixture of $C_{2v}$–$C_{78}$ and $D_3$–$C_{78}$ with two equivalents of diethyl 2-bromomalonate yielded eight tris-adducts and two $C_1$-symmetrical bisadducts.[113] Furthermore, the retro-cyclopropanation reaction can be used to isolate pure $C_{2v}$–$C_{78}$ and a $C_2$-symmetric bisadduct.[17] Simple bromination of a fullerene mixture containing only 0.67% of $C_{78}$ led to the isolation of crystals containing $C_{78}Br_{18}$, the structure of which was studied by X-ray diffraction. This study showed a highly symmetrical addition of 18 Br atoms characterized by the presence of conjugated double-bond systems and isolated double bonds in the carbon cage. Separation of the $C_{2v}$ and $D_3$ occurred in the course of the bromination.[114]

Recently, Echegoyen and co-workers[115] described the synthesis and characterization of a monoadduct derivative of $Sc_3N@C_{80}$, an N-ethylpyrrolidinofullerene, by a 1,3-dipolar cycloaddition reaction between an azomethine ylide and $Sc_3N@C_{80}$. The addition occured regioselectively on a [5,6] double bond on the surface of the icosahedral symmetric $Sc_3N@C_{80}$ in the same position as described previously for a Diels–Alder adduct of the same compound, synthesized by reaction of $Sc_3N@C_{80}$ with an excess of 6,7-dimethoxyisochroman-3-one.[116]

A retro-Bingel[17] reaction was applied in the separation of enantiomers of $C_{76}$ using an optically active malonate as a chiral auxiliary. This approach was also used recently to separate constitutional isomers and enantiomers of $C_{84}$.

## References

1. W. Krätschmer, L.D. Lamb, K. Fotstiropoulos and D.R. Huffmann, *Nature*, 1990, **347**, 354.
2. R.S. Ruoff, D.S. Tse, R. Malhotra and D.C. Lorents, *J. Phys. Chem.*, 1993, **97**, 3379.
3. A. Hirsch and M. Brettreich, in *Fullerenes: Chemistry and Reactions*, (ed), Wiley, New York, 2005.
4. M.A. Yurovskaya and I.V. Trushkov, *Russ. Chem. Bull., Int. Ed.*, 2002, **51**, 367.
5. T. Suzuki, Q. Li, C. Khemani, F. Wudl and Ö. Almarsson, *Science*, 1991, **254**, 1186.
6. F. Diederich, L.I. saacs and D. Philp, *Chem. Soc. Rev.*, 1994, **23**, 243.
7. J.F. Nierengarten, *Synthesis of methanofullerenes for materials science and biological applications*, in *Fullerenes: From Synthesis to Optoelectronic Properties*, Kluwer, Dordrecht, Chapter 4, 2002, 51.
8. C. Bingel, *Chem. Ber.*, 1993, **126**, 1957.
9. (*a*) J.-F. Nierengarten, V. Gramlich, F. Cardullo and F. Diederich, *Angew. Chem. Int. Ed. Engl.*, 1996, **35**, 2101; (*b*) J.-F. Nierengarten and J.-F. Nicoud, *Tetrahedron Lett.*, 1997, **38**, 1595.
10. X. Camps and A. Hirsch, *J. Chem. Soc., Perkin Trans. 1*, 1997, **11**, 1595.
11. S. Taillemite and D. Fichou, *Eur. J. Org. Chem.*, 2004, 4981.
12. S. Gónzalez, N. Martín and D.M. Guldi, *J. Org. Chem.*, 2003, **68**, 779.
13. E. Allard, F. Oswald, B. Donnio, D. Guillon, J.L. Delgado, F. Langa and R. Deschenaux, *Org. Lett.*, 2005, **7**, 383.

14. (a) T. Chuard and R. Deschenaux, *Helv. Chim. Acta*, 1996, **79**, 736; (b) R. Deschenaux, M. Even and D. Guillon, *Chem. Commun.*, 1998, 537; (c) B. Dardel, R. Deschenaux, M. Even and E. Serrano, *Macromolecules*, 1999, **32**, 5193; (d) N. Tirelli, F. Cardullo, T. Habicher, U.W. Ster and F. Diederich, *J. Chem. Soc. Perkin Trans.*, 2000, 193.
15. D. Bonifazi, G. Accorsi, N. Aramaroli, F. Song, A. Palkar, L. Echegoyen, M. Scholl, P. Seiler, B. Jaun and F. Diederich, *Helv. Chim. Acta*, 2005, **88**, 1839.
16. Y. Nakamura, S. Minami, K. Iizuka and J. Nishimura, *Angew. Chem. Int. Ed. Engl.*, 2003, **42**, 3158.
17. M.A. Herranz, F. Diederich and L. Echegoyen, *Eur. J. Org. Chem.*, 2004, 2299.
18. R. Kessinger, J. Crassous, A. Hermann, M. Rüttimann, L. Echegoyen and F. Diederich, *Angew. Chem. Int. Ed. Engl.*, 1998, **37**, 1919.
19. (a) R. Kessinger, M. Gómez-López, C. Boudon, J.-P. Gisselbrecht, M. Gross, L. Echegoyen and F. Diederich, *J. Am. Chem. Soc.*, 1998, **120**, 8545; (b) L.E. Echegoyen, F.D. Djojo, A. Hirsch and L. Echegoyen, *J. Org. Chem.*, 2000, **65**, 4994.
20. N.N.P. Moonen, C. Thilgen, L. Echegoyen and F. Diederich, *Chem. Commun.*, 2000, 335.
21. M.A. Herranz, L. Echegoyen, M.W.J. Beulen, J.A. Rivera, N. Martín, B. Illescas and M.C. Díaz, *Proc. Electrochem. Soc.*, 2002, **12**, 307.
22. R. Kessinger, N.S. Fender, L.E. Echegoyen, C. Thilgen, L. Echegoyen and F. Diedierich, *Chem. Eur. J.*, 2000, **6**, 2184.
23. G.A. Burley, P.A. Keller, S.G. Pyne and G.E. Ball, *J. Org. Chem.*, 2002, **67**, 8316.
24. F. Langa, P. de la Cruz, A. de la Hoz, A. Díaz-Ortiz and E. Diez-Barra, *Contemp. Org. Synth.*, 1997, 373.
25. (a) V.M. Rotello, J.B. Howard, T. Yadav, M.M. Conn, E. Viani, L.M. Giovane and A.L. Lafleur, *Tetrahedron Lett.*, 1993, **34**, 1561; (b) M. Tsuda, T. Ishida, T. Nogami, S. Kurono and M. Ohashi, *J. Chem. Soc., Chem. Commun.*, 1993, 1296.
26. (a) M.F. Meidine, A.G. Avent, A.D. Darwish, H.W. Kroto, O. Ohashi, R. Taylor and D.R.M. Walton, *J. Chem. Soc., Perkin Trans. 2*, 1994, 1189; (b) H. Takeshita, J.-F. Liu, N. Kato, A. Mori and R. Isobe, *Chem. Lett.*, 1995, **24**, 377; (c) S.R. Wilson, M.E. Yurchenko, D.I. Schuster, A. Khong and M. Saunders, *J. Org. Chem.*, 2000, **65**, 2619.
27. (a) J.A. Schlüter, J.M. Seaman, S. Taha, H. Cohen, K.R. Lykke, H.H. Wang and J.M. Williams, *J. Chem. Soc., Chem. Commun.*, 1993, 972; (b) K. Komatsu, Y. Murata, N. Sugita, K. Takeuchi and T.S.M. Wan, *Tetrahedron Lett.*, 1993, **34**, 8473.
28. (a) R. Kraütler, T. Müller and A. Duarte-Ruiz, *Chem. Eur. J.*, 2001, **2**, 3223; (b) N. Chronakis and M. Organopoulus, *Tetrahedron Lett.*, 2001, **42**, 1201; (c) I. Lamparth, C. Maichle-Mössmer and A. Hirsch, *Angew. Chem. Int. Ed. Engl.*, 1995, **34**, 1607.

29. M.F. Meidine, R. Roers, G.J. Langley, A.G. Avent, A.D. Darwish, S. Firth, H.W. Kroto, R. Taylor and D.R.M. Walton, *J. Chem. Soc., Chem. Commun.*, 1993, 1342.
30. (a) Y. Takaguchi, T. Tajima, K. Ohta, J. Motoyoshiya, H. Aoyama, T. Wakahara and T. Akasaka, *Angew. Chem. Int. Ed. Engl.*, 2002, **41**, 817; (b) Y. Takaguchi, Y. Sako, Y. Yanagimoto, S. Tsunoi, J. Motoyoshiya, H. Aoyama, T. Wakahara and T. Akasaka, *Tetrahedron Lett.*, 2003, **44**, 5777.
31. M.A. Herranz, N. Martín, J. Ramey and D.M. Guldi, *Chem. Commun.*, 2002, 2968.
32. P. de la Cruz, A. de la Hoz, F. Langa, B. Illescas and N. Martín, *Tetrahedron Lett.*, 1997, **38**, 2599.
33. Y. Murata, N. Kato, K. Fujiwara and K. Komatsu, *J. Org. Chem.*, 1999, **64**, 3483.
34. K. Mikami, S. Mtsumoto, T. Tonoi, Y. Okubo, T. Suenobu and S. Fukuzumi, *Tetrahedron Lett.*, 1998, **39**, 3733.
35. (a) B. Kraütler and M. Puchberger, *Helv. Chim. Acta*, 1993, **76**, 1626; (b) M. Ohno, Y. Shirakawa and S. Eguchi, *Synthesis*, 1998, **12**, 1812.
36. M. Diekers, A. Hirsch, S. Pyo, J. Rivera and L. Echegoyen, *Eur. J. Org. Chem.*, 1998, 1111.
37. J.L. Segura and N. Martín, *Chem. Rev.*, 1999, **99**, 3199.
38. (a) P. Belik, A. Gügel, J. Spickerman and K. Müllen, *Angew. Chem. Int. Ed. Engl.*, 1993, **32**, 78; (b) Y. Nakamura, T. Minowa, S. Tobita, H. Shizuka and J. Nishimura, *J. Chem. Soc., Perkin Trans. 2*, 1995 **1**, 2351; (c) M. Taki, S. Sugita, Y. Nakamura, E. Kasashima, E. Yashima, Y. Okamoto and J. Nishimura, *J. Am. Chem. Soc.*, 1997, **119**, 926.
39. (a) A. Kraus, A. Gügel, P. Belik, M. Walter and K. Müllen, *Tetrahedron*, 1995, **51**, 9927; (b) M. Iyoda, F. Sultana, S. Sasaki and M. Yoshida, *J. Chem. Soc., Chem. Commun.*, 1994, 1929; (c) M. Iyoda, S. Sasaki, F. Sultana, M. Yoshida, Y. Kuwatani and S. Nagase, *Tetrahedron Lett.*, 1996, **37**, 7987; (d) H. Tomioka and K. Yamamoto, *J. Chem. Soc., Chem. Commun.*, 1995, 1961.
40. F. Effenberg and G. Grube, *Synthesis*, 1998, **9**, 1372.
41. (a) B.M. Illescas, N. Martín, C. Seoane, E. Ortí, P.M. Viruela, R. Viruela and A. de la Hoz, *J. Org. Chem.*, 1997, **62**, 7585; (b) J.-H. Liu, A.-T. Wu, M.-H. Huang, C.-W. Wu and W.-S. Chung, *J. Org. Chem.*, 2000, **65**, 3395.
42. (a) U.M. Fernandez-Paniagua, B. Illescas, N. Martín, C. Seoane, P. de la Cruz, A. de la Hoz and F. Langa, *J. Org. Chem.*, 1997, **62**, 3705; (b) M. Ohno, N. Koide, H. Sato and S. Eguchi, *Tetrahedron*, 1997, **53**, 9075; (c) F. Effenberger and G. Grube, *Synthesis*, 1998, **9**, 1372.
43. (a) M.A. Herranz and N. Martín, *Org. Lett.*, 1999, **1**, 2005; (b) C. Boulle, J.M. Rabreau, P. Hudhomme, M. Cariou, J. Orduna and J. Garín, *Tetrahedron Lett.*, 1997, **38**, 3909; (c) J. Llacay, M. Mas, E. Molins, J. Veciana, D. Powell and C. Rovira, *Chem. Commun.*, 1997, 659.

44. A.S.D. Sandanayaka, K.-I. Ikeshita, Y. Araki, N. Kihara, Y. Furusho, T. Takata and O. Ito, *J. Mater. Chem.*, 2005, **15**, 2276.
45. B. Illescas, N. Martín, C. Seoane, P. de la Cruz, F. Langa and F. Wudl, *Tetrahedron Lett.*, 1995, **36**, 8307.
46. C.-C. Chi, I.-F. Pai and W.-S. Chung, *Tetrahedron*, 2004, **60**, 10869.
47. G.P. Miller and M.C. Tetreau, *Org. Lett.*, 2000, **2**, 3091.
48. (*a*) M. Maggini, G. Scorrano and M. Prato, *J. Am. Chem. Soc.*, 1993, **115**, 9798; (*b*) M. Prato and M. Maggini, *Acc. Chem. Res.*, 1998, **31**, 519; (*c*) N. Tagmatarchis and M. Prato, *Synlett*, 2003, **6**, 768.
49. (*a*) K.G. Thomas, V. Biju, M.V. George, D.M. Guldi and P.V. Kamat, *J. Phys. Chem. A*, 1998, **102**, 5341; (*b*) A. Bianco, M. Maggini, G. Scorrano, C. Toniolo, G. Marconi, C. Villani and M. Prato, *J. Am. Chem. Soc.*, 1996, **118**, 4072; (*c*) A. Bianco, F. Gasparrini, M. Maggini, D. Misiti, A. Polese, M. Prato, G. Scorrano, C. Toniolo and C. Villani, *J. Am. Chem. Soc.*, 1997, **119**, 7550.
50. (*a*) P. de la Cruz, A. de la Hoz, L.M. Font, F. Langa and M.C. Pérez-Rodríguez, *Tetrahedron Lett.*, 1998, **39**, 6053; (*b*) F. Langa, P. de la Cruz, A. de la Hoz, E. Espildora, F.P. Cosssio and B. Lecea, *J. Org. Chem.*, 2000, **65**, 2499; (*c*) J. Zhang, W. Yang, S. Wang, P. He and S. Zhu, *Synth. Commun.*, 2005, **35**, 89.
51. G.-W. Wang, T.-H. Zhang, E.-H. hao, L.-J. Jiao, Y. Murata and K. Komatsu, *Tetrahedron*, 2002, **59**, 55.
52. (*a*) F. D'Souza, M.E. Zandler, P.M. Smith, G.R. Deviprasad, K. Arkady, M. Fujitsuka and O. Ito, *J. Phys. Chem. A*, 2002, **106**, 649; (*b*) D.M. Guldi, M. Maggini, G. Scorrano and M. Prato, *J. Am. Chem. Soc.*, 1997, **119**, 974; (*c*) V. Mamane and O. Riant, *Tetrahedron*, 2001, **57**, 2555.
53. (*a*) J.J. Oviedo, P. de la Cruz, J. Garín, J. Orduna and F. Langa, *Tetrahedron Lett.*, 2005, **46**, 4781; (*b*) J.J. Oviedo, M.E. El-Khouly, P. de la Cruz, L. Pérez, J. Garín, J. Orduna, Y. Araki, F. Langa and O. Ito, *New J. Chem.*, 2006, **30**, 93.
54. (*a*) R. Fong, D.I. Schuster and S.R. Wilson, *Org. Lett.*, 1999, **1**, 729; (*b*) H. Imahori, H. Yamada, Y. Nishimura, I. Yamazaki and Y. Sakata, *J. Phys. Chem. B*, 2000, **104**, 2099; (*c*) H. Imahori, H. Yamada, S. Ozawa, Y. Sakata and K. Ushida, *Chem. Commun.*, 1999, 1165; (*d*) H. Imahori, H. Norteda, H. Yamada, Y. Nishimura, I. Yamazaki, Y. Sakata and S. Fukuzumi, *J. Am. Chem. Soc.*, 2001, **123**, 100; (*e*) C. Luo, D.M. Guldi, H. Imahori, K. Tamaki and Y. Sakata, *J. Am. Chem. Soc.*, 2000, **122**, 6535; (*f*) A.S.D. Sandanayaka, K.-I. Ikeshita, Y. Araki, N. Kihara, Y. Furusho, T. Takata and O. Ito, *J. Mater. Chem.*, 2005, **15**, 2276.
55. D. González-Rodríguez, T. Torres, D.M. Guldi, J. Rivera, M.A. Herranz and L. Echegoyen, *J. Am. Chem. Soc.*, 2004, **126**, 6301.
56. (*a*) N. Martín, L. Sánchez, M.A. Herranz and D.M. Guldi, *J. Phys. Chem. A*, 2000, **104**, 4648; (*b*) M.C. Diaz, M.A. Herranz, B.M. Illescas, N. Martín, N. Godbert, M.R. Bryce, C. Luo, A. Swartz, G. Anderson and

D.M. Guldi, *J. Org. Chem.*, 2003, **68**, 7711; (*c*) F. Giacalone, J.L. Segura, N. Martin, J. Ramey and D.M. Guldi, *Chem. Eur. J.*, 2005, **11**, 4819.
57. (*a*) T. Chuard and R. Deschenaux, *J. Mater. Chem.*, 2002, **12**, 1944; (*b*) Y. Rio, J.-F. Nicoud, J.L. Rehspringer and J.-F. Nierengarten, *Tetrahedron Lett.*, 2000, **41**, 10207; (*c*) S. Campidelli, E. Vazquez, D. Milic, M. Prato, J. Barbera, D.M. Guldi, M. Marcaccio, D. Paolucci, F. Paolucci and R. Deschenaux, *J. Mater. Chem.*, 2004, **14**, 1266; (*d*) S. Campidelli, J. Lenoble, J. Barbera, F. Paolucci, M. Marcaccio, D. Paolucci and R. Deschenaux, *Macromolecules*, 2005, **38**, 7915.
58. (*a*) T. Gu and J.F. Nierergarten, *Tetrahedron Lett.*, 2001, **42**, 3175; (*b*) J.-F. Eckert, J.-F. Nicoud, J.-F. Nierengarten, S.-G. Liu, L. Echegoyen, F. Barigelletti, N. Armaroli, L. Ouali, V. Krasnikov and G. Hadziioannou, *J. Am. Chem. Soc.*, 2000, **122**, 7467; (*c*) C. Martineau, P. Blanchard, D. Rondeau, J. Delaunay and J. Roncali, *Adv. Mater.*, 2002, **14**, 283.
59. V. Mamane and O. Riani, *Tetrahedron*, 2001, **57**, 2555.
60. P.A. Troshin, A.S. Peregudov, D. Muehlbacher and R.N. Lyubovskaya, *Eur. J. Org. Chem.*, 2005, 3064.
61. S.H. Kang, H. Ma, M.-S. Kang, K.-S. Kim, A.K.-Y. Jen, M.H. Zareie and M. Sarikaya, *Angew. Chem. Int. Ed. Engl.*, 2004, **43**, 1512.
62. (*a*) T. Da Ros, M. Prato, F. Novello and M. Magginiand E. Banfi, *J. Org. Chem.*, 1996, **61**, 9070; (*b*) S. Bosi, L. Feruglio, D. Milic and M. Prato, *Eur. J. Org. Chem.*, 2003, 4741; (*c*) B.M. Illescas, R. Martinez-Alvarez, J. Fernandez-Gadea and N. Martin, *Tetrahedron*, 2003, **59**, 6569; (*d*) J.L. Segura, E.M. Priego, N. Martin, C. Luo and D.M. Guldi, *Org. Lett.*, 2000, **2**, 4021.
63. P. Sofou, Y. Elemes, E. Panou-Pomonis, A. Stavrakoudis, V. Tsikaris, C. Sakarellos, M. Sakarellos-Daitsiotis, M. Maggini, F. Formaggio and C. Toniolo, *Tetrahedron*, 2004, **60**, 2823.
64. (*a*) T. Da Ros, M. Prato, M. Carano, P. Ceroni, F. Paolucci and S. Roffia, *J. Am. Chem.*, 1998, **120**, 11645; (*b*) T. Mashino, D. Nishikawa, K. Takahashi, N. Usui, T. Yamori, M. Seki, T. Endo and M. Mochizukia, *Bioorg. Med. Chem. Lett.*, 2003, **13**, 4395.
65. (*a*) A. Bagno, S. Claeson, M. Maggini, M.L. Martini, M. Prato and G. Scorrano, *Chem. Eur. J.*, 2002, **8**, 1015; (*b*) F. D'Souza, M.E. Zandler, G.R. Deviprasad and W. Kutner, *J. Phys. Chem. A*, 2000, **104**, 6887.
66. (*a*) A. Polese, S. Mondini, A. Bianco, C. Toniolo, G. Scorrano, D.M. Guldi and M. Maggini, *J. Am. Chem. Soc.*, 1999, **121**, 3446; (*b*) M. Maggini, A. Karlsson, L. Pasimeni, G. Scorrano, M. Prato and L. Valli, *Tetrahedron Lett.*, 1994, **35**, 2985; (*c*) F. D'Souza, M.E. Zandler, P.M. Smith, G.R. Deviprasad, K. Aracady, M. Fujitsuka and O. Ito, *J. Phys. Chem. A*, 2002, **106**, 649; (*d*) M.A. Herranz, B. Illescas, N. Martín, C. Luo and D.M. Guldi, *J. Org. Chem.*, 2000, **65**, 5728.
67. (*a*) P. de la Cruz, A. de la Hoz, F. Langa, N. Martín, M.C. Pérez-Rodríguez and L. Sánchez, *Eur. J. Org. Chem.*, 1999, **1**, 3433; (*b*) G.R. Deviprasad, M.S. Rahman and F. D'Souza, *Chem. Commun.*, 1999, 849.

68. N. Martín, M. Altable, S. Filipone, A. Martín-Domenech, L. Echegoyen and C.M. Cardona, *Angew. Chem. Int. Ed. Engl.*, 2006, **45**, 110.
69. (*a*) J.C. Hummelen, B. Knight, J. Pavlovich, R. González and F. Wudl, *Science*, 1995, **269**, 1554; (*b*) F. Zhou, G.J.V. Berkel and B.T. Donovan, *J. Am. Chem. Soc.*, 1994, **116**, 5485; (*c*) N. Liu, H. Touhara, Y. Morio, D. Komichi, F. Okino and S. Kawasaki, *J. Electrochem. Soc.*, 1996, **143**, 214.
70. C.K. Mathews, P.R.V. Rao, R. Ragunathan, P. Maruthamuthu and S. Muthu, *Tetrahedron Lett.*, 1994, **35**, 1763.
71. (*a*) Y. Matsubara, H. Tada, S. Nagase and Z. Yoshida, *J. Org. Chem.*, 1995, **60**, 5372; (*b*) Y. Matsubara, H. Muraoka, H. Tada and Z. Yoshida, *Chem. Lett.*, 1996, **25**, 373.
72. P. de la Cruz, A. Díaz-Ortiz, J.J. García, M.J. Gómez-Escalonilla, A. de la Hoz and F. Langa, *Tetrahedron Lett.*, 1999, **40**, 1587.
73. E. Espíldora, J.L. Delgado, P. de la Cruz, A. de la Hoz, V. López-Arza and F. Langa, *Tetrahedron*, 2002, **58**, 5821.
74. N. Armaroli, G. Accorsi, J.P. Gisselbrecht, M. Gross, V. Krasnikov, D. Tsamouras, G. Hadziioannou, M.J. Gómez-Escalonilla, F. Langa, J.F. Eckert and J.F. Nierengarten, *J. Mater. Chem.*, 2002, **12**, 2077.
75. F. Langa, M.J. Gómez-Escalonilla, E. Díez-Barra, J.C. García-Martínez, A. de la Hoz, J. Rodríguez-Lopéz, A. González-Cortés and V. López-Arza, *Tetrahedron Lett.*, 2001, **42**, 3435.
76. J.L. Delgado, P. de la Cruz, V. López-Arza, F. Langa, D.B. Kimball, M.M. Haley, Y. Araki and O. Ito, *J. Org. Chem.*, 2004, **69**, 2661.
77. M.J. Gómez-Escalonilla, F. Langa, J.M. Rueff, L. Oswald and J.F. Nierengarten, *Tetrahedron Lett.*, 2002, **43**, 7507.
78. F. Langa and F. Oswald, *C. R. Chimie*, 2006, **9**, 1058.
79. M.V. Reinov, M.A. Yurovskaya, D.V. Davydov and A.V. Streletskii, *Chem. Heterocycl. Comp.*, 2004, **40**, 188.
80. G.W. Wang, Y.J. Li, R.F. Peng, Z.H. Liang and Y.C. Liu, *Tetrahedron*, 2004, **60**, 3921.
81. W.H. Powell, F. Cozzi, G.P. Moss, C. Thilgen, R.J.R. Hwu and A. Yerin, *Pure Appl. Chem.*, 2002, **74**, 629.
82. A. Hirsch, I. Lamparth and H.R. Karfunkel, *Angew. Chem. Int. Ed.*, 1994, **33**, 437.
83. (*a*) K.D. Kampe, N. Egger and M. Vogel, *Angew. Chem.*, 1993, **105**, 1203; (*b*) J.M. Hawkins, A. Meyer, T.A. Lewis, U. Bunz, R. Nunlist, G.E. Ball, T.W. Wbbesen and K. Tanigaki, *J. Am. Chem. Soc.*, 1992, **114**, 7954; (*c*) C.C. Henderson, C.M. Rohlfing, R.A. Assink and P.A. Cahill, *Angew. Chem. Int. Ed. Engl.*, 1994, **33**, 786; (*d*) Y. Nakamura, N. Takano, T. Nishimura, E. Yashima, M. Sato, T. Kudo and J. Nishimura, *Org. Lett.*, 2001, **3**, 1193.
84. X. Camps and A. Hirsch, *J. Chem. Soc., Perkin Trans. 1*, 1997, 1595.
85. (*a*) K. Kordatos, S. Bosi, T. Da Ros, A. Zambon, V.V. Lucchini and M. Prato, *J. Org. Chem.*, 2001, **66**, 2802; (*b*) Q. Lu, D.I. Schuster and S.R. Wilson, *J. Org. Chem.*, 1996, **61**, 4764.

86. L. Isaacs, R.F. Haldimann and F. Diederich, *Angew. Chem. Int. Ed.*, 1994, **33**, 2339.
87. (*a*) A. Hirsch and O. Vostrowsky, *Eur. J. Org. Chem.*, 2001, **1**, 829; (*b*) F. Diederich and R. Kessinger, *Templated Organic Synthesis*, Wiley-VCH, Weinheim, 2000, 189.
88. For reviews on tether-directed synthesis: (a) F. Diederich and R. Kessinger, *Acc. Chem. Res.*, 1999, 32, 537; (b) Z. Zhou and S.R. Wilson, *Curr. Org. Chem.*, 2005, 9, 789.
89. J.-P. Bourgeois, L. Echegoyen, M. Fibbioli, E. Pretsch and F. Diederich, *Angew. Chem. Int. Ed.*, 1998, **37**, 2118.
90. J.-F. Nierengarten, N. Armaroli, G. Accorsi, Y. Rio and J.-F. Eckert, *Chem. Eur. J.*, 2003, **9**, 37.
91. (*a*) J. Tröger, *J. Prakt. Chem.*, 1887, **36**, 225; (*b*) V. Prelog and P. Wieland, *Helv. Chim. Acta*, 1944, **27**, 1127; (*c*) E. Kim, S. Paliwal and C.S. Wilcox, *J. Am. Chem. Soc.*, 1998, **120**, 11192.
92. S. Sergeyev and F. Diederich, *Angew. Chem. Int. Ed.*, 2004, **43**, 1738.
93. (*a*) U. Reuther, F. Hampel and A. Hirsch, *Chem. Eur. J.*, 2002, **8**, 2833; (*b*) N. Chronakis and A. Hirsch, *C. R. Chimie*, 2006, **9**, 862.
94. F. Beuerle, N. Chronakis and A. Hirsch, *Chem. Commun.*, 2005, 3676.
95. N. Chronakis and A. Hirsch, *Chem. Commun.*, 2005, 3709.
96. P.M. Smith, A.L. McCarty, N.Y. Nguyen, M.E. Zandler and F. D'Souza, *Chem. Commun.*, 2003, 1754.
97. Z. Zhou, D.I. Schuster and S.R. Wilson, *J. Org. Chem.*, 2006, **71**, 1545.
98. (*a*) Y. Rubin, *Chem. Eur. J.*, 1997, **3**, 1009; (*b*) W. Quian and Y. Rubin, *Angew. Chem. Int. Ed. Engl.*, 1999, **38**, 2356; (*c*) W. Quian and Y. Rubin, *J. Org. Chem.*, 2002, **67**, 7683.
99. M. Taki, S. Sugita, Y. Nakamura, E. Kasashima, E. Yashima, Y. Okamoto and J. Nishimura, *J. Am. Chem. Soc.*, 1997, **119**, 926.
100. R. Schwenninger, T. Müller and B. Kräutler, *J. Am. Chem. Soc.*, 1997, **119**, 9317.
101. F. Diederich and R.L. Whetten, *Acc. Chem. Res.*, 1992, **25**, 119.
102. C. Thilgen, A. Herrmann and F. Diederich, *Angew. Chem., Int. Ed.Engl.*, 1997, **36**, 2269.
103. (*a*) H.R. Karfunkel and A. Hirsch, *Angew. Chem.*, 1992, **104**, 1529; (*b*) C.C. Henderson, C.M. Rohlfing, K.T. Gillen and P.A. Cahill, *Science*, 1994, **264**, 397.
104. (*a*) C. Thilgen and F. Diederich, *Top. Curr. Chem.*, 1999, **199**, 135; (*b*) C. Thilgen, I. Gosse and F. Diederich, *Top. Stereochem.*, 2003, **23**, 1.
105. A. Hermann, F. Diederich, C. Thilgen, H.-U. ter Meer and W.H. Müller, *Helv. Chim. Acta*, 1994, **77**, 1689.
106. S.R. Wilson and Q. Lu, *J. Org. Chem.*, 1995, **60**, 6496.
107. F. Langa, P. de la Cruz, A. de la Hoz, E. Espíldora, F.P. Cossío and B. Lecea, *J. Org. Chem.*, 2000, **65**, 2499.
108. M.M. Wienk, J.M. Kroon, W.J.H. Verhees, J. Knol, J.C. Hummelen, P.A. van Hal and R.A.J. Janssen, *Angew. Chem. Int. Ed.*, 2003, **42**, 3371.

109. X. Wang, E. Perzon, F. Oswald, F. Langa, S. Admassie, M.R. Andersson and O. Inganäs, *Adv. Funct. Mater.*, 2005, **15**, 1665.
110. A. Herrmann, F. Diederich, C. Thilgen, H.-U. ter Meer and W.H. Müller, *Helv. Chim. Acta*, 1994, **77**, 1689.
111. A. Herrmann and F. Diederich, *Helv. Chim. Acta*, 1996, **79**, 1741.
112. A.D. Darwish, N. Martsinovich and R. Taylor, *Org. Biomol.Chem.*, 2004, **2**, 1364.
113. A. Herrmann and F. Diederich, *J. Chem. Soc., Perkin Trans. 2*, 1997, 1679.
114. S.I. Troyanov and E. Kemnitz, *Eur. J. Org. Chem.*, 2003, 3916.
115. C.M. Cardona, A. Kitaygorodskiy, A. Ortiz, A.A. Herranz and L. Echegoyen, *J. Org. Chem.*, 2005, **70**, 5092.
116. E.B. Iezzi, J.C. Duchamp, K. Harich, T.E. Glass, H. Man Lee, M.M. Olmstead, A.L. Balch and H.C. Dorn, *J. Am. Chem. Soc.*, 2002, **124**, 524.

CHAPTER 3
# Three Electrodes and a Cage: An Account of Electrochemical Research on $C_{60}$, $C_{70}$ and their Derivatives[†]

MAURIZIO CARANO[‡], MASSIMO MARCACCIO AND FRANCESCO PAOLUCCI

INSTM, Unit of Bologna, Dipartimento di Chimica G. Ciamician, University of Bologna, Alma Mater Studiorum, Via Selmi 2, 40126 Bologna, Italy

## 3.1 Introduction

After quite a few years since their discovery[1] fullerenes are now a well-established subject of research in many areas of chemistry.[2] Years of studies have made researchers aware of both potentials and limitations of this class of compounds. The first spread of ideas and the more or less plausible foreseen applications have now converged and focused in what can be viewed as an integrated field of science. Fullerene science plays a noticeable role in fields, such as Nanotechnology, Supramolecular Chemistry and Materials Chemistry. One of the most promising territories of investigation involves modification of the fullerene cage (mainly $C_{60}$) to improve and/or enhance its qualities for various applicative purposes. In this perspective one of the first proposed applications, drug delivery, is still an open research territory once it will be clear how to choose the appropriate ligands to be attached onto the outside of the carbon cage. Biological applications in a broader meaning are still and constantly under study.[3] Some biological activities based on their unique physical properties and chemical reactivities have been reported. Fullerenes have also shown the ability to block the HIV virus from attacking cells under certain conditions,[4] DNA scissions and oxidation of biological materials depend on photo-induced active oxygen production by fullerene and enzyme-inhibition

---
[†] Dedicated to Prof. Sergio Roffia on his 70th birthday.
[‡] On leave from Department of Chemistry, Monash University, Victoria, Australia.

activities depend on the high hydrophobicity of fullerene. Many groups have reported antioxidant activities of fullerene, which are thought to depend on high reactivity for radicals.[5] Last but not least, the use of fullerenes in solar cells, sensors and other devices is still a strong driving force for much research.

Since the very beginning of the fullerene era, with the observation of the six reversible one-electron reductions of $C_{60}$ (and $C_{70}$), electrochemistry has been among the most popular and powerful tools to explore the rather unique electronic and chemical properties of $C_{60}$ and higher fullerenes and of their derivatives. Methods and concepts borrowed from molecular electrochemistry,[6] with its wealth of experimental techniques, methodologies and theoretical knowledge accumulated over a period of many decades, were largely applied to the new class of molecules. Then, a comprehensive and unifying description of the redox properties and of the typical electrochemically induced reactivity of fullerenes and of their derivatives was achieved over a period of few years.

Fullerenes were soon recognized, for their unique electronic and structural properties, as ideal candidates for the development of molecular or supramolecular devices, such as, for instance, Donor–Spacer–Acceptor systems where the fullerenes can play successfully the role of electron/energy acceptor, or in electrochemical sensors in which electronic transduction is carried out by the fullerene. In such a context, electrochemistry has also played a fundamental role in the investigation of the subtle and sometimes elusive effects that the supramolecular environment exerts on the redox properties of the molecular components, of both the fullerene moieties and their partners. The electrochemical study, supported by spectroscopic, photochemical and photoelectrochemical techniques, may provide fundamental information (i) on the spatial organisation of the redox sites within the molecular or supramolecular structure, along with the energy location of the corresponding redox orbitals; (ii) on the entity of the interactions between units in the supramolecular framework and therefore on the thermodynamical feasibility of such processes as intramolecular electron/energy transfers and electrochemically induced isomerization and conformational changes; (iii) on the kinetic stability of reduced/oxidized and charge-separated species; and (iv) on the interaction of such species with conducting and semiconducting materials in view of the realization of electrochemical, photoelectrochemical and photovoltaic devices.

The organic functionalization of fullerenes,[7] is driven by the possibility of combining some of the outstanding properties of the fullerenes with those of other interesting materials, such as photoactive and/or electroactive units. Another useful task of the organic modification of the fullerenes is the increased solubility of the fullerene derivatives in polar solvents or aqueous media where $C_{60}$, $C_{70}$ and the other fullerenes are not soluble or only sparingly soluble.

A comprehensive coverage of all the aspects of the electrochemistry of fullerenes and fullerenes derivatives is beyond the scopes of the authors. The selected examples are taken from their own contribution to this research field or otherwise reflect their personal preferences. Excellent and comprehensive reviews and chapters in either this or other books[8] cover many other aspects of this fascinating area, and the reader is directed to them for further information.

## 3.2 The Electrochemical Properties of Fullerenes

The modulation of electron accepting and/or donor properties and the understanding of electro-induced reactivity are naturally the reign of electrochemistry. The knowledgeable and controlled fine-tuning of fullerenes properties has represented for several years the main research focus for many groups involved in the field. Improved electron affinity, complexing and sensing capabilities, electronic communication in molecular and supramolecular units and assemblies and many others can often be accurately described by cyclic voltammetry experiments. These techniques greatly benefit from the use of ultramicroelectrodes and low temperatures and implies a careful control of all the experimental conditions[§] for the study of ultra-fast kinetics and electro-generated intermediates with micro and sub-microseconds lifetimes.

### 3.2.1 Increasing the Electronegativity of Fullerenes

From the earliest days of fullerene chemistry, the electrochemistry of both the parent cages and their derivatives has been of considerable interest; the topic has been reviewed recently.[8] In particular the effect of successive additions to the cages has been analyzed in terms of changing hybridization of the cage carbons, and electron donation (or withdrawal) by the addends.

In general, change in hybridization of carbons from $sp^2$ to the less electronegative $sp^3$ state reduces electron withdrawal. It then becomes more difficult to add electrons and so reduction becomes harder. Thus addition of most addends causes a decrease in the electrochemical potential. A notable exception to this is the addition of two (electron-withdrawing) phenyl groups to the fullerene which makes reduction easier, though addition of further groups then makes subsequent reduction harder.[9] If however, a large number of strongly electron-withdrawing addends become attached to the cage, then the effect of hybridization change can be outweighed. This however has only been achieved in fluorination, where the small size of the addends allows substantial polyaddition to take place. The consequence is that the reduction potentials of $C_{60}$, $C_{60}F_{18}$, $C_{60}F_{36}$ and $C_{60}F_{48}$, are (in dichloromethane, vs. SCE) −0.59, 0.04, 0.24 and 0.79 V, respectively.[10] As a consequence of the saturation of its double bonds, most of $C_{60}$ mono and polyadducts have decreased electronegativity with respect to pristine fullerene. $C_{60}F_{18}$ (Figure 1) is a rather unique member of the fluorofullerene family because it possesses a flattened hemisphere comprising an aromatic face surrounded by a fluorinated crown; and a curved "normal" hemisphere akin to its all-carbon parent.[11] Substitution chemistry to $C_{60}F_{18}$ allows the attachment of a plethora of functionalities in three precise locations on the fullerene surface leading to derivatives of general formula $C_{60}F_{15}X_3$.[12]

---

[§](That is particular attention to purity and strictly anhydrous experimental conditions have characterized the activity of our Laboratory).

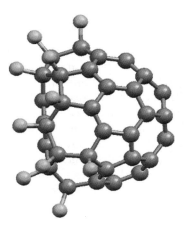

**Figure 1**   $C_{60}F_{18}$

### 3.2.2 Fulleropyrrolidines: Mono and Bis-Adducts

We have recently shown that the standard potential relative to reduction of fulleropyrrolidines can be modulated by alkylating the nitrogen atom in the fullerene-fused heterocycle (Scheme 1).[13,14] These molecules have been shown to possess enhanced electron-accepting properties with respect to both the parent pyrrolidine derivatives and $C_{60}$

The first reduction of **2** (Figure 2) is in fact located at a less negative potential than the corresponding process in its non-ionic version **1** and $C_{60}$ by 180 and 60 mV respectively. Furthermore, CV measurements performed at low temperatures and fast scan rates, using ultramicroelectrodes, allowed the observation, for the first time in a fullerene derivative, of six $C_{60}$-centered reductions.

As a probable consequence of the stabilizing effect of the two positive charges in the methylated *bis*-adduct (from *cis*-1 to *trans*-1, 8 different stereo-isomers are possible for such a molecule), the first reductions of 4-*trans*-**2** and 4-*trans*-**1** are even easier (by 20 and 30 mV, respectively) than those in monoadduct **2**. Incidentally, this places these charged isomers among the strongest reversible electron-accepting $C_{60}$ derivatives, being anticipated with respect to $C_{60}$ itself by 80 and 90 mV, respectively.

Furthermore, the study evidenced that, in both **3** and **4**, the CV pattern (see for instance Figure 3 for the *trans*-1 and *cis*-3 isomers), and in particular the potential separation between the second and third reductions, changes significantly with the addition pattern (Figure 4).

Both empirical considerations and quantum chemical AM1 calculations failed to reproduce the experimental differences that could instead be simply explained with a sequential π-electron model whose main features have been used in the past to predict the halogen addition pattern in $C_{60}X_6$ (X = Cl, Br) and $C_{70}Cl_{10}$.

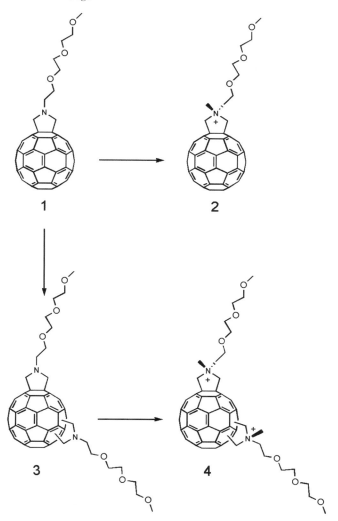

**Scheme 1** *Some derivatization possibilities using alkylation reaction on fulleropyrrolidines*

### 3.2.3 Polyphenylated Derivatives

In the perspective of studying the influence of multiple addition also on $C_{70}$, several aryl-fullerenes ($C_{70}Ph_{2n}$, with $n = $ 1–5) were characterized (Figure 5).[10]

The investigation, made using CV, evidenced electrochemical properties, which are the result of a competition between the electron-withdrawing effect exerted by the phenyl groups and the destabilization of redox orbitals deriving from the saturation of double bonds. As a result, $C_{70}Ph_2$ is easier to reduce than

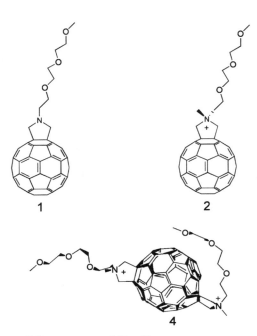

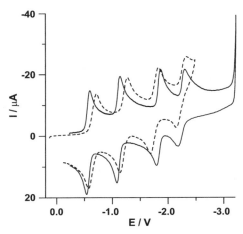

**Figure 2** *Fulleropyrrolidines: mono and bisadducts*

**Figure 3** *CV curves of 0.5 mM (full line) 3-trans-1 and (dashed line) 3-cis-3 THF solution (0.05 M TBAH). Scan rate: 0.5 V/s. Working electrode: Pt wire. T=−60 °C*

$C_{70}$, a feature possibly shared with $C_{70}Ph_4$ that makes these derivatives of $C_{70}$ among the best electron acceptor molecules so far discovered. In the rest of the series destabilization prevails, and reductions become increasingly more cathodic as the number of phenyls added to the fullerene cage goes up.

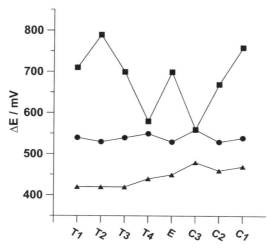

**Figure 4**  Potential gaps ($\Delta E$, absolute values) between subsequent redox processes in bisfulleropyrrolidine 3. triangles: $E_{1/2}(II)$-$E_{1/2}(I)$; squares: $E_{1/2}(III)$-$E_{1/2}(II)$; circles: $E_{1/2}(IV)$-$E_{1/2}(III)$. Data obtained at $-60°C$

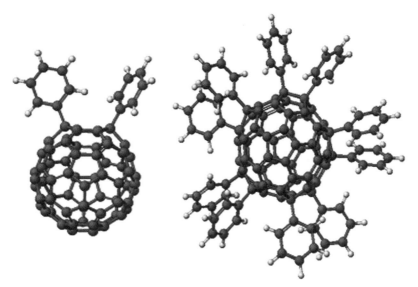

**Figure 5**  $C_{70}Ph_2$ and $C_{70}Ph_{10}$

### 3.2.4  $C_{60}Ph_5Cl$, a Very Special Polyphenylated Fullerene Derivative

In $C_{60}Ph_5Cl$ (Figure 6) the presence of six substituents (1 Cl and 5 Ph) on the $C_{60}$ unit should, at least in principle, shift the reduction processes toward more negative potentials.[15]

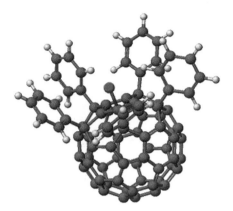

**Figure 6**  *$C_{60}Ph_5Cl$, the electrochemical yo-yo*

At the same time the presence of the phenyls groups is supposed to increase the electron accepting properties *via* an inductive effect. The chlorine atom sits on one of the carbons of the pentagon surrounded by the phenyls to effectively form a chlorocyclopentadiene fragment.[16] $C_{60}Ph_5Cl$ has electrochemical properties strikingly different from pristine $C_{60}$: the cyclic voltammetric curve (Figure 7) shows that, in the first reduction peak, two electrons are sequentially and irreversibly injected at the same potential.

The kinetic analysis of the voltammetric response suggested that, concomitantly with the electron transfer, the carbon-chlorine bond is severed and that back-oxidation of the resulting fragments regenerates the pristine fullerene derivative. The cleavage and formation of the C-Cl bond is therefore driven by the electron flow (Scheme 2): the different oxidation states of this derivative of $C_{60}$ repel and attract the chlorine atom, a property that further adds to the well-known special character of fullerene anions. $C_{60}Ph_5Cl$ therefore presents very interesting chemical features,[17] brought together in a single molecule by an elegant architecture.

## 3.3 Electrochemically Induced Reactivity in Fullerene Derivatives

The first example of reactivity induced in fullerene derivatives by electron transfer processes was the bond-isomerization reaction that, upon the addition of 3-4 electrons (depending on the temperature and/or potential scan rate), converts the kinetically favored -but less stable- [5,6] fulleroid **5** in the corresponding more stable [6,6] methanofullerene **6** (Scheme 3).

Such a reaction, that may also occur upon either thermal or photochemical activation, was discovered in the early 1990s[18,19] and dominated the electrochemical scene for a few years. The various mechanisms proposed for this reaction all agree about a first step in which the breaking of one of the bonds

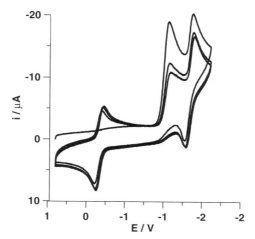

**Figure 7** *Cyclic voltammetric curve of $C_{60}Ph_5Cl$ in multiple-scan CV curve of a 0.5 mM $C_{60}Ph_5Cl$, 0.05 M $Bu_4NPF_6$, $CH_2Cl_2$ solution. Scan rate=0.5 V s−1; T=−50 °C; working electrode: Pt. The initial scan starts toward negative potentials and the successive scans are performed without the renewal of the diffusion layer*

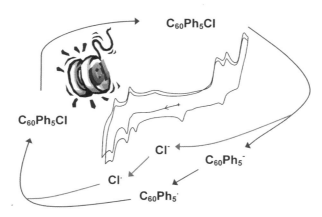

**Scheme 2** *Pictorial representation of the reversible electrochemically induced C-Cl bond breaking and reformation mechanism in $C_{60}Ph_5Cl$*

between the fullerene cage and the bridging carbon atom occurs following the electron transfer reaction. It was also proposed that the electronic structure of the added group may play a fundamental role in the bond-breaking reaction (*periconjugation*).[20] An interesting example of fulleroid-methanofullerene conversion upon injection of a single electron was reported for species **7** (Figure 8).[21] The mechanism was confirmed by performing the digital simulation of the CV curves and the intervention of the nitroxide group in the stabilization of the open form (intermediate), with the formation of a covalent bond *via* a radical–radical coupling between nitroxide oxygen and a bridgehead carbon atom, was invoked.

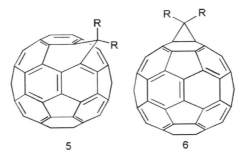

**Scheme 3**  *Structures of [5,6] fulleroid and [6,6] methanofullerene*

**Figure 8**  *2,2,6,6-tetramethylpiperidine-1-oxyl (TEMPO) fulleroid*

### 3.3.1 The Electrochemically Induced Retro-cyclopropanation Reaction

While the presence of an electrochemically induced reactivity may represent a complication in the characterization of fullerene-based devices, the possibility of switching on and off such a reactivity upon the transfer of electrons can profitably be used for specific purposes. This is exemplified for instance by the recent work of Echegoyen *et al.*, who developed very elegant regioselective synthetic routes to complex supramolecular architectures that, in view of the unique electronic properties of the fullerenes, would allow the development of novel molecular electronic devices. This topic was recently reviewed.[22] Such an approach is largely based on the extensive use of the (Bingel[23]) cyclopropanation/electrolytic retro-cyclopropanation strategy, the so-called retro-Bingel reaction, *i.e.* the electrochemical removal of addends from malonate fullerene derivatives – the Bingel derivatives (Figure 9).[24] In this context, digital simulation[25] of the voltammetric experiments proved to be a very useful tool for the description at both a qualitative and quantitative level of the overall electrode mechanism.[26]

## 3.3.2 The Role of Digital Simulation of the CV Experiments in Mechanistic Studies

The electrochemical removal of methano-adducts is not limited to the malonate derivatives of fullerenes (Bingel adducts, Figure 9) but is common to all compounds bearing a cyclopropane ring fused to the carbon cage.[27] Figure 10 shows an example of the typical voltammetric pattern that is observed with such compounds.

Important results, in this respect, obtained by the CV investigations, are (i) that the reaction responsible for the irreversible behavior shown in Figure 10 is only triggered by the injection of at least three (at room temperature) or four (at

**Figure 9** *Fullerene malonate (Bingel adduct)*

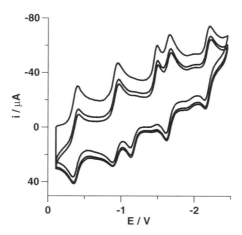

**Figure 10** *Cyclic voltammetric curve of 0.5 mM $C_{60}$ malonate in tetrahydrofuran (0.05 M tetrabutylammonium hexafluorophosphate). $T = 25°C$, scan rate=0.2 V/s, working electrode Pt, potential vs. SCE*

low temperature) electrons; and (ii) that the reaction coupled to the multiple reduction of the fullerene derivatives is reversed upon their re-oxidation, using relatively low scan rates. This implies that the transient chemical species generated by the reduction of malonate derivatives, presumably to be identified with the corresponding fullerene derivatives in which one of the cyclopropane bonds has been cleaved, are stable in the relatively long time scale of CV. Such is not the case when controlled potential electrolysis is carried out. At potentials that correspond to the second fullerene-centered reduction of the malonate derivative, the formation of pristine $C_{60}$ is observed: the species undergoes the so-called retro-Bingel reaction, that consists of the irreversible removal, under bulk electrolysis conditions, of *bis*(alkoxycarbonyl)methanoadducts to give the parent $C_{60}$.

A systematic study to determine the kinetic parameters for the electrochemically induced retro-cyclopropanation reaction carried out using digital simulations evidenced also the role of the substituents in the electro-induced reactivity of cyclopropane $C_{60}$ derivatives. The study was conducted with three different $C_{60}$ derivatives (Figure 11) in which the addends result in different electrochemical responses. They range from the non-electroactive addend in **8** to the electroactive and highly interacting unit in compound **10**, while in **9** the anthraquinone moiety is an electroactive species that interacts poorly with the fullerene cage. This wide range of characteristics explains the differences between the proposed mechanisms for the compounds. Comparing the rate constants at the same reduction stage of the $C_{60}$ core, that is, when the same number of electrons has been transferred to the fullerene cage shows that the rate for the cleavage of the cyclopropane ring increases in going from **8** to **10**. Digital simulations confirmed the proposed mechanism and allowed the estimation of the rate constants of the chemical reactions. Quantum mechanical calculation well supported both experimental evidence and the mechanism hypothesis.[27]

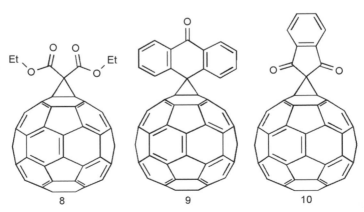

**Figure 11** *Bingel derivatives used in the CV investigation of the retro cyclopropanation reaction*

## 3.4 Fullerenes as Active Components of Molecular Devices

### 3.4.1 Photoactive Dyads

Photoactive dyads are molecular devices designed to perform, through the generation of charge separated states, the conversion of light energy to chemical energy, in analogy to photosynthesis (Figure 12). Fullerenes have widely been used as redox and photoactive units in D-S-A photoactive dyads, triads and polyads.[28]

In contrast with their poor ability to form stable cations, fullerenes and their derivatives exhibit good electron acceptor properties: $C_{60}$ and its derivatives can accept reversibly up to 6 electrons in aprotic media,[9,14] the first reduction potential being similar to that of quinones, the biological redox couples involved in natural photosynthetic processes. Fullerene derivatives have been coupled to electron donor groups in dyads where the fullerene also acts as the primary photosensitizer. However, the poor absorption properties of fullerenes in the visible region has prompted to the development of D-S-$C_{60}$ dyads (and triads) where the main chromophoric function is associated to the D unit rather than to the fullerene. Porphyrins and metalloporphyrins have been preferentially used as donor/chromophores in such systems, for the unique combination of favorable photophysical and redox properties of these compounds. Furthermore, these compounds, particularly appealing for their similarity with natural chromophores, are stable and allow high synthetic availability to many structural variations and therefore great flexibility in the design of D-S-A systems. Since 1994, a great number of different porphyrin-C60 dyads, triads and more complex systems, also including supramolecular assemblies, have been investigated (see Figure 13 for some examples) and many examples are collected in a recent special issue of the *Journal of Materials Chemistry* dedicated to *Functionalized Fullerene Materials*.[28]

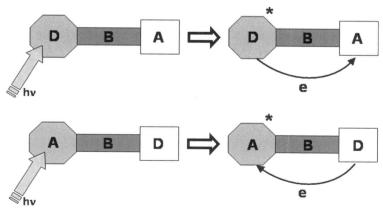

**Figure 12** *Photoactive (electron) donor–spacer–(electron) acceptor dyads*

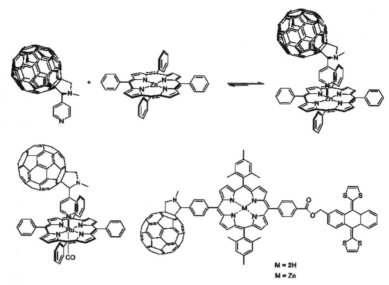

**Figure 13** *Porphyrin-$C_{60}$ photoactive dyads*

Investigation of photoinduced intramolecular ET processes in such systems have in general shown that, compared to analogous porphyrin/quinone systems (i) charge separation occurs with higher efficiency and (ii) charge-separated states have longer lifetimes, Figure 14. Such a behavior is the result of the combination of two favorable characteristics of fullerenes compared to quinones and other normal donor units, namely (i) the curvature of fullerene surface that allows a better electronic coupling between the acceptor unit and the hydrocarbon bridge and, ultimately with the D unit,[29] and (ii) the smaller reorganization energy associated to fullerenes, with respect to most acceptors, as a consequence of their large size and rigid framework in either the ground, excited or reduced states.

Most photoinduced charge separation processes are only slightly exoergonic, *i.e.* the process occurs generally in the normal Marcus region; a smaller reorganization energy therefore implies, for a given driving force, a faster process. Vice-versa, charge recombination is often a highly exoergonic process; this places the process in the Marcus inverted region and a smaller reorganization energy makes, for a given driving force, the kinetic constant of the charge recombination process smaller (Figure 14).[3]

Electron-transfer quenching reactions may involve either the oxidation or the reduction of the excited state (Figure 12).

$$D^* + A \rightarrow D^+ + A^- \quad \text{oxidative quenching}$$

$$D + A^* \rightarrow D^+ + A^- \quad \text{reductive quenching}$$

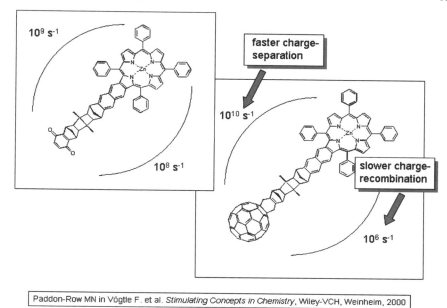

**Figure 14** *Rates of photoinduced charge separation and subsequent charge recombination in metalloporphyrin-based dyads (benzonitrile)*

The thermodynamic driving force of such reactions ($\Delta G°$) is estimated (in eV) from the relationship

$$-\Delta G^0_{CR} = E_{1/2}(D^+/D) - E_{1/2}(A/A^-) + \Delta G_S \quad (1)$$

where $E_{0\text{-}0}$ is the 0-0 energy of the chromophore excited state, estimated from the emission spectrum, $E_{1/2}(D^+/D)$ and $E_{1/2}(A/A^-)$ are the halfwave (~ standard) potentials for the oxidation of the donor and the reduction of the acceptor respectively, as conveniently obtained for instance by CV, and $\Delta G_s$ is the correction term for the effects of ion-solvent interaction for the charge-separated (CS) species. Such a term accounts both for ion-pair electrostatic interaction in the CS state and for energy correction when electrochemical and photophysical data respectively were obtained in different solvents.

## 3.4.2 Fullerene-Based Photoactive Liquid Crystals

The search for new fullerene-based materials, whose properties could be of interest in nanotechnology (*e.g.* molecular switches, solar cells), has prompted the investigation of fullerene-containing thermotropic photoactive liquid crystals.[30] Liquid-crystalline fullerenes were designed according to two different approaches: in the first approach, $C_{60}$ was functionalized with liquid-crystalline malonates by applying the Bingel reaction (leading to mesomorphic methanofullerenes, Figure 15), and, in the second one, $C_{60}$ was functionalized with

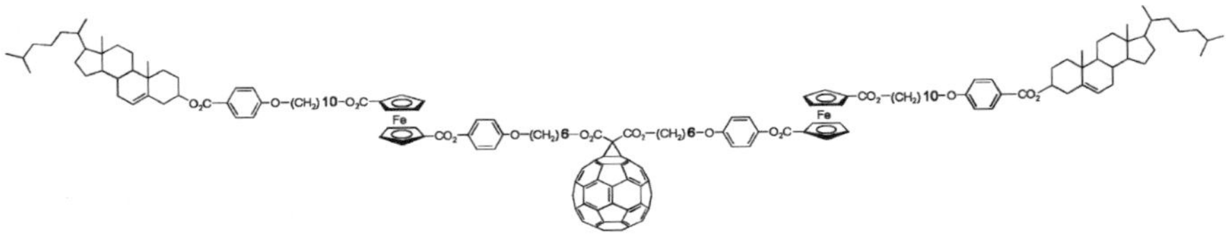

**Figure 15** *A mesomorphic methanofullerene*

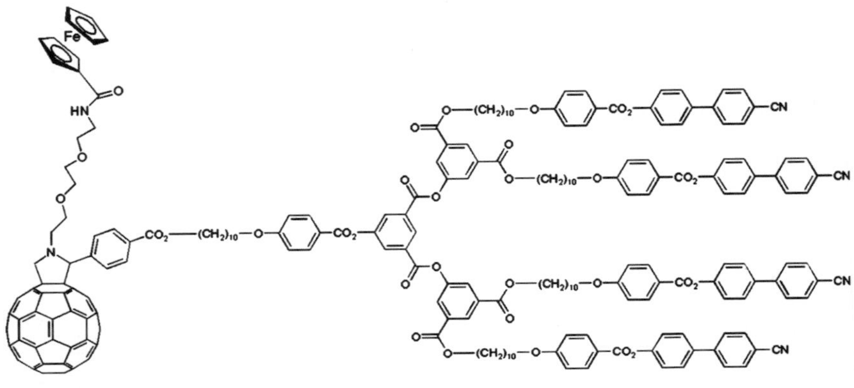

**Figure 16** *A mesomorphic fulleropyrrolidine*

liquid-crystalline aldehydes and *N*-methylglycine or an amino acid derivative by applying the 1,3-dipolar addition reaction (leading to mesomorphic fulleropyrrolidines, Figure 16). The two classes of compounds differ at a fundamental level from the electrochemical point of view: whereas methanofullerenes undergo retro-Bingel reaction upon chemical[23] and electrochemical[24,26] reduction, fulleropyrrolidines lead to stable reduced species. Dendritic liquid-crystalline fulleropyrrolidines would represent an interesting family of electroactive macromolecules, as they would combine the electrochemical behavior of $C_{60}$ with the rich mesomorphism found for dendrimers.

### 3.4.3 Fullerene-$[Ru(bpy)_3]^{2+}$ Systems

Ru *tris*-bipyridine complexes have been widely used in photoactive dyads to study photoinduced electron- and energy transfer processes for their unique photophysical and redox properties.[31] Furthermore, the chemistry of bipyridine is well developed and allows a wide range of possible functionalized derivatives.[32] By combining the standard potential for the Ru(II)-based oxidation process ($E° \sim +1.3$ V) and its $^3$MLCT excited state energy ($E_{00} \sim 2.0$ eV), the standard potential of excited state $E°[Ru(III)/*Ru(II)]$ is $\sim -0.7$ V. This latter may therefore ignite the photoinduced electron transfer to fullerene derivatives ($E° \sim -0.4$ V).

The nature and strength of the electronic interaction between a fulleropyrrolidine and the $[Ru(bpy)_3]^{2+}$ moiety, and the influence of the bridging spacer on such an interaction, were investigated by electrochemistry, UV-Vis-NIR absorption spectroscopy and steady state and time-resolved emission spectroscopy in a series of dyads where the units were covalently linked to each other by spacers showing different flexibility, *i.e.* either a rigid androstane bridge that suppresses conformational freedom in the dyad **11** (Figure 17),[33] a hexapeptide[34] that allows variable dimensional freedom upon temperature and/or solvent changes, and finally a highly flexible triethylene glycol spacer.[35]

Complex CV pattern are usually observed in the case of heteroleptic coordination compounds, in polynuclear complexes and supramolecular systems, where the various subunits may have different and also partly superimposing redox patterns. In general, the analysis of the voltammetric curves, also by the use of digital simulation techniques, allows to obtain the redox standard potentials for overlapping processes in multielectron waves. The localization of all the redox process, *i.e.* their attribution to the various electroactive subunits present in the supramolecular structure, is based on the comparison with the redox properties of suitable model compounds with the construction of *genetic diagrams*, *i.e.* diagrams in which each redox potential for the various processes observed in the supramolecular system is correlated to that of an equivalent process observed in an appropriate model. Additionally, the absorption characteristics of the reduced/oxidized species generated during spectroelectrochemical experiments and the results of quantum chemical calculations may support the attribution of the redox processes.

Steady-state absorption and CV (Figure 18) evidenced in the case of dyad **11**, no significant interaction between the C60 and $[Ru(bpy)3]^{2+}$ units in the ground state.

A reversible one-electron transfer at +1.34 V was observed, assigned to the metal-centered Ru(II)/Ru(III) oxidation. The complex cathodic pattern comprises ten successive reversible one-electron transfer, whose $E_{1/2}$ values were obtained also by digital simulation of the CV curves. The electrochemical behavior of this dyad is the sum of that of its two subunits, the $C_{60}$ core and the Ru(II) *tris*-bipyridyl complex, thus excluding any significant electronic interaction between the two moieties in the ground state. The first reduction processes (at −0.47 V) involves the fullerene moiety. On the other hand, the

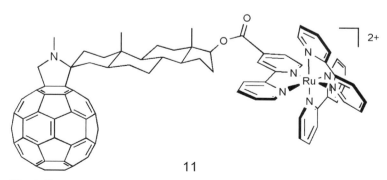

**Figure 17** *Fulleropyrrolidine-androstane-Ru(bpy)3 dyad*

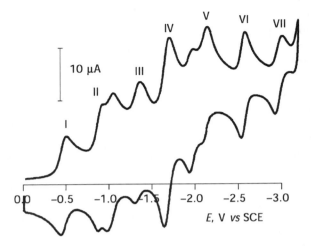

**Figure 18**  *CV curve of dyad 11, 0.5 mM in tetrahydrofuran (0.05 M tetrabutylammonium hexafluorophosphate). T=−60°C, scan rate=0.2 V/s, working electrode: Pt*

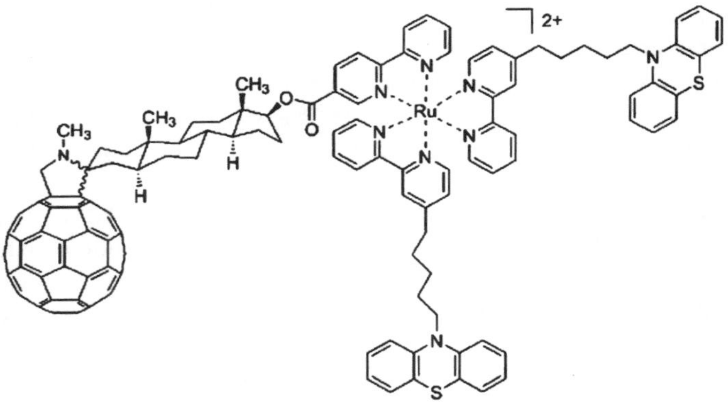

**Figure 19**  *Fulleropyrrolidine-Androstane-Ru(bpy)3 Phenothiazine triad*

driving force for the photoinduced ET process, from Ru $^3$MLCT to fulleropyrrolidine, calculated according to Equation (1), confirmed the hypothesis that electron transfer from the ruthenium complex to the fulleropyrrolidine may quench the photoexcited $[Ru(bpy)_3]^{2+}$ MLCT state. Also expectedly, the excited state dynamics was strongly affected by the solvent polarity. Charge-separated state is therefore formed from the Ru $^3$MLCT excited state, *via* a through-bond mechanism, with intramolecular rate constants of $0.69 \times 10^9$ s$^{-1}$ and $5.1 \times 10^9$ s$^{-1}$ in CH$_2$Cl$_2$-toluene (1:1, v/v) and CH$_3$CN respectively.

As a conceptual development of the above system, the triad shown in Figure 19 was obtained where the strong electron donor phenothiazine (PTZ) was covalently linked to the $[Ru(bpy)_3]^{2+}$ moiety to act as secondary electron donor capable to stabilize the charge-separated state.

Upon photoexcitation of the [Ru(bpy)$_3$]$^{2+}$ chromophore, the PTZ-Ru(III)-C60.- pair develops, with similar intramolecular kinetics (2.8 × 10$^9$ s$^{-1}$ in CH$_3$CN) as that observed in **11**, followed by formation of the the PTZ$^+$-Ru(II)-C$_{60}$$^-$ species. Importantly in view of the practical exploitation of photogenerated chemical potential, lifetimes of charge-separated state are significantly longer than those observed for **11**, 1.29 μs and 304 ns in CH$_2$Cl$_2$ and CH$_3$CN respectively.[35]

### 3.4.4 Fullerenes Containing Crown Ethers

There are a number of known instances in which fullerenes have been connected to crown ethers to exploit the well-known complexing properties of these subunits.[36] It therefore also seemed interesting to prepare methanofullerenes from cyclic malonic esters and to explore their properties. Crown ether derivatives incorporating a malonate unit and functionalized with π-electron-donating conjugated substituents to form ylidenemalonate moieties are able to form 1:1 complexes with Lewis acid-like metal cations, such as Mg$^{2+}$ and Eu$^{3+}$, in organic solutions.[37] The complexation of the metal cation is essentially directed toward the conjugated 1,3-dicarbonyl system, but the ethylene glycol ether structure, depending on its chemical structure and overall dimensions, gives additional stabilisation to the 1:1 complex. The synthesis and intercomparison of several methanofullerenes each bearing a macrocyclic crown ether attached to the C$_{60}$-fullerene, their complexing abilities toward primary ammonium ions, their electrochemical behavior, and their analysis by EPR techniques was reported.[38] The study illustrates the synthesis and characterisation of fullerene derivatives bearing macrocyclic malonates directly attached to the C$_{60}$ core through a cyclopropane moiety. The CV investigation carried out on the species shown in Figure 20 has shown that the injection of three (at 25 °C) or four (at −60 °C) electrons triggers a follow-up reaction responsible for irreversible CV behavior; such a reaction is, however, reversed upon re-oxidation with relatively low scan rates. This implies that the transient chemical species generated by the reduction of the above species, and presumably to be identified with the corresponding fullerene derivatives in which one of the cyclopropane bonds has been cleaved, a step believed to be preliminary to the retro-Bingel reaction, are stable on the relatively long timescales of CV.

The binding properties of such molecules and the variation of their electrochemical and spectrolectrochemical characteristics upon binding are currently under study.

### 3.4.5 Trannulenes

The recent discovery of the family of trannulenes comprising an 18 π annulene in the all-*trans* configuration (Figure 21) provided a new opportunity for the design of organic-based photovoltaic devices.[39]

Trannulenes have high electron affinities, enabling them to stabilize charged entities more effectively than conventional fullerene derivatives. Additionally

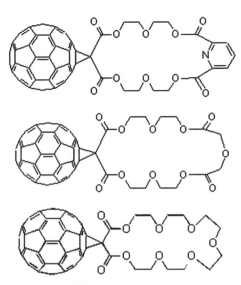

**Figure 20** *Crown ethers methanofullerenes*

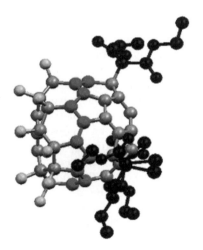

**Figure 21** *Trannulenes*

trannulenes have rich visible absorptions attributed to the diatropic 18 $\pi$ annulene substructure. Coupled with a mild preparative methodology producing radial three-dimensional architectures, these molecules could be a useful molecular building block for tailor-made components for optoelectronics, molecular-scale logic gates, and sensor design. Preliminary electron-transfer investigations of an extended tetrathiafulvalene trannulene dyad confirmed the ability of these remarkable molecules to accept a charged entity and store the photoexcitation energy in a longlived charge-separated state. Key to the use of trannulenes is their outstanding electron-accepting ability with first reduction potentials at around $-0.1$ V *vs.* SCE in tetrachloroethane.[40] Depending on the relative energies of the lowest excited state or of the charge-separated state, the outcome of the excited-state chromophore deactivation is either an energy- (*i.e.* $C_{60}F_{15}$-pyrene) or an electron-transfer (*i.e.* $C_{60}F_{15}$-perylene and $C_{60}F_{15}$-ferrocene) scenario, which proceed with quantum efficiencies of 99.4% ($C_{60}F_{15}$-perylene), 95.3% ($C_{60}F_{15}$-pyrene), and 71.5% ($C_{60}F_{15}$-ferrocene).[41]

### 3.4.6 Electropolymerizable $C_{60}$ Derivatives

Incorporation of fullerenes in a conducting polymer has been pursued with the aim of exploiting the many interesting properties of fullerenes in materials science applications.[7] By plain mixing of $C_{60}$ with polymers, a material suitable for surface coating can be obtained with optical limiting properties[42] or capable of performing light-to-electricity[43] conversion in photoconducting devices.[44] A great appeal is also shown by the direct synthesis of polymers from fullerene-containing monomers (Figure 22). Figure 23 shows the electropolymerization of a bithiophene covalently linked to a $C_{60}$ derivative.[45] The resulting conducting film still retained the redox properties of the pristine fullerene derivative.

### 3.4.7 $C_{60}$ as a Mediator in Bio Electrochemical Sensors

The unique electronic and/or structural properties of fullerenes and fullerene derivatives have been exploited for the development of chemical and biochemical devices. The electrocatalytic properties and sensor applications of fullerenes and carbon nanotubes have recently been reviewed.[46]

A novel biosensor for the amperometric detection of glutathione was proposed in which glutathione reductase is coupled to the fulleropyrrolidine *bis*-adduct **4** shown in Figure 2a that plays the role of redox mediator between electrode and enzyme.[47] The adoption of a simple procedure for realizing the

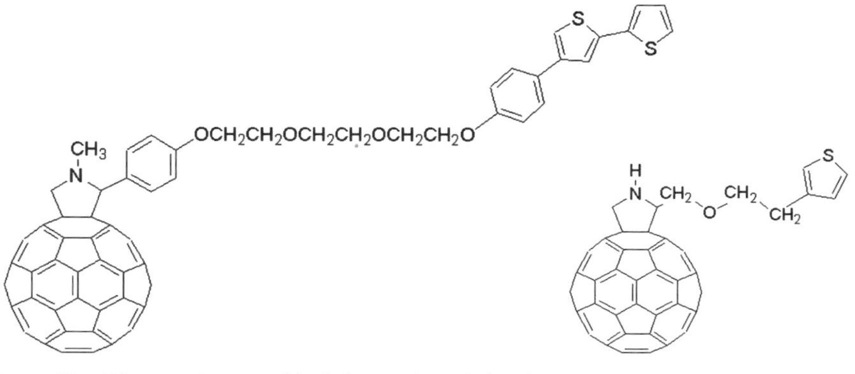

**Figure 22** *Electropolymerizable fullerene-based thiophene monomers*

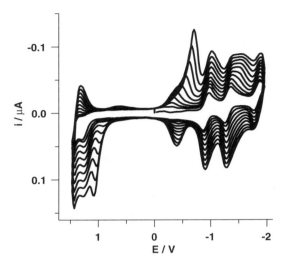

**Figure 23** *Potentiodynamic polymerization of a fullerene-based thiophene monomer: growth curves*

immobilization of the redox mediator together with the enzyme, within a polypyrrole film cast onto a glassy carbon electrode, was made possible by the amphiphilic properties of the fullerene derivative (Figure 24). The biosensor showed a fast and reproducible response to glutathione and, because of its lower sensitivity, a wider linear range with respect to a similar device based on a diffusional redox mediator.

## 3.5 Back to the Basics: The Oxidative Electrochemistry of $C_{60}$

The facile reduction of fullerenes, exemplified by the electrochemical stepwise reversible addition of up to six electrons to $C_{60}$ and $C_{70}$,[48] contrasts with their difficult oxidation.[49] Furthermore, in condensed media, fullerenium radical cations – the first all-carbon carbocations – react immediately with any nucleophile present in solution leading to decomposition.[50] This was observed in early CV experiments with the generation of multi-electron oxidation peaks, as the decomposition products undergo further oxidation. In 1993, Echegoyen *et al.*, employing scrupulously dried tetrachloroethane (TCE), tetrabutylammonium hexafluorophosphate (TBAPF$_6$) as supporting electrolyte, and low temperatures, reported the first observation of electrochemical reversible one-electron oxidation of $C_{60}$ at $E_{1/2}$=1.26 V (*vs.* Fc$^+$/Fc).[51] Higher fullerenes are easier to oxidize than $C_{60}$, and the second electrochemical oxidation (although irreversible) of $C_{70}$, $C_{76}$ and $C_{78}$ in TCE/TBAPF$_6$ was reported.[9] By contrast, the further oxidation of $C_{60}^+$ was disallowed likely by the too narrow positive potential window available in such a solvent, and because $C_{60}^{2+}$ is an exceedingly reactive species.

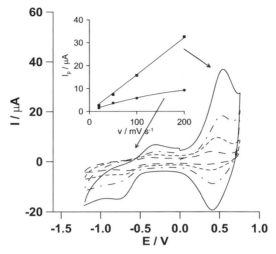

**Figure 24** Cyclic voltammetric curves relative to a polypyrrole/4 film electrogenerated onto a glassy carbon electrode (r ~ 1.5 mm) in 0.1 M aqueous $LiClO_4$ solution. Scan rate: 0.02 (——), 0.05 (---), 0.1 (.–.) and 0.2 V/s (——). Inset: dependence of peak current on scan rate, for the polypyrrole oxidation (■) and 4-centred reduction (●)

**Table 1** $E_{1/2}$ (Volts vs. $Fc^+/Fc$) of oxidation/reduction of $C_{60}$

| Solvent | Oxidations | | | Reductions | | | | | |
|---|---|---|---|---|---|---|---|---|---|
| | I | II | III | I | II | III | IV | V | VI |
| $CH_2Cl_2$[a] | 1.27 | 1.74 | 2.14 | −1.06 | −1.46 | −1.89 | | | |
| THF[b] | | | | −0.90 | −1.47 | −2.05 | −2.54 | −3.07 | −3.53 |

[a] Supporting electrolyte: tetrabutylammonium hexafluoroarseniate (0.05 M), T = −60°C. W: Pt.
[b] Supporting electrolyte: tetrabutylammonium hexafluorophosphate (0.05 M), T = −60°C. W: Pt.

We recently succeeded in obtaining the cyclic voltammetric reversible generation of $C_{60}^{2+}$ and $C_{60}^{3+}$ (Table 1).[51] This was allowed by the adoption of suitable experimental conditions that comprise ultra-dry solvents and electrolytes with very high oxidation resistance and low nucleophilicity. Our experiments were carried out using as solvent dichloromethane, a solvent widely used in fullerene electrochemistry, and avoiding supporting electrolytes containing nucleophilic counter anions such as $ClO_4^-$, $BF_4^-$, and $PF_6^-$, that are known to attack $C_{60}^+$ in such a solvent. $AsF_6^-$ was instead chosen which is known for its high oxidation resistance and low nucleophilicity.[52]

$C_{60}$ is therefore a single molecular unit, *i.e.* not made up of independent subcomponents, that may exist in solution in up to 10 different redox states, energetically spanning a potential range larger than 5 V. This property adds to the many other unique properties of such a fascinating molecule.

## 3.6 Final Remarks

During the 1990s, $C_{60}$ was chemists' superstar, but what does it mean today? Despite it has not (so far) found any major applications, the influence of $C_{60}$ is now pervasive in chemistry and beyond.[53] $C_{60}$ is a kind of ideal nanoscale building block that can be picked up and manipulated with nanotechnological tools. Its curved, hollow structure has made us familiar with another view of carbon materials, different and complementary to that of flat sheets of carbon atoms in graphite. Ultimately, all the "noise" around these carbon allotropes has generated new drive for research and has introduced the perhaps most notable representatives of the present nanoworld, carbon nanotubes.

The ability of some SWNT to encapsulate various atoms and molecules into their one-dimensional nanocavity is well known. So fullerenes, endo- and exo-hedral metallofullerenes, exohedral fullerene derivatives[54] have been successfully incorporated into the nanocavity of SWNT and these hybrid nanomaterials have been christened with the common name of peapods. By using a different approach, we have recently used a $C_{60}$-pyrene derivative in the presence of an homogeneous dispersion of purified HiPCO SWNT. Cyclic voltammetry experiments, photophysical and microscopic characterization of the dispersed material supports the view that π-π interactions between SWNT and the pyrene moiety govern the intermolecular association, resulting in a sort of *topological mutant* of peapods (Figure 25).[55] Use of this novel systems for applications in photovoltaics and electronics is currently underway.

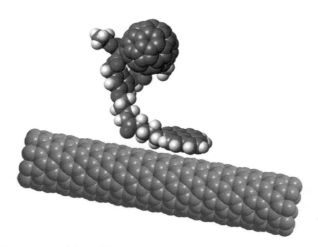

**Figure 25** *Semiempirical (PM3) optimized geometry of pyrene-fulleropyrrolidine in the presence of a (9,0) SWNT*

# References

1. H.W. Kroto, J.R. Heath, S.C. O'Brien, R.F. Curl and R.E. Smalley, *Nature*, 1985, **318**, 162.
2. T.Y. Zakharian, A. Seryshev, B. Sitharaman, B.E. Gilbert, V. Knight and L.J. Wilson, *J. Am. Chem. Soc.*, 2005, **127**, 12508; E. Perzon, X. Wang, F. Zhang, W. Mammo, J.L. Delgado, P. de la Cruz, O. Inganas, F. Langa and M. Andersson, *Synth. Metals*, 2005, **154**, 53; H. Imahori, A. Fujimoto, S. Kang, H. Hotta, K. Yoshida, T. Umeyama, Y. Matano and S. Isoda, *Adv. Mater.*, 2005, **17**, 1727; Q.Sun, Q. Wang, P. Jena and Y. Kawazoe, *J. Am. Chem. Soc.*, 2005, **127**, 14582; Y. Liu, P. Liang, Y. Chen, Y.M. Zhang, J.Y. Zheng and H. Yue, *Macromolecules*, 2005, 38, 9095; S. Pekker, E. Kovats, G. Ozlanyi, G. B'Enyei, G.O. Klupp, G Bortel, I. Jalsovszky, E. Jakab, F. Borondics, K. Kamaras, M. Bokor, G. Kriza, K. Tompa and G. Faigel, *Nature Mat.*, 2005, **4**, 764; M. Prato and M. Maggini, *Acc. Chem. Res.*, 1998, **31**, 519.
3. T. Mashino, N. Usui, K. Okuda, T. Hirota and M. Mochizukia, *Bioorg. Med. Chem.*, 2003, **11**, 1433.
4. S.H. Friedman, D.L. DeCamp, R.P. Sijbesma, G. Srdanov, F. Wudl and G.L. Kenyon, *J. Am. Chem. Soc.*, 1993, **115**, 6506.
5. N. Gharbi, M. Pressac, M. Hadchouel, H. Szwarc, S.R. Wilson and F. Moussa, *Nano Letters*, 2005, **5**, 2578; Y. -L. Laia, P. Murugan and K.C. Hwang, *Life Sciences*, 2003, **72**, 1271.
6. A.J.L. Pombeiro and C. Amatore (eds), *Trends in Inorganic Molecular Electrochemistry*, FontisMedia-Marcel Dekker, Lausanne, 2004.
7. A. Hirsch, The Chemistry of the Fullerenes, Thieme, Stuttgart, 1994; F. Diederich, L. Isaacs and D. Philp, *Chem. Soc. Rev.*, 1994, **23**, 243; R. Taylor, *The Chemistry of Fullerenes*, World Scientific, Singapore, 1995; M. Prato, *J. Mater. Chem.*, 1997, **7**, 1097.
8. L. Echegoyen and L.E. Echegoyen, *Acc. Chem. Res.*, 1998, **31**, 593; N. Armaroli, G. Accorsi, F.Y. Song, A. Palkar, L. Echegoyen, D. Bonifazi and F. Diederich, *Chem. Phys. Chem.*, 2005, **6**, 732; P.L. Boulas, M. Gomez Kaifer and L. Echegoyen, *Angew. Chem. Int. Ed.*, 1998, **37**, 216; L. Echegoyen and L.E. Echegoyen in H. Lund and O. Hammerich (eds) *Organic Electrochemistry*, 4$^{th}$ Edition, Marcel Dekker, 2001, p. 323–339.
9. A.G. Avent, P.R. Birkett, M. Carano, A.D. Darwish, H.W. Kroto, J.O. Lopez, F. Paolucci, S. Roffia, R. Taylor, N. Wachter, D.R.M. Walton and F. Zerbetto, *J. Chem. Soc. Perkin Trans.*, 2001, **2**, 140.
10. F. Zhou, G.F. Van Berkel and B.T. Donovan, *J. Am. Chem. Soc.*, 1994, **116**, 5485; N. Liu, Y. Morio, F. Okino, H. Touhara, O.V. Boltalina and V.K. Pavlovich, *Synth. Metals*, 1997, **96**, 2289; K. Okhubo, R. Taylor, O.V. Boltalina, S. Ogo and S. Fukuzumi, *Chem. Commun.*, 2002, 1952; D. Paolucci, F. Paolucci, M. Marcaccio, M. Carano and R. Taylor, *Chem. Phys. Lett.*, 2004, 400, 389.
11. X.W. Wei, A.D. Darwish, O.V. Boltalina, P.B. Hitchcock, J.M. Street and R. Taylor, *Angew. Chem. Int. Ed.*, 2001, **40**, 2989.

12. (a) X.W. Wei, A.G. Avent, O.V. Boltalina, A.D. Darwish, P.W. Fowler, J.P.B. Sandall, J.M. Street and R. Taylor, *J. Chem. Soc., Perkin Trans.*, 2002, **2**, 41; (b) A.D. Darwish, I.V. Kuvytchko, X.W. Wei, O.V. Boltalina, I.V. Gol'dt, J.M. Street and R. Taylor, *J. Chem. Soc. Perkin Trans.*, 2002, **2**, 1118.
13. T. Da Ros, M. Prato, M. Carano, P. Ceroni, F. Paolucci and S. Roffia, *J. Am. Chem. Soc.*, 1998, **120**, 11645.
14. M. Carano, T. Da Ros, M. Fanti, K. Kordatos, M. Marcaccio, F. Paolucci, M. Prato and S. Roffia, *J. Am. Chem. Soc.*, 2003, **125**, 7139.
15. P.R. Birkett, R. Taylor, N.K. Wachter, M. Carano, F. Paolucci, S. Roffia and F. Zerbetto, *J. Am. Chem. Soc.*, 2000, **122**, 4209.
16. A.G. Avent, P.R. Birkett, J.D. Crane, A.D. Darwish, G.J. Langley, H.W. Kroto, R. Taylor and D.R.M. Walton, *J. Chem. Soc. Chem. Comm.*, 1994, 1463.
17. H. Iikura, S. Mori, M. Sawamura and E. Nakamura, *J. Org. Chem.*, 1997, **62**, 7912.
18. M. Eiermann, F. Wudl, M. Prato and M. Maggini, *J. Am. Chem. Soc.*, 1994, **116**, 8364.
19. T. Suzuki, Q. Li, K.C. Khemani and F. Wudl, *J. Am. Chem. Soc.*, 1992, **114**, 7301; F. Diederich, L. Isaacs and D. Philp, *Chem. Soc. Rev.*, 1994, **23**, 243; F. Paolucci, M. Marcaccio, S. Roffia, G. Orlandi, F. Zerbetto, M. Prato, M. Maggini and G. Scorrano, *J. Am. Chem. Soc.*, 1995, **117**, 6572.
20. M. Eiermann, R.C. Haddon, B. Knight, Q. Li, M. Maggini, N. Martin, T. Ohno, M. Prato, T. Suzuki and F. Wudl, Angew. Chem., *Int. Ed. Engl.*, 1995, **34**, 1591.
21. P. Ceroni, F. Conti, C. Corvaja, M. Maggini, F. Paolucci, S. Roffia, G. Scorrano and A. Toffoletti, *J. Phys. Chem. A.*, 2000, **104**, 156.
22. M.Á. Herranz, F. Diederich and L. Echegoyen, *Eur. J. Org. Chem.*, 2004, **11**, 2299.
23. C. Bingel, *Chem. Ber.*, 1993, **126**, 1957.
24. M. Ocafrain, M.A. Herranz, L. Marx, C. Thilgen, F. Diederich and L. Echegoyen, *Chem. Eur. J.*, 2003, **9**, 4811; V.V. Yanilkin, N.V. Nastapova, V.P. Gubskaya, V.I. Morozov, L.S. Berezhnaya and I.A. Nuretdinov, *Russian Chem. Bull.*, 2002, **51**, 72.
25. D. Britz, Digital Simulation in Electrochemistry, Lecture Notes in Physics Springer-Verlag GmbH, Berlin, 2005; S.W. Feldberg, in *Electroanalytical Chemistry*, vol 3, A.J. Bard (ed), Marcel Dekker, New York, 1969; F. Alden, R.G. Hutchinson, R. Compton, *J. Phys. Chem. B.*, 1997, **101**, 949.
26. M. Carano and L. Echegoyen, *Chem. Eur. J.*, 2003, **9**, 1974.
27. M.W.J. Beulen and L. Echegoyen, *Chem. Commun.*, 2000, 1065; M.W.J. Beulen, J.A. Rivera, M.A. Herranz, A. Martin-Domenech, N. Martin and L. Echegoyen, *Chem. Commun.*, 2001, 407.
28. M.D. Meijer, G.P.M. van Klink and G. van Koten, *Coord. Chem. Rev.*, 2002, **230**, 141; N. Martín, L. Sánchez, B. Illescas and I. Pérez, *Chem. Rev.*, 1998, **98**, 2527; F. Diederich and M. Gómez-López, *Chem. Soc. Rev.*, 1999, **28**, 263; H. Imahori and Y. Sakata, *Adv. Mater.*, 1997, **9**, 537; M. Prato

(ed), *J. Mater. Chem.*, 2002, **12** (Special Issue on Functionalised Fullerene Materials); S. Fukuzumi and D.K. Guldi, in *Electron Transfer in Chemistry*, vol 2, V. Balzani, Wiley-VCH, 2001, 270.
29. M.N. Paddon-Row, in *Stimulating Concepts in Chemistry*, F. Vögtle, J.F. Stoddart and M. Shibasaki (eds), Wiley-VCH, Weinheim, 2000; M.N. Paddon-Row, in *Electron Transfer in Chemistry*, vol 3, V. Balzani (ed), Wiley-VCH, 2001, 179.
30. T. Chuard and R. Deschenaux, *Helv. Chim. Acta*, 1996, **79**, 736; S. Campidelli and R. Deschenaux, *Helv. Chim. Acta*, 2001, **84**, 589; T. Chuard and R. Deschenaux, *J. Mater. Chem.*, 2002, **12**, 1944; S. Campidelli, R. Deschenaux, J-F. Eckert, D. Guillon and J. -F. Nierengarten, *Chem. Commun.*, 2002, 656; S. Campidelli, E. Vázquez, D. Milic, M. Prato, J. Barberá, D.M. Guldi, M. Marcaccio, D. Paolucci, F. Paolucci and R. Deschenaux, *J. Mater. Chem.*, 2004, **14**, 1266; S. Campidelli, J. Lenoble, J. Barbera, F. Paolucci, M. Marcaccio, D. Paolucci and R. Deschenaux, *Macromolecules*, 2005, **38**, 7915.
31. A. Juris, V. Balzani, F. Barigelletti, S. Campagna, P. Belser and A. Zelewsky, *Coord. Chem. Rev.*, 1988, **84**, 85; J.P. Sauvage, J.P. Collin, J.C. Chambron, S. Gillerez, C. Coudret, V. Balzani, F. Barigelletti, L. De Cola and L. Flamigni, *Chem. Rev.*, 1994, **94**, 993; V. Balzani, A. Juris, M. Venturi, S. Campagna and S. Serroni, *Chem. Rev.*, 1996, 96, 759.
32. C. Kaes, A. Katz and M.W. Hosseini, *Chem. Rev.*, 2000, **100**, 3553; F. Kröhnke, *Synthesis*, 1976, **1**.
33. M. Maggini, D.M. Guldi, S. Mondini, G. Scorrano, F. Paolucci, P. Ceroni and S. Roffia, *Chem. Eur. J.*, 1998, **4**, 1992.
34. A. Polese, S. Mondini, A. Bianco, C. Toniolo, G. Scorrano, D.M. Guldi and M. Maggini, *J. Am. Chem. Soc.*, 1999, **121**, 3456.
35. D.M. Guldi, M. Maggini, E. Menna, G. Scorrano, P. Ceroni, M. Marcaccio, F. Paolucci and S. Roffia, *unpublished results*.
36. R. Wilson and Y. Wu, *Chem. Commun.*, 1993, 784; U. Jonas, F. Cardullo, P. Belik, F. Diederich, A. Gügel, E. Hart, A. Herrmann, L. Isaacs, K. Müllen, H. Ringsdorf, C. Thilgen, P. Uhlmann, A. Vasella, C.A.A. Waldraff and M. Walter, *Chem. Eur. J.*, 1995, **1**, 243; Z. Guo, Y. Li, J. Yan, F. Bai, F. Li, D. Zhu, J. Si and P. Ye, *Appl. Phys. B., Lasers and Optics*, 2000, **257**, 60.
37. D. Pasini, P.P. Righetti and V. Rossi, *Org. Lett.*, 2002, **4**, 23; L. Garlaschelli, I. Messina, D. Pasini and P.P. Righetti, *Eur. J. Org. Chem.*, 2002, 3385.
38. M. Carano, C. Corvaja, L. Garlaschelli, M. Maggini, M. Marcaccio, F. Paolucci, D. Pasini, P.P. Righetti, E. Sartori and A. Toffoletti, *Eur. J. Org. Chem.*, 2003, **2**, 374.
39. X.W. Wei, A.D. Darwish, O.V. Boltalina, P.B. Hitchcock, J.M. Street and R. Taylor, *Angew. Chem., Int. Ed.*, 2001, **40**, 2989; X.W. Wei, A.G. Avent, O.V. Boltalina, A.D. Darwish, P.W. Fowler, J.P.B. Sandall, J.M. Street and R. Taylor, *J. Chem. Soc., Perkin Trans.*, 2002, **2**, 41; A.D. Darwish, I.V. Kuvytchko, X.W. Wei, O.V. Boltalina, I.V. Gold't, J.M. Street and R.

Taylor, *J. Chem. Soc., Perkin Trans.*, 2002, **2**, 1118; G.A. Burley, A.G. Avent, O.V. Boltalina, T. Drewello, I.V. Gol'dt, M. Marcaccio, F. Paolucci, D. Paolucci, J.M. Street and R. Taylor, *Org. Biomol. Chem.*, 2003, **1**, 2015; G.A. Burley, A.G. Avent, I.V. Goldt, P.B. Hitchcock, H. Al-Matar, D. Paolucci, F. Paolucci, P.W. Fowler, A. Soncini, J.M. Street and R. Taylor, *Org. Biomol. Chem.*, 2004, **2**, 319.
40. G.A. Burley, A.G. Avent, O.V. Boltalina, I.V. Gol'dt, D.M. Guldi, M. Marcaccio, D. Paolucci, F. Paolucci and R. Taylor, *Chem. Commun.*, 2003, 148.
41. D.M. Guldi, M. Marcaccio, F. Paolucci, D. Paolucci, J. Ramey, R. Taylor and G.A. Burley, *J. Phys. Chem. A.*, 2005, **109**, 9723.
42. C. Li, C. Liu, F. Li and Q. Gong, *Chem. Phys. Lett.*, 2003, **380**, 201; A. Kost, L. Tutt, M.B. Klein, T.K. Dougherty and W.E. Elias, *Opt. Lett.*, 1993, **18**, 334.
43. H. Imaori, A. Fujimoto, S. Kang, H. Hotta, K. Yoshida, T. Umeyama, Y. Matano and S. Isoda, *Adv. Mater.*, 2005, **17**, 1727.
44. P.V. Kamat, M. Haria and S. Hotchandani, *J. Phys. Chem. B.*, 2004, **108**, 5166; A. Capobianchi and M. Tucci, *Thin Solid Films*, 2004, **451**, 33; N.S. Sariciftci, L. Smilowitz, A.J. Heeger and F. Wudl, *Science*, 1993, 258, 1474.
45. T. Benincori, E. Brenna, F. Sannicoló, L. Trimarco, G. Zotti, and P. Sozzani, *Angew, Chem, Int, Ed. Engl.*, 1996, **35**, 648; A. Cravino and N.S. Sariciftci, *Nature Mat.*, 2003, **2**, 360.
46. B.S. Sherigara, W. Kutner and F. D'Souza, *Electroanalysis*, 2003, **15**, 753.
47. M. Carano, S. Cosnier, K. Kordatos, M. Marcaccio, M. Margotti, F. Paolucci, M. Prato and S. Roffia, *J. Mat. Chem.*, 2002, **12**, 1996.
48. Q. Xie, E. Perez-Cordero and L. Echegoyen, *J. Am. Chem. Soc.*, 1992, **114**, 3977; Y. Oshawa and T. Saji, *J. Chem. Soc., Chem. Commun.*, 1992, 781.
49. Q. Xie, F. Arias and L. Echegoyen, *J. Am. Chem. Soc.*, 1993, **115**, 9818.
50. C.A. Reed, K-C. Kim, R.D. Bolskar and L.J. Mueller, *Science*, 2000, **289**, 101.
51. C. Bruno, I. Doubitski, M. Marcaccio, F. Paolucci, D. Paolucci and A. Zaopo, *J. Am. Chem. Soc.*, 2003, **125**, 15738.
52. E. Garcia, J. Kwak and A.J. Bard, *J. Am. Chem. Soc.*, 1988, **27**, 4377; A.J. Bard, E. Garcia, S. Kucharenko and V.V. Stretlets, *Inorg. Chem.*, 1993, 32, 3528; P. Ceroni, F. Paolucci, C.Paradisi, A. Juris, S. Roffia, S. Serroni, S. Campagna and A.J. Bard, *J. Am. Chem. Soc.*, 1998, **120**, 5480.
53. P. Ball, *Chemistry World*, 2005, **2**(12), 28.
54. N. Nakashima, Y. Tanaka, Y. Tomonari, H. Murakami, H. Kataura, T. Sakaue and K. Yoshikawa, *J. Phys. Chem. B.*, 2005, **109**, 13076; Y. Maeda, S. Kimura, Y. Hirashima, M. Kanda, Y. Lian, T. Wakahara, T. Akasaka, T. Hasegawa, H. Tokumoto, T. Shimizu, H. Kataura, Y. Miyauchi, S. Maruyama, K. Kobayashi and S. Nagase, *J. Phys. Chem. B*, 2004, **108**, 18395; Y.-B. Sun, Y. Sato, K. Suenaga, T. Okazaki, N. Kishi, T. Sugai, S. Bandow, S. Iijima and S. Shinohara, *J. Am. Chem. Soc.*, 2005, **127**, 17972.
55. D.M. Guldi, E. Menna, M. Maggini, M. Marcaccio, D. Paolucci, F. Paolucci, S. Campidelli, M. Prato, G.M. Aminur Rahman and S. Schergna, *Chem. Eur. J.*, 2006, **12**, 3975.

CHAPTER 4
# Light-Induced Processes in Fullerene Multicomponent Systems

NICOLA ARMAROLI AND GIANLUCA ACCORSI

Molecular Photoscience Group, Istituto per la Sintesi Organica e la Fotoreattività, Consiglio Nazionale delle Ricerche, Via Gobetti 101, 40129, Bologna, Italy

## 4.1 Introduction

Light-induced processes in supramolecular and multicomponent systems containing $C_{60}$ fullerenes have been the object of intensive research since mid-1990s.[1–5] Such a great interest is related to special electronic and excited state properties of the $C_{60}$ carbon cage[6] which, upon a suitable choice of the partner units, can give rise to a variety of complex architectures featuring electron- and/or energy-transfer processes.[7,8] Indeed, the discovery of buckminsterfullerene $C_{60}$ coincided somewhat fortuitously with the very beginnings of the area of supramolecular chemistry itself.[3] The fields of fullerene and supramolecular chemistry have since matured and overlapped considerably and today the literature displays many wonderful examples, where fullerene $C_{60}$ has been employed in such assemblies. An obvious question, and one which has preoccupied chemists interested in harnessing the potential of fullerene $C_{60}$, is what are the most appropriate electron/energy donors to be coupled with $C_{60}$ in complex architectures. There are many criteria which must be met which include reasonably long-lived excited state lifetimes, good absorption in the UV/Vis and photostability. Many candidates are possible such as porphyrins,[1–2,4] phtalocyanines,[5,9] organic conjugated oligomers[10–14] and polymers,[15] coordination compounds,[16] and other molecular subunits.[17–22] Supramolecular architectures[23] range from fullerodendrimers[3,24] to host–guest systems,[25–27] to hydrogen-bonded motifs,[28–30] whereas the approach to materials science[31] has been attempted by means of nanoparticles,[32,33] monolayers,[34–39] and liquid crystals[40,41] just to mention a few. The large majority of these systems are designed to take advantage of the electron-accepting properties of fullerene $C_{60}$ and make prototype light energy conversion devices.[1,15,42] However, other potential applications of photoactive fullerene arrays can be envisaged in the field of optical limiters[36,43] and, thanks to the singlet oxygen sensitization capability of $C_{60}$ and its functionalized derivatives,[44] medicinal chemistry.[45]

During the last decade an impressive amount of research work has been carried out in the field of light-induced processes within multicomponent and supramolecular fullerene arrays by a number of research groups, and it is out of the scope of this chapter to give a comprehensive description of the field. The attention will be focused to a few selected classes of systems recently investigated in our research group.

## 4.2 Photophysical Properties of $C_{60}$ and its Derivatives

The photophysical properties of $C_{60}$ and its derivatives have been extensively investigated, whereas $C_{70}$,[46-49] higher fullerenes,[50-53] and open cage fullerenes[54] have been studied to a much lesser extent. All of these carbon cages are characterized by low lying excited electronic levels.[55] The lowest singlet ($^1C_{60}^*$) and triplet ($^3C_{60}^*$) excited states of pristine $C_{60}$ are located, respectively, at about 1.7[56] and 1.5[57] eV above the ground state. $^1C_{60}^*$ is partly deactivated to the ground state *via* a radiative process giving rise to a weak fluorescence band centred around 700 nm.[48,56,58] However, the main deactivation pathway is intersystem crossing to give $^3C_{60}^*$ with high yield (>99%).[57] $^3C_{60}^*$ can be monitored by means of a variety of techniques including EPR,[59] photoacoustic calorimetry[57] and transient absorption (TA) spectroscopy.[60] A few authors have also reported phosphorescence spectra of $C_{60}$, but only in heavy atom-containing solvents to promote singlet to triplet intersystem crossing.[61,62]

The profile of the electronic absorption spectrum of functionalized fullerenes is substantially modulated relative to pristine $C_{60}$, leading to a variety of colours of the corresponding solutions in organic solvents (Figure 1). This reflects substantial modification of transition probabilities of electronic excited states upon cage functionalization. The fluorescence quantum yield of $C_{60}$ is lower than that of most of its derivatives (Figure 2). However, the position of the lowest electronically singlet and triplet states is very similar in all cases, as can be inferred from the onset of fluorescence spectra depicted in Figure 2.[63]

$C_{60}$ generates excited singlet molecular oxygen $O_2(^1\Delta_g)$ with high efficiency.[64] The photosensitization process occurs *via* a triplet–singlet electron-exchange energy-transfer process, which is spin allowed thanks to the triplet nature of ground state molecular oxygen.[65] The reaction can be schematized as follows:

$$C_{60} \xrightarrow{h\nu} {}^1C_{60}^* \xrightarrow{\text{i.s.c.}} {}^3C_{60}^* \xrightarrow{O_2} C_{60} + C_{60} + {}^1O_2^* \qquad (1)$$

where i.s.c. indicates intersystem crossing, $^1C_{60}^*$ and $^3C_{60}^*$ denote the lowest singlet and triplet fullerene excited states, and $^1O_2^*$ stands for $O_2(^1\Delta_g)$,

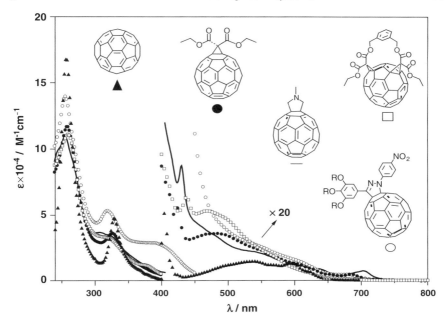

**Figure 1** *Electronic absorption spectra of pristine $C_{60}$ and some of its functionalized derivatives in dichloromethane solution. The visible spectral window ($\lambda > 400$ nm) is multiplied by a factor of 20. Pristine $C_{60}$ (▲), methanofullerene (●), bismethanofullerene (□), fulleropyrrolidine (–), and fulleropyrazoline (○)*

commonly named "singlet oxygen". $^1O_2^*$ deactivates back to the ground state giving rise to a characteristic near-infrared luminescence band centred at 1268 nm.[66,67] This signal is the most powerful tool to monitor and study the photosensitization process[68,69] which, when no other $^3C_{60}^*$ decay pathways are available, can be described by a series of important parameters such as (i) the quantum yield of fullerene triplet formation ($\Phi_T$), (ii) the fraction of oxygen quenching reaction that leads to $^1O_2^*$ ($S_\Delta$), (iii) the quantum yield of singlet oxygen sensitization ($\Phi_\Delta$). For $C_{60}$ it is found that $\Phi_T (=\Phi_{i.s.c.}) = 1$ and $S_\Delta = 1$. Therefore $\Phi_\Delta = 1$, as measured in toluene and benzene, with various techniques.[67]

## 4.3 Dyads with Oligophenylenevinylenes

Molecular dyads made of electron acceptor (*e.g.* $C_{60}$ fulleropyrrolidine) and electron donor (oligophenylenevinylene, OPV) subunits, proved to be useful to make prototype photovoltaic devices.[11,69–71] They open an alternative way to the assembling of plastic solar cells, in competition with the "classical" approach,[72] in which the photoactive material is a blend of fullerene and poly(*p*-phenylenevinylene).[42,73,74] In particular, the molecular approach may

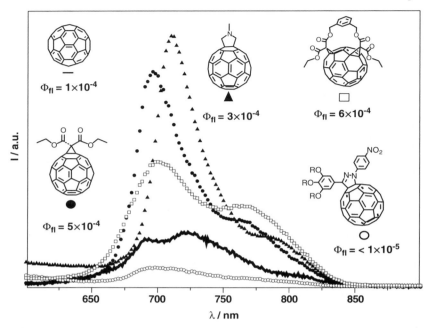

**Figure 2** *Fluorescence spectra of pristine $C_{60}$ and some of its functionalized derivatives at room temperature. The spectral intensities reflect the luminescence quantum yields ($\lambda_{exc} = 330$ nm, $OD = 0.2$ in dichloromethane solution for all samples). Pristine $C_{60}$ (–), methanofullerene (●), bismethanofullerene (□), fulleropyrrolidine (▲), and fulleropyrazoline (○)*

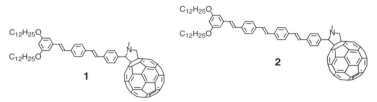

**Figure 3** *OPV-fulleropyrrolidine dyads 1 and 2*

avoid phase separation between the donor and the acceptor that occurs under working conditions in classical devices where the two materials are simply blended.

The light-to-current efficiency of the photovoltaic devices made of thin films of **1** and **2** hybrids (Figure 3) is poor, *i.e.* 0.01 and 0.03%, respectively, under illumination at $\lambda = 400$ nm.[11]

In principle, excitation of the OPV subunit in the two dyads can lead to both charge separation and fullerene singlet sensitization. Photophysical studies in $CH_2Cl_2$ solution show that, upon excitation of the OPV moiety, intramolecular sensitization of $C_{60}$ fluorescence is observed.[11] Calculations on the OPV→$C_{60}$ singlet–singlet energy-transfer process, according to the Förster treatment, give a rate constant value of at least $10^{12}$ s$^{-1}$, suggesting that effective competition by electron transfer is unlikely. On the other hand, once the lowest fullerene singlet excited state is populated (1.71 eV), the charge separated OPV$^+$-$C_{60}^-$ state can no longer be reached since located at 2.01 and 1.86 eV for **1** and **2**, respectively. All these data indicate that the preferential way of OPV quenching by the attached $C_{60}$ moiety is energy transfer rather than electron transfer. As a consequence, in the photovoltaic device, the chance of intramolecular charge separation events is extremely low thus leading to poor photocurrent yields.[11]

Photophysical investigations on the fulleropyrrolidine/OPV system **3** (Figure 4), analogous to **1** and **2**, showed that excitation of the OPV moiety sensitizes the lowest fullerene singlet and then (*via* intersystem crossing) triplet excited states, as revealed by means of fluorescence and transient absorption spectroscopy, both in dichloromethane and in more polar benzonitrile.[75,76]

These results confirm that OPV→$C_{60}$ energy transfer successfully compete with electron transfer in OPV-$C_{60}$ dyads. Also, they suggest that higher yields of photocurrent might be obtained by improving the electron-accepting character of the carbon sphere and/or the electron-donating ability of the oligomeric chain, so as to favour charge separation over energy transfer. Indeed, the fact that the photovoltaic performances of **2** are three times better that those of **1** are likely to be related to the better electron-donating properties of the longer tetrameric 4PV fragment.[11] Importantly, this suggests that understanding photoinduced processes at the molecular level in solution allows to obtain structure–activity relationships and devise strategies for the implementation of the device performances.

The electron-donating properties of the OPV moieties in **1** and **2** can be improved by adding another $-OC_{12}H_{25}$ residue in the *para* position of the terminal phenylene group. In this way the first oxidation potential of the OPV unit in **4** (Figure 5) is shifted by 280 mV to less positive potentials compared to

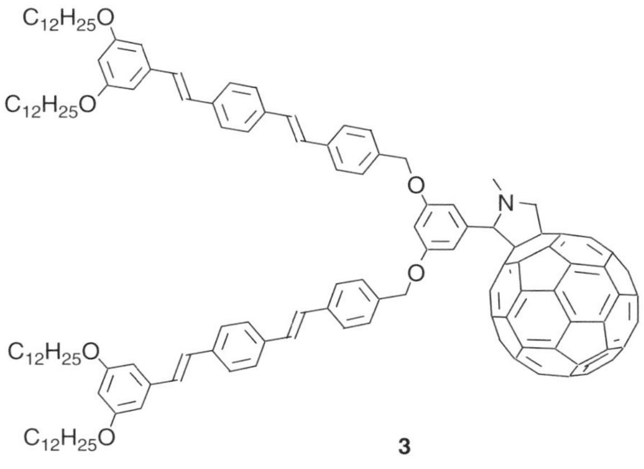

**Figure 4** *OPV-fulleropyrrolidine dyad 3*

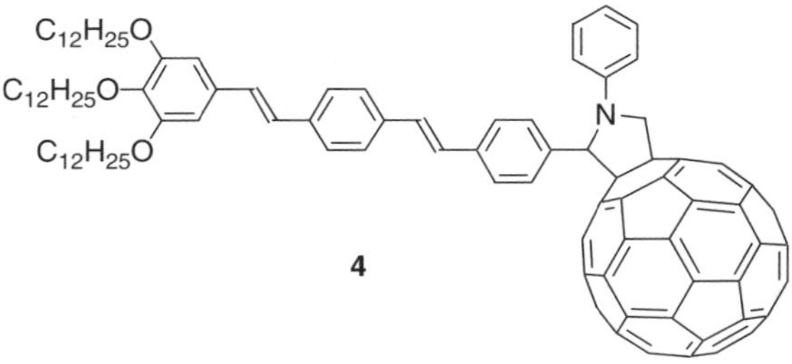

**Figure 5** *OPV-fulleropyrrolidine dyad* **4**

**1**, whereas the reduction potential of the fulleropyrrolidine moiety is practically identical in the two systems.

Detailed investigations on **4**[69] and on other similar arrays[10,75,77] have confirmed that ultrafast OPV→$C_{60}$ Förster-type energy transfer occurs in solution. In principle direct OPV→$C_{60}$ electron transfer may also take place but it is highly exergonic and located in the Marcus inverted region, thus it cannot compete with the energy-transfer process.[10,69] Therefore, electron transfer may only be originated from the lowest-singlet excited state of the fullerene moiety and the OPV fragment simply acts as an antenna unit. This might be the intimate pattern of photoinduced processes also in solid-state devices, even though it has been suggested that photovoltaic effect is likely to be the consequence of "material" rather than "molecular" processes.[78] Electron transfer from the fullerene singlet in OPV/$C_{60}$ arrays suffers from competition of internal deactivation and can be promoted by solvent polarity, which can conveniently lower the energy of the OPV$^+$-$C_{60}^-$ charge separated state. Thus, only in polar benzonitrile OPV→$C_{60}$ electron transfer is detected for **4**. All the above effects are illustrated in the energy diagram of Figure 6, which summarizes the cascade of photoinduced processes in **4** following excitation of the OPV moiety.

Several sophisticated OPV/$C_{60}$ arrays have been prepared in recent years.[69,79–83] In **5** (Figure 7), the fullerene is provided with both an energy (OPV) and an electron donor (pyrazoline) unit.[69]

Detailed studies on the dependence of photoinduced processes on solvent polarity, addition of acid (which leads to protonation of the pyrazoline nitrogen), and temperature reveal that this compound can be considered a fullerene-based molecular switch. The switchable parameters are photoinduced processes, namely OPV→$C_{60}$ energy transfer and pyrazoline→$C_{60}$ electron transfer.[69] The complex pattern of light-induced properties of **5** is sketched in Figure 8.

Interestingly, the incorporation of **5** in photovoltaic devices results in very low light to current efficiency since charge separation involving the fullerene moiety and the pyrazoline N atom is not able to contribute to the photocurrent and, instead, the pyrazoline unit can act as an electron trap. Nevertheless, the design principle of multicomponent arrays featuring an antenna unit (like OPV) and a charge separation module (like pyrazoline-$C_{60}$) is very appealing for the construction of devices for charge separation and light energy conversion. More recently, two new dumbbell pyrazoline-fullerene arrays **6** and **7** have

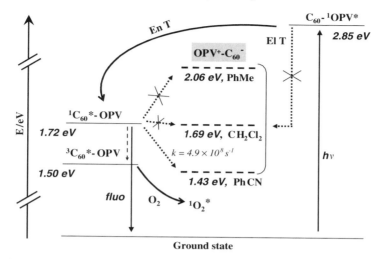

**Figure 6** *Energy-level diagram describing excited state intercomponent processes following light excitation of the OPV subunit in **4**. Charge separated states are indicated as dashed lines, whereas full lines represent localized singlet and triplet electronic excited levels. EnT and ElT stand for energy-transfer and electron-transfer, respectively. $C_{60}$ and OPV represent the fullerene and the oligophenylenevinylene moieties in **4**, respectively*

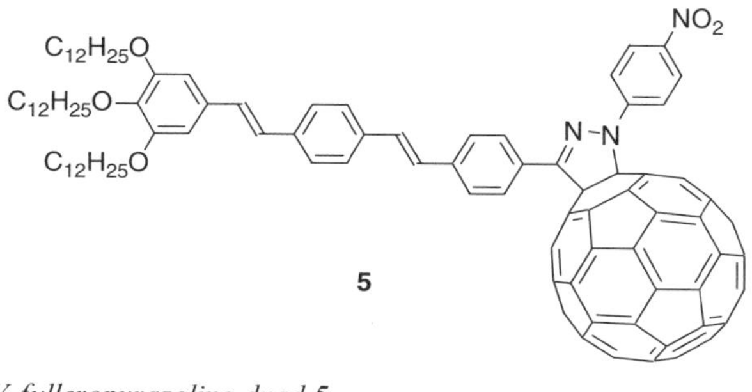

**Figure 7** *OPV-fulleropyrazoline dyad **5***

been prepared and their properties compared to those of the corresponding non-dumbbell systems **8–9** (Figure 9).[80]

The trend in photoinduced processes for **6–9** is very similar to that observed for **5** and described above (Figure 8) The experimental rates of singlet OPV → $C_{60}$ energy transfer have been found to fit the expectations based on the Förster dipole–dipole mechanism. As depicted in Figure 9 the energy-transfer process is the fastest for **6** which has a short central OPV rod and two terminal fullerene units, and the slowest for **9** having the long five-membered OPV and just one $C_{60}$ unit. Such two extreme behaviours correspond to the largest and smallest overlap between the emission spectrum of the OPV unit and the absorption spectrum of the fullerene part(s) of the molecules thus driving the energy-transfer

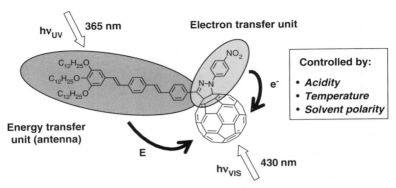

**Figure 8**  *OPV fulleropyrazoline system **5** where the fullerene unit acts as energy or electron acceptor for the OPV and the pyrazoline moiety, respectively. Switching of photoinduced processes can be obtained by operating on different parameters, namely excitation wavelength, proton concentration, solvent polarity, and temperature.*

rate to its maximum and minimum value. These results corroborates the occurrence of Förster energy transfer between the organic conjugated fragment and the carbon sphere in OPV-$C_{60}$ arrays postulated earlier.[11,75]

The occurrence of ultrafast energy transfer between OPV and $C_{60}$ units has prompted the synthesis of arrays with increasingly large OPV moieties acting as light harvesting "antenna" units, such as **10**, **11** (Figure 10) and **12** (Figure 11).[24,84] **10–12** belong to the family of fullerodendrimers, which will be discussed in more detail on the next section.

Molar absorptivities $\varepsilon$ ($M^{-1}$ $cm^{-1}$) at the OPV-type band maximum (394 nm) are as follows: 134,800 for **10**, 255,100 for **11**, and 730,400 for **12**. Thus, the relative amount of light captured by the organic conjugated moiety compared to the fullerene unit ($\varepsilon = 7600$ $M^{-1}$ $cm^{-1}$, 394 nm) is progressively increased along the series reaching a 99:1 ratio for **12**, where incident light is virtually captured only by the OPV giant fragment (antenna effect). Despite virtually identical thermodynamic driving forces for OPV→$C_{60}$ electron transfer, substantial differences in the extent of electron transfer are found as a function of solvent polarity. This suggests subtle dendritic effects on the rates and relative importance of light-induced phenomena.[85]

As far as the construction of $C_{60}$-conjugated oligomer/polymer arrays for photovoltaic applications is concerned, two further approaches have been proposed besides the molecular one described above. The first one is based on the so-called double cable concept.[15] Fullerene units are grafted to conjugated polymeric backbones so as to obtain intrinsically bipolar double cable polymers in which the negative charge carriers (fullerenes) are spatially close to each other and covalently linked to the positive charge carrier (polymeric backbone). In this way, the effective donor–acceptor interfacial area is maximized and positive effects on device efficiency and duration are expected.[15]

The second approach is termed "supramolecular" and is aimed at creating morphological organization in the active layer of photovoltaic cells *via*

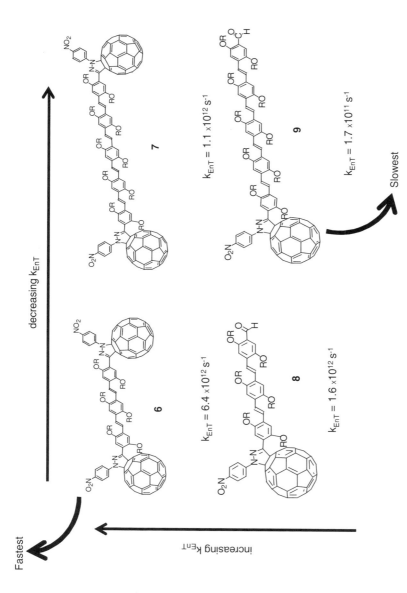

**Figure 9** *OPV fulleropyrazoline dumbbell systems 6–7 and their reference dyads 8–9 ($R = C_{12}H_{25}$). Experimental rates and trends of energy transfer are also indicated*

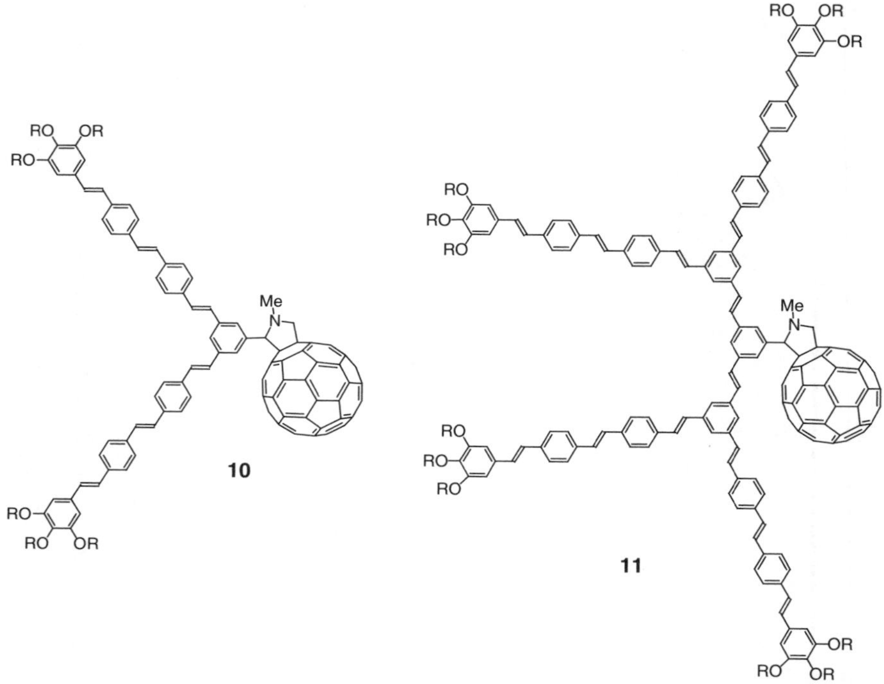

**Figure 10** *OPV-fulleropyrrolidine systems **10** and **11** ($R = C_{12}H_{25}$). The increasing OPV branched moiety acts as light harvesting "antenna"*

spontaneous supramolecular organization.[86] Recently, photophysical investigation on adducts between a methanofullerene and an OPV molecule, both provided with self-complementary 2-ureido-4[1H]-pyrimidinone units have been carried out. In low polarity solvents, the self-association constant is very high and the supramolecular adduct exhibits OPV → $C_{60}$ singlet energy transfer, with no evidence of electron transfer even in polar solvents.[87]

Finally, it is worth pointing out that not only OPV/$C_{60}$ arrays but also oligophenyleneethynylene/$C_{60}$[88] and oligothiophene/$C_{60}$[89–93] systems are intensively investigated in order to elucidate their photophysical properties and test them for photovoltaic applications. Photocurrents have been also generated in devices including gold electrodes modified with fullerene-linked oligothiophenes.[94]

## 4.4 Fullerodendrimers

When first prepared in the late 1970s, and for some years thereafter, dendrimers were basically considered a scientific curiosity. In the past 20 years, however, they have attracted a great deal of attention since potentially interesting in several fields of application such as drugs and diagnostics,[95–97] sensors,[98]

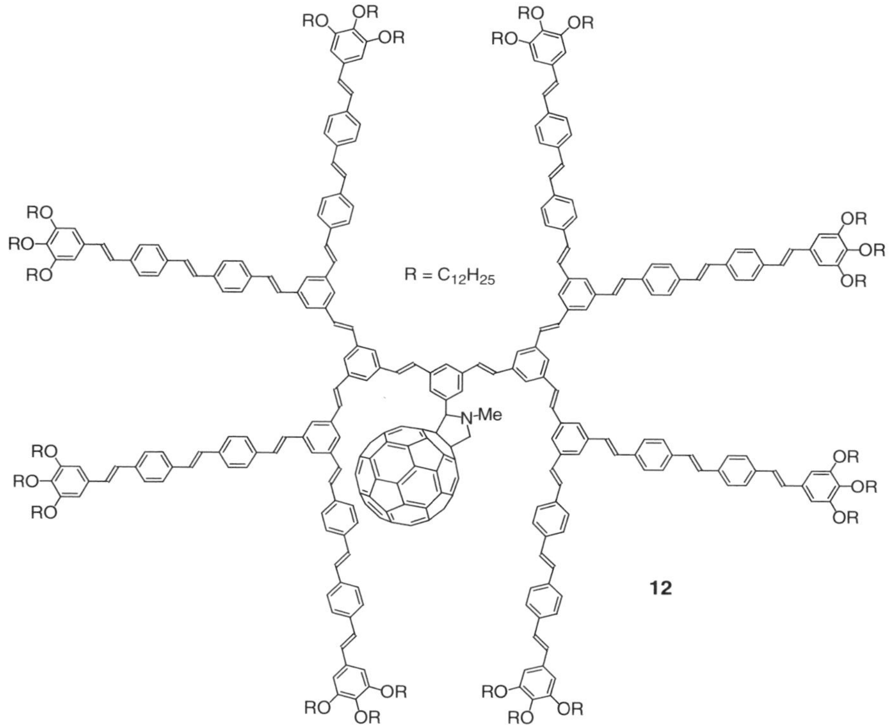

**Figure 11** *OPV-fulleropyrrolidine giant light harvesting array* **12**

biotechnology,[99] light harvesting,[100,101] luminescence,[102] catalysis,[103] polymer science,[104] *etc*. This interest in mainly related to the capability of dendritic architectures to generate specific properties as a result of their unique molecular structures. For example, their hollow cavities within the branching structures have been used to hold metal nanoparticles, drugs and imaging agents.

Owing to its physical and chemical properties, $C_{60}$ is a convenient building block in dendrimer chemistry.[105,106] In particular, its spherical shape makes it a natural candidate as a central core unit, also thanks to the possibility of multiple addition patterns, from mono- to exadducts, through a variety of substituents. This *fullerene-inside* approach is complemented by the *fullerene-outside* concept, where the carbon sphere is employed as terminal unit of dendritic structures. In the following sections examples of photoactive fullerodendrimers with inside and outside design will be presented.

### 4.4.1 Fullerene Inside

Figures 12 and 13 report two series of dendrimers having a fullerene core where one (**13–16**) or two (**17–20**) dendritic branches have been attached to $C_{60}$. Their

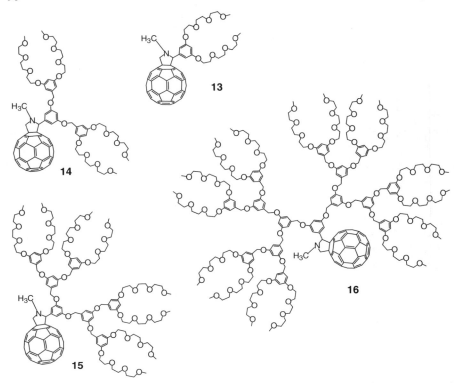

**Figure 12** *Fullerodendrimers 13, 14, 15, and 16 with a fulleropyrrolidine central core*

photophysical properties have been systematically investigated in different solvents and the changes observed along the series suggest an increasing isolation of the central chromophore towards the exterior. Notably, external dendritic shells are able to affect the process of singlet oxygen sensitization.

### 4.4.1.1 Fulleropyrrolidine Dendrimers 13–16

The electronic absorption spectra of **13–16** in toluene, dichloromethane, and acetonitrile (hereafter indicated as, respectively, PhCH$_3$, CH$_2$Cl$_2$, and CH$_3$CN) are reported in Figures 14–16.

The absorption spectra in CH$_2$Cl$_2$ and CH$_3$CN reveal the increasing contribution of the poly(aryl ether) dendritic branches above 250 nm and around 280 nm (shoulder) in passing from **13** to **16** (Figures 14–16), in agreement with the dendrimers structures. In all solvents changes of the absorption spectral shapes are observed for **13–16**. In particular, it can be noticed: (a) an intensity decrease of the fullerene bands in the 250–350 nm region, quite remarkable for **16**; (b) a progressive loss of spectral resolution in the Vis spectral region, particularly evident from the marked intensity reduction of the diagnostic fulleropyrrolidine

**Figure 13** *Fullerodendrimers 17, 18, 19, and 20 with a bismethanofullerene central core*

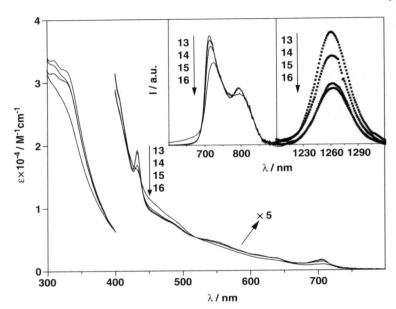

**Figure 14** *Absorption spectra of **13–16** in toluene solution at 298 K; above 400 nm a multiplying factor of 5 is applied. Inset: fluorescence (left, $\lambda_{exc} = 370$ nm, OD = 0.30) and sensitized singlet oxygen luminescence spectra (right, $\lambda_{exc} = 480$ nm, OD = 0.20) of **13–16***

absorption peaks at 431 and 706 nm. This might reflect a modification of the $C_{60}$ chromophore solvation environment as a consequence of a tighter contact with the external dendritic arms containing aromatic units possibly promoted by favourable electronic donor–acceptor interactions. The effect is the strongest in $CH_3CN$ (Figure 16), where the high solvent polarity promotes intramolecular stacking between the dendritic wedges and the hydrophobic central core leading to a more compact molecular structure.

The fluorescence spectra of **13–16** (Figures 14–16) exhibit a steady red-shift of the band maxima (up to 10–15 nm) as a function of dendrimer size in all solvents. Little or no variation of the fullerene singlet lifetime and fluorescence quantum yield are detected.

The very short singlet lifetime of fullerenes makes it unsuited to probe dendritic effects of protection, since the very fast intrinsic singlet excited state deactivation of fullerenes (around 1.5 ns) is expected to be much faster than the time needed for bimolecular-quenching processes. Thus, in order to test protective effects, the study of the much longer triplet lifetimes (hundreds of ns in air-equilibrated solutions) can be more helpful, also considering the high intersystem crossing yield of fulleropyrrolidines ($\geq 90\%$).[107] The shape and position of triplet–triplet transient absorption (TA) spectra are virtually identical for **13–16** in all solvents. The spectrum of **16** in $CH_2Cl_2$ (Figure 17), is quite comparable to those of simple fulleropyrrolidines. By contrast, the triplet

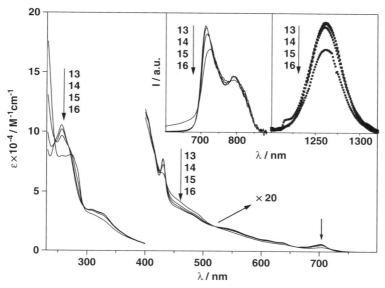

**Figure 15** *Absorption spectra of **13–16** in $CH_2Cl_2$ solution at 298 K; above 400 nm a multiplying factor of 20 is applied. Inset: fluorescence (left, $\lambda_{exc} = 370$ nm, OD = 0.30) and sensitized singlet oxygen luminescence spectra (right, $\lambda_{exc} = 480$ nm, OD = 0.20) of **13–16***

lifetimes are steadily increased on passing from **13** to **16**, as displayed in the inset of Figure 17 and summarized in Table 1.

The differences between triplet lifetimes measured in air equilibrated solutions (AER) with respect to the corresponding values in air-free samples (DEA) is attributable to oxygen quenching [Equation (1)], The occurrence of this bimolecular energy-transfer process has been probed by monitoring the diagnostic $^1O_2^*$ luminescence band centred at 1269 nm. whose relative intensity in a given solvent can be taken to evaluate the relative yields of singlet oxygen generation (Figures 14–16).

For **13–16** interesting trends can be obtained from the analysis of triplet lifetimes in air-equilibrated solutions (Table 1) and of the relative yields of formation of singlet oxygen, as derived from the of the $^1O_2^*$ NIR luminescence band intensities (Figures 14–16). A steady increase of lifetimes is found by increasing the dendrimer size in all solvents, suggesting that dendritic wedges are able to partially shield the fullerene core from contacts with dioxygen molecules. This hypothesis can be supported by the fact that the increase is particularly marked for polar $CH_3CN$, where a better shielding of the fullerene chromophore is expected (see above); in this case a 45% lifetime prolongation is found in passing from **14** to **16** (23 and 28% only for $PhCH_3$ and $CH_2Cl_2$, respectively). It must be emphasized that triplet lifetimes of **16** in the three solvents are rather different from each other, likely reflecting specific solvent–fullerene interactions that affect excited state deactivation rates. This suggests

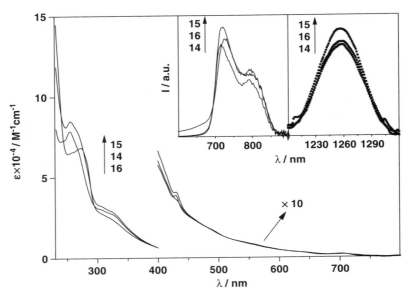

**Figure 16** *Absorption spectra of **14–16** in $CH_3CN$ solution at 298 K; above 400 nm a multiplying factor of 10 is applied. Inset: fluorescence (left, $\lambda_{exc} = 370$ nm, $OD = 0.30$) and sensitized singlet oxygen luminescence spectra (right, $\lambda_{exc} = 480$ nm, $OD = 0.20$) of **14–16**. **13** is not soluble in this solvent*

that, albeit a dendritic effect is evidenced, even the largest wedges are not able to provide a complete shielding of the central fulleropyrrolidine core. In any case, the dendritic wedges are able to reduce the singlet oxygen sensitization capability of the fullerene chromophore, and **16** exhibits the lowest singlet oxygen sensitization yield in all the investigated solvents (Figures 14–16). Notably, in all the investigated solvents, the triplet lifetimes are constantly increased with the dendrimers size also in oxygen-free samples (Table 1). This demonstrates that other quenchers of fullerene triplet states are present in solution. The observed effect is attributed to solvent impurities such as stabilizers and/or paramagnetic ions that are commonly present in spectroscopic grade solvents.

### 4.4.1.2  Bis-Methanofullerene Dendrimers 17–20

The UV–VIS absorption spectra of **17–20** in toluene, dichloromethane, and acetonitrile ($PhCH_3$, $CH_2Cl_2$, and $CH_3CN$, respectively) are reported in Figures 18–20. In $CH_2Cl_2$ and $CH_3CN$, the increasing contribution of the poly(aryl ether) dendritic branches above 250 nm and around 280 nm is observed in passing from **17** to **20** (Figures 18–20). A substantial increase of absorption intensity with dendrimer generation is recorded in all solvents in the region between 380 and 500 nm. The spectrum of **17**, the smallest representative of the series, is virtually unaffected by change of the solvent polarity throughout

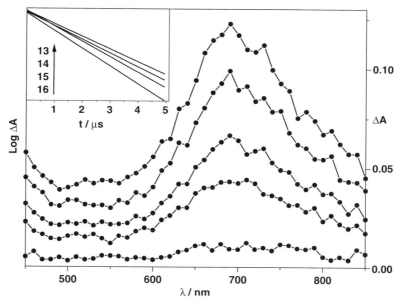

**Figure 17** *Triplet transient absorption spectrum of **16** at 298 K in $CH_2Cl_2$ air equilibrated solution upon laser excitation at 355 nm (energy = 3 mj pulse$^{-1}$). The spectra were recorded at delays of 100, 300, 600, 900, and 2000 ns following excitation. The inset shows the different decay time profiles of $\Delta A$ (700 nm) for the whole series*

**Table 1** *Triplet lifetimes as determined by transient absorption spectroscopy ($\lambda$ = 700 nm) at 298 K in air-equilibrated (AER) and oxygen-free solutions (DEA)*

|    | $PhCH_3$ | | $CH_2Cl_2$ | | $CH_3CN$ | |
|----|----------|-----------|------------|-----------|----------|-----------|
|    | AER (ns) | DEA (μs)  | AER (ns)   | DEA (μs)  | AER (ns) | DEA (μs)  |
| 13 | 279      | 34.0      | 598        | 26.3      | [a]      | [a]       |
| 14 | 304      | 37.9      | 643        | 27.4      | 330      | 14.0      |
| 15 | 318      | 39.4      | 732        | 30.1      | 412      | 21.7      |
| 16 | 374      | 42.5      | 827        | 34.6      | 605      | 25.5      |

[a] Not soluble.

the UV–VIS spectral range and is virtually identical to that of other bismethanofullerenes, showing a diagnostic peak at 438 nm. Thus, spectral variations observed upon increase of dendrimer shell cannot be attributed to changes of the fullerene/solvent interactions but must be related to increasing intramolecular interaction between the fullerene moiety and the external branches, most likely driven by electronic donor–acceptor attractions. Notably, a similar trend of absorption spectra has been observed for other families of

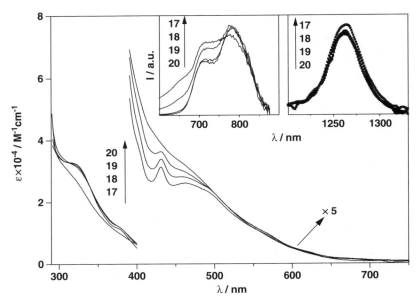

**Figure 18** *Absorption spectra of **17–20** in toluene solution at 298 K; above 400 nm a multiplying factor of 5 is applied. Inset: fluorescence (left, $\lambda_{exc} = 370$ nm, $OD = 0.20$) and sensitized singlet oxygen luminescence spectra (right, $\lambda_{exc} = 480$ nm, $OD = 0.20$) of **17–20***

fullerodendrimers[108,109] and upon complexation of fullerene molecules by dendrimeric hosts.[25] The fluorescence spectra of **17–20** in the three solvents are displayed in Figures 18–20.

By increasing the dendrimer size, fluorescence band broadening is observed in all cases on the high-energy side regardless of the excitation wavelength. However, little or no variations for singlet lifetimes or fluorescence quantum yields were measured along the series in any solvents, as for the fulleropyrrolidine dendrimer family **13–16**.

The shape of triplet–triplet TA spectra is substantially identical for **17–20** in all the investigated solvents, with a wide absorption band peaked at 705 nm.[108] Triplet lifetime values determined by TA spectroscopy, are gathered in Table 2.

Differences found for the non-dendrimeric **17** in the three solvents (see also **13** in the previous section) indicate that solvent interactions strongly affect the deactivation rate constants of fullerene-excited states. For **17–20** a steady increase of air-equilibrated triplet lifetimes is found by increasing the dendrimers size in all solvents (Table 2). Since, the fullerene active core is identical along the series one has to conclude that dendritic wedges are able to partially shield the fullerene core from interactions with dioxygen molecules. Similar protective effects towards dioxygen quenching have been already observed, besides the **13–16** series presented in the previous section, for other dendrimer cores exhibiting long-lived electronic excited states.[110–112]

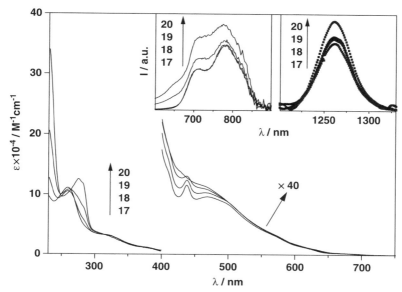

**Figure 19** *Absorption spectra of **17–20** in $CH_2Cl_2$ solution at 298 K; above 400 nm a multiplying factor of 40 is applied. Inset: fluorescence (left, $\lambda_{exc} = 370$ nm, $OD = 0.20$) and sensitized singlet oxygen luminescence spectra (right, $\lambda_{exc} = 480$ nm, $OD = 0.20$) of **17–20***

By inspecting the data of air-equilibrated triplet lifetimes in Table 2 interesting trends can be emphasized. For instance, although the lifetime of **17** is substantially different in $CH_2Cl_2$ and $CH_3CN$ (611 and 314 ns, respectively) an identical lifetime, within experimental uncertainties, is measured for **20** ($\approx 1100$ ns). This suggests that in $CH_2Cl_2$ and $CH_3CN$ the fullerene core of **20** is buried inside the dendronic cage, with negligible interactions with solvent molecules. Notably, in toluene and acetonitrile the lifetime value of **17** is shorter than in $CH_2Cl_2$, but in both solvents an increment of over 200% is observed for **20**, pointing to a very effective shielding of the dendronic cage in solvents of any polarity.

The triplet lifetime data of **17–20** (Table 2) can be interestingly compared to those of **13–16** (Table 1). For the latter less marked lifetime increase has been observed in passing from the smallest to the largest dendrimer (only 34% in toluene, for instance), pointing to a much less effective protection of the fullerene active core for the fulleropyrrolidine relative to the bis-methanofullerene dendrimers. This difference can be rationalized by comparing the structure of **20** with that of **16**. When the dendronic moiety is anchored at two different points of the carbon sphere, as in **20**, a more effective wrapping can be reasonably expected in comparison to fulleropyrrolidine analogues.

Triplet lifetimes of **17–20** are also constantly increased, by enlarging the dendrimers size, also in oxygen-free solutions (Table 2). As pointed out above this is attributed to quenching by solvent impurities such as stabilizers and/or

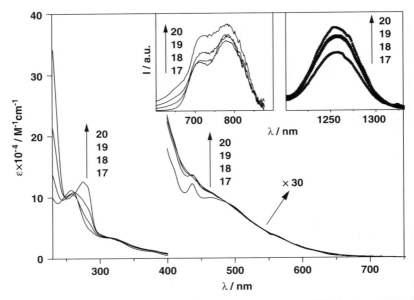

**Figure 20** *Absorption spectra of **17–20** in $CH_3CN$ solution at 298 K; above 400 nm a multiplying factor of 30 is applied. Inset: fluorescence (left, $\lambda_{exc} = 370$ nm, $OD = 0.20$) and sensitized singlet oxygen luminescence spectra (right, $\lambda_{exc} = 480$ nm, $OD = 0.20$) of **17–20***

**Table 2** *Triplet lifetimes as determined by transient absorption spectroscopy ($\lambda = 700$ nm) at 298 K in air equilibrated (AER) and oxygen-free solutions (DEA) solutions*

|  | $PhCH_3$ | | $CH_2Cl_2$ | | $CH_3CN$ | |
|---|---|---|---|---|---|---|
|  | AER (ns) | DEA (μs) | AER (ns) | DEA (μs) | AER (ns) | DEA (μs) |
| **17** | 288 | 21.3 | 611 | 19.0 | 314 | 13.6 |
| **18** | 317 | 23.0 | 742 | 20.0 | 380 | 14.7 |
| **19** | 448 | 26.8 | 873 | 25.9 | 581 | 20.0 |
| **20** | 877 | 34.7 | 1103 | 31.7 | 1068 | 27.9 |

paramagnetic ions that are commonly present in organic solvents. This trend in triplet lifetimes of air-purged samples is taken as a further proof of a protective dendritic effect along the series **17–20**. Notably, in fullerodendrimers with more rigid dendronic wedges no prolongation of triplet lifetimes has been observed.[113,114]

The less effective quenching of fullerene triplet states by increasing the dendrimer size can be assigned to one or more of the following factors: (a) decrease in the diffusion rate of $O_2$ inside the peripheral wedges, with increasing volume of the compound; (b) lower solubility of $O_2$ in the interior of the dendrimers; (c) preferential solvation of the $C_{60}$ core by the dendrimer branches, hindering suitable orbital overlap for singlet oxygen sensitization.

Detailed studies suggest that (a) is the most likely explanation.[44] Furthemore singlet oxygen lifetimes, which are strongly affected by the chemical environment,[115–118] indicate that for **20** in $CH_3CN$ oxygen trapping can occur. In other words, the structure of the largest dendrimers in the most polar medium is so compact that trapping of very small molecules is made possible.

### 4.4.2 Fullerene Outside

In Figures 21 and 22 are reported three dendrimers with the same pentameric OPV core and two, four, or eight bis-methanofullerenes as terminal units. In these complex architectures a progressive isolation of the internal hydrophobic core by the bulky external carbon spheres can be anticipated by increasing the dendrimers size and the solvent polarity. Since **21–23** exhibit exactly the same OPV and $C_{60}$ moieties, modulation in the pattern of the typical OPV→$C_{60}$ photoinduced processes (see Section 4.3) can be ascribed solely to structural

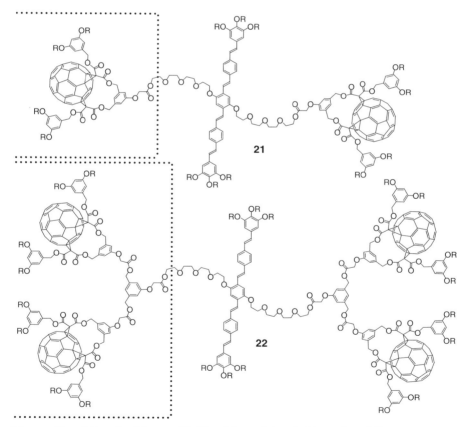

**Figure 21** *Fullerodendrimers **21–22** with peripheral bis-methanofullerene moieties ($R = C_{12}H_{25}$). In the areas delimited by dotted lines the fullerene molecules used as reference in photophysical investigations are highlighted (see text)*

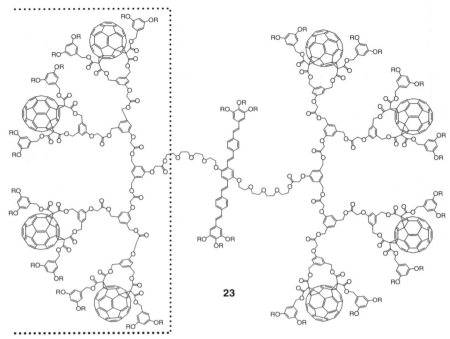

**Figure 22** *Fullerodendrimer **23** with peripheral bis-methanofullerene moieties ($R = C_{12}H_{25}$). In the area delimited by dotted lines the fullerene dendron used as reference in photophysical investigations is highlighted (see text)*

"dendritic" factors. In order to probe these effects the photophysical properties of **21–23** have been systematically investigated in different solvents.

The absorption and fluorescence spectra and excited state lifetime of **21–23** in CH$_2$Cl$_2$ solution indicate that (i) absorption profiles do not match the spectra obtained by summing the component units (dendronic fragments and OPV); (ii) the OPV strong fluorescence ($\Phi_{FL} = 0.52$) is quenched by over a factor of 100 in all fullerodendrimers; (iii) excitation spectra taken at 800 nm (fullerene fluorescence) exhibit the intense fullerene band at 260 nm but also the contribution of the OPV absorption at 425 nm. These results show that in **21–23** strong ground state interaction between OPV and C$_{60}$ moieties take place (thanks to floppy structures favouring close contacts) and that singlet–singlet $^1$OPV* → $^1$C$_{60}$* energy transfer occurs, in line with previous reports on OPV-C$_{60}$ hybrids.

As discussed in Section 4.3, photoinduced electron transfer in OPV-C$_{60}$ arrays has been obtained in solution from the lowest fullerene singlet excited state ($\sim 1.7$ eV), directly or indirectly populated following light absorption. On the contrary, population of upper lying OPV singlet levels ($> 2.8$ eV) does not result in electron transfer since competitive ultrafast OPV → C$_{60}$ energy transfer prevails. Thus electron transfer in OPV-C$_{60}$ systems can be conveniently signaled by the quenching of $C_{60}$ fluorescence and can be observed or not, depending on the solvent polarity (see Figure 6).[10,69]

In Figures 23–25 (top) are reported the fluorescence spectra of **21–23** in solvent of increasing polarity, compared to those of the corresponding reference fullerene dendrons (see Figures 21 and 22) under the same conditions. Toluene (PhMe), dichloromethane ($CH_2Cl_2$), and benzonitrile (PhCN) have been used, which exhibit static relative permittivity $\varepsilon$ of 2.4, 8.9, and 25.2, respectively.

In PhMe solution (Figure 23), the fullerene-type fluorescence is unquenched relative to the model compounds for the whole **21–23** family. This signals the

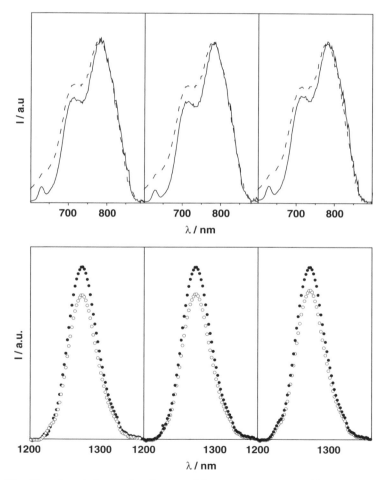

**Figure 23** *Top, from left to right: Corrected fluorescence spectra of **21–23** (dashed line) and their dendronic fullerene-only reference compounds (full line) in $PhCH_3$; $\lambda_{exc} = 425$ nm or $\lambda_{exc} = 530$ nm. Bottom, in the same order: Sensitized singlet oxygen emission in $PhCH_3$ of **21–23** (empty circles) and their reference compounds (full circles); $\lambda_{exc} = 500$ nm. $A = 0.30$ for all compounds*

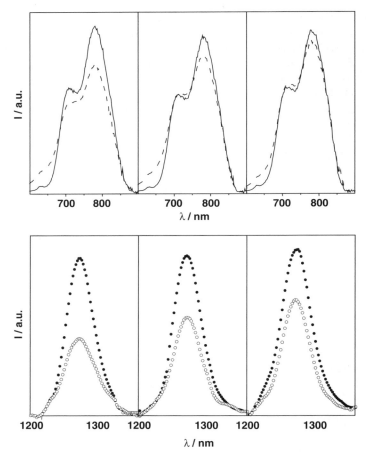

**Figure 24** *Top, from left to right: Corrected fluorescence spectra of **21–23** (dashed line) and their dendronic fullerene-only reference compounds (full line) in $CH_2Cl_2$; $\lambda_{exc} = 425$ nm or $\lambda_{exc} = 530$ nm. Bottom, in the same order: Sensitized singlet oxygen emission in $CH_2Cl_2$ of **21–23** (empty circles) and their reference compounds (full circles); $\lambda_{exc} = 500$ nm. $A = 0.30$ for all compounds*

absence of OPV→$C_{60}$ electron transfer upon population of the fullerene singlet state in apolar toluene, as observed in many other OPV-$C_{60}$ arrays.[10,69] It is interesting to note that, in any solvent, the quantum yield of singlet oxygen sensitization of the dendrimers is about 15% lower than that of the corresponding reference compounds. This is not a consequence of triplet depletion by electron transfer from the OPV unit, since this process is thermodynamically not allowed and no fullerene triplet lifetime quenching is measured. The decrease of the singlet oxygen sensitization yield is attributed to peculiar intramolecular contacts occurring in the dendrimer structures, which decrease the surface of the fullerene cages available for dioxygen bimolecular interactions.[68]

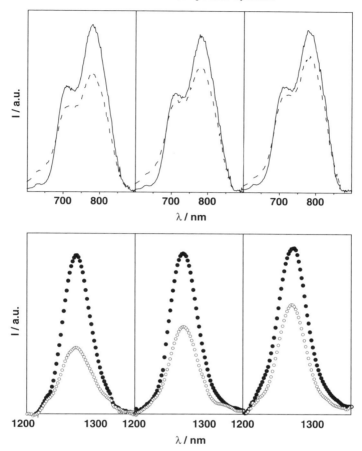

**Figure 25** *Top, from left to right: Corrected fluorescence spectra of **21–23** (dashed line) and their dendronic fullerene-only reference compounds (full line) in PhCN; $\lambda_{exc} = 425$ nm or $\lambda_{exc} = 530$ nm. Bottom, in the same order: Sensitized singlet oxygen emission in PhCN of **21–23** (empty circles) and their reference compounds (full circles); $\lambda_{exc} = 500$ nm. $A = 0.30$ for all compounds*

In $CH_2Cl_2$, the energy of the $OPV^+$-$C_{60}^-$ charge separated state of **21** is estimated to be 1.68 eV from electrochemical potentials, *i.e.* almost isoenergetic to the fullerene singlet. Similar energetics do not allow electron transfer in rigid OPV-$C_{60}$ arrays, owing to an activation barrier of about 0.2 eV.[69] Clearly, in the present flexible systems where electron transfer partners can get in tighter vicinity, such barrier is lower and electron transfer can occur, as suggested by the decrease of fullerene fluorescence for **21** (Figure 24, top left). A more marked effect on fullerene fluorescence in polar PhCN supports this interpretation (Figure 25, top left). The singlet lifetime of the fullerene moiety of **21** is 750 ps, compared to 1600 ps of the reference moiety **24**, corresponding to an electron transfer rate constant of $7.1 \times 10^8$ s$^{-1}$ in benzonitrile.

Very importantly, for larger dendrimers **22** and **23** there is recovery of fluorescence intensity in both $CH_2Cl_2$ and PhCN, suggesting a decreased electron transfer efficiency, despite the fact that electron transfer partners and thermodynamics are identical to **21**. Practically, no electron transfer from fullerene singlet occurs for **23** in $CH_2Cl_2$ (Figure 24, top right), whereas some of it is still detected in the more polar PhCN (Figure 25, top right). These trends can be rationalized by considering increasingly compact dendrimer structures in more polar solvents, as observed for the fullerene-inside systems of the previous section. This implies that the actual polarity experienced by the involved electron transfer partners, particularly the central OPV, is no longer that of the bulk solvent. This strongly affects electron transfer thermodynamics which, being reasonably located in the normal region of the Marcus parabola,[10,69] becomes less exergonic and thus slower and less competitive towards intrinsic deactivation of the fullerene singlet state. This dendritic effect is in line with the molecular dynamics studies, which suggest that the central OPV unit is more and more protected by the dendritic branches when the generation number is increased. Actually, the calculated structure of **23** shows that the two dendrons of third generation are able to fully cover the central OPV core.[119]

The trend of sensitized singlet oxygen luminescence resembles that of fullerene fluorescence, since the detected signal is more intense by increasing the dendrimer size, confirming the above-mentioned polarity effects. However, as in the case of PhMe solvent, the amount of fullerene triplet (indirectly monitored *via* NIR singlet oxygen emission)[68,69] is comparably lower than that of singlet (probed by fullerene fluorescence). This behaviour is in contrast to what was observed in rigid OPV-$C_{60}$ arrays where the relative amount of fullerene fluorescence intensity and sensitized singlet oxygen are identical.[69] Similarly to what discussed above on PhMe solvent it has to be excluded that any electron transfer from the triplet state occurs, since triplet lifetimes of **21–23** are unchanged relative to the fullerene models. Thus reduced singlet oxygen sensitization is attributed to structural factors which, in polar solvents where the compactness of the dendrimers structure is probably enhanced, are expected to play an even major role.

In summary, the photophysical properties of **21–23** investigated in different solvents reveal significant polarity effects resulting from the dendritic structure. While the photophysical properties of the first-generation compound are similar to those already reported for related OPV-fullerene systems, the two largest structures evidence interesting effects. In simple OPV-$C_{60}$ dyads photoinduced electron transfer originating from the fullerene singlet excited state are found to be strongly solvent dependent,[10,69] because the energy of the charge-separated state can be finely tuned around the energy value exhibited by the fullerene singlet excited state ($\sim 1.7$ eV). Practically, solvent polarity can affect the electron transfer energetics and transform an endoergonic process in a moderately exergonic one, by switching from apolar (*e.g.* PhMe) to more polar media (*e.g.* PhCN). For the highest generation fullerodendrimer described here, the strong solvent effects on the OPV-$C_{60}$ charge separation processes are severely limited. For a given solvent, the extent of electron transfer is reduced with dendrimers size because progressive isolation of the central OPV electron donor by the surrounding dendrons tends to disfavour solvent-induced stabilization of the transient ionic species. The dendritic architecture is therefore not only able to isolate the central core unit but it can also influence dramatically its properties.

## 4.5 $C_{60}$: Metal Complex Arrays

The relatively long-lived metal-to-ligand-charge-transfer (MLCT) excited states characterizing complexes of Ru(II),[120] Re(I),[121] and Cu(I)[122] with 2,2′-bipyridine or 1,10-phenanthroline ligands have been widely exploited in the design of supramolecular molecular architectures featuring photoinduced energy- and electron-transfer processes.[122–125] The MLCT excited states of these metal complexes have a marked reducing character that, in principle, make them ideal partners for $C_{60}$ fullerene oxidants in the construction of donor–acceptor arrays for photoinduced electron transfer. Indeed, several examples of multicomponent arrays containing $C_{60}$ fullerenes and coordination compounds can be found in the literature which feature interesting photoinduced processes.[126–132,68,108]

The popular $[Ru(bpy)_3]^{2+}$ chromophore has been coupled with fullerene subunits in supramolecular architectures and electron transfer has been observed.[126] On the other hand, photophysical investigations on the systems reported in Figure 26 have shown that electron transfer in **24** is rapidly followed by fast and quantitative charge recombination to the low-lying fullerene triplet;

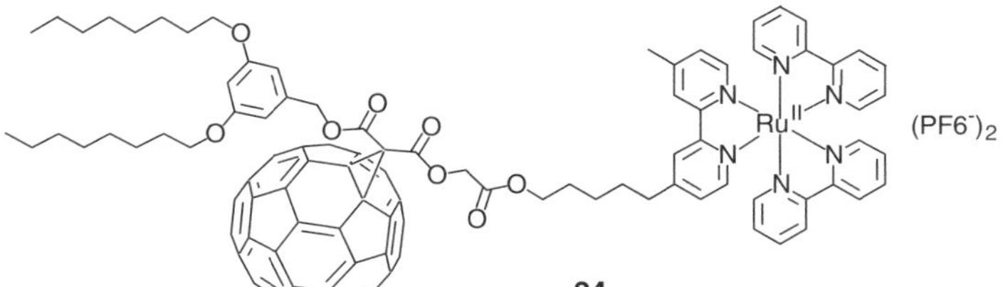

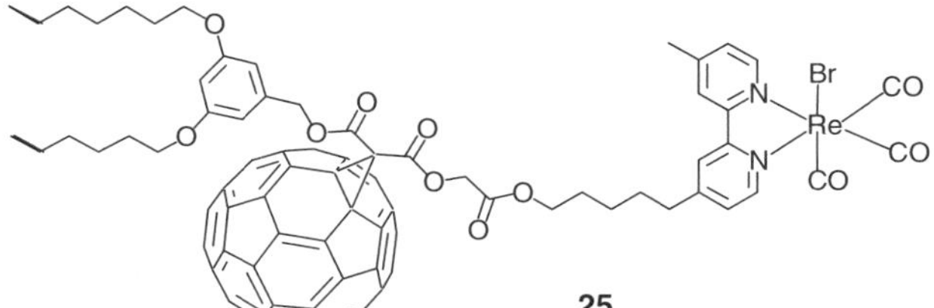

**Figure 26** Hybrid dyads **24** and **25**, containing a $C_{60}$ methanofullerene moiety and a Ru(II) or Re(I) complex, respectively

the same applies to the Re(I) analogue **25**.[68] This suggests that these hybrid systems are probably not suited for the generation of long-lived and highly exergonic charge separated states.

In a rotaxane made of a Cu(I)-bisphenanthroline core ($[Cu(NN)_2]^+$) and two $C_{60}$ terminal units (**26**) (Figure 27) the typical low-lying excited states of each moiety are strongly quenched, namely MLCT emission of the core, $C_{60}$ fluorescence, and $C_{60}$ triplet absorption. Also, the singlet oxygen sensitization observed for both (separated) subunits, is dramatically reduced. All these findings are a consequence of the fact that, in the supramolecular complex **26**, a low-energy charge-separated state is made available. Excitation of the central inorganic chromophore causes direct $(Cu(I) \rightarrow C_{60})$ electron transfer; importantly, this process is preceded by a $C_{60} \rightarrow [Cu(NN)_2]^+$ energy-transfer step when the light input is addressed to the fullerene chromophores.[129]

A fascinating family of fullerodendrimers with a Cu(I)-bisphenanthroline central core and a variable number (4, 8, or 16) of terminal bis-methanofullerene units has been reported.[108] The smallest representative of the series (**27**), with only four buckyball terminals, is depicted in Figure 28. Upon selective excitation of the peripheral $C_{60}$ units at 600 nm, fullerene fluorescence is observed for all dendrimers and the corresponding emission quantum yield and excited state lifetimes are identical, within experimental error, to those of the corresponding dendronic reference subunits.

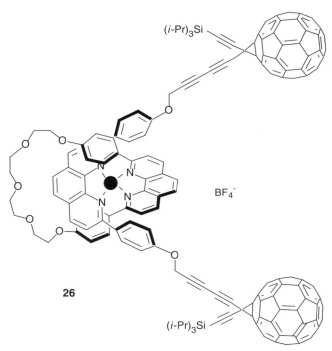

**Figure 27** *A rotaxane containing a Cu(I)-bisphenanthroline core and two methanofullerene stoppers*

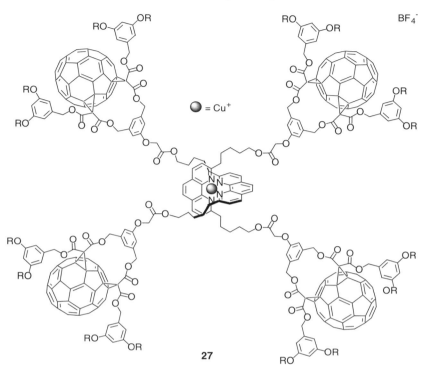

**Figure 28** *A fullerodendrimer with a Cu(I)-complexed core and four terminal bismethanofullerenes units ($R = -C_8H_{17}$)*

Despite the fact that selective excitation of the Cu-(bisphenanthroline) central core is not possible, clear evidence for the quenching of the luminescence of this moiety can be obtained. The quenching of the MLCT excited state of the Cu(I)-complexed moiety is tentatively attributed to energy transfer to the $C_{60}$ singlet or triplet states, even though electron transfer cannot be completely ruled out. Unfortunately detailed transient absorption studies on this family of fullerodendrimers upon selective or prevalent excitation of the metal complexed core is impossible due to the unfavourable light partitioning with the overwhelmingly absorbing fullerene moieties.

By comparing the results of the light-induced processes in **26** and **27**, an interesting point is that the fullerene moiety is quenched by the Cu(I) centre in the first sample but not in the second. The former contains methanofullerene fragments, the latter bismethano derivatives. In more recent studies involving fullerohelicates made of an identical central dinuclear Cu(I) complex and two external mono-(**28**) or bis-(**29**) methano derivatives (Figure 29), this trend has been confirmed.[127,128]

Therefore, investigations on several Cu(I)-phenanthroline/$C_{60}$ arrays with topologies ranging from rotaxanes,[129] to dendrimers,[108] to sandwich-type triads,[127,128] have shown that the fullerene moieties are invariably quenched

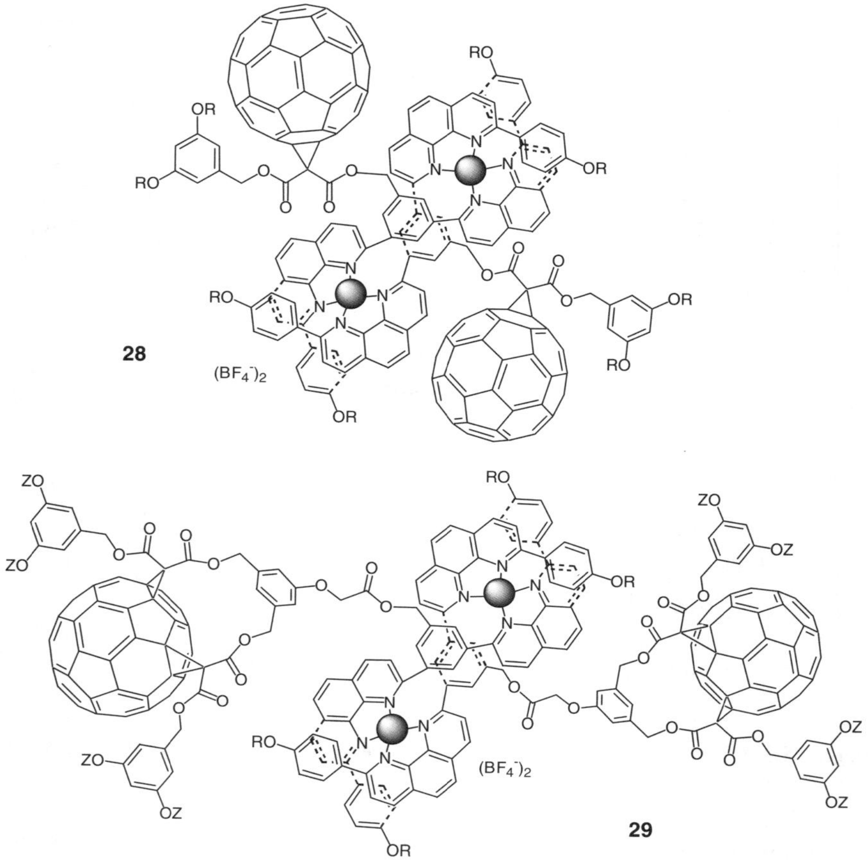

**Figure 29** Cu(I)-complexed fullerohelicates **28** and **29** ($R = -C_{12}H_{25}$)

(electron or energy transfer) by the metal-complexed unit when methanofullerenes are involved whereas quenching does not occur for bismethanofullerene arrays. This has to do with the inherently different electronic structure of the two-fullerene derivatives. By means of an analysis of their fluorescence spectra, which are substantially different, it was possible to conclude that the singlet excited state of methanofullerenes is more prone to undergo electron transfer than that of bismethanofullerenes, thanks to the associated smaller internal reorganization energy.[133] In addition, methanofullerenes are slightly easier to reduce than bismethanofullerenes, giving also a thermodynamic advantage for electron transfer in multicomponent arrays containing the monofunctionalized derivative. The combined effect of these two factors (kinetic and thermodynamic) can explain the different and somehow unexpected trend in photoprocesses of multicomponent arrays containing Cu(I)-phenanthrolines linked to methanofullerenes *vs.* bismethanofullerenes, which has been found in a variety of molecular architectures. Notably, when excitation is addressed to the Cu(I) complexed unit, Cu(I)→$C_{60}$ electron transfer is always observed for all supramolecular architectures, whatever the substitution pattern of the carbon sphere. This is related to the CT nature of the involved excited state that facilitates the electron-transfer process due to kinetic factors.[127–128,133]

## 4.6 $C_{60}$-Porphyrin Assemblies

The first report on light-induced processes in fullerene/porphyrin arrays dates back to late 1994, when Gust, Moore, Moore *et al.*[134] demonstrated that in a zinc porphyrin/fullerene array photoinduced electron transfer from the inorganic to the organic moiety occurs. This prompted a huge amount of synthetic work aiming at the construction of increasingly sophisticated arrays containing fullerene and porphyrin moieties. Mainly covalently linked arrays have been prepared;[1,2,4,135] however, arrays relying on weaker interactions, *e.g.* coordinative binding to metal porphyrins, are increasingly popular.[5,136,137] Many of these systems have been designed in order to get artificial models featuring the fundamental acts of natural photosynthetic systems, namely light harvesting and charge separation.[2] To this end impressive results have been obtained.[1]

An interesting aspect of the chemistry of fullerenes and porphyrins is that they are spontaneously attracted to each other, as a result of electronic donor–acceptor interactions.[138] This can be observed both in the solid state[139] and in solution.[26] For instance, Aida *et al.*[26] have successfully achieved the complexation of $C_{60}$ inside rigid porphyrin cages and this approach has been elegantly applied to the separation of mixtures of fullerenes of different size.[140] The triad (**30**) depicted in Figure 30, regardless of solvent polarity, adopts a conformation in which one carbon sphere is tangential to the porphyrin plane, as shown by NMR investigations.[141] This spontaneous attraction can also be monitored

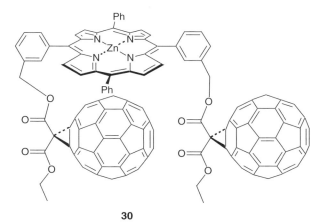

**30**

**Figure 30** *A fullerene/porphyrin array, characterized by spontaneous attractions of the two chromophores through donor–acceptor and van der Waals interactions*

photochemically since ground state charge transfer absorption bands (CT) are recorded, which is not the case for reference solutions containing the three molecular subunits unlinked. Quite remarkably, the CT states are luminescent in the near infrared region ($\lambda_{max}$ = 890 nm) and exhibit a lifetime of 720 ps.

The photochemistry and photophysics of face-to-face and porphyrin–fullerene arrays has further developed recently.[142–144] A fairly interesting case concerns the two dyads **31** and **32** reported in Figure 31.[145–147] The porphyrin moieties are meso–meso linked (**31**) or triply fused (**32**) dimers. In the former, the two heterocyclic rings are perpendicular to each other in the latter they are on the same plane and form a small tape. With the same concept Osuka et al.[148–150] have prepared molecular ribbons and tapes with tens of porphyrins subunits.

The absorption spectrum of **31** is depicted in Figure 32, along with the spectrum obtained by summing the corresponding molecular subunits (two methanofullerenes + one porphyrin dimer).

The porphyrin Soret absorption band around 420 nm (split by exciton coupling, typical for porphyrin oligomers) is substantially decreased in intensity, compared to the reference mathematical spectrum. Furthermore a new, weak absorption band above 700 nm is recorded. These two findings unambiguously point to close vicinity between the porphyrins plane and the carbon sphere, leading to charge transfer interactions. A new low-lying charge transfer state is established in the supramolecular array, which is populated upon excitation of any chromophore. The CT level partially deactivates *via* radiative paths originating a charge-transfer emission in the NIR region above 950 nm in toluene solution (Figure 32). This band exhibits spectral red-shift, intensity decrease, and shorter lifetimes when solvent of greater polarity are used (diethyl ether and THF). The CT excited state lifetimes can also be determined with transient absorption (TA) spectroscopy, and the data obtained are in excellent agreement with luminescence measurements. TA spectra evidence the formation of the porphyrin radical cation at *ca.* 630 nm (Figure 33), confirming the

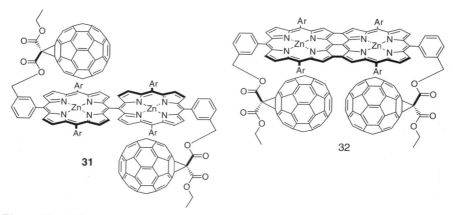

**Figure 31** *Fullerene/porphyrin arrays with a meso–meso linked (**31**) and triply fused (**32**) porphyrin core*

# Light-Induced Processes in Fullerene Multicomponent Systems 111

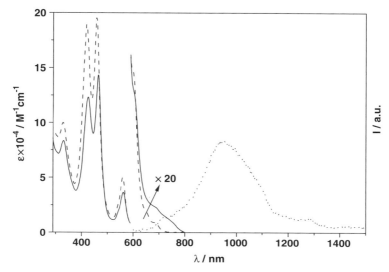

**Figure 32** *Absorption (full line) and luminescence (dotted line) spectra of **31** in toluene. The dashed lines represents the mathematical profile obtained by summing the spectra of the molecular subunits of the supramolecular array. The absorption spectral region at $\lambda > 600$ nm is multiplied by a factor of 20*

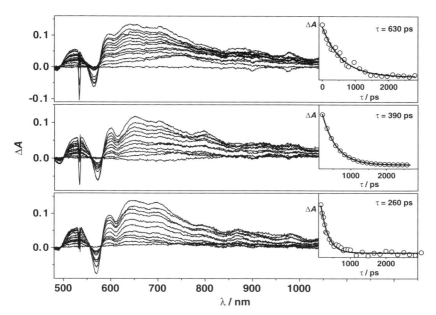

**Figure 33** *Picosecond transient absorption spectra of **31** in solvents of increasing polarity at 298 K: toluene (top), Et2O (centre), and THF (bottom). The explored time window is 3000 ps, and spectra are taken at intervals of 33 or 66 ps. In the inset are reported the spectral decay profiles of the porphyrin radical cation at $\lambda = 640$ nm with the corresponding lifetime determined via monoexponential fitting*

CT nature of the luminescent excited state. The faster charge recombination in more polar media is attributable to the energy-gap law[151] (lower CT energy, faster non-radiative deactivation rates) and/or to the fact that the CR process of the porphyrin–fullerene tight pairs is highly exergonic and lying in the Marcus inverted region.[2] Under this regime, charge recombination processes are expected to be faster in more polar media (THF) owing to a decrease of the activation energy for charge recombination.[152]

The electronic absorption spectrum of **32** is depicted in Figure 34, along with the spectrum obtained by summing the corresponding molecular subunits (two methanofullerenes + one porphyrin fused dimer). In this case, there is an excellent agreement between the experimental spectrum and the profile obtained by summing the component units. The spectrum of **32** is dominated by the porphyrin features with a strongly split Soret band in the range 400–650 nm and the $Q$ bands shifted in the NIR region above 800 nm. Absorption transitions centred on the fullerene core are prevailing only below 400 nm. The emission spectrum of **32** is not characterized by a new band as in the case of **31**, but the porphyrin emission profile is recorded (Figure 34). Notably, the fullerene luminescence above 700 nm is quenched and sensitization of the porphyrin moiety is observed (Figure 34, inset). This means that the low-lying

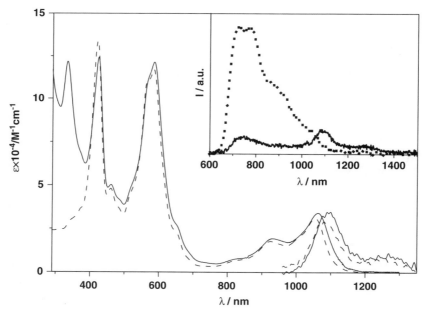

**Figure 34** *Absorption spectrum of **32** (full line) and that of the central triply fused porphyrin dimer alone (dashed line) in toluene. The low energy profiles with maxima at ca. 1100 nm are the corresponding fluorescence spectra ($\lambda_{exc} = 420$ nm). Inset: fluorescence spectra of a methanofullerene reference compound (dotted line) and of **32** (full line), upon excitation on the fullerene moieties ($\lambda_{exc} = 335$ nm, OD = 0.3) in toluene*

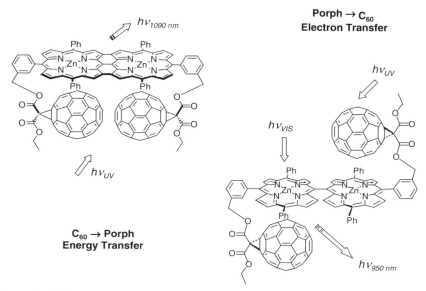

**Figure 35** *Schematic representation of light-induced processes occurring in* **31** *and* **32**

and very short-lived (4.5 ps) porphyrin singlet level offers an extremely competitive deactivation pathway and acts as a sink for all the higher energy electronic levels. Notably, this is the first case of a fullerene–porphyrin array where the carbon sphere acts as energy donor in a multicomponent array. This is possible because the triply fused dimer is characterized by a low-lying singlet level around 1 eV, *i.e.* well below the $C_{60}$ lowest singlet and triplet states.

In conclusion, despite very similar chemical structures, the photoinduced process in **31** and **32** are completely different. In the case of **31** a porphyrin → $C_{60}$ electron-transfer mechanism is active, while **32** is characterized by a $C_{60}$ → porphyrin energy-transfer process. Very interestingly, both molecules are NIR emitters. In Figure 35 is depicted a scheme summarizing the trend of light-induced processes in both **31** and **32**. An extensive descriptions on the photophysics of fullerene arrays with metal complexes and porphyrins can be found elsewhere.[1,2,4]

## 4.7 Conclusions

The electronic properties of fullerenes make them extremely attractive from the photochemical and photophysical point of view. In the presence of thoroughly chosen partners, they may undergo energy and/or electron-transfer processes upon light irradiation. The two phenomena are often in competition and the prevalence of one of the two is strictly related to the reaction thermodynamics involved which, in their turn, affect the reaction kinetics of electron transfer and/or energy-transfer processes,[153] according to the Marcus theory.[154] The solvent polarity is the simplest way of addressing photoprocesses in the desired direction, although the scarce intrinsic solubility of fullerenes does not allow a wide choice of solvents. In toluene solution, energy transfer is the main or exclusive photoinduced process; by switching to more polar benzonitrile, electron transfer often emerges as a strongly competing quenching mechanism.[8]

Here, we have described light-induced processes in selected fullerene multicomponent systems, which belong to some of the most extensively investigated classes in this area of research, namely fullerodendrimers, arrays with organic conjugated oligomers, coordination compounds, and porphyrins.

OPV-$C_{60}$ systems undergo energy and electron-transfer processes; the extent and competition between them are finely tuned by solvent polarity or little chemical modifications of the OPV donor moiety, which increases its electron-donating capability. These multicomponent systems have been tested in prototype photovoltaic devices, and the observed poor light to current efficiencies can be related to the trend of photoinduced processes at the molecular level, highlighting structure–activity relationships. Notably, the combined electronic properties of molecular architectures with OPV and $C_{60}$ subunits can be exploited to design fullerene molecular switches.

Hybrid systems combining a carbon cage and a coordination compound allow to join subunits with $\pi\pi$ (organic moiety) and MLCT (metal complex fragment) electronic levels. The intrinsically different nature of such excited states causes a quite different outcome of light stimulation of either moieties, which is also strongly influenced by the substitution pattern of the carbon sphere. The study of these systems has shed light on several subtle aspects of photoinduced electron-transfer in fullerene systems and has shown that Cu(I) complexes promote the generation of low-lying charge separated states. By contrast, in Ru(II) complexes, the fullerene triplet level is the final sink of all upper-lying excited states.

Face-to-face fullerene arrays with meso-linked porphyrins are extremely interesting since exhibit low-lying NIR emitting charge transfer states, very rarely observed in supramolecular photochemistry. Similar triply fused systems are the only example reported to date in which one can observe a fullerene → porphyrin energy transfer due to a very low-lying porphyrin singlet state.

Finally, in fullerodendrimers, substantial protective effects of fullerene cores by peripheral dendronic branches have been evidenced by studying oxygen-quenching processes (fullerene-inside approach). On the other hand, when $C_{60}$s are used as external decorating fragments in dendritic architectures they can affect the partitioning between energy and electron transfer processes from the OPV central core, thanks to structure-induced polarity effects (fullerene-outside approach).

More than 20 years after their discovery, fullerenes still keep a great potential for large-scale practical applications and this has prompted industrial production on the ton scale in order to decrease the initially prohibitive price.[155] Possible extensive utilization of fullerenes, as well as of carbon nanotubes, can be envisioned in the fields of solar batteries and fuel cells, gas storage materials (*e.g.* $H_2$, $CH_4$, $O_2$), additives for plastics or rubber, pharmacological treatments. As far as light-induced properties are concerned, the spin-off of current research could concern three classes of useful materials: optical limiters, photodynamic therapy agents, and plastic solar cells. The latter are probably receiving the highest attention, also in light of the mounting problem of world energy supply in the decades to come.[156] The efficiency of plastic solar cells based on fullerenes has substantially increased in the past 10 years but the current record on prototype devices (*ca.* 5%)[157] is still far from that of the best commercial silicon modules (approaching 17%). So far, only a restricted number of $C_{60}$ derivatives has been tested for this application. It is reasonable to assume that when a larger number of carbon materials (*e.g.* higher fullerenes, endohedral fullerenes, nanotubes, *etc.*) will be available on a wider scale further improvements in photovoltaic efficiency will be achieved. This could be accomplished also through implemented energy donors and a better understanding of the device nanomorphology. The unique electronic properties of fullerene and related materials are still largely unexploited and, probably, the only limit to new discoveries and applications of their light-induced properties will be only the scientists' ingenuity.

## Acknowledgements

We thank the Italian National Research Council-CNR (commessa PM-P04-ISTM-C1-ISOF-M5, Componenti Molecolari e Supramolecolari o Macromolecolari con Proprietà Fotoniche ed Optoelettroniche) and EU (RTN Contract "FAMOUS", HPRN-CT-2002-00171) for the financial support to our research work over the years.

## References

1. D. Gust, T.A. Moore and A.L. Moore, Mimicking photosynthetic solar energy transduction, *Acc. Chem. Res.*, 2001, **34**, 40–48.
2. D.M. Guldi, Fullerene–porphyrin architectures; photosynthetic antenna and reaction center models, *Chem. Soc. Rev.*, 2002, **31**, 22–36.
3. N. Armaroli, From metal complexes to fullerene arrays: exploring the exciting world of supramolecular photochemistry fifteen years after its birth, *Photochem. Photobiol. Sci.*, 2003, **2**, 73–87.
4. H. Imahori, Porphyrin–fullerene linked systems as artificial photosynthetic mimics, *Org. Biomol. Chem.*, 2004, **2**, 1425–1433.
5. M.E. El-Khouly, O. Ito, P.M. Smith and F. D'Souza, Intermolecular and supramolecular photoinduced electron transfer processes of fullerene-porphyrin/phthalocyanine systems, *J. Photochem. Photobiol. C-Photochem. Rev.*, 2004, **5**, 79–104.
6. D.M. Guldi, Fullerenes: three dimensional electron acceptor materials, *Chem. Commun.*, 2000, 321–327.

7. P.J. Bracher and D.I. Schuster, Electron transfer in fuctionalized fullerenes, in *Fullerenes: From Synthesis to Optoelectronic Properties*, D.M. Guldi and N. Martin (eds), Kluwer Academic Publishers, Dordrecht, The Netherlands, 2002, 163–212.
8. N. Armaroli, Photoinduced energy transfer processes in fuctionalized fullerenes, in *Fullerenes: From Synthesis to Optoelectronic Properties*, D.M. Guldi and N. Martin (eds), Kluwer Academic Publishers, Dordrecht, The Netherlands, 2002, 137–162.
9. D.M. Guldi, I. Zilbermann, A. Gouloumis, P. Vazquez and T. Torres, Metallophthalocyanines: versatile electron-donating building blocks for fullerene dyads, *J. Phys. Chem. B*, 2004, **108**, 18485–18494.
10. E. Peeters, P.A. van Hal, J. Knol, C.J. Brabec, N.S. Sariciftci, J.C. Hummelen and R.A.J. Janssen, Synthesis, photophysical properties, and photovoltaic devices of oligo(p-phenylene vinylene)-fullerene dyads, *J. Phys. Chem. B*, 2000, **104**, 10174–10190.
11. J.F. Eckert, J.F. Nicoud, J.F. Nierengarten, S.G. Liu, L. Echegoyen, F. Barigelletti, N. Armaroli, L. Ouali, V. Krasnikov and G. Hadziioannou, Fullerene-oligophenylenevinylene hybrids: synthesis, electronic properties, and incorporation in photovoltaic devices, *J. Am. Chem. Soc.*, 2000, **122**, 7467–7479.
12. J. Roncali, Linear π-conjugated systems derivatized with $C_{60}$-fullerene as molecular heterojunctions for organic photovoltaics, *Chem. Soc. Rev.*, 2005, **34**, 483–495.
13. J.L. Segura, N. Martin and D.M. Guldi, Materials for organic solar cells: the $C_{60}/\pi$-conjugated oligomer approach, *Chem. Soc. Rev.*, 2005, **34**, 31–47.
14. L. Sanchez, M.A. Herranz and N. Martin, $C_{60}$-based dumbbells: connecting $C_{60}$ cages through electroactive bridges, *J. Mater. Chem.*, 2005, **15**, 1409–1421.
15. A. Cravino and N.S. Sariciftci, Double-cable polymers for fullerene based organic optoelectronic applications, *J. Mater. Chem.*, 2002, **12**, 1931–1943.
16. J.N. Clifford, G. Accorsi, F. Cardinali, J.-F. Nierengarten and N. Armaroli, Photoinduced electron and energy transfer processes in fullerene $C_{60}$ – Metal complex hybrid assemblies, *C. R. Chim.*, 2006, **9**, 1005–1013.
17. A.S.D. Sandanayaka, H. Sasabe, Y. Araki, Y. Furusho, O. Ito and T. Takata, Photoinduced electron-transfer processes between $C_{60}$ fullerene and triphenylamine moieties tethered by rotaxane structures. Through-space electron transfer via excited triplet states of [60]fullerene, *J. Phys. Chem. A*, 2004, **108**, 5145–5155.
18. H. Nishikawa, S. Kojima, T. Kodama, I. Ikemoto, S. Suzuki, K. Kikuchi, M. Fujitsuka, H.X. Luo, Y. Araki and O. Ito, Photophysical study of new methanofullerene-TTF dyads: an obvious intramolecular charge transfer in the ground states, *J. Phys. Chem. A*, 2004, **108**, 1881–1890.
19. T. Makinoshima, M. Fujitsuka, M. Sasaki, Y. Araki, O. Ito, S. Ito and N. Morita, Competition between intramolecular electron-transfer and

energy-transfer processes in photoexcited azulene-$C_{60}$ dyad, *J. Phys. Chem. A*, 2004, **108**, 368–375.
20. H. Kanato, K. Takimiya, T. Otsubo, Y. Aso, T. Nakamura, Y. Araki and O. Ito, Synthesis and photophysical properties of ferrocene–oligothiophene–fullerene triads, *J. Org. Chem.*, 2004, **69**, 7183–7189.
21. K.G. Thomas, V. Biju, P.V. Kamat, M.V. George and D.M. Guldi, Dynamics of photoinduced electron-transfer processes in fullerene-based dyads: effects of varying the donor strength, *ChemPhysChem*, 2003, **4**, 1299–1307.
22. M. Fujitsuka, N. Tsuboya, R. Hamasaki, M. Ito, S. Onodera, O. Ito and Y. Yamamoto, Solvent polarity dependence of photoinduced charge separation and recombination processes of Ferrocene-$C_{60}$ dyads, *J. Phys. Chem. A*, 2003, **107**, 1452–1458.
23. F. Diederich and M. Gomez-Lopez, Supramolecular fullerene chemistry, *Chem. Soc. Rev.*, 1999, **28**, 263–277.
24. J.F. Nierengarten, N. Armaroli, G. Accorsi, Y. Rio and J.F. Eckert, [60]Fullerene: a versatile photoactive core for dendrimer chemistry, *Chem. Eur. J.*, 2003, **9**, 37–41.
25. J.F. Eckert, D. Byrne, J.F. Nicoud, L. Oswald, J.F. Nierengarten, M. Numata, A. Ikeda, S. Shinkai and N. Armaroli, Polybenzyl ether dendrimers for the complexation of [60]fullerenes, *New J. Chem.*, 2000, **24**, 749–758.
26. J.Y. Zheng, K. Tashiro, Y. Hirabayashi, K. Kinbara, K. Saigo, T. Aida, S. Sakamoto and K. Yamaguchi, Cyclic dimers of metalloporphyrins as tunable hosts for fullerenes: a remarkable effect of rhodium(III), *Angew. Chem. Int. Ed.*, 2001, **40**, 1858–1861.
27. T. Kawase, K. Tanaka, N. Fujiwara, H.R. Darabi and M. Oda, Complexation of a carbon nanoring with fullerenes, *Angew. Chem. Int. Ed.*, 2003, **42**, 1624–1628.
28. L. Sanchez, M.T. Rispens and J.C. Hummelen, A supramolecular array of fullerenes by quadruple hydrogen bonding, *Angew. Chem. Int. Ed.*, 2002, **41**, 838–840.
29. J.J. Gonzalez, S. Gonzalez, E.M. Priego, C.P. Luo, D.M. Guldi, J. de Mendoza and N. Martin, A new approach to supramolecular $C_{60}$-dimers based on quadruple hydrogen bonding, *Chem. Commun.*, 2001, 163–164.
30. M.T. Rispens, L. Sanchez, J. Knol and J.C. Hummelen, Supramolecular organization of fullerenes by quadruple hydrogen bonding, *Chem. Commun.*, 2001, 161–162.
31. F. Wudl, Fullerene materials, *J. Mater. Chem.*, 2002, **12**, 1959–1963.
32. P.K. Sudeep, B.I. Ipe, K.G. Thomas, M.V. George, S. Barazzouk, S. Hotchandani and P.V. Kamat, Fullerene-functionalized gold nanoparticles. A self-assembled photoactive antenna-metal nanocore assembly, *Nano Lett.*, 2002, **2**, 29–35.
33. T. Hasobe, H. Imahori, P.V. Kamat and S. Fukuzumi, Quaternary self-organization of porphyrin and fullerene units by clusterization with gold

nanoparticles on $SnO_2$ electrodes for organic solar cells, *J. Am. Chem. Soc.*, 2003, **125**, 14962–14963.
34. N. Papageorgiou, M. Grätzel, O. Enger, D. Bonifazi and F. Diederich, Lateral electron transport inside a monolayer of derivatized fullerenes anchored on nanocrystalline metal oxide films, *J. Phys. Chem. B*, 2002, **106**, 3813–3822.
35. H. Yamada, H. Imahori, Y. Nishimura, I. Yamazaki, T.K. Ahn, S.K. Kim, D. Kim and S. Fukuzumi, Photovoltaic properties of self-assembled monolayers of porphyrins and porphyrin–fullerene dyads on ITO and gold surfaces, *J. Am. Chem. Soc.*, 2003, **125**, 9129–9139.
36. J.F. Nierengarten, Chemical modification of $C_{60}$ for materials science applications, *New J. Chem.*, 2004, **28**, 1177–1191.
37. H. Imahori, M. Kimura, K. Hosomizu, T. Sato, T.K. Ahn, S.K. Kim, D. Kim, Y. Nishimura, I. Yamazaki, Y. Araki, O. Ito and S. Fukuzumi, Vectorial electron relay at ITO electrodes modified with self-assembled monolayers of ferrocene-porphyrin–fullerene triads and porphyrin–fullerene dyads for molecular photovoltaic devices, *Chem. Eur. J.*, 2004, **10**, 5111–5122.
38. H. Imahori, M. Kimura, K. Hosomizu and S. Fukuzumi, Porphyrin and fullerene-based photovoltaic devices, *J. Photochem. Photobiol. A-Chem.*, 2004, **166**, 57–62.
39. D.M. Guldi, I. Zilbermann, G.A. Anderson, K. Kordatos, M. Prato, R. Tafuro and L. Valli, Langmuir-Blodgett and layer-by-layer films of photoactive fullerene–porphyrin dyads, *J. Mater. Chem.*, 2004, **14**, 303–309.
40. M. Even, B. Heinrich, D. Guillon, D.M. Guldi, M. Prato and R. Deschenaux, A mixed fullerene–ferrocene thermotropic liquid crystal: synthesis, liquid-crystalline properties, supramolecular organization and photoinduced electron transfer, *Chem. Eur. J.*, 2001, **7**, 2595–2604.
41. T. Chuard and R. Deschenaux, Design, mesomorphic properties, and supramolecular organization of [60]fullerene-containing thermotropic liquid crystals, *J. Mater. Chem.*, 2002, **12**, 1944–1951.
42. J.K.J. van Duren, X.N. Yang, J. Loos, C.W.T. Bulle-Lieuwma, A.B. Sieval, J.C. Hummelen and R.A.J. Janssen, Relating the morphology of poly(p-phenylenevinylene)/methanofullerene blends to solar-cell performance, *Adv. Funct. Mater.*, 2004, **14**, 425–434.
43. G. Brusatin and R. Signorini, Linear and nonlinear optical properties of fullerenes in solid state materials, *J. Mater. Chem.*, 2002, **12**, 1964–1977.
44. Y. Rio, G. Accorsi, H. Nierengarten, C. Bourgogne, J.-M. Strub, A. Van Dorsselaer, N. Armaroli and J.-F. Nierengarten, A fullerene core to probe dendritic shielding effects, *Tetrahedron*, 2003, **59**, 3833–3844.
45. T. Da Ros and M. Prato, Medicinal chemistry with fullerenes and fullerene derivatives, *Chem. Commun.*, 1999, 663–669.
46. J.W. Arbogast and C.S. Foote, Photophysical properties of $C_{70}$, *J. Am. Chem. Soc.*, 1991, **113**, 8886–8889.
47. R.R. Hung and J.J. Grabowski, $C_{70}$ Intersystem crossing and singlet oxygen production, *Chem. Phys. Lett.*, 1992, **192**, 249–253.

48. D.H. Kim, M.Y. Lee, Y.D. Suh and S.K. Kim, Observation of fluorescence emission from solutions of $C_{60}$ and $C_{70}$ and measurement of their excited-state lifetimes, *J. Am. Chem. Soc.*, 1992, **114**, 4429–4430.
49. R.V. Bensasson, M. Schwell, M. Fanti, N.K. Wachter, J.O. Lopez, J.-M. Janot, P.R. Birkett, E.J. Land, S. Leach, P. Seta, R. Taylor and F. Zerbetto, Photophysical properties of the ground and triplet state of four multiphenylated [70]fullerene compounds, *ChemPhysChem*, 2001, **2**, 109–114.
50. K. Komatsu, K. Fujiwara and Y. Murata, The fullerene cross-dimer $C_{130}$: synthesis and properties, *Chem. Commun.*, 2000, **1**, 1583–1584.
51. M. Fujitsuka, H. Takahashi, T. Kudo, K. Tohji, A. Kasuya and O. Ito, Photophysical and photochemical properties of $C_{120}O$, a $C_{60}$ dimer linked by a saturated furan ring, *J. Phys. Chem. A*, 2001, **105**, 675–680.
52. S.M. Bachilo, A.F. Benedetto and R.B. Weisman, Triplet state dissociation of $C_{120}$, the dimer of $C_{60}$, *J. Phys. Chem. A*, 2001, **105**, 9845–9850.
53. M. Fujitsuka, O. Ito, N. Dragoe, S. Ito, H. Shimotani, H. Takagi and K. Kitazawa, Photophysical and photochemical processes of an unsymmetrical fullerene dimer, $C_{121}$, *J. Phys. Chem. B*, 2002, **106**, 8562–8568.
54. R. Stackow, G. Schick, T. Jarrosson, Y. Rubin and C.S. Foote, Photophysics of open $C_{60}$ derivatives, *J. Phys. Chem. B*, 2000, **104**, 7914–7918.
55. G. Orlandi and F. Negri, Electronic states and transitions in $C_{60}$ and $C_{70}$ fullerenes, *Photochem. Photobiol. Sci.*, 2002, **1**, 289–308.
56. S.P. Sibley, S.M. Argentine and A.H. Francis, A photoluminescence study of $C_{60}$ and $C_{70}$, *Chem. Phys. Lett.*, 1992, **188**, 187–193.
57. R.R. Hung and J.J. Grabowski, A Precise determination of the triplet energy of $C_{60}$ by photoacoustic calorimetry, *J. Phys. Chem.*, 1991, **95**, 6073–6075.
58. Y.P. Sun, P. Wang and N.B. Hamilton, Fluorescence spectra and quantum yields of buckminsterfullerene ($C_{60}$) in room-temperature solutions. No excitation wavelength dependence, *J. Am. Chem. Soc.*, 1993, **115**, 6378–6381.
59. M.R. Wasielewski, M.P. O'Neil, K.R. Lykke, M.J. Pellin and D.M. Gruen, Triplet states of fullerenes $C_{60}$ and $C_{70}$ – Electron paramagnetic resonance spectra, photophysics, and electronic structures, *J. Am. Chem. Soc.*, 1991, **113**, 2774–2776.
60. R.J. Sension, C.M. Phillips, A.Z. Szarka, W.J. Romanow, A.R. McGhie, J.P. McCauley, A.B. Smith and R.M. Hochstrasser, Transient absorption studies of $C_{60}$ in solution, *J. Phys. Chem.*, 1991, **95**, 6075–6078.
61. F. Prat, C. Marti, S. Nonell, X.J. Zhang, C.S. Foote, R.G. Moreno, J.L. Bourdelande and J. Font, $C_{60}$ fullerene-based materials as singlet oxygen $O_2(^1\Delta_g)$ photosensitizers: a time-resolved near-IR luminescence and optoacoustic study, *Phys. Chem. Chem. Phys.*, 2001, **3**, 1638–1643.
62. F. Prat, R. Stackow, R. Bernstein, W.Y. Qian, Y. Rubin and C.S. Foote, Triplet-state properties and singlet oxygen generation in a homologous series of functionalized fullerene derivatives, *J. Phys. Chem. A*, 1999, **103**, 7230–7235.

63. D.M. Guldi and K.D. Asmus, Photophysical properties of mono- and multiply-functionalized fullerene derivatives, *J. Phys. Chem. A*, 1997, **101**, 1472–1481.
64. J.W. Arbogast, A.P. Darmanyan, C.S. Foote, Y. Rubin, F.N. Diederich, M.M. Alvarez, S.J. Anz and R.L. Whetten, Photophysical properties of $C_{60}$, *J. Phys. Chem.*, 1991, **95**, 11–12.
65. A. Gilbert and G. Baggott, *Essentials of Molecular Photochemistry*, Blackwell Scientific Publications, Oxford, UK, 1991.
66. F. Wilkinson, W.P. Helman and A.B. Ross, Rate constants for the decay and reactions of the lowest electronically excited singlet-state of molecular-oxygen in solution – An expanded and revised compilation, *J. Phys. Chem. Ref. Data*, 1995, **24**, 663–1021.
67. R.W. Redmond and J.N. Gamlin, A compilation of singlet oxygen yields from biologically relevant molecules, *Photochem. Photobiol.*, 1999, **70**, 391–475.
68. N. Armaroli, G. Accorsi, D. Felder and J.F. Nierengarten, Photophysical properties of the Re(I) and Ru(II) complexes of a new $C_{60}$-substituted bipyridine ligand, *Chem. Eur. J.*, 2002, **8**, 2314–2323.
69. N. Armaroli, G. Accorsi, J.P. Gisselbrecht, M. Gross, V. Krasnikov, D. Tsamouras, G. Hadziioannou, M.J. Gomez-Escalonilla, F. Langa, J.F. Eckert and J.F. Nierengarten, Photoinduced processes in fullerenopyrrolidine and fullerenopyrazoline derivatives substituted with an oligophenylenevinylene moiety, *J. Mater. Chem.*, 2002, **12**, 2077–2087.
70. J.F. Nierengarten, J.F. Eckert, J.F. Nicoud, L. Ouali, V. Krasnikov and G. Hadziioannou, Synthesis of a $C_{60}$-oligophenylenevinylene hybrid and its incorporation in a photovoltaic device, *Chem. Commun.*, 1999, 617–618.
71. J.F. Nierengarten, J.F. Eckert, D. Felder, J.F. Nicoud, N. Armaroli, G. Marconi, V. Vicinelli, C. Boudon, J.P. Gisselbrecht, M. Gross, G. Hadziioannou, V. Krasnikov, L. Ouali, L. Echegoyen and S.G. Liu, Synthesis and electronic properties of donor-linked fullerenes towards photochemical molecular devices, *Carbon*, 2000, **38**, 1587–1598.
72. N.S. Sariciftci, L. Smilowitz, A.J. Heeger and F. Wudl, Photoinduced electron-transfer from a conducting polymer to buckminsterfullerene, *Science*, 1992, **258**, 1474–1476.
73. C.J. Brabec, N.S. Sariciftci and J.C. Hummelen, Plastic solar cells, *Adv. Funct. Mater.*, 2001, **1**, 15–26.
74. H. Hoppe, M. Niggemann, C. Winder, J. Kraut, R. Hiesgen, A. Hinsch, D. Meissner and N.S. Sariciftci, Nanoscale morphology of conjugated polymer/fullerene-based bulk-heterojunction solar cells, *Adv. Funct. Mater.*, 2004, **14**, 1005–1011.
75. N. Armaroli, F. Barigelletti, P. Ceroni, J.F. Eckert, J.F. Nicoud and J.F. Nierengarten, Photoinduced energy transfer in a fullerene-oligophenylenevinylene conjugate, *Chem. Commun.*, 2000, 599–600.
76. N. Armaroli, F. Barigelletti, P. Ceroni, J.F. Eckert and J.F. Nierengarten, A fulleropyrrolidine with two oligophenylenevinylene substituents:

77. P.A. van Hal, R.A.J. Janssen, G. Lanzani, G. Cerullo, M. Zavelani-Rossi and S. De Silvestri, Two-step mechanism for the photoinduced intramolecular electron transfer in oligo(p-phenylenevinylene)-fullerene dyads, *Phys. Rev. B*, 2001, **64**, 0752061–0752067.
78. P.A. van Hal, J. Knol, B.M.W. Langeveld-Voss, S.C.J. Meskers, J.C. Hummelen and R.A.J. Janssen, Photoinduced energy and electron transfer in fullerene–oligothiophene–fullerene triads, *J. Phys. Chem. A*, 2000, **104**, 5974–5988.
79. M. Elhabiri, A. Trabolsi, F. Cardinali, U. Hahn, A.M. Albrecht-Gary and J.F. Nierengarten, Cooperative recognition of $C_{60}$-ammonium substrates by a ditopic oligophenylenevinylene/crown ether host, *Chem. Eur. J.*, 2005, **11**, 4793–4798.
80. F. Langa, M.J. Gomez-Escalonilla, J.M. Rueff, T.M.F. Duarte, J.F. Nierengarten, V. Palermo, P. Samori, Y. Rio, G. Accorsi and N. Armaroli, Pyrazolino[60]fullerene-oligophenylenevinylene dumbbell-shaped arrays: synthesis, electrochemistry, photophysics, and self-assembly on surfaces, *Chem. Eur. J.*, 2005, **11**, 4405–4415.
81. E.E. Neuteboom, P.A. van Hal and R.A.J. Janssen, Donor–acceptor polymers: a conjugated oligo(p-phenylenevinylene) main chain with dangling perylene bisimides, *Chem. Eur. J.*, 2004, **10**, 3907–3918.
82. A.M. Ramos, S.C.J. Meskers, P.A. van Hal, J. Knol, J.C. Hummelen and R.A.J. Janssen, Photoinduced multistep energy and electron transfer in an oligoaniline-oligo(p-phenylenevinylene)-fullerene triad, *J. Phys. Chem. A*, 2003, **107**, 9269–9283.
83. M. Gutierrez-Nava, H. Nierengarten, P. Masson, A. Van Dorsselaer and J.F. Nierengarten, A supramolecular oligophenylenevinylene-$C_{60}$ conjugate, *Tetrahedron Lett.*, 2003, **44**, 3043–3046.
84. G. Accorsi, N. Armaroli, J.F. Eckert and J.F. Nierengarten, Functionalization of [60]fullerene with new light-collecting oligophenylenevinylene-terminated dendritic wedges, *Tetrahedron Lett.*, 2002, **43**, 65–68.
85. N. Armaroli, G. Accorsi, J.N. Clifford, J.F. Eckert and J.-F. Nierengarten, *Chem. Asian J.*, 2006, **1**, in press.
86. L. Schmidt-Mende, A. Fechtenkotter, K. Mullen, E. Moons, R.H. Friend and J.D. MacKenzie, Self-organized discotic liquid crystals for high-efficiency organic photovoltaics, *Science*, 2001, **293**, 1119–1122.
87. E.H.A. Beckers, P.A. van Hal, A. Schenning, A. El-ghayoury, E. Peeters, M.T. Rispens, J.C. Hummelen, E.W. Meijer and R.A.J. Janssen, Singlet-energy transfer in quadruple hydrogen-bonded oligo(p-phenylenevinylene)-fullerene dyads, *J. Mater. Chem.*, 2002, **12**, 2054–2060.
88. T. Gu, D. Tsamouras, C. Melzer, V. Krasnikov, J.P. Gisselbrecht, M. Gross, G. Hadziioannou and J.F. Nierengarten, Photovoltaic devices from fullerene oligophenyleneethynylene conjugates, *ChemPhysChem*, 2002, **3**, 124–127.

89. N. Negishi, K. Takimiya, T. Otsubo, Y. Harima and Y. Aso, Synthesis and photovoltaic effects of oligothiophenes incorporated with two [60]fullerenes, *Chem. Lett.*, 2004, **33**, 654–655.
90. M. Maggini, G. Possamai, E. Menna, G. Scorrano, N. Camaioni, G. Ridolfi, G. Casalbore-Miceli, L. Franco, M. Ruzzi and C. Corvaja, Solar cells based on a fullerene-azothiophene dyad, *Chem. Commun.*, 2002, 2028–2029.
91. C. Martineau, P. Blanchard, D. Rondeau, J. Delaunay and J. Roncali, Synthesis and electronic properties of adducts of oligothienylenevinylenes and fullerene $C_{60}$, *Adv. Mater.*, 2002, **14**, 283–287.
92. E.H.A. Beckers, P.A. van Hal, A. Dhanabalan, S.C.J. Meskers, J. Knol, J.C. Hummelen and R.A.J. Janssen, Charge transfer kinetics in fullerene–oligomer–fullerene triads containing alkylpyrrole units, *J. Phys. Chem. A*, 2003, **107**, 6218–6224.
93. F. Padinger, R.S. Rittberger and N.S. Sariciftci, Effects of postproduction treatment on plastic solar cells, *Adv. Funct. Mater.*, 2003, **13**, 85–88.
94. D. Hirayama, K. Takimiya, Y. Aso, T. Otsubo, T. Hasobe, H. Yamada, H. Imahori, S. Fukuzumi and Y. Sakata, Large photocurrent generation of gold electrodes modified with [60]fullerene-linked oligothiophenes bearing a tripodal rigid anchor, *J. Am. Chem. Soc.*, 2002, **124**, 532–533.
95. U. Boas and P.M.H. Heegaard, Dendrimers in drug research, *Chem. Soc. Rev.*, 2004, **33**, 43–63.
96. C.Z.S. Chen and S.L. Cooper, Recent advances in antimicrobial dendrimers, *Adv. Mater.*, 2000, **12**, 843–846.
97. S.E. Stiriba, H. Frey and R. Haag, Dendritic polymers in biomedical applications: from potential to clinical use in diagnostics and therapy, *Angew. Chem. Int. Ed.*, 2002, **41**, 1329–1334.
98. D. Astruc, M.C. Daniel and J. Ruiz, Dendrimers and gold nanoparticles as exo-receptors sensing biologically important anions, *Chem. Commun.*, 2004, 2637–2649.
99. M.W. Grinstaff, Biodendrimers: new polymeric biomaterials for tissues engineering, *Chem. Eur. J.*, 2002, **8**, 2838–2846.
100. A. Adronov and J.M.J. Frechet, Light-harvesting dendrimers, *Chem. Commun.*, 2000, 1701–1710.
101. S. Campagna, C. Di Pietro, F. Loiseau, B. Maubert, N. McClenaghan, R. Passalacqua, F. Puntoriero, V. Ricevuto and S. Serroni, Recent advances in luminescent polymetallic dendrimers containing the 2,3-bis(2′-pyridyl)pyrazine bridging ligand, *Coord. Chem. Rev.*, 2002, **229**, 67–74.
102. V. Balzani, P. Ceroni, M. Maestri, C. Saudan and V. Vicinelli, Luminescent dendrimers. Recent advances, *Top. Curr. Chem.*, 2003, **228**, 159–191.
103. L.J. Twyman, A.S.H. King and I.K. Martin, Catalysis inside dendrimers, *Chem. Soc. Rev.*, 2002, **31**, 69–82.
104. S. Förster and M. Konrad, From self-organizing polymers to nano- and biomaterials, *J. Mater. Chem.*, 2003, **13**, 2671–2688.

105. J.F. Nierengarten, Fullerodendrimers: a new class of compounds for supramolecular chemistry and materials science applications, *Chem. Eur. J.*, 2000, **6**, 3667–3670.
106. J.F. Nierengarten, Fullerodendrimers: fullerene-containing macromolecules with intriguing properties, *Top. Curr. Chem.*, 2003, **228**, 87–110.
107. K. Kordatos, T. Da Ros, M. Prato, S. Leach, E.J. Land and R.V. Bensasson, Triplet state properties of N-mTEG[60]fulleropyrrolidine mono and bisadduct derivatives, *Chem. Phys. Lett.*, 2001, **334**, 221–228.
108. N. Armaroli, C. Boudon, D. Felder, J.P. Gisselbrecht, M. Gross, G. Marconi, J.F. Nicoud, J.F. Nierengarten and V. Vicinelli, A copper(I) bisphenanthroline complex buried in fullerene-functionalized dendritic black boxes, *Angew. Chem. Int. Ed.*, 1999, **38**, 3730–3733.
109. Y. Murata, M. Ito and K. Komatsu, Synthesis and properties of novel fullerene derivatives having dendrimer units and the fullerenyl anions generated therefrom, *J. Mater. Chem.*, 2002, **12**, 2009–2020.
110. J. Issberner, F. Vögtle, L. De Cola and V. Balzani, Dendritic bipyridine ligands and their tris(bipyridine)ruthenium(II) chelates – Syntheses, absorption spectra, and photophysical properties, *Chem. Eur. J.*, 1997, **3**, 706–712.
111. F. Vögtle, M. Plevoets, M. Nieger, G.C. Azzellini, A. Credi, L. De Cola, V. De Marchis, M. Venturi and V. Balzani, Dendrimers with a photoactive and redox-active [Ru(bpy)$_3$]$^{2+}$-type core: photophysical properties, electrochemical behaviour, and excited-state electron-transfer reactions, *J. Am. Chem. Soc.*, 1999, **121**, 6290–6298.
112. X.L. Zhou, D.S. Tyson and F.N. Castellano, First generation light-harvesting dendrimers with a [Ru(bpy)$_3$]$^{2+}$ core and aryl ether ligands functionalized with coumarin 450, *Angew. Chem. Int. Ed.*, 2000, **39**, 4301–4305.
113. M. Schwell, N.K. Wachter, J.H. Rice, J.P. Galaup, S. Leach, R. Taylor and R.V. Bensasson, Coupling a dendrimer and a fullerene chromophore: a study of excited state properties of C$_{61}$(poly(aryl)acetylene)$_2$, *Chem. Phys. Lett.*, 2001, **339**, 29–35.
114. R. Kunieda, M. Fujitsuka, O. Ito, M. Ito, Y. Murata and K. Komatsu, Photochemical and photophysical properties of C$_{60}$ dendrimers studied by laser flash photolysis, *J. Phys. Chem. B*, 2002, **106**, 7193–7199.
115. A.A. Gorman, A.A. Krasnovsky and M.A.J. Rodgers, Singlet oxygen infrared luminescence – Unambiguous confirmation of a solvent-dependent radiative rate-constant, *J. Phys. Chem.*, 1991, **95**, 598–601.
116. R.D. Scurlock, S. Nonell, S.E. Braslavsky and P.R. Ogilby, Effect of solvent on the radiative decay of singlet molecular-oxygen (a$^1\Delta_g$), *J. Phys. Chem.*, 1995, **99**, 3521–3526.
117. R. Schmidt and H.-D. Brauer, Radiationless deactivation of singlet oxygen ($^1\Delta_g$) by solvent molecules, *J. Am. Chem. Soc.*, 1987, **109**, 6976–6981.
118. F. Wilkinson, W.P. Helman and A.B. Ross, Quantum yields for the photosensitized formation of the lowest electronically excited singlet-state

of molecular-oxygen in solution, *J. Phys. Chem. Ref. Data*, 1993, **22**, 113–262.
119. M. Gutierrez-Nava, G. Accorsi, P. Masson, N. Armaroli and J.F. Nierengarten, Polarity effects on the photophysics of dendrimers with an oligophenylenevinylene core and peripheral fullerene units, *Chem. Eur. J.*, 2004, **10**, 5076–5086.
120. A. Juris, V. Balzani, F. Barigelletti, S. Campagna, P. Belser and A. von Zelewsky, Ru(II)-polypyridine complexes: photophysics, photochemistry, electrochemistry, and chemiluminescence, *Coord. Chem. Rev.*, 1988, **84**, 85–277.
121. V.W.W. Yam, Luminescent carbon-rich rhenium(I) complexes, *Chem. Commun.*, 2001, 789–796.
122. N. Armaroli, Photoactive mono- and polynuclear Cu(I)-phenanthrolines. A viable alternative to Ru(II)-polypyridines?, *Chem. Soc. Rev.*, 2001, **30**, 113–124.
123. K.S. Schanze and K.A. Walters, Photoinduced electron transfer in metal-organic dyads, , in *Organic and Inorganic Photochemistry*, V. Ramamurthy and K.S. Schanze (eds), Marcel Dekker, New York, 1998, 75–127.
124. F. Scandola, C. Chiorboli, M.T. Indelli and M.A. Rampi, Covalently linked systems containing metal complexes, in *Electron Transfer in Chemistry*, Vol. 3, V. Balzani (ed), Wiley-VCH, Weinheim, Germany, 2001, 337–408.
125. F. Barigelletti and L. Flamigni, Photoactive molecular wires based on metal complexes, *Chem. Soc. Rev.*, 2000, **29**, 1–12.
126. M.D. Meijer, G.P.M. van Klink and G. van Koten, Metal-chelating capacities attached to fullerenes, *Coord. Chem. Rev.*, 2002, **230**, 141–163.
127. F. Cardinali, H. Mamlouk, Y. Rio, N. Armaroli and J.F. Nierengarten, Fullerohelicates: a new class of fullerene-containing supermolecules, *Chem. Commun.*, 2004, 1582–1583.
128. Y. Rio, G. Enderlin, C. Bourgogne, J.F. Nierengarten, J.P. Gisselbrecht, M. Gross, G. Accorsi and N. Armaroli, Ground and excited state electronic interactions in a bis(phenanthroline) copper(I) complex sandwiched between two fullerene subunits, *Inorg. Chem.*, 2003, **42**, 8783–8793.
129. N. Armaroli, F. Diederich, C.O. Dietrich-Buchecker, L. Flamigni, G. Marconi, J.F. Nierengarten and J.P. Sauvage, A copper(I)-complexed rotaxane with two fullerene stoppers: synthesis, electrochemistry, and photoinduced processes, *Chem. Eur. J.*, 1998, **4**, 406–416.
130. D.M. Guldi, T. Da Ros, P. Braiuca and M. Prato, A topologically new ruthenium porphyrin–fullerene donor–acceptor ensemble, *Photochem. Photobiol. Sci.*, 2003, **2**, 1067–1073.
131. D.M. Guldi, M. Maggini, E. Menna, G. Scorrano, P. Ceroni, M. Marcaccio, F. Paolucci and S. Roffia, A photosensitizer dinuclear ruthenium complex: intramolecular energy transfer to a covalently linked fullerene acceptor, *Chem. Eur. J.*, 2001, **7**, 1597–1605.

132. T. Da Ros, M. Prato, D.M. Guldi, M. Ruzzi and L. Pasimeni, Efficient charge separation in porphyrin–fullerene-ligand complexes, *Chem. Eur. J.*, 2001, **7**, 816–827.
133. M. Holler, F. Cardinali, H. Mamlouk, J.-F. Nierengarten, J.-P. Gisselbrecht, M. Gross, Y. Rio, F. Barigelletti and N. Armaroli, Synthesis of fullerohelicates and fine tuning of the photoinduced processes by changing the number of addends on the fullerene subunits, *Tetrahedron*, 2006, **62**, 2060–2073.
134. P.A. Liddell, J.P. Sumida, A.N. Macpherson, L. Noss, G.R. Seely, K.N. Clark, A.L. Moore, T.A. Moore and D. Gust, Preparation and photophysical studies of porphyrin-$C_{60}$ dyads, *Photochem. Photobiol.*, 1994, **60**, 537–541.
135. D.I. Schuster, Synthesis and photophysics of new types of fullerene–porphyrin dyads, *Carbon*, 2000, **38**, 1607–1614.
136. N. Armaroli, F. Diederich, L. Echegoyen, T. Habicher, L. Flamigni, G. Marconi and J.F. Nierengarten, A new pyridyl-substituted methanofullerene derivative. Photophysics, electrochemistry and self-assembly with zinc(II) meso-tetraphenylporphyrin (ZnTPP), *New J. Chem.*, 1999, **23**, 77–83.
137. F. D'Souza and O. Ito, Photoinduced electron transfer in supramolecular systems of fullerenes functionalized with ligands capable of binding to zinc porphyrins and zinc phthalocyanines, *Coord. Chem. Rev.*, 2005, **249**, 1410–1422.
138. P.D.W. Boyd and C.A. Reed, Fullerene–porphyrin constructs, *Acc. Chem. Res.*, 2005, **38**, 235–242.
139. D. Sun, F.S. Tham, C.A. Reed and P.D.W. Boyd, Extending supramolecular fullerene–porphyrin chemistry to pillared metal-organic frameworks, *Proc. Natl. Acad. Sci. USA*, 2002, **99**, 5088–5092.
140. Y. Shoji, K. Tashiro and T. Aida, Selective extraction of higher fullerenes using cyclic dimers of zinc porphyrins, *J. Am. Chem. Soc.*, 2004, **126**, 6570–6571.
141. N. Armaroli, G. Marconi, L. Echegoyen, J.P. Bourgeois and F. Diederich, Charge-transfer interactions in face-to-face porphyrin–fullerene systems: solvent-dependent luminescence in the infrared spectral region, *Chem. Eur. J.*, 2000, **6**, 1629–1645.
142. V. Chukharev, N.V. Tkachenko, A. Efimov, D.M. Guldi, A. Hirsch, M. Scheloske and H. Lemmetyinen, Tuning the ground-state and excited-state interchromophore interactions in porphyrin–fullerene-stacks, *J. Phys. Chem. B*, 2004, **108**, 16377–16385.
143. N.V. Tkachenko, H. Lemmetyinen, J. Sonoda, K. Ohkubo, T. Sato, H. Imahori and S. Fukuzumi, Ultrafast photodynamics of exciplex formation and photoinduced electron transfer in porphyrin–fullerene dyads linked at close proximity, *J. Phys. Chem. A*, 2003, **107**, 8834–8844.
144. V. Vehmanen, N.V. Tkachenko, H. Imahori, S. Fukuzumi and H. Lemmetyinen, Charge-transfer emission of compact porphyrin–fullerene

dyads analyzed by Marcus theory of electron-transfer, *Spectrochim. Acta A*, 2001, **57**, 2229–2244.
145. N. Armaroli, G. Accorsi, F.Y. Song, A. Palkar, L. Echegoyen, D. Bonifazi and F. Diederich, Photophysical and electrochemical properties of meso, meso-linked oligoporphyrin rods with appended fullerene terminals, *ChemPhysChem*, 2005, **6**, 732–743.
146. D. Bonifazi, G. Accorsi, N. Armaroli, F.Y. Song, A. Palkar, L. Echegoyen, M. Scholl, P. Seiler, B. Jaun and F. Diederich, Oligoporphyrin arrays conjugated to [60]fullerene: preparation, NMR analysis, and photophysical and electrochemical properties, *Helv. Chim. Acta*, 2005, **88**, 1839–1884.
147. D. Bonifazi, M. Scholl, F.Y. Song, L. Echegoyen, G. Accorsi, N. Armaroli and F. Diederich, Exceptional redox and photophysical properties of a triply fused diporphyrin-$C_{60}$ conjugate: novel scaffolds for multicharge storage in molecular scale electronics, *Angew. Chem. Int. Ed.*, 2003, **42**, 4966–4970.
148. N. Aratani, A. Osuka, H.S. Cho and D. Kim, Photochemistry of covalently-linked multi-porphyrinic systems, *J. Photochem. Photobiol. C: Photochem. Rev.*, 2002, **3**, 25–52.
149. D. Kim and A. Osuka, Photophysical properties of directly linked linear porphyrin arrays, *J. Phys. Chem. A*, 2003, **107**, 8791–8816.
150. A. Tsuda and A. Osuka, Fully conjugated porphyrin tapes with electronic absorption bands that reach into infrared, *Science*, 2001, **293**, 79–82.
151. R. Englman and J. Jortner, The energy gap law for radiationless transitions in large molecules, *Mol. Phys.*, 1970, **18**, 145–164.
152. M. Koeberg, M. de Groot, J.W. Verhoeven, N.R. Lokan, M.J. Shephard and M.N. Paddon-Row, U-shaped donor [bridge] acceptor systems with remarkable charge transfer fluorescent properties: an experimental and computational investigation, *J. Phys. Chem. A*, 2001, **105**, 3417–3424.
153. V. Balzani, F. Bolletta and F. Scandola, Vertical and "non-vertical" energy transfer processes. A general classical treatment, *J. Am. Chem. Soc.*, 1980, **102**, 2152–2163.
154. R.A. Marcus, Electron-transfer reactions in chemistry – theory and experiment (Nobel Lecture), *Angew. Chem. Int. Ed.*, 1993, **32**, 1111–1121.
155. J.F. Tremblay, Fullerenes by the ton, *Chemical & Engineering News*, 2003, **81**, 13–14.
156. V. Smil, *Energy at the Crossroads*, MIT Press, Cambridge, MA, 2003.
157. W.L. Ma, C.Y. Yang, X. Gong, K. Lee and A.J. Heeger, Thermally stable, efficient polymer solar cells with nanoscale control of the interpenetrating network morphology, *Adv. Funct. Mater.*, 2005, **15**, 1617–1622.

CHAPTER 5

# Encapsulation of [60]Fullerene into Dendritic Materials to Facilitate their Nanoscopic Organization

JEAN-FRANÇOIS NIERENGARTEN,[1] NATHALIE SOLLADIE[2] AND ROBERT DESCHENAUX[3]

[1] Groupe de Chimie des Fullerènes et des Systèmes Conjugués, Laboratoire de Chimie de Coordination du CNRS, 205 route de Narbonne, 31077 Toulouse Cedex 4, France
[2] Groupe de Synthèse de Systèmes Porphyriniques, Laboratoire de Chimie de Coordination du CNRS, 205 route de Narbonne, 31077 Toulouse Cedex 4, France
[3] Institut de Chimie, Université de Neuchâtel, Avenue de Bellevaux 51, Case Postale 158, Neuchâtel 2009, Switzerland

## 5.1 Introduction

Since [60]fullerene ($C_{60}$) became available in macroscopic quantities in 1990,[1] the properties of this fascinating carbon cage have been intensively investigated. Among the most spectacular findings, $C_{60}$ was found to behave like an electronegative molecule able to reversibly accept up to six electrons,[2] to become a supraconductor in $M_3C_{60}$ species (M = alkali metals)[3] or to be an interesting material for photovoltaic applications.[4] However, this new molecular material aggregates very easily and is insoluble or only sparingly soluble in most solvents.[5] Therefore, pristine $C_{60}$ is difficult to handle. This serious obstacle for practical applications can be, at least in part, overcome with the help of organic modifications of $C_{60}$. Indeed, recent developments in the functionalization of fullerenes allow the preparation of soluble $C_{60}$ derivatives which are easier to handle.[6] In this respect, the functionalization of $C_{60}$ with a controlled number of dendritic branches is of particular interest. The solubility of the fullerenes is thus dramatically improved. In addition, the surrounding dendritic shell provides a compact insulating layer around the carbon sphere. The latter effect can be particularly useful for the design of fullerene derivatives with mesomorphic properties and for the preparation of amphiphilic $C_{60}$ compounds capable of forming stable Langmuir films at the air–water interface. The present chapter illustrates the current state of the art of the self-organization of fullerene-containing molecular materials. The aim of this chapter is not to present an exhaustive review but to describe some of the most significant examples.

## 5.2 Langmuir-Blodgett Films with Fullerene-Containing Dendrimers

Since the incorporation of fullerenes into thin films is required for the preparation of many optoelectronic devices, the past several years have seen a considerable growth in the use of fullerene-based derivatives at surfaces and interfaces.[7] A possible approach toward structurally ordered $C_{60}$ assemblies is the preparation of Langmuir films at the air–water interface and their subsequent transfer onto solid substrates.[7] However, all the studies on the spreading behavior of $C_{60}$ at the air–water interface revealed the formation of collapsed films due to the non-amphiphilic nature of these compounds and to aggregation phenomena resulting from strong $C_{60}$–$C_{60}$ interactions.[8] Furthermore, all attempts to obtain well-defined Langmuir-Blodgett (LB) films have failed.

Two approaches have been used to overcome these problems. The first one consists in preventing $C_{60}$–$C_{60}$ interactions by incorporating fullerene into a matrix of an amphiphilic compound to produce mixed Langmuir films. Fatty acids or long-chain alcohols have been used;[9] however, the expected protection is not always very effective and fullerene aggregation remains a problem. Amphiphilic molecules containing a cavity, which is able to incorporate fullerene, such as azacrowns[10] or calixarenes[11] have been found to be the most suitable matrices for the preparation of fullerene-containing composite Langmuir films of good quality. The second approach is achieved by chemical modification of $C_{60}$, in general by covalent attachment of a hydrophilic head-group onto the fullerene core to obtain an adduct with an amphiphilic character.[12–13] By attaching a hydrophilic head-group to the fullerene core, significant improvement of the spreading behavior have been reported. The polar head-group is responsible for attractive interactions with the aqueous subphase, thus preventing three-dimensional aggregation and allowing the preparation of monolayers at the air–water interface. However, in most of the cases, once the fullerene cores are in contact with each other in the compressed Langmuir films, they irreversibly aggregate, and the monolayer does not expand back to the initial state. The resulting Langmuir films are also usually rigid and, as a result, their transfer onto solid substrates is difficult. As part of this research, the best results have been obtained with compounds combining the advantages of these two approaches, meaning a $C_{60}$ derivative functionalized with a polar head-group to obtain an adduct with an amphiphilic character and bulky substituents to prevent $C_{60}$–$C_{60}$

interactions.[14,15] For example, substitution of an amphiphilic cyclic fullerene *bis*-adduct substructure with cholesterol groups is a convenient way for the preparation of suitable derivatives for efficient incorporation in Langmuir films.[15] The amphiphilic fullerene *bis*-adducts 1 and 2 substituted with two and four cholesterol residues, respectively, have been studied (Figure 1). The molecular area, extrapolated to zero surface pressure, $170 \pm 5$ Å$^2$ for **1**, is in agreement with the value which can be estimated by molecular modeling. This observation suggests that a monomolecular film has been obtained. Unfortunately, the isotherm still shows poor reversibility, even when the highest pressure is kept below the collapse value.

Brewster Angle Microscope (BAM) images show that the Langmuir film obtained with compound **1** is homogeneous at the end of the compression (Figure 2). In addition to the observation of the expected molecular area for **1**, the high-quality film observed by BAM clearly indicates the formation of an homogeneous monomolecular layer. As shown in Figure 2, during the decompression, the film broke, hence the lack of reversibility of the isotherms. The pressure–area isotherm for compound **2** is shown in Figure 1. Good-quality films are obtained and the BAM picture of the film at the end of the compression (not shown) looks like the picture shown in Figure 2 for **1**. When compared to **1**, one immediately sees (Figure 1) that the additional two cholesterol groups increase the molecular area which becomes $270 \pm 10$ Å$^2$. It can also be noticed that the shape of the isotherm is different, the surface pressure starts to increase slowly at a molecular area of 400 Å$^2$, before rising in a steeper way around 250 Å$^2$. This liquid-expanded phase between 400 and 250 Å$^2$, not seen for

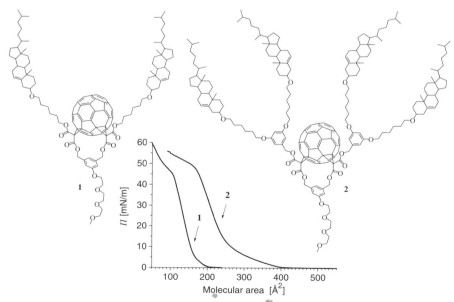

**Figure 1** *Pressure–area isotherms for amphiphilic fullerene derivatives 1 and 2*

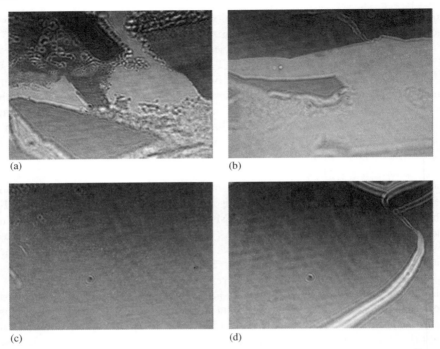

**Figure 2** *Brewster angle microscopy images for **1** at $A = 300$ Å$^2$ (a) $A = 200$ Å$^2$ (b) $A = 162$ Å$^2$ (c) and during the decompression (d)*

compound **1**, indicates long-range intermolecular interactions in the Langmuir film. Further evidence of a better film cohesion is confirmed by the hysteresis curve, showing a better reversibility of the isotherm. By increasing the number of cholesterol units, the encapsulation of the carbon sphere within its addend is more effective, thus limiting $C_{60}$–$C_{60}$ interactions and aggregation phenomena.

The Langmuir films obtained from amphiphilic fullerene derivative **2** have been transferred onto solid substrates with the LB technique. A great number of layers could be deposited. The excellent quality of the LB films prepared with these amphiphilic fullerene *bis*-adducts is deduced from the plot of their UV/Vis absorbance as a function of the layer number, which results in straight lines, indicating an efficient stacking of the layers. The main feature of the UV/Vis spectra is the broadening of the absorption in the LB films when compared to the solution.[15] The latter observation is indicative for $C_{60}$–$C_{60}$ interactions within the LB films. Owing to the presence of the long alkyl chains around the $C_{60}$ subunits within a layer, we believe that these $C_{60}$–$C_{60}$ interactions may be the result of the contact of carbon spheres from neighboring layers rather than within the layers.

Diederich and co-workers have also shown that the encapsulation of a fullerene moiety in the middle of a dendritic structure is an efficient strategy

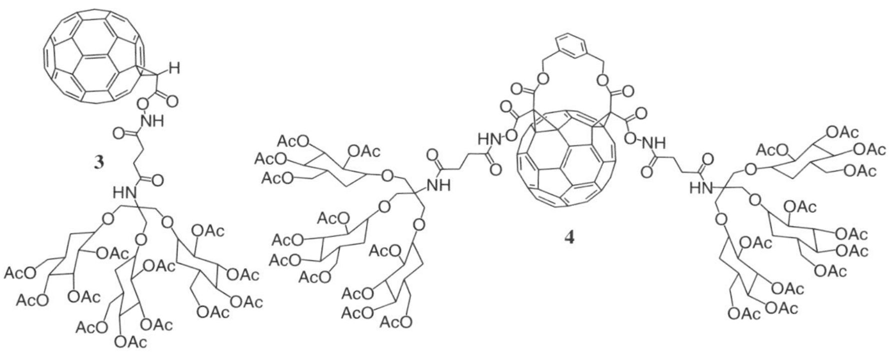

**Figure 3** *Dendrimers with a fullerene core and peripheral acylated glucose units*

to obtain amphiphilic derivatives with good spreading characteristics.[16] Indeed, functionalization of $C_{60}$ with a controlled number of dendrons provides a compact insulating layer around the carbon sphere capable of preventing contact between neighboring fullerenes when the film is compressed, thus the irreversible aggregation usually observed for amphiphilic fullerene derivatives cannot occur. Dendrimers with a fullerene core and peripheral acylated glucose units were prepared and incorporated in Langmuir films (Figure 3).

Fullerene amphiphiles **3** and **4** form stable and ordered monomolecular layers at the air–water interface. The isotherm for monoadduct **3** changed from a liquid-expanded phase to a liquid-condensed phase at $\Pi < 4 \text{ mN m}^{-1}$. The value of $A_0 = 96 \text{ Å}^2 \text{ molecule}^{-1}$ for the limiting area per molecule was extrapolated from the linear part of the curve and is in agreement with the expectations (87 Å$^2$ molecule$^{-1}$) for a 2D close-packed monolayer of the fullerene with the hydrophilic head-group oriented toward the water phase. For *bis*-adduct **2**, the monolayer changed from a liquid-expanded phase to a liquid-condensed phase at $\Pi = 10 \text{ mN m}^{-1}$ and extrapolation to zero surface pressure led to $A_0 = 220 \text{ Å}^2$ molecule$^{-1}$. The $A_0$ value obtained for *bis*-adduct **4** clearly indicates that the bulky dendrons at the two malonate bridges limit the packing and thus keep the fullerene spheres at a distance that is larger than that corresponding to a close contact between the $C_{60}$ spheres. Successive compression and expansion cycles were performed on both different monolayers at the air–water interface in order to explore the reversibility of their formation. After an initial compression to $\Pi = 30 \text{ mN m}^{-1}$, the measured surface pressure during the decompression followed a curve that is very close to that recorded during the compression process. Recompression yielded curves, which were close to the original ones, with only a slight decrease of the measured molecular area, indicating that the formed films are very stable. The hydrophilic carbohydrate-based dendrons are polar enough to establish a strong amphiphilic character of the fullerene derivatives, yet bulky enough to avoid irreversible aggregation of the fullerene centers.

The synthesis of diblock globular fullerodendrimers **5–7** with hydrophobic chains on one hemisphere and hydrophilic groups on the other one was reported (Figure 4).[17] The hydrophobic/hydrophilic balance of these

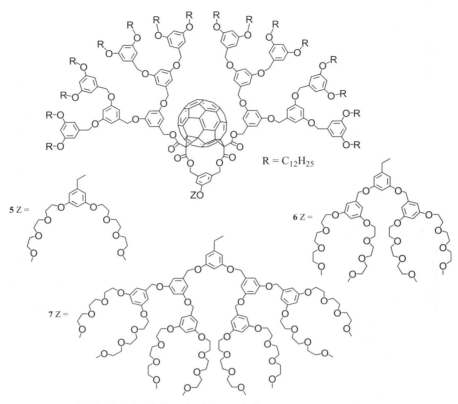

**Figure 4** *Diblock globular fullerodendrimers 5–7*

dendrimers was systematically modified by changing the size of the polar head-group in order to investigate the role of the amphiphilicity at the air–water interface and during the deposition onto solid substrates.

Langmuir films have been obtained for all dendrimers **5–7**. Regardless of the size, the polar head-group attached to these compounds is responsible for an attractive interaction with the aqueous subphase, thus forcing the molecules toward the water surface in a two-dimensional arrangement. The pressure/area ($\Pi/A$) isotherms of fullerodendrimers **5–7** are shown in Figure 5. Compounds **5–7** exhibit a similar behavior: the surface pressure $\Pi$ increases smoothly at molecular areas between 400 and 500 Å$^2$ before taking a sharper rise between 250 and 350 Å$^2$, depending on the compound. The general shape of the isotherms indicates that the films are first in a liquid-condensed phase. Final molecular areas $A_0$ extrapolated at zero surface pressure are 320 ± 3 (**5**), 295 ± 3 (**6**) and 280 ± 3 Å$^2$ (**7**) and are in agreement with the value which can be estimated by molecular modeling. Interestingly, the $A_0$ values decrease when the size of the polar head-group increases. The latter observation could be ascribed to a conformational change in the dendritic structure when anchoring on the water surface is stronger. The repulsion of the $C_{12}H_{25}$ terminated

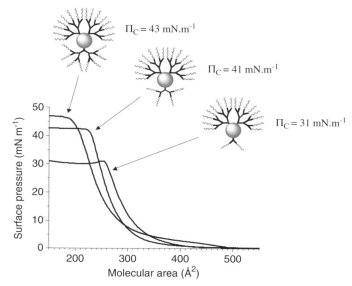

**Figure 5** *Pressure–area isotherms of compounds 5–7 taken on pure water at 20°C. The collapse pressure increases with the size of the polar head-group*

dendrons from the water surface must be more effective in the case of **6** and **7** and as a result, the molecules adopt a more compact structure. In other words, the two $C_{12}H_{25}$-terminated dendritic branches are forced to wrap around the $C_{60}$ core. In contrast, the dendritic structure of compound **5** may be less densely packed around the central fullerene core due to weaker anchoring on the water surface. This model could also be supported by observations done on the LB films prepared from the Langmuir films of compounds **5–7** (see below).

Langmuir films of **5–7** were successfully transferred onto solid substrates to obtain multilayers. In agreement with previous observations, the transfer is more efficient upon increasing the size of the polar head-group. The excellent quality of the LB films prepared from these amphiphilic dendrimers is deduced from the plot of their UV/Vis absorbance as a function of the number of layers, which results in straight lines, indicating that the films obtained from **5** to **7** grow regularly. As discussed above for the LB films prepared from **2**, the main feature of the UV/Vis spectra of the LB film of **5** is the broadening of the film absorption when compared to the solution spectra. This observation is indicative of $C_{60}$–$C_{60}$ interactions resulting from contacts of carbon spheres among neighboring layers. Remarkably, a clear evolution can be seen on going from **5** to **7**: the broadening of the absorption spectrum seen for **5** is almost vanished for **6** and **7**. Actually, the UV/Vis spectra of LB films of **6** and **7** are close to those recorded in $CH_2Cl_2$ solutions suggesting limited $C_{60}$–$C_{60}$ interactions within the LB film. This observation is in agreement with the more compact structures proposed for **6** and **7** at the air–water interface when compared to **5**. In summary, the Langmuir studies of fullerodendrimers **5–7** have revealed a conformational change in the dendritic structure with the size of the polar-head group. Owing to better anchoring onto the water surface, the compounds with the largest polar-head groups adopt a more compact structure and the dendritic branches are forced to wrap around the fullerene core. This model is confirmed by the amount of $C_{60}$–$C_{60}$ interactions within the LB films deduced from their absorption spectra. On the one hand, the results obtained in this systematic study show some of the fundamental architectural requirements for obtaining stable Langmuir films with amphiphilic dendrimers. On the other, it is worth noting that the $C_{60}$ chromophores are almost completely isolated from external contacts by the dendritic structure, thus paving the way toward ordered thin films of isolated functional molecular units. This appears to be an important finding for future nanotechnological applications, in particular for data storage at a molecular level.

The Langmuir and LB film investigations of the amphiphilic dendrimers with peripheral fullerene units **8–10** has also been reported (Figure 6).[18]

Compounds **8–10** form good-quality Langmuir films at the air–water interface and the isotherms taken at 20°C are shown in Figure 7. Fullerodendrimers

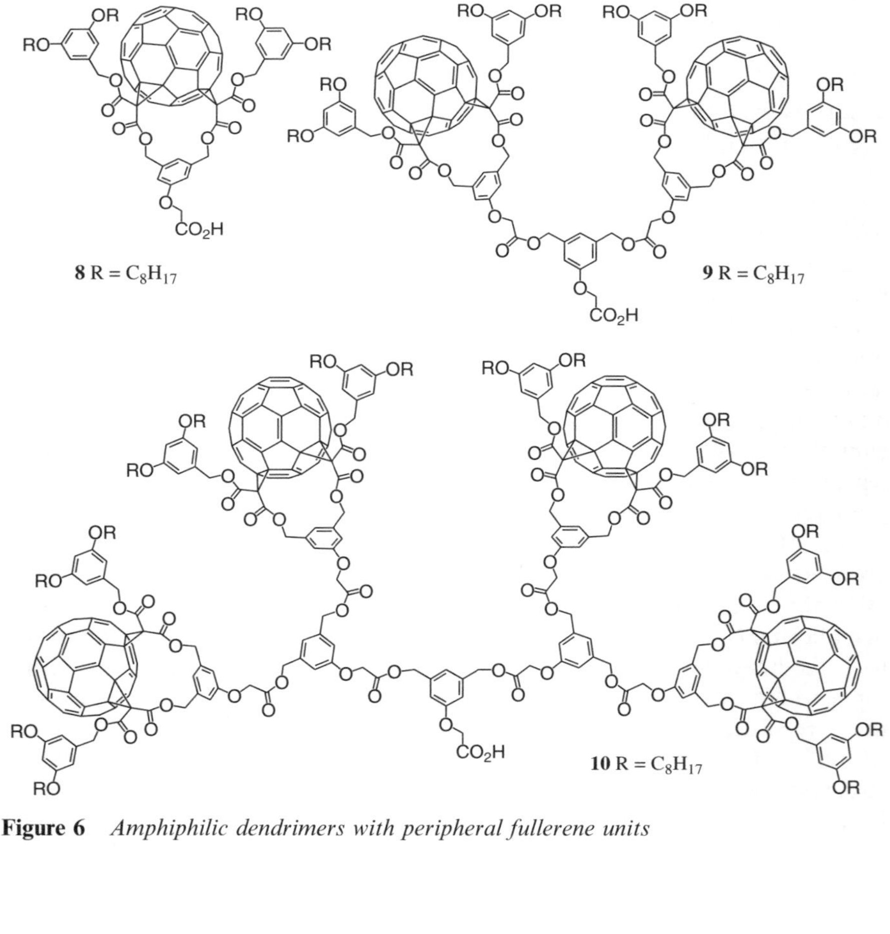

**Figure 6** *Amphiphilic dendrimers with peripheral fullerene units*

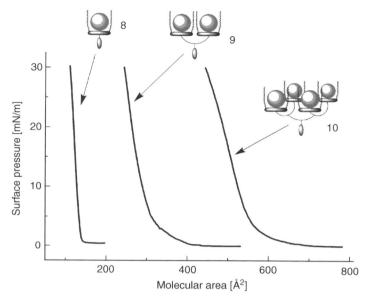

**Figure 7** *Pressure–area isotherms for amphiphilic fullerodendrimers 8–10*

**8** and **10** can withstand pressures up to $\Pi \approx 20$ mN m$^{-1}$, the collapse of the films being indicated by a change of compressibility, while the film obtained from **9** begins to collapse around 30 mN m$^{-1}$. These isotherms are perfectly reversible upon multiple compression-decompression cycles and no hysteresis is observed as long as the collapse pressure is not exceeded, even for **10**. The molecular area for **8**, **9** and **10** extrapolated at zero pressure are $140 \pm 7$, $310 \pm 15$ and $560 \pm 30$ Å$^2$, respectively. They are in the expected 1:2:4 proportion given the structure of the dendrimers and are in agreement with values which can be estimated from molecular modeling. The spreading behavior of **8–10** shows no significant dependence on compression rate or temperature up to 50°C. Observation of the Langmuir films by BAM shows good-quality films for all the compounds. When the surface pressure reaches $\Pi \approx 10$ mN m$^{-1}$ and as long as the films do not enter the collapse regime, at $\Pi \approx 20$ mN m$^{-1}$, only homogeneous surfaces could be observed.

Monolayers of **8** are easily transferred onto silicon or glass substrates covered with a monolayer of octadecyltrichlorosilane (OTS). The transfer ratio (TR) corresponding to the first layer was $1.0 \pm 0.05$, but subsequent layers usually had a slightly smaller TR (0.9). Since the Langmuir films were stable, this should mean that defects were present. Nevertheless, all LB films had a very good appearance, with a uniform brown color for the multilayer ones. Figure 8 shows the grazing X-ray patterns for two LB films of **8**, together with the fits. The monomolecular film shows only two Kiessig fringes, the best fit giving a thickness of $21.0 \pm 0.5$ Å and a roughness of about 4 Å. For the multilayer films, since all layers have the same quality given the almost constant TR (0.9),

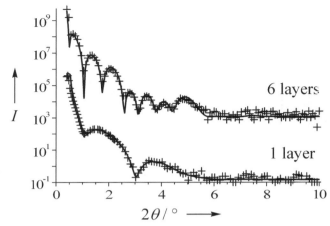

**Figure 8** *Grazing incidence X-ray patterns for LB films (6 layers: upper curve; 1 layer: lower curve) of **8**, together with the best fit (continuous lines) to the data. Overall thickness and film roughness, as deduced from the fits are $21.0 \pm 0.5$ and 4 Å for the 1 layer film, $104 \pm 1$ and 6 Å for the 6-layer film (see text). Not all data points are plotted and upper curve has been shifted for clarity*

one would expect an average layer thickness of 21 Å. However, the average thickness for multilayers film was smaller. This apparent contradiction is solved if one measures the overall thickness of multilayer films as a function of the number of layers. One then finds that odd layers (downstroke) increase the film thickness by 21 Å, whereas even layers (upstroke) only bring 17 Å. This result should indicate that the molecules of even layers do not deposit atop odd ones, but partially wedge themselves in between, like hard spheres piling up on hard spheres (Figure 9). Studies based on molecular modeling show that this model is indeed reasonable, and suggests that intermolecular hydrogen bonds could take place as schematically depicted in Figure 9, producing additional stabilization of the films. The IR spectra of the multilayer films of 8 showing a signal at 3340 cm$^{-1}$ is in good agreement with the presence of such hydrogen bonds.

Owing to the difference in size between the hydrophobic and hydrophilic groups, the preparation of multilayered films was found to be difficult with **9** and **10**. As discussed above, the peripheral substitution of a diblock globular dendrimer with hydrophobic chains on one hemisphere and hydrophilic groups on the other one provides a perfect hydrophobic/hydrophilic balance allowing the formation of stable Langmuir films. In order to incorporate efficiently such fullerene-rich nanostructures in LB films, a dendron with peripheral long alkyl chains and containing five $C_{60}$ units in the branching shell has been prepared and attached to a Fréchet-type dendron functionalized with ethyleneglycol chains (Figure 10).[19]

Compounds **11** and **12** form Langmuir films at the air–water interface. The general shapes of the isotherms obtained at 25°C for both **11** and **12** are similar

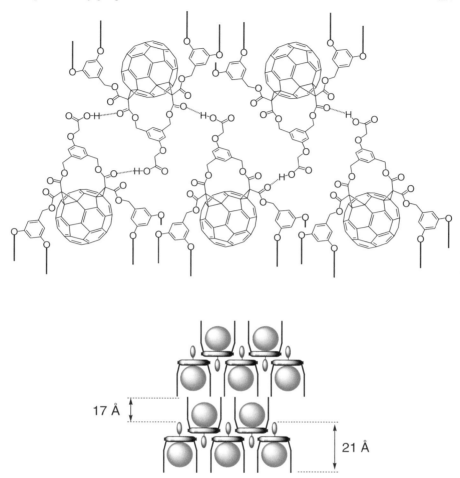

**Figure 9** *Schematic model of the structure of the multilayer LB films obtained from* **8**

and the molecular areas extrapolated to zero pressure (610 ± 30 Å$^2$ for **11** and 680 ± 30 Å$^2$ for **12**) in good agreements with the values estimated by molecular modeling. The stability of both films and their behavior upon successive compression/decompression cycles is however different. Indeed, for compound **11**, the films are not very stable and the isotherm shows poor reversibility, even when the highest pressure is kept below the collapse value. In contrast, Langmuir films of **12** are stable and a perfect reversibility has been observed in successive compression/decompression cycles (Figure 11). It is noteworthy that the five C$_{60}$ units of **11** or **12** are buried in the middle of the dendritic structure, which is capable of providing a compact insulating layer around the carbon spheres, thus preventing the irreversible three-dimensional aggregation resulting from strong C$_{60}$–C$_{60}$ interactions as usually observed with amphiphilic C$_{60}$ derivatives. For both **11** and **12**, the polar head-group causes an

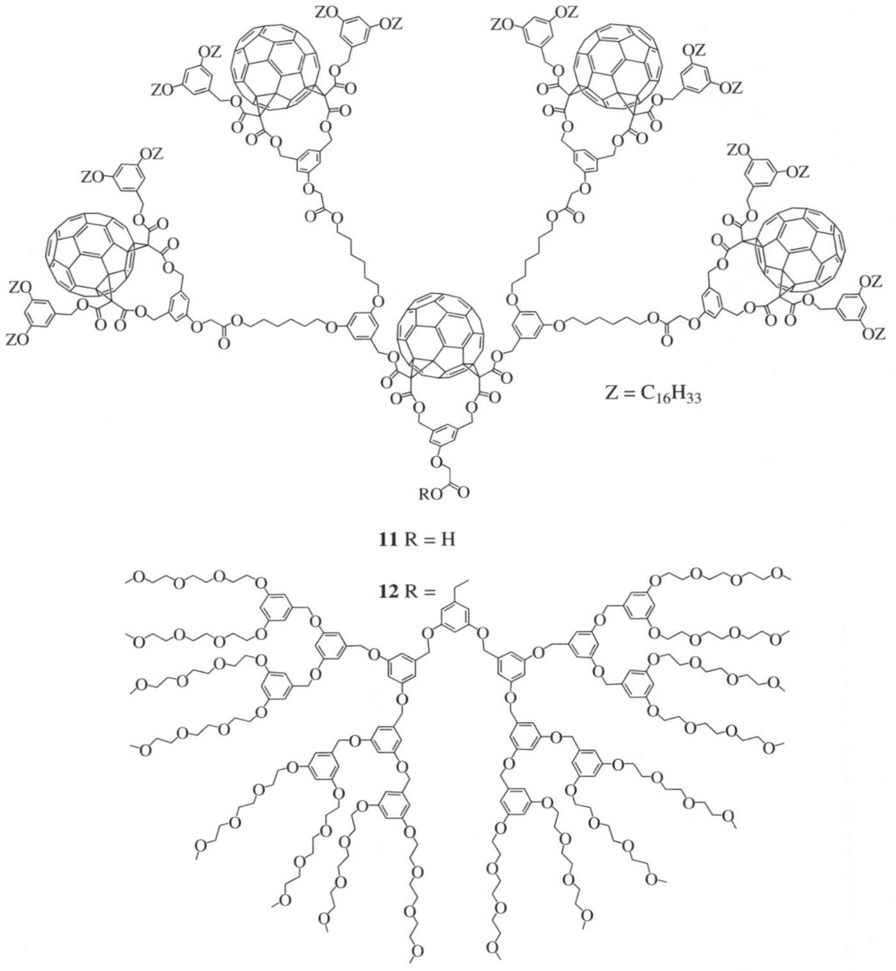

**Figure 10** *Amphiphilic fullerene-rich dendrimers*

attractive interaction with the aqueous subphase, forcing the molecules to adopt a two-dimensional arrangement on the water surface. However, for compound **11**, due to the small size of the polar head-group relative to the large hydrophobic moiety, the interaction with the aqueous layer is certainly not strong enough to stabilize the film. In contrast, a perfect hydrophilic/hydrophobic balance is provided by the 16 peripheral long alkyl chains and 16 triethylene glycol units in the diblock dendrimer **12** allowing the preparation of stable Langmuir films. As already discussed for compounds **5–7**, the functionalization of the branching shell of a diblock dendritic structure is an efficient strategy for the incorporation of functional groups in thin ordered films.

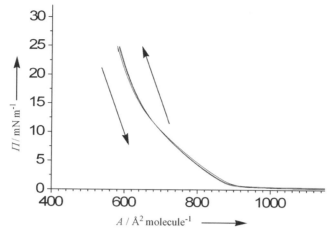

**Figure 11** *Four successive compression/expansion cycles with a monolayer of 12 (25°C, compression/decompression rate: 5 mm min$^{-1}$) showing the perfect reversibility of the process*

Transfer experiments of the Langmuir films onto solid substrates and the preparation of LB films were investigated for **11** and **12**. The deposition of films of **12** occurred regularly on quartz slides or silicon wafers with a TR of 1 ± 0.05. The diblock structure of dendrimer **12** providing the perfect hydrophilic/hydrophobic balance appeared also crucial for efficient transfers of the Langmuir films in order to obtain well-ordered multilayered LB films. Indeed, the transfer of the Langmuir films of dendrimer **11** with the small polar head-group was found to be difficult with a TR of about 0.5–0.7. The structural quality of mono- and multilayered films of **11** and **12** was investigated by grazing incidence X-ray diffraction. The quality of the LB films obtained with **11** was not so good and only allowed an estimation of their thickness, the roughness being always in the 3 Å range. Monolayers of **11** were *ca.* 20 Å thick and the average value of the layer thickness was found to be somewhat smaller than expected for the multilayer films (*ca.* 18 Å). This smaller value is probably the result of partial interpenetration of the successive layers within the film as seen for compound **8**. For LB films of **12**, the presence of low-angle Kiessig fringes in the grazing X-ray patterns indicates that the overall quality of the films is good. The best fit of the grazing X-ray pattern obtained for a monomolecular film gives a thickness of 36 ± 1 Å and a roughness of about 2 Å. For the multilayer films, the average layer thickness was found to be *ca.* 36 Å, indicating no or little interpenetration of successive layers. The excellent quality of the LB films prepared with **12** is also deduced from the plot of their UV/Vis absorbance as a function of the number of layers, which results in a straight line, indicating an efficient stacking of the layers. It is worth stressing out the quality of the stacking and, as a consequence, the quality of the multilayered films obtained with such a megamolecule.

## 5.3 Fullerene-Containing Thermotropic Liquid Crystals

Fullerene-containing liquid crystals represent an interesting class of supramolecular materials. However, liquid-crystalline ordering of fullerenes is difficult since $C_{60}$ does not behave as a mesogenic unit. Two approaches have been developed for the preparation of fullerene-containing liquid-crystalline derivatives:[20] in the first one (covalent approach), $C_{60}$ is functionalized with liquid-crystalline addends, whereas in the second one (non-covalent approach), a supramolecular complex is formed between $C_{60}$ and mesogenic receptors. Obviously, the covalent and non-covalent concepts demonstrate the numerous possibilities for $C_{60}$ to give rise to mesomorphic materials when it is placed in the adequate environment.[21]

### 5.3.1 Non-Covalent Fullerene-Containing Liquid Crystals

The non-covalent approach is interesting since it does not require the development of specific reactions to make $C_{60}$ mesomorphic. A disadvantage of such inclusion compounds, when compared to covalent fullerene derivatives, is their limited thermal stability at elevated temperature where $C_{60}$ might be expelled from the complex. Two different supramolecular fullerene derivatives with liquid-crystalline properties have been described so far. The first one is an inclusion complex obtained from a cyclotriveratrylene (CTV) derivative[22] and $C_{60}$; the second one is based on a dendritic porphyrin hosting $C_{60}$.[23]

The CTV core is appropriate to induce mesophases when functionalized; indeed, several liquid-crystalline CTV derivatives have already been described.[24] Furthermore, recent studies have shown that the CTV macrocycle is a suitable candidate for the formation of inclusion complexes with $C_{60}$.[25] Therefore, a CTV substituted with long-alkyl chains is virtually programmed for self-assembly in a discrete inclusion complex with $C_{60}$, and for self-organization of the resulting system into an extended lattice with liquid-crystalline properties. Inclusion complex **14** was obtained by slow evaporation of a benzene solution containing liquid-crystalline CTV derivative **13** and $C_{60}$ (Figure 12).[22] Observation by polarized optical microscopy (POM) of the brown product obtained after preparation revealed a fluid birefringent phase at room temperature. When the sample was heated above 70°C, the birefringence of the texture disappeared and the X-ray diffraction pattern was transformed into a pattern characteristic of a cubic phase. It is important to highlight that complete encapsulation of $C_{60}$ in the inclusion complex is essential for the stability of the supramolecular system in the mesophase. Indeed, analogous assemblies have been obtained from $C_{60}$ and CTV derivatives bearing less alkyl chains and irreversible phase separation has been observed when the temperature reached the melting point. Indeed, if $C_{60}$ is not completely buried in the middle of the assembly, $C_{60}$–$C_{60}$ interactions are possible, and crystallization of $C_{60}$ occurs.

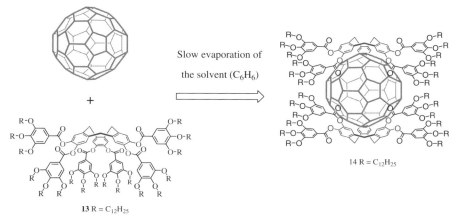

**Figure 12** *Preparation of inclusion complex **14***

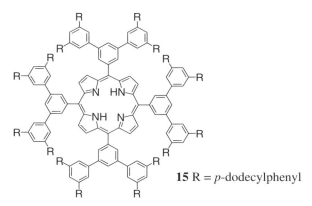

**Figure 13** *Dendritic porphyrin **15** used for the preparation of a supramolecular fullerene-containing liquid crystal*

The second example of mesomorphic fullerene-containing supramolecular derivative has been described by Saito.[23] A dendrimer bearing a central porphyrin unit and carrying 16 peripheral long alkyl chains has been prepared (Figure 13). This compound self-organized to form liquid-crystalline phases by the occurrence of phase segregation between the rigid interior and the peripheral alkyl chains. The thin film obtained from a 1:1 mixture of **15** and $C_{60}$ was homogeneous and showed no irregular aggregates of $C_{60}$ by means of POM and atomic force microscopy (AFM). The complex exhibited two reversible transitions at 99 and 250°C, and POM observations revealed a needle-like texture in the range of 100–250°C. The XRD pattern of the 1:1 mixture of **15** and $C_{60}$ differed from that of **15** alone, due to the fact that the inclusion of $C_{60}$ within the nanospace of **15** strongly affects the mesophase structure in the thermotropic liquid-crystalline phase.

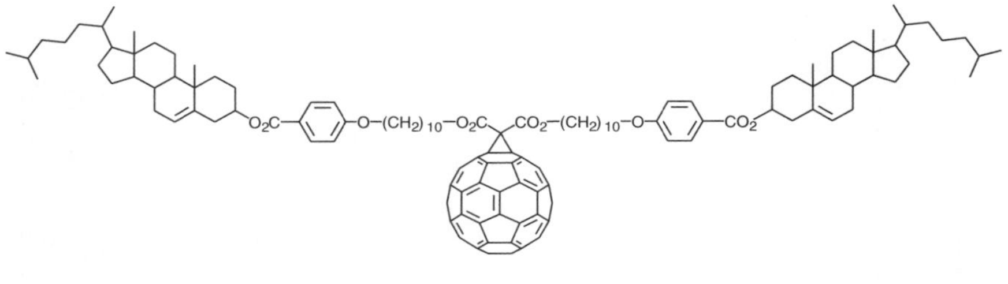

**16**

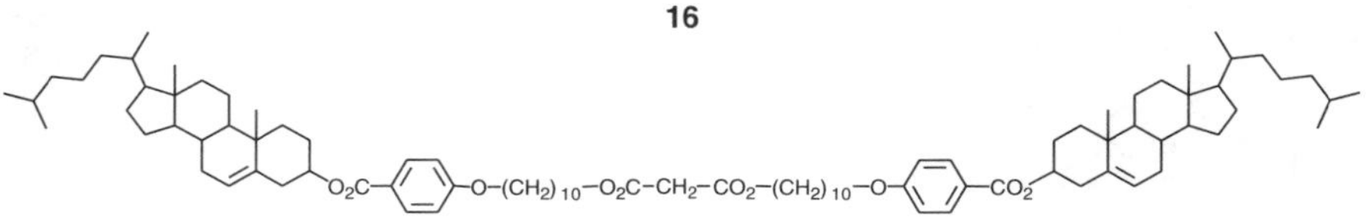

**17**

**Figure 14** *First fullerene-containing liquid crystal*

### 5.3.2 Covalent Fullerene-Containing Thermotropic Liquid Crystals

The first fullerene-containing liquid crystal **16** was obtained by the functionalization of $C_{60}$ with malonic ester **17** bearing two mesogenic cholesterol units (Figure 14).[26] A monotropic smectic A phase formed for **16** at 190°C when the sample was cooled from the isotropic melt. Malonate precursor **17** (Cr → SmA: 112°C; SmA → N*: 214°C; N* → BP: 224°C; BP → I: 225°C)[†] showed smectic A and chiral nematic (cholesteric) phases as well as a blue phase. Therefore, grafting of $C_{60}$ onto the malonate led to a drastic change of the liquid-crystalline properties. This observation indicated that $C_{60}$ disrupts the liquid-crystalline organization, as a consequence of its size and shape.

The concept developed to elaborate fullerene-containing thermotropic liquid crystals, *i.e.* addition reaction of a mesomorphic malonate derivative to $C_{60}$, was successfully extended to dendritic architectures.[27] The mixed [60]fullerene-ferrocene liquid-crystalline dendrimer of second generation **18** ($T_g$: not detected; SmA → I: 157°C)[‡] (Figure 15) and the corresponding malonate addend (structure not shown) showed similar mesomorphic and thermal properties: both compounds gave rise to a smectic A phase, and only a moderate difference was noticed between their clearing temperatures (157°C for **18**; 169°C for the malonate). Taking into account their respective size, it is understandable that $C_{60}$ has only little influence on the thermal properties of the dendrimer. This is the first observation of the beneficial effect of dendrimers when they are used to functionalize $C_{60}$ in order to generate thermotropic liquid-crystalline phases.

Dendritic liquid-crystalline methanofullerenes **19** (first generation; $T_g$: 48°C; SmA → I: 179°C), **20** (second generation; $T_g$: not detected; SmA → N: 183°C;

---

[†] Cr, crystal; SmA, smectic A phase, N*, chiral nematic (cholesteric) phase; BP, blue phase; I, isotropic liquid.
[‡] $T_g$: glass transition temperature.

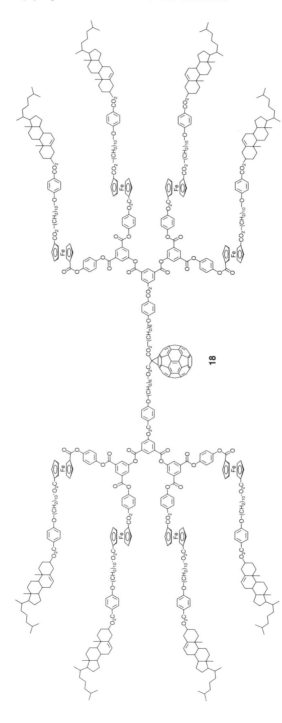

**Figure 15** *Mixed [60]fullerene-ferrocene liquid-crystalline dendrimer*

N → I: 184°C),§ **21** (third generation; $T_g$: not detected; SmA → 212°C) and **22** (fourth generation; $T_g$: not detected; SmA → 252°C) (Figure 16) were synthesized from their corresponding malonates by applying the Bingel reaction.[28]

The malonate precursors (structures not shown) gave rise to nematic (first generation), smectic A (third and fourth generations) or smectic A and nematic (second generation) phases. The clearing point of both the malonate precursors and methanofullerene derivatives increased with the dendrimer generation. A striking difference between the fullerenes and the malonates of low generation resides in the liquid-crystalline phases which were observed: the fullerene derivatives displayed only smectic A phases, with the exception of the second-generation dendrimer **20**, which showed nematic and smectic A phases. This result is an indication that the sphere-like $C_{60}$ unit governs, at least in part, the supramolecular organization of the mesomorphic units within the liquid-crystalline state. This behavior is clearly emphasized by comparing the mesomorphic behavior of the first-generation malonate (nematic phase) with that of its fullerene analogue **19** (smectic A phase).

The molecular organization within the smectic A layers was investigated by XRD. For the first-generation dendrimer **19**, the organization is determined by steric factors: the supramolecular organization depends on the cross-sectional area of the fullerene moiety (90–100 Å$^2$) and that of the mesogenic groups (22–25 Å$^2$ per mesogenic unit) to fill the space in an optimized way. For the high dendrimer generations (**20**, **21** and **22**), the mesogenic groups impose a microphase organization. Owing to the lateral extension of the branching part, the cyano mesogenic groups arrange in a parallel fashion as in classical smectic A phases. The almost constant value of the layer spacing found for the high dendrimer generations indicates that the dendritic core expands laterally with respect to the plane of the layers (Figure 17).

Dendritic liquid-crystalline fulleropyrrolidines **23** (first generation; Cr → I: 178°C), **24** (second generation; $T_g$: 44°C; SmA → I: 168°C), **25** (third generation; $T_g$: 51°C; SmA → I: 196°C) and **26** (fourth generation; $T_g$: 36°C; SmA → I: 231°C) were synthesized from the corresponding aldehyde-based dendrimers and sarcosine by applying the 1,3-dipolar cycloaddition reaction (Figure 18).[29] With the exception of the first generation dendrimer (**23**), which was found to be non-mesomorphic, all fullerene-based dendrimers gave rise to a smectic A phase. The fact that **23** did not show mesomorphism was unexpected taking into account that the corresponding methanofullerene carrying also two mesogenic units exhibited a smectic A phase. Two structural factors could have prevented **23** from being mesomorphic: (1) the bent structure induced by the fulleropyrrolidine moiety, and (2) the benzene nucleus located next to the nitrogen atom of the pyrrolidine ring, the presence of which increases the bulkiness around the fullerene. As a consequence, the mesomorphic part in **23** is not large enough to thwart the steric effects generated by the fulleropyrrolidine framework.

---

§N: nematic phase.

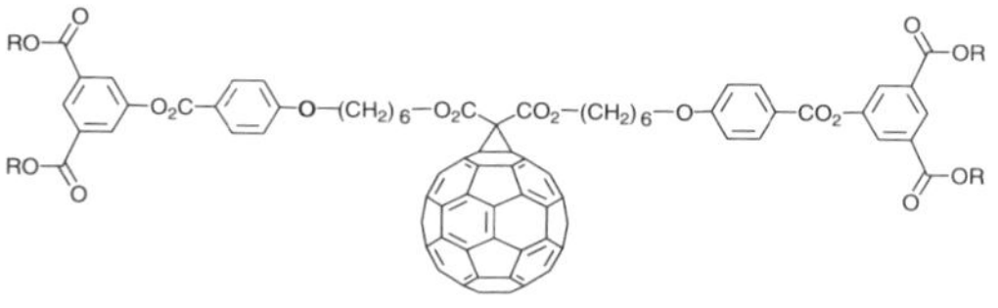

**Figure 16** *Dendritic liquid-crystalline methanofullerenes* **19–22**

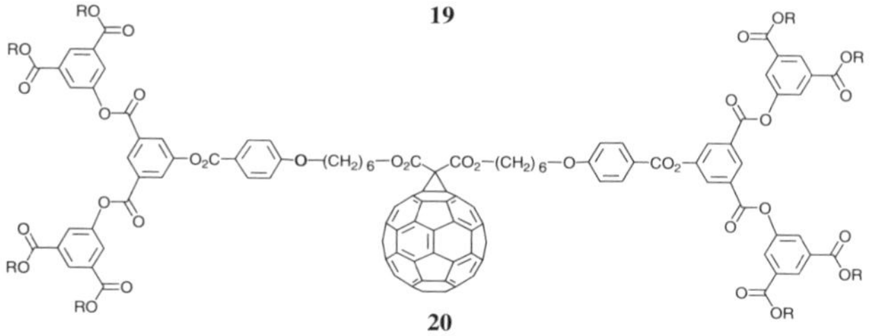

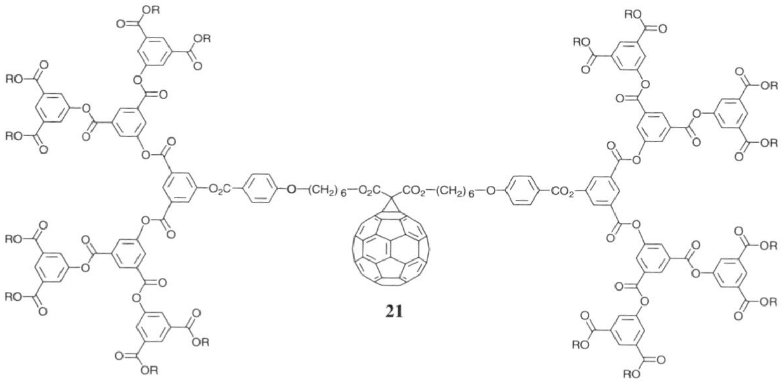

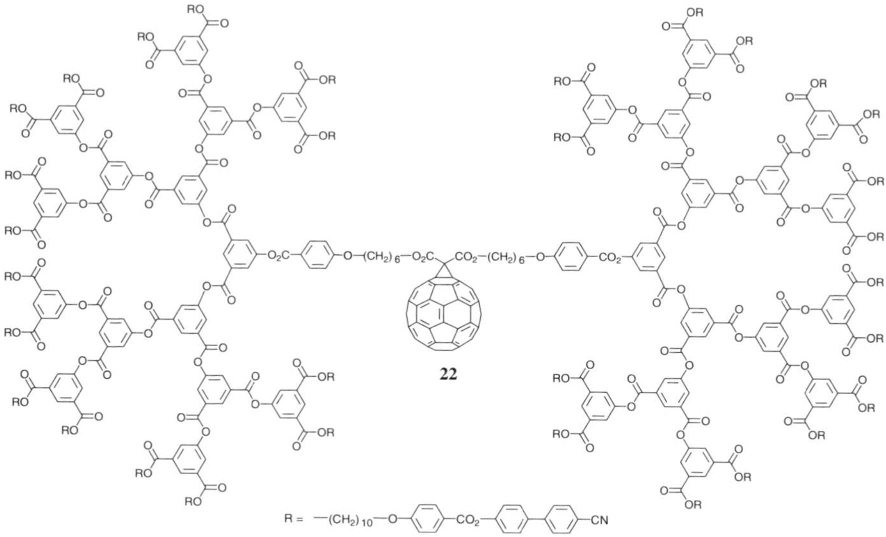

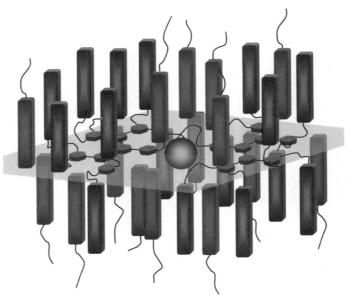

**Figure 17** *Postulated supramolecular organization of **21** (as an illustrative example) within the smectic A phase. The interdigitation is illustrated by the red and green cyanobiphenyl units: the red units belong to the dendrimer which is displayed on the drawing, and the green units belong to dendrimers of adjacent layers*

The $d$-layer spacings determined by XRD were found to be nearly independent of temperature, which is in agreement with the nature of the smectic A phase. For the dendrimer of second generation (**24**), for which the $d$-layer spacing is larger than its molecular length, the organization within the smectic layers requires the adequacy between the cross-sectional area of $C_{60}$ (90–100 Å$^2$) and that of four mesogens (22–25 Å$^2$ per mesogenic unit). For efficient space filling, the molecules are oriented in a head-to-tail fashion within the layers, and for each molecule the cyanobiphenyl units point in the same direction interdigitating with the cyanobiphenyl groups of the adjacent layer (Figure 19a). This model is consistent with the tendency of the cyano-substituted mesogens to interdigitate in an antiparallel way promoted by dipole–dipole interactions. For **24** the supramolecular organization is thus governed by steric constraints.

For the dendrimers of third (**25**) and fourth (**26**) generations, for which the $d$-layer spacings are significantly smaller than their molecular lengths, the dendritic core extends laterally, parallel to the layer planes. With such an orientation, the $d$-layer spacing does not increase (practically) when the size of the dendrimer increases. The mesogenic units are oriented above and below the dendritic core and, as for **24**, interdigitation occurs from one layer to the adjacent one. In this case, $C_{60}$ is hidden in the dendritic core and has no influence on the supramolecular organization. Therefore, from the third

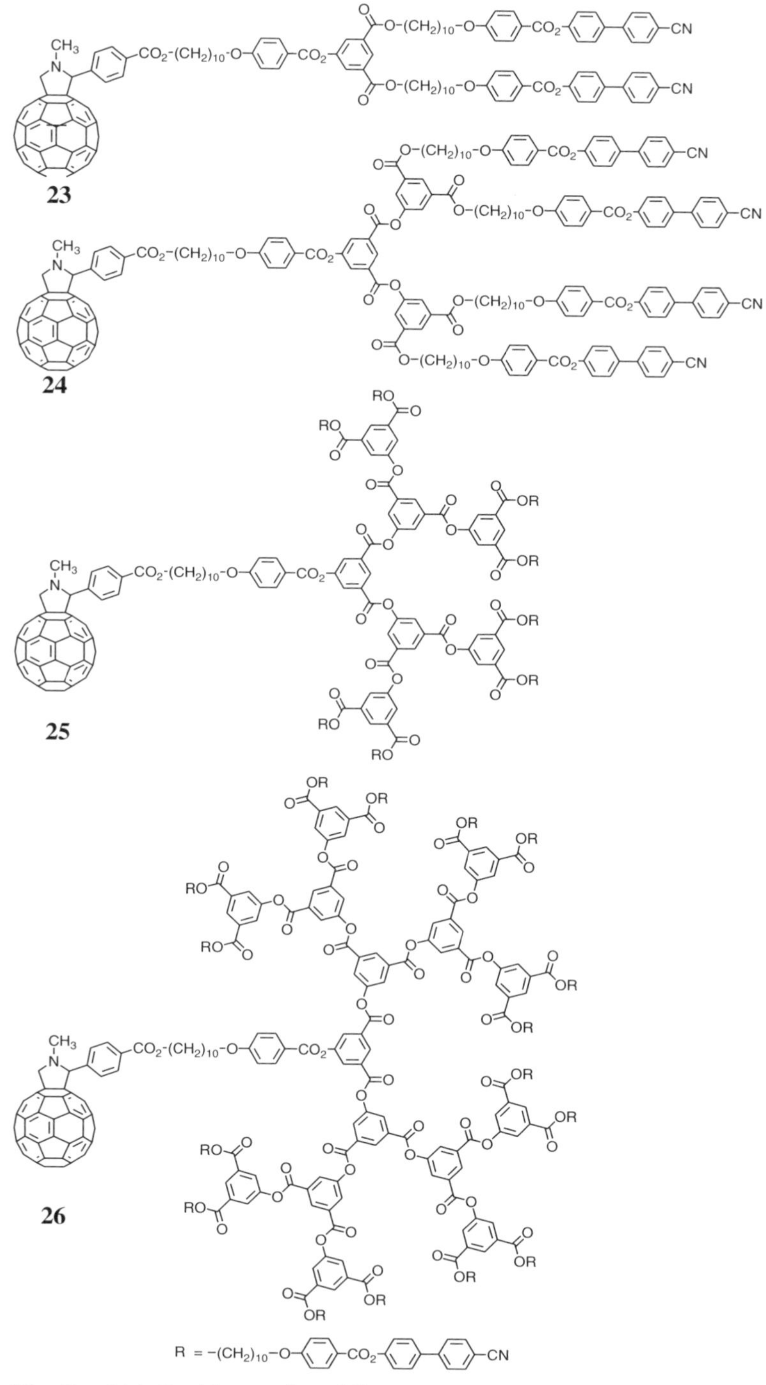

**Figure 18** Dendritic liquid-crystalline fulleropyrrolidines *23–26*

**Figure 19** *Postulated supramolecular organization of fulleropyrrolidines 24 (a), 25 (b) and 26 (c) within the smectic A phases*

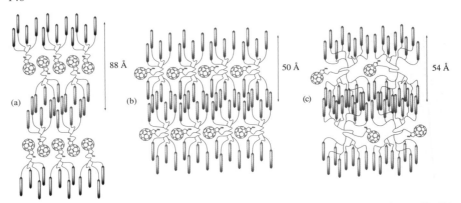

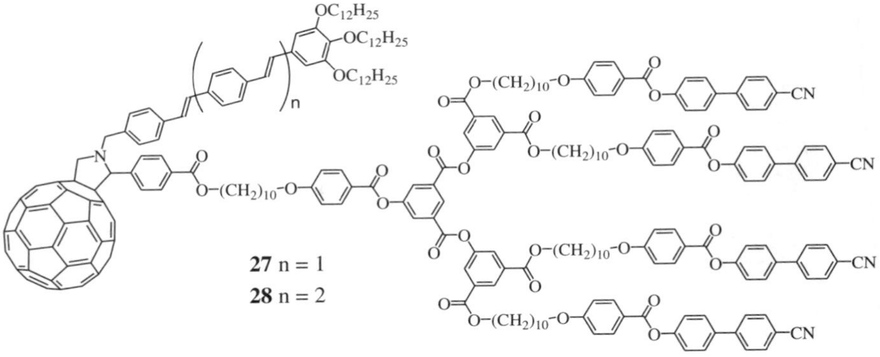

**Figure 20** *Liquid-crystalline $C_{60}$–OPV conjugates*

generation, the supramolecular organization is governed by the nature and structure of the mesogenic units (Figure 19b and c). This result opens avenues for the design of fullerene-containing liquid-crystalline dendrimers with tailored mesomorphic properties: fine tuning of the liquid-crystalline properties could be achieved when the adequate mesomorphic promoter (polar, apolar, chiral) is associated with the appropriate dendrimer (stiff, flexible). Both series of dendrimers (methanofullerenes and fulleropyrrolidines) gave rise to similar behavior in agreement with the nature of the dendritic core and mesogenic units.

The versatility of the 1,3-dipolar cycloaddition reaction[30] is an ideal platform for the introduction of various functional groups onto mesomorphic fullerenes.[31] For example, this strategy was applied to the design of liquid-crystalline fullerenes bearing oligophenylenevinylene (OPV) subunits **27** and **28** (Figure 20).[32] Fullerene derivatives **27** ($T_g$: 50°C; SmA → I: 171°C) and **28** ($T_g$: 50°C; SmA → I: 169°C) gave rise to smectic A phases. Compounds

**27** and **28** exhibit similar results indicating that the overall behavior is mainly governed by the second-generation dendritic framework in agreement with results obtained for the above-described fullerene-containing liquid-crystalline dendrimers. The functionalisation of $C_{60}$-OPV conjugates with a dendritic mesogenic group allows the liquid-crystalline ordering of such donor-acceptor systems which present all the characteristics required for photovoltaic applications.[4] Indeed, the use of liquid-crystalline $C_{60}$-OPV conjugates could be of particular interest since such materials would spontaneously form ordered assemblies that could then be oriented and lead to high-performance thin films.

## 5.4 Conclusions

The recent progresses in the chemistry of $C_{60}$ allow to prepare a large variety of fullerene derivatives for various applications in materials science. In the first part of this chapter, we have shown some of the fundamental architectural requirements needed for the design of amphiphilic fullerene derivatives capable of forming stable Langmuir films. The encapsulation of $C_{60}$ in dendritic branches is an efficient strategy to prevent the irreversible aggregation resulting from the strong $C_{60}$–$C_{60}$ interactions usually observed for amphiphilic $C_{60}$ derivatives at the air–water interface. Some of these compounds have also been successfully incorporated in LB films. In the second part, we have described different concepts to design liquid-crystalline fullerenes. In the non-covalent approach, a supramolecular complex is formed from $C_{60}$ and mesogenic receptors, whereas in the covalent approach, $C_{60}$ is functionalized with liquid-crystalline addends. In the latter case, $C_{60}$ can be functionalized with liquid-crystalline malonates by applying the Bingel reaction thus leading to mesomorphic methanofullerenes or with liquid-crystalline aldehydes and $N$-substitutedglycine derivative by applying the 1,3-dipolar addition reaction thus leading to mesomorphic fulleropyrrolidines. A great variety of liquid crystals was obtained, including photochemical molecular devices combining the fullerene acceptor with various donor moieties. With the view to construct supramolecular fullerene materials, the properties of which could be of interest in nanotechnology (*e.g.* molecular switches, solar cells), fullerene-containing thermotropic liquid crystals could play a major role owing to their self-organization capabilities.

## References

1. W. Krätschmer, L.D. Lamb, K. Fostiropoulos and D.R. Huffman, *Nature*, 1990, **347**, 354.
2. L. Echegoyen and L.E. Echegoyen, *Acc. Chem. Res.*, 1998, **31**, 593.
3. (*a*) R.C. Haddon, A.F. Hebard, M.J. Rosseinski, D.W. Murphy, S.J. Duclos, K.B. Lyons, B. Miller, J.M. Zahurak, R. Tycko, G. Dabbagh and F.A. Thiel, *Nature*, 1991, **350**, 320; (*b*) K. Holczer, O. Klein, S.-M.

Huang, R.B. Kaner, K.-J. Fu, R.L. Whetten and F. Diederich, *Science*, 1991, **252**, 1154.
4. (*a*) C.J. Brabec, N.S. Sariciftci and J.C. Hummelen, *Adv. Funct. Mater.*, 2001, **11**, 15; (*b*) J.-F. Nierengarten, *Solar Energy Mater Solar Cells*, 2004, **83**, 187.
5. R.S. Ruoff, D.S. Tse, R. Malhotra and D.C. Lorents, *J. Phys. Chem.*, 1993, **97**, 3379.
6. A. Hirsch, *The Chemistry of Fullerenes*, Thieme, New York, 1994.
7. C.A. Mirkin and W.B. Caldwell, *Tetrahedron* 1996, **52**, 5113, and references therein.
8. For examples, see: (a) Y.S. Obeng and A.J. Bard, *J. Am. Chem. Soc.*, 1991, **113**, 6279; (b) T. Nakamura, H. Tachibana, M. Yamura, M. Matsumoto, R. Azumi, M. Tanaka and Y. Kawabata, *Langmuir*, 1992, **8**, 4; (c) P. Wang, M. Shamsuzzoha, X.-L. Wu, W.-J. Lee and R.M. Metzger, *J. Phys. Chem.*, 1992, **96**, 9025.
9. (*a*) J.A. Milliken, D.D. Dominguez, H.H. Nelson and W.R. Barger, *Chem. Mater.*, 1992, **4**, 252; (*b*) C. Ewins and B. Steward, *J. Chem. Soc. Farraday Trans.*, 1994, **90**, 969; (*c*) S.S. Shiratori, M. Shimizu and K. Ikezaki, *Thin Solid Films*, 1998, **327–329**, 655.
10. F. Diederich, J. Effing, U. Jonas, L. Jullien, T. Plesnivy, H. Ringsdorf, C. Thilgen and D. Weinstein, *Angew. Chem. Int. Ed.*, 1992, **31**, 1599.
11. (*a*) P.L. Nostro, A. Casnati, L. Bossoletti, L. Dei and P. Baglioni, *Colloids Surfaces A Physicochem. Eng. Aspects*, 1996, **116**, 203; (*b*) Z.I. Kazantseva, N.V. Lavrik, A.V. Nabok, O.P. Dimitriev, B.A. Nesterenko, V.I. Kalchenko, S.V. Vysotsky, L.N. Markovskiy and A.A. Marchenko, *Supramolecular Science*, 1997, **4**, 341; (*c*) L. Dei, P.L. Nostro, G. Capuzzi and P. Baglioni, *Langmuir*, 1998, **14**, 4143.
12. J.-F. Nierengarten, *New J. Chem.*, 2004, **28**, 1177.
13. U. Jonas, F. Cardullo, P. Belik, F. Diederich, A. Gügel, E. Harth, A. Herrmann, L. Isaacs, K. Müllen, H. Ringsdorf, C. Thilgen, P. Uhlmann, A. Vasella, C.A.A. Waldraff and M. Walter, *Chem.-Eur. J.*, 1995, **1**, 243.
14. (*a*) J.-F. Nierengarten, C. Schall, J.-F. Nicoud, B. Heinrich and D. Guillon, *Tetrahedron Lett.*, 1998, **39**, 5747; (*b*) D. Felder, M. Gutiérrez Nava, M. del Pilar Carreon, J.-F. Eckert, M. Luccisano, C. Schall, P. Masson, J.-L. Gallani, B. Heinrich, D. Guillon and J.-F. Nierengarten, *Helv. Chim. Acta*, 2002, **85**, 288.
15. D. Felder, M. del Pilar Carreon, J.-L. Gallani, D. Guillon, J.-F. Nierengarten, T. Chuard and R. Deschenaux, *Helv. Chim. Acta*, 2001, **84**, 1119.
16. F. Cardulo, F. Diederich, L. Echegoyen, T. Habicher, N. Jayaraman, R.M. Leblanc, J.F. Stoddart and S. Wang, *Langmuir*, 1998, **14**, 1955.
17. S. Zhang, Y. Rio, F. Cardinali, C. Bourgogne, J.-L. Gallani and J.-F. Nierengarten, *J. Org. Chem.*, 2003, **68**, 9787.
18. D. Felder, J.-L. Gallani, D. Guillon, B. Heinrich, J.-F. Nicoud and J.-F. Nierengarten, *Angew. Chem. Int. Ed.*, 2000, **39**, 201.
19. J.-F. Nierengarten, J.-F. Eckert, Y. Rio, M.P. Carreon, J.-L. Gallani and D. Guillon, *J. Am. Chem. Soc.*, 2001, **123**, 9743.

20. T. Chuard and R. Deschenaux, *J. Mater. Chem.*, 2002, **12**, 1944.
21. (*a*) R. Deschenaux, M. Even and D. Guillon, *Chem. Commun.*, 1998, 537; (*b*) T. Chuard, R. Deschenaux, A. Hirsch and H. Schönberger, *Chem. Commun.*, 1999, 2103; (*c*) M. Even, B. Heinrich, D. Guillon, D.M. Guldi, M. Prato and R. Deschenaux, *Chem. Eur. J.*, 2001, **7**, 2595; (*d*) S. Campidelli, C. Eng, I.M. Saez, J.W. Goodby and R. Deschenaux, *Chem. Commun.*, 2003, 1520; (*e*) E. Allard, F. Oswald, B. Donnio, D. Guillon, J.L. Delgado, F. Langa and R. Deschenaux, *Org. Lett.*, 2005, **7**, 383; (*f*) S. Campidelli and R. Deschenaux, *Helv. Chim. Acta*, 2001, **84**, 589; (*g*) M. Sawamura, K. Kawai, Y. Matsuo, K. Kanie, T. Kato and E. Nakamura, *Nature*, 2002, **419**, 702; (*h*) Y. Matsuo, A. Muramatsu, R. Hamasaki, N. Mizoshita, T. Kato and E. Nakamura, *J. Am. Chem. Soc.*, 2004, **126**, 432; (*i*) N. Tirelli, F. Cardullo, T. Habicher, U.W. Suter and F. Diederich, *J. Chem. Soc., Perkin Trans.*, 2000, **2**, 193; (*j*) R.J. Bushby, I.W. Hamley, Q. Liu, O. R. Lozman and J.E. Lydon, *J. Mater. Chem.*, 2005, **15**, 4429; (*k*) D. Felder-Flesch, L. Rupnicki, C. Bourgogne, B. Donnio and D. Guillon, *J. Mater. Chem.*, 2006, **16**, 304.
22. D. Felder, B. Heinrich, D. Guillon, J.-F. Nicoud and J.-F. Nierengarten, *Chem. Eur. J.*, 2000, **6**, 3501.
23. M. Kimura, Y. Saito, K. Ohta, K. Hanabusa, H. Shirai and N. Kobayashi, *J. Am. Chem. Soc.*, 2002, **124**, 5274.
24. J. Malthête and A. Collet, *J. Am. Chem. Soc.*, 1987, **109**, 7544.
25. J.-F. Nierengarten, *Fullerenes, Nanotubes Carbon Nanostruct.*, 2005, **13**, 229.
26. T. Chuard and R. Deschenaux, *Helv. Chim. Acta*, 1996, **79**, 736.
27. B. Dardel, R. Deschenaux, M. Even and E. Serrano, *Macromolecules*, 1999, **32**, 5193.
28. B. Dardel, D. Guillon, B. Heinrich and R. Deschenaux, *J. Mater. Chem.*, 2001, **11**, 2814.
29. S. Campidelli, J. Lenoble, J. Barberá, F. Paolucci, M. Marcaccio, D. Paolucci and R. Deschenaux, *Macromolecules*, 2005, **38**, 7915.
30. M. Prato and M. Maggini, *Acc. Chem. Res.*, 1998, **31**, 519.
31. S. Campidelli, E. Vázquez, D. Milic, M. Prato, J. Barberá, D.M. Guldi, M. Marcaccio, D. Paolucci, F. Paolucci and R. Deschenaux, *J. Mater. Chem.*, 2004, **14**, 1266.
32. S. Campidelli, R. Deschenaux, J.-F. Eckert, D. Guillon and J.-F. Nierengarten, *Chem. Commun.*, 2002, 656.

CHAPTER 6
# Hydrogen Bonding Donor–Acceptor Carbon Nanostructures

M. ÁNGELES HERRANZ, FRANCESCO GIACALONE, LUIS SÁNCHEZ AND NAZARIO MARTÍN

Departamento de Química Orgánica, Facultad de Química, Universidad Complutense, E-28040, Madrid, Spain

## 6.1 Introduction

The success of Nature to convert inexpensive, non-polluting and inexhaustible sunlight into energy has stimulated the investigation on the natural photosynthesis in order to emulate it. The photosynthetic process starts with the absorption of light by plants and bacteria and continues with cascades of energy (ET) and electron (eT) transfer events. This complex process results in the conversion of $CO_2$ and $H_2O$ into glucose and energy (ATP).[1] Energy and electron transfer processes are also key events in molecular-scale optoelectronics (for instance in plastic solar cells), photonics, sensors and some other emerging areas of nanoscience and nanotechnology.[2]

A remarkable feature is that in Nature, all these events are produced in a rigid matrix and, hence, in a well-ordered media of supramolecular ensembles. Whereas a precise control on the structure as well as on the function of the materials created from molecular components can be efficiently achieved by using the vast tools of covalent chemistry, to reach a similar control with supramolecular principles involves a higher degree of complexity. To face this challenge, Mother Nature provides excellent and plentiful examples of nanometric multifunctional materials in which size, shape and function are regulated by non-covalent interactions.[3] The non-covalent forces used to create multi-molecular arrays comprise ion–ion, ion–dipole, dipole–dipole, π–π stacking and, finally, hydrogen bonds. Using these weak interactions, whose binding energies range from a few kJ $mol^{-1}$ to several hundreds of kJ $mol^{-1}$, supramolecular growth is driven by thermodynamic forces or is the result of kinetic processes.[4]

The binding energies of hydrogen bonding interactions, which can reach up to 120 kJ $mol^{-1}$, have prompted their use as a versatile supramolecular methodology. It has been demonstrated that H-bonding specificity and directionality governs the 3D structures in novel chemical and biological systems.[5] Nonetheless, and due to the characteristic weakness of a single H-bond, to create stable architectures and arrays of multiple H-bonds or the combination of H-bonds with additional supramolecular interactions, such as hydrophobic or electrostatic forces, is necessary (Figures 1a and b).[6] Whereas association constant ($K_a$) values of $\sim 10$ M$^{-1}$ are achieved in the simplest DH•A arrays, built upon one H-donor (DH) and one H-acceptor (A) site, $K_a$ values as large as $10^2$–$10^3$ M$^{-1}$ are determined for triple H-bonding motifs (see Figure 1c).[7] Remarkably, Meijer et al. have demonstrated that the concurrence of H-bonds together with four attractive secondary interactions in 2-ureido-4-pyrimidinones (UP) shed extraordinary high $K_a$ values, higher than $10^7$ M$^{-1}$ (see Figure 1d).[8]

A major goal in the field of *Supramolecular Chemistry* is the search for new artificial photosynthetic models, in which biomimetic principles can be used for implementation in molecular electronic devices. Thus, such biomimetic models should fulfill the following requisites: (i) the presence of an electron donor (D) connected to an electron acceptor (A); (ii) the ability of one of the components to absorb visible light, thus acting as antenna, and (iii) an organizational principle that controls their electronic interactions (and therefore the rates and yields of eT).[1] The structural and electronic features of fullerenes and carbon nanotubes (*i.e.*, their extended and delocalized π-electron system and their good electron acceptor properties) offer new possibilities in the quest for biomimetic model systems as well as in the construction of photoactive devices.[9]

Despite the importance of non-covalent motifs, and in particular H-bonding in the design of supramolecular architectures, their application to nanoscale carbon-based materials, such as fullerenes and carbon nanotubes, has been largely neglected. Only recently the combination of such family of compounds and non-covalent motifs started to develop into an interdisciplinary field.[10] In particular, we and others became interested in applying such weak intermolecular forces as a mean to modulate fullerene-based architectures and to control their function.[11]

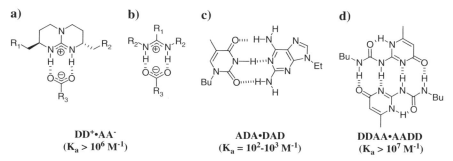

**Figure 1** *$K_a$ values for different H-bonding motifs [measured in Toluene/DMSO (99/1) for (a) and (b) and in CHCl$_3$ for (c) and (d)]*

The aim of the present chapter is threefold: (i) showing that electronic communication in $C_{60}$-based donor–acceptor ensembles – connected through hydrogen bonds – is, at least, as efficient as that found in covalently connected systems, (ii) sketching that hydrogen-bonding fullerene chemistry is a versatile concept to construct supramolecular polymers with exciting and singular features and, (iii) highlighting the promising applications of non covalently bonded carbon nanotubes in donor–acceptor supramolecular architectures.

## 6.2 Hydrogen Bonded $C_{60}$•Donor ($C_{60}$•D) Ensembles

It is well-established how a thermodynamically driven charge separation process is generated by light irradiation in electron donor/electron acceptor ensembles.[12] In $C_{60}$-based derivatives, such charge-separated radical-ion pairs generated upon irradiation show lifetimes ranging from picoseconds to seconds. Most of these donor–acceptor compositions are based on covalent linkages between donor and acceptor units. However, much less is known on H-bonded model systems,[13] and only recently, examples of H-bonded organofullerene dyads have been reported. Such non-covalent $C_{60}$•D conjugates will be summarized next attending to their electron–donor moiety nature.

### 6.2.1 H-bonding interfaced metallomacrocycles•$C_{60}$ dyads

To the best of our knowledge, the first H-bonded interfaced $C_{60}$•D dyad (**1**) was reported in 2002 by Guldi, *et al.*[14] using a zinc phthalocyanine as electron-donor moiety (Figure 2). In this pseudorotaxane-like complex, the corresponding radical-pair species (*i.e.*, $C_{60}^{•-}$•$ZnPc^{•+}$) evolve as a product of an efficient intracomplex electron-transfer starting from the excited state of the ZnPc

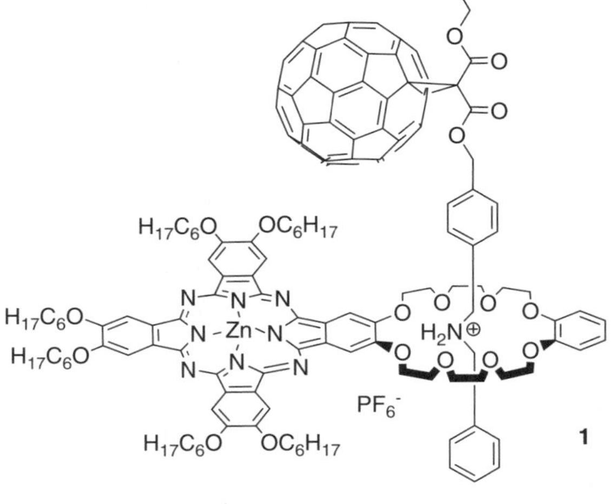

**Figure 2** *H-bonded $C_{60}$•ZnPc dyad **1***

fragment. Interestingly, the reported lifetime was in the range of microseconds. This value is three orders of magnitude higher than that reported for the related covalently bounded $C_{60}$-ZnP dyads.[15]

The similarity of porphyrins (P) with natural electron-donor centers has made this kind of electroactive chromophore the main choice for the preparation of a wide variety of covalent $C_{60}$-P dyads.[16] Nevertheless, only very recently H-bonded $C_{60}$•P dyads have been reported in the literature.[11]

Takata et al. reported on a rotaxane-like $C_{60}$•ZnP supramolecular dyad (**2a**) (Figure 3).[17] Laser irradiation on such composition induced a charge-separation process that commences with the initial photoexcitation of the ZnP moieties and that yields the $C_{60}{}^{•-}$•ZnP${}^{•+}$ radical ion pair whose lifetime (180 ns) is comparable to those reported for some other covalently linked $C_{60}$-P dyads ($\tau_{CS} = 770$ ns).[18]

Very recently, these authors have also reported on the synthesis and photophysical study of a series of $C_{60}$-based rotaxanes (**2b-c**) in which the length of the axle has been systematically modified (Figure 4).[19] A photoinduced electron transfer process takes place after laser irradiation, thus generating the corresponding radical pair. Interestingly, the rate constants and the quantum yields of the radical pairs decrease with the length of the axle. However, they are not sensitive to solvent polarity. In addition, the lifetime of the charge-separated (CS) state increases from 250 ns for **2b** to 650 ns for **2c**. The axle length also conditions the excited state from which the radical pair is formed. Thus, for **2a** this CS state is formed via singlet-excited state of the porphyrin moiety, whereas in the case of **2c** the CS state is formed through the triplet state. All

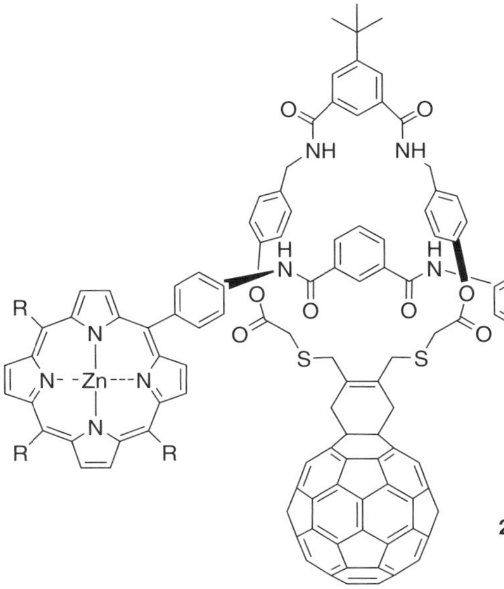

**Figure 3** *Rotaxane-like $C_{60}$•ZnP supramolecular dyad **2a***

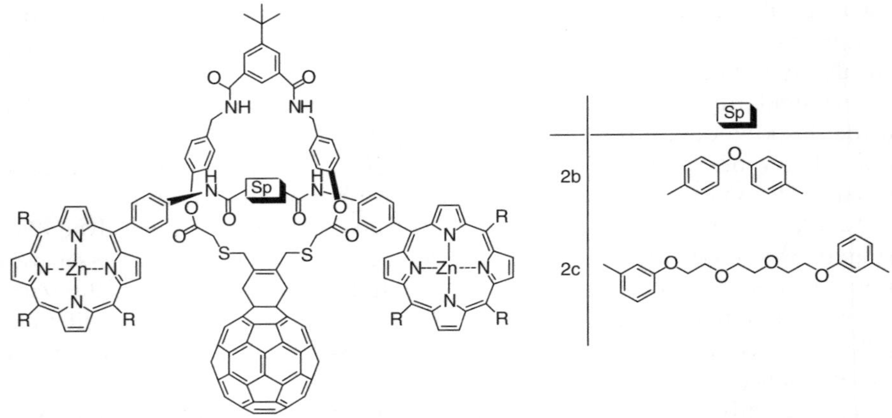

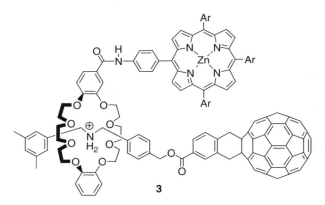

**Figure 4** *Structures of rotaxane-like $C_{60}$•ZnP supramolecular dyads **2b–c***

**Figure 5** *End-capped $C_{60}$•ZnP rotaxane dyad **3***

these phenomena clearly point out to a through-space eT mechanism in such ensembles.

Following the end-capping methodology, Takata *et al.* have reported the synthesis of rotaxane ensembles **3** in which the [60]fullerene cage makes up the axle of the supramolecular structure.[20] In [2]rotaxane **3**, the acceleration of the end-capping process, which enhances its overall efficiency, has been ascribed to an attractive interaction between the two chromophores, *i.e.*, $C_{60}$ and ZnP (Figure 5).

The so-called "two-point" binding strategy is a very successful methodology to assembly fullerenes and porphyrins. First used by D'Souza, Ito and co-workers, this methodology involves axial coordination of the Zn atom from the P and H-bonding interactions.[21] Following this strategy, highly stable supramolecular $C_{60}$•ZnP complexes (**4**) (Figure 6), led to the creation of well-defined distanced and orientated non-covalent ensembles.[22] Time-resolved emission and nanosecond transient absorption studies have revealed efficient charge

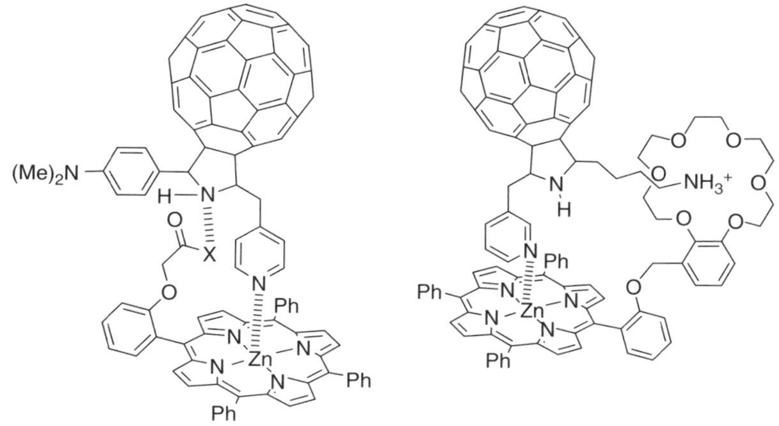

**Figure 6** *"Two-point" assembled $C_{60}\bullet ZnP$ compositions*

separation processes, with rates of $6.3 \times 10^7$ and $3.1 \times 10^9$ s$^{-1}$ in **4a** and **4b**, respectively.

The effect of different non-covalent forces (axial ligation or π–π interactions) on the photophysical properties of "two-point" bound supramolecular $C_{60}\bullet ZnP$ compositions, have recently been reported by the same authors.[23] The different fullerene hosts (**5a,b**) and porphyrin guests (**6a-c**) are depicted in Figure 7. In all these "two-point" ensembles, and regardless of the type of non-covalent bonds, an electron transfer process induced by light is observed. These eT events give rise to radical pairs with lifetimes ranging from 50 to 500 ns depending upon the host and the guest. In the case of P-crown ether **6a**, longer lifetimes are detected for the corresponding CS states independently of the $C_{60}$-host used.

Gathering $C_{60}$ with a porphyrin moiety to form array **7a** showed a remarkable impact on the lifetime of the photogenerated radical ion pair. Thus, the lifetime of the radical ion pair in **7a**, in which $C_{60}$ and ZnP are tethered by means of a guanosine-cytidine scaffold[13b,24] – one of the Watson–Crick pairs – is 2.02 ms. The beneficial effect of the H-bonds, as a part of the whole bond framework, stems from the fact that this value is higher than those reported for related covalently linked $C_{60}$-ZnP dyads, mentioned above.[18] In addition, the comparison between **7a** and **7b**[24] points to a remarkable increase in the lifetime of the radical ion pair state due to the unique combination of $C_{60}$ and ZnP (Figure 8).

Nierengarten's and Soladié's groups have recently described the synthesis of supramolecular cup-and-ball $C_{60}$-P conjugates (**10** in Scheme 1) by direct mixing of P-crown ethers **8** and $C_{60}$-ammonium host **9**.[25] In conjugates **10**, two different non-covalent interactions, ammonium-crown ether and π–π stacking between the fullerene surface and the planar P, exert a dramatic effect on the recognition interaction between the two redox centers. The $K_a$ value, obtained by fluorescence

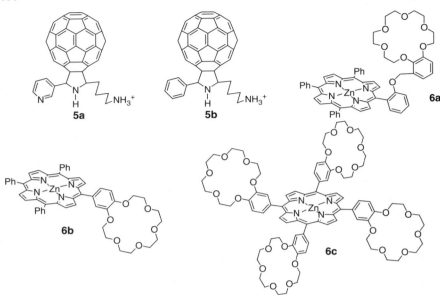

**Figure 7** *Structure of the $C_{60}$-hosts (**5a,b**) and porphyrin guests (**6a–c**) used to create "two-point" supramolecular conjugates*

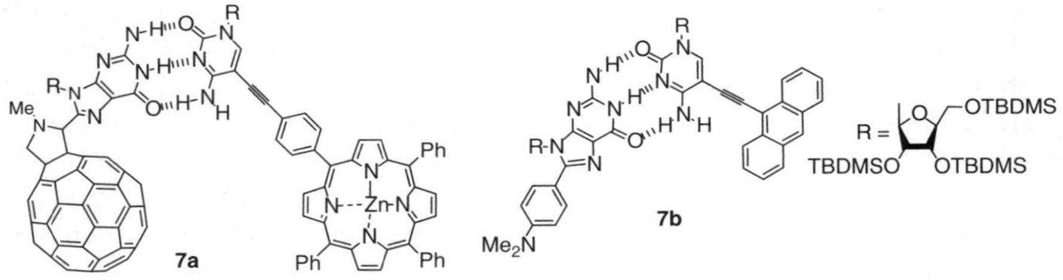

**Figure 8** *Watson–Crick H-bonded D•A dyads*

titration, for conjugate **10** was of 375,000 $M^{-1}$. This value is two orders of magnitude higher than those reported for related ammonium-crown ether complexes formed by using **8** with some other crown ether receptors.[26] Interestingly, the NMR studies carried out for compositions **10** showed that after the initial fast ammonium-crown ether complexation, which gives rise to conformer **10A**, a slow exchange takes place to give rise to conformer **10B** (Scheme 1).

### 6.2.2 H-Bonding Tethered π-Conjugated Oligomer•$C_{60}$ Dyads

Fullerenes and π-conjugated oligomers/polymers are being widely used as building blocks for optoelectronic devices – light-emitting diodes (LEDs),[27] and plastic photovoltaic devices.[28] It is well-known how the morphology of the active layer in photovoltaic cells conditions the final efficiency of the

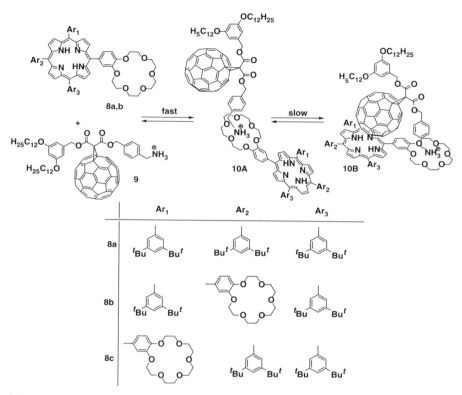

**Scheme 1** *Complexation process between porphyrin receptors **8** and fullerene ammonium host **9**. Schematic representation of the formation of both conformers of conjugate **10***

devices. A potential way to attain better morphologies comprises the supramolecular assembly of their constituents. The orientation of the electroactive components could give rise to an enhanced charge separation and charge carriers movement.[29]

In this context and in order to control the organization in active layers of photovoltaic cells, conjugate **11**[30] and its covalent analogue **12**[31] were synthesized. In both cases, a strong quenching of the oligophenylene vinylene (OPV) fluorescence was observed, being a singlet-energy transfer from the excited OPV to the fullerene responsible for the quenching. Surprisingly, whereas in the covalently bound **12** an ultrafast energy transfer is followed by an intramolecular electron transfer, this sequence does not occur in the H-bonded **11**. This is primarily a consequence of the low electronic coupling in **11** between the electroactive units (Figure 9).[32]

The combination of an UP oligomer (see below) with OPVs endowed with UPs allowed the formation of heterodimers linked through quadruple H-bonds (**13**). This strategy resulted in a new supramolecular donor–acceptor–donor

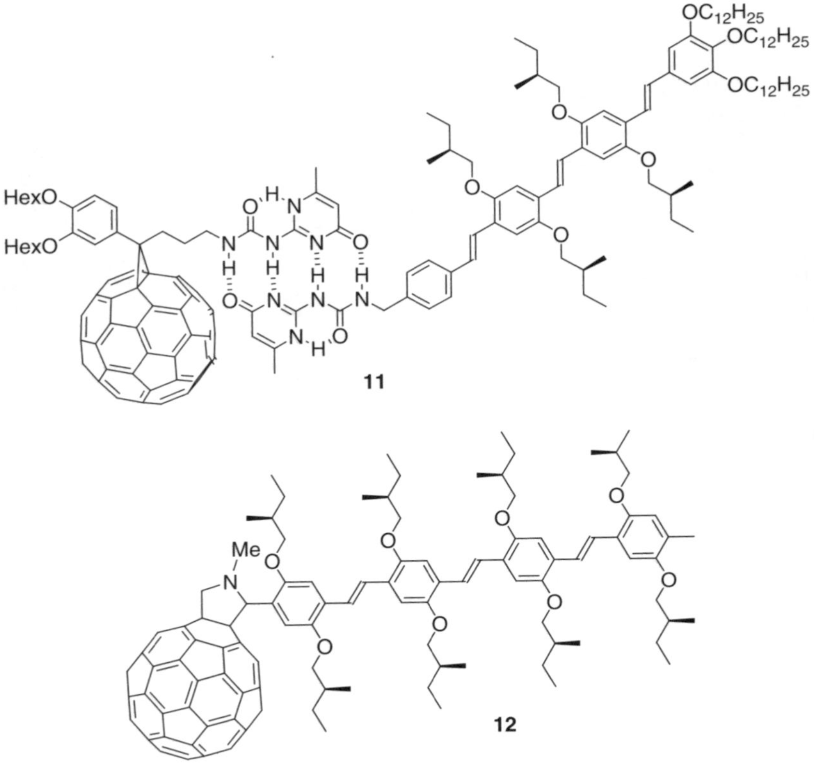

**Figure 9** H-bonding interfaced (**11**) and covalent (**12**) $C_{60}$-OPV dyads

triad (Figure 10).[33] Remarkably, in contrast to previous examples in which the self-complementary nature of such H-bonding motifs leads to a statistical mixture of homo- and hetero-dimers, triad **13** represents the first experimental observation of preferential formation of functional supramolecular heterodimers linked by UP moieties.

It has already been mentioned in the previous section how the cooperative effect between ammonium hosts – crown ethers guests and π-stacking increases the $K_a$ value in comparison with some other related complexes (see Figure 6). This beneficial effect over the association process has also been observed in analogous examples in which the π-conjugated OPV **14** is mixed with $C_{60}$ host **15**. Again, the binding studies carried out in the complex **14•15** demonstrates a beneficial effect on the association constant stemming from the π–π interaction between the fullerene and the OPV moieties (Figure 11).[34] As occurs in $C_{60}$•ZnP complexes, a strong stabilization of about two orders of magnitude is observed for $K_a$ in comparison with many other examples of complexes formed between crown ether receptors and ammonium, alkylammonium or arylammonium hosts.[35]

Remarkably, and despite the expected steric hindrance between the two fullerene-based hosts, when a ditopic OPV-type crown ether guest (**16**) is used

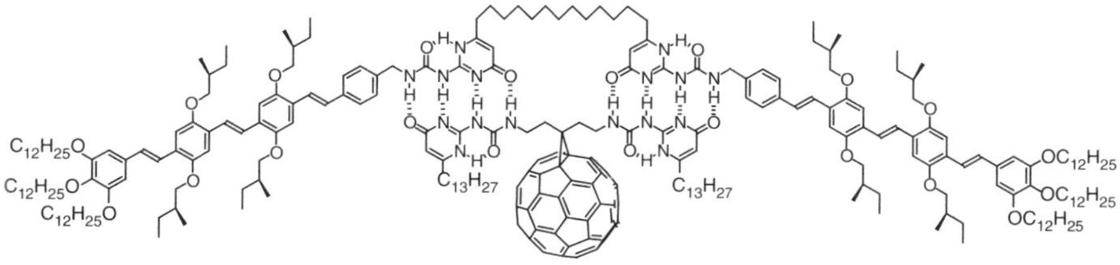

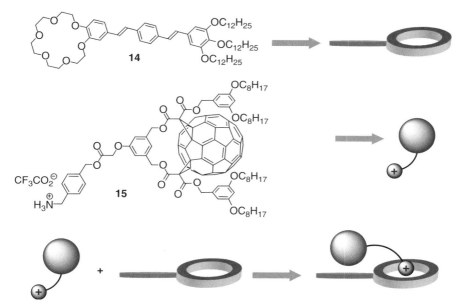

**Figure 10** *Supramolecular OPV•C$_{60}$•OPV triad, 13*

**Figure 11** *Complexation of OPV guest 14 and fullerene ammonium host 15*

to form the corresponding supramolecular complex, similar values of $K_a$ to that measured for complex **14•15** are observed.[34] These findings have been accounted for by the sum of secondary weak interactions, like π–π stacking, between host **15** and receptor **16**. In addition, the flexible character of the spacer seems to be the reason why the expected steric repulsion between the two hosts **15** is avoided and, therefore, the complexation of two C$_{60}$-ammonium hosts with just one receptor **16** is observed. The complex thus obtained can adopt either a *syn* or *anti* conformation (Figure 12). The authors claim that the thermodynamic data point out to a preferential *syn* aggregation as a consequence of the strong π–π interaction between the two C$_{60}$ cages.

When *bis*(ammonium) fullerene ligand **17** is complexed with the previously mentioned ditopic OPV receptor **16**, a perfect complementarity between them is observed.[36] This methodology represents a powerful tool to control the

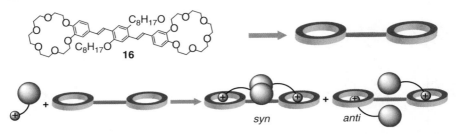

**Figure 12** *Syn/anti complexation of ditopic OPV guest 16 and fullerene ammonium host 15*

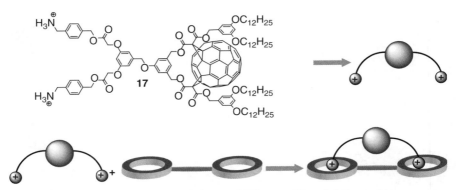

**Figure 13** *"Click" complexation of ditopic OPV guest 16 and fullerene bis-ammonium host 17*

formation of new supramolecular macrocyclic arrays. Analogously to the concept of click chemistry coined by Sharpless *et al.*[37] Dorsselaer, Albrecht-Gary and Nierengarten have named this perfect fit as "supramolecular click chemistry". In fact, *bis*-cationic host **17** "clicks" onto the OPV receptor **16** giving rise to the non-covalently bonded macrocyclic complex **16●17** (Figure 13) with a binding constant of log $K_1 = 6.3 \pm 0.4$ M$^{-1}$. This $K_a$ value is three orders of magnitude higher than those previously reported for other analogous complexes formed by crown ethers and ammonium derivatives.[26]

In the search of H-bonded assemblies capable to undergo an intermolecular ground-state electron transfer process, Bassani, De Cola and co-workers have used the C$_{60}$-based barbituric acid host (**18**) to form a C$_{60}$–oligothienylenevinylene conjugate (**18●19**) showing a $K_a$ value of 5500 M$^{-1}$ (in *o*-DCB).[38] Steady-state fluorescence emission studies showed a strong quenching for the supramolecular conjugate **18●19** when compared with the free species **18** and **19**. A diagnostic absorption band at 560 nm is observed in the transient absorption spectra of the **18●19** pair which could be assigned to the formation of the radical-cation for guest **19**. These findings clearly support an efficient electron transfer process between the two redox centres in composition **18●19** (Figure 14).

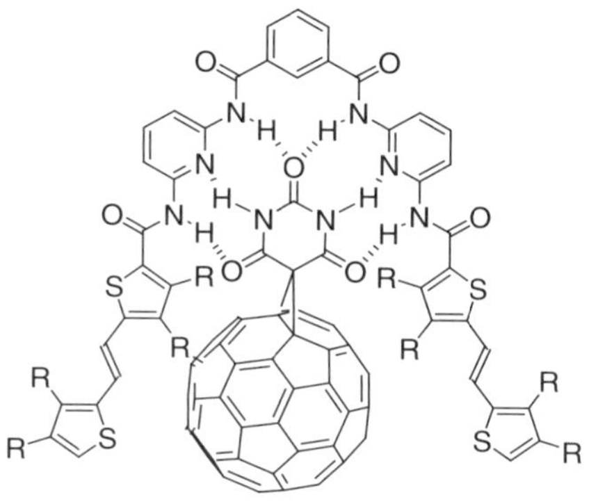

**Figure 14** *Structure of the H-bonded barbituric acid-oligothienylenevinylene pair 18•19*

Within this supramolecular oligomer•$C_{60}$ approach, another remarkable example is the fullerene barbituric acid (**18**)-pentathienylmelamine (**20**) conjugate reported by Bassani and co-workers.[39] The directionality of the H-bonding framework in complex **18•20** is a useful tool to control the exact geometries of the electron donor and acceptor beyond the molecular level. It is well-known that one of the main drawbacks of organic solar cells is the incomplete utilization of the incident light due to the poor match between the absorption spectrum of the materials and the incident solar irradiation.[28c] Interestingly, the conjugation of the melamine **20** allows its absorption in the visible region of the spectrum, which supposes that it is not necessary to synthesize longer oligomers in order to increase this property. In order to evaluate the suitability of these H-bonded materials for their application in the construction of solar cells, simple photovoltaic devices were fabricated. For comparison, devices from oligothiophene-melamine (**20**) and from mixtures **18•20** and **20•$C_{60}$** were prepared. The photovoltaic response of these three devices is depicted in Figure 15. The incorporation of $C_{60}$ (device B in Figure 15b) onto the device induces two-fold gain in the performance of the device in comparison with device A (fabricated from pure oligothiophene **20**). In device C, in which the active layer is the conjugate **18•20**, the enhancement of the photocurrent is fivefold, thus suggesting a higher organization (Figure 15a) degree in the blend which leads to an improved charge separation process.

## 6.3 Other Electron Donor Moieties H-Bonding Interfaced with [60]fullerene

Tetrathiafulvalene, TTF, has demonstrated to be a useful building block in macrocyclic and supramolecular chemistry.[40] Owing to its non-aromatic 14π-electron character, TTF and its π-extended derivatives undergo a reversible oxidation process leading to aromatic and planar cation and/or dication species respectively.[41] These effects have been successfully used to improve significantly

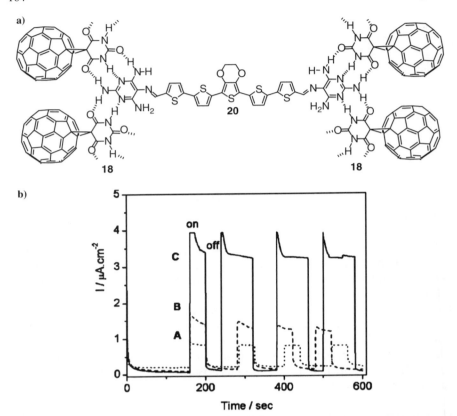

**Figure 15** *(a) Tapelike structure of melamine-barbituric acid pair 18•20; (b) Photovoltaic response of devices A (solid line), B (dashed line) and C (dotted line). The applied potential was −100 mV*
(Reprinted with permission from ref. [39]. Copyright (2005) American Chemical Society)

the radical ion pair lifetimes of the species formed upon visible light irradiation.[42] In collaboration with Mendoza's group, we have recently synthesized a series of H-bonded $C_{60}$•TTF ensembles **21**,[43] in which the TTF unit acts as electron donor. The photo- and redox-active units are held together not only through complementary H-bonds, but also by strong electrostatic interactions through guanidinium and carboxylate ion pairs (Figure 16). Two chemical spacers of different lengths (*i.e.*, phenyl *vs.* biphenyl) as well as two functional groups (*i.e.*, ester *vs.* amide) have been used in order to modulate the molecular architectures.

In these supramolecular dyads, the flexible nature of the spacer results in through-space electron-transfer processes. The lifetime measured for the radical ion pair states, *i.e.*, $C_{60}^{•-}$•$TTF^{•+}$, are in the range of hundred of nanoseconds, thus being several orders of magnitude higher than those reported for covalently linked $C_{60}$-TTF dyads.[42a–c]

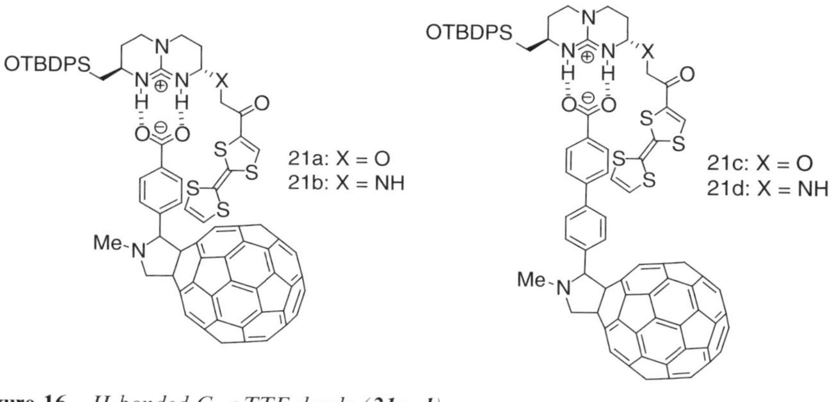

**Figure 16** H-bonded $C_{60}$•TTF dyads (**21a–d**)

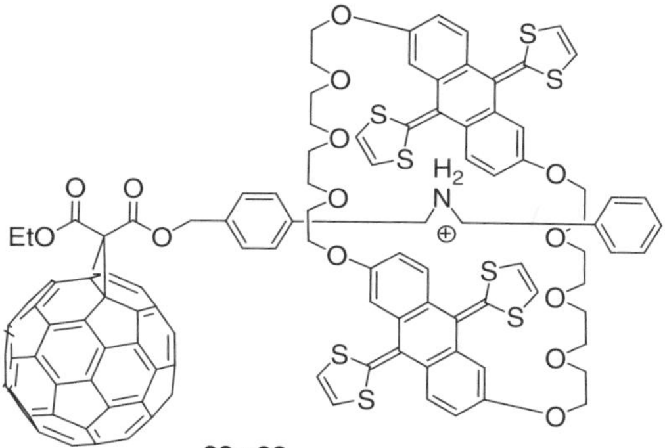

**Figure 17** $C_{60}$•exTTF pseudo-rotaxane triad **22•23**

We have recently reported a supramolecular triad in which the redox-active species are a pseudo-rotaxane crown-ether receptor endowed with two π-extended TTF units (**22**) and a $C_{60}$-ammonium host (**23**). $K_a$ values around 50 M$^{-1}$ have been determined by $^1$H NMR (in CDCl$_3$/CD$_3$CN) and fluorescence (in o-DCB) experiments (Figure 17).[44]

Unlike previous reported π-extended TTF systems covalently linked to $C_{60}$ in which just one oxidation wave involving two electrons are seen,[42f-i] macrocycle **22** shows two closely-spaced oxidation waves. These findings have been accounted for by the flexibility of the polyether chains, which originates weak interactions between the two electron donor moieties.

The donor ability of the nitrogen atom in triphenylamine (TPA) units together with their ability to transport positive charges *via* their radical cations[45] has prompted the use of such molecules in different research fields within molecular electronics.[46] These donor systems have also been used to obtain H-bonding dyads by combining them with $C_{60}$ in a rotaxane fashion.

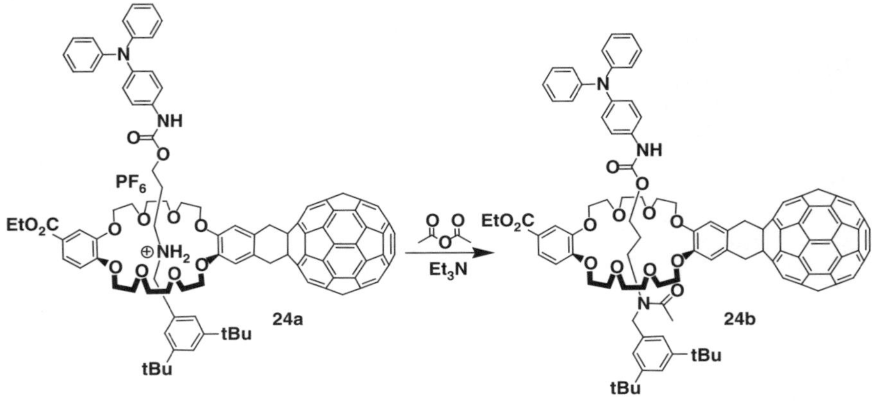

**Scheme 2** $C_{60}$-TPA ensembles tethered by rotaxane structures

Rotaxane **3** was decorated with a $C_{60}$ unit in its axle (Figure 5). In contrast, rotaxanes **24a,b** make up [60]fullerene in their wheel (Scheme 2).[47] In these rotaxanes, non-covalent interactions are modulated by the positive charge that is placed at the nitrogen atom in the middle of the axle (**24a**). The neutral amide group in **24b** evoked a shifting of the axle with respect to the $C_{60}$-wheel. As in the previous examples, in **24a-b**, a through-space intra-rotaxane photoinduced electron transfer gives rise to a long-lived CS state ($C_{60}^{\bullet-}$•TPA$^{\bullet+}$), whose lifetimes range between 170 and 300 ns. A beneficial effect stemming from the supramolecular framework is observed since the lifetime of the $C_{60}^{\bullet-}$•TPA$^{\bullet+}$ radical pair in rotaxanes **24a,b** is longer than those measured for the corresponding covalently bounded $C_{60}$-TPA dyad.[47]

## 6.4 H-Bonded Supramolecular $C_{60}$-Based Polymers

In the previous section we have presented a wide variety of H-bonding donor–acceptor assemblies endowed with $C_{60}$. However, fullerene containing supramolecular polymers have been scarcely studied and, despite their huge potential applicability, only a few examples of donor–acceptor structures are known so far. In this section we will concentrate on the scope of $C_{60}$-based polymer chemistry to develop unprecedented architectures at a supramolecular level, with special emphasis on the more relevant examples including H-bonding architectures and the complementary interactions between pristine $C_{60}$ and ditopic concave guests.

On the basis of the DDAA principles, Hummelen and co-workers prepared the supramolecular polymer **25** from a self-complementary monomer able to form quadruple hydrogen bonds (Figure 18).[48] In polymer **25**, the presence of the UP units – as the molecular recognition motif – confers high $K_a$ values. The dynamic behaviour of **25** was investigated by $^1$H-NMR spectroscopy. When the spectra were analyzed at low concentrations (10 mM), different sets of multiple signals appeared. A likely rationale infers polymeric and low molecular weight

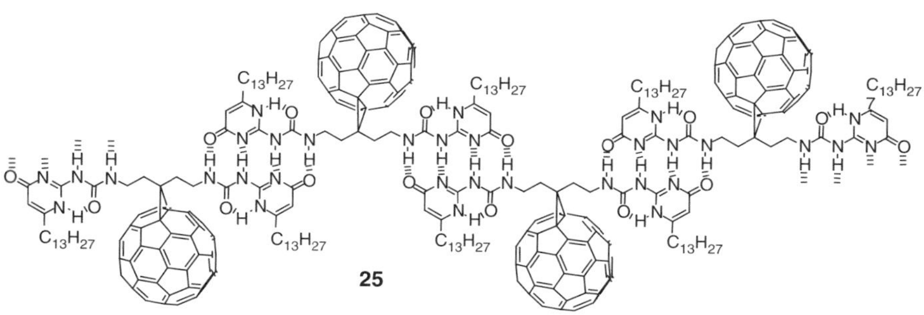

**Figure 18** *Supramolecular $C_{60}$-based polymer 25*

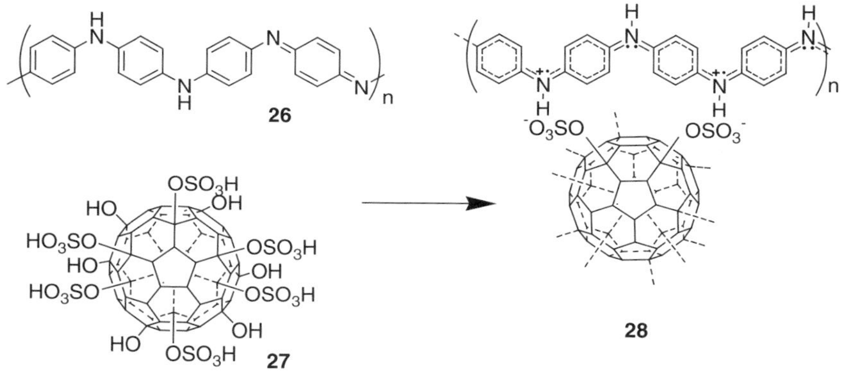

**Figure 19** *Doped polyaniline emeraldine base chains (26) with the hydrogen sulfonated fullerenol derivatives containing multiple – $OSO_3H$ groups (27)*

cyclic aggregates, as it has been proposed in related systems.[49] A fairly high association constant of $6 \times 10^7$ M$^{-1}$ was determined under these conditions.[8,50]

An interesting class of supramolecular binding between $C_{60}$ derivatives and polymers involves multisubstituted fullerenes, which can interact with a properly functionalized polymers. For this purpose, Dai, in 1998, described the behaviour of polyaniline emeraldine base (PANI-EB) **26** doped with a $C_{60}$ structure (**27**) endowed with an average number of hydrogenosulfated and hydroxylic groups of six (Figure 19).[51]

After doping a PANI-EB film with **27**, the conductivity increased up to 11 orders of magnitude, reaching values of *ca.* 106 S cm$^{-1}$ at 250 K. This value was $10^6$ times higher than the typical value for fullerene-doped conducting polymers.[52] Furthermore, dedoping could be achieved after exposure to $NH_3$ vapours, recovering both optical and electronic properties and these processes could be repeatedly reproduced. This huge enhancement of the conductivity has been attributed to the doping-induced "uncoiling" of the PANI-EB chains when **28** is formed. The unravelling of polymeric chains leads to enhanced intrachains carrier mobility but, at the same time, results in an improvement of the interchain ordering augmenting the final conductance. An analogue

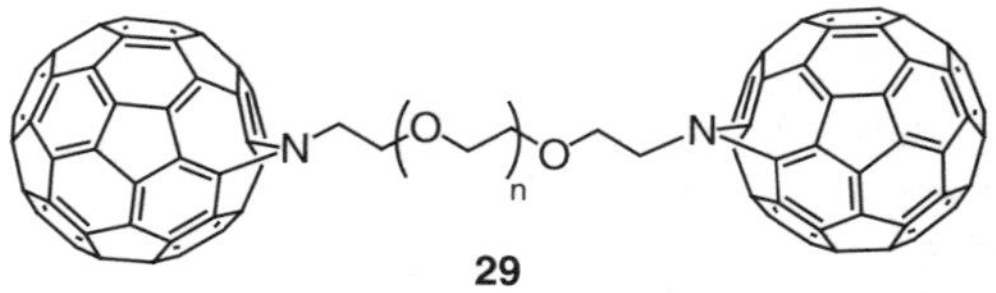

**Figure 20** *Poly(ethelyeneglycol)-based polymer* **29**

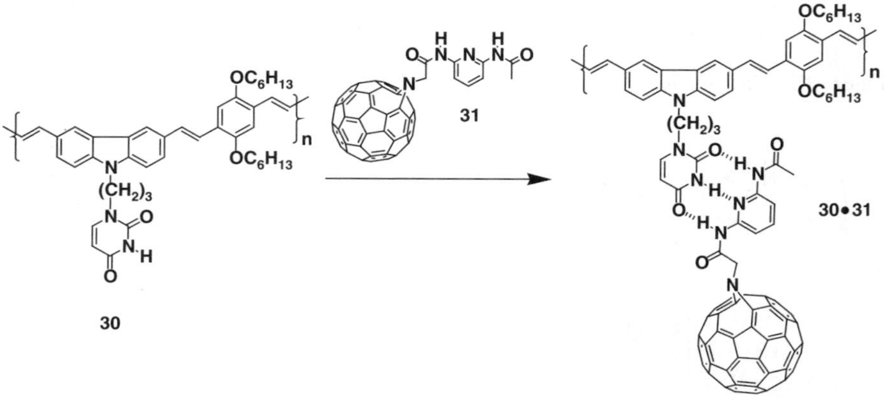

**Scheme 3** *Synthesis of the three-point H-bonded $C_{60}$•PPV assembly* **30•31**

attempt to complex **27** with the accepting H-bonding poly-(4-vinylpyridine) (P4VPy), demonstrates the ionic nature of the interactions in the complexes.[53]

A number of polymers carrying $C_{60}$ as a side substituent have been prepared and their interpolymer complexes, when mixed together with P4VPy and poly(1-vinylimidazole) as proton-acceptors, investigated.[54] $C_{60}$-end-capped polyethelyeneglycol such as **29** (Figure 20) also form interpolymer complexes with different H-donating polymers such as poly(p-vinylphenol),[55] poly(vinylchloride),[56] poly(methacrylic acid),[57a] or poly(acrylic acid).[57b] Furthermore, by melt blending of poly(methyl methacrylate) (PMMA) and **29**, pseudo-semi-interpenetrating polymer networks have been prepared showing a storage modulus as high as 42 kPa, 16 times larger than that of PMMA.[58]

Li and co-workers have prepared an original H-bonding assembly between $C_{60}$ and a π-conjugated polymer (**30•31**) that at the same time avoids high $C_{60}$-contents in the polymer, which prevents the structure from becoming insoluble or not processable.[59] The synthetic procedure involves the preparation of a poly-p-phenylenevinylenecarbazole endowed with an uracil moiety (**30**) able to connect a complementary 2,6-diacylaminopyridine-$C_{60}$ derivative (**31**) through a threefold hydrogen bonding (Scheme 3). Fluorescence experiments are symptomatic of strong interactions between uracil-PPV (**30**) and diaminopyridine-$C_{60}$ (**31**).

By using random-coil diblock polystyrene-poly(4-vinylpyridine) polymers (**PS-b-P4VP, 32**) as hosts and N-methylfulleropyrrolidine carboxylic acid (**33**)

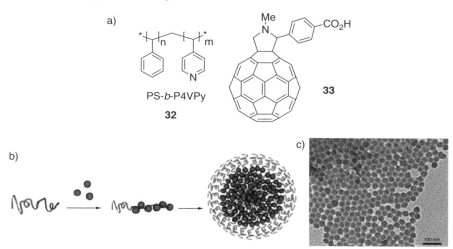

**Figure 21** (a) Structure of the Ps-b-P4PV host **32** and the $C_{60}$-based carboxylic acid guest **33**, (b) schematic representation of the formation of supramolecular rod–coil polymers leading to the generation of micelles (c) TEM image of polymer **32●33** composite
(Reprinted from ref. 60 with permission from Wiley-VCH)

as guest, Shinkai *et al.* have reported the formation of spherical $C_{60}$-based nanoclusters bearing controlled size and morphology (Figure 21).[60] The authors claimed the formation of micelle-like superstructures in which the P4VPy blocks of the random-coil structure of **32** adopts a rod-like rigid conformation by interacting with **33**. This P4VPy-**33** complex possesses a poor solubility that forces the highly soluble PS blocks to orientate in the outside of the micelle, thus forming a shell, as it is shown in Figures 21(b) and 21(c). This practical methodology, carried out just by mixing the respective components, represents a novel approach to nanomaterials chemistry and could be a milestone for the controlled construction of organic nanoparticles.

More recently, Reynolds and Schanze described several photovoltaic devices based on an active layer formed by electrostatic layer-by-layer (LBL) deposition of bilayers of PPE-SO$_3^-$ (**34**), PPE-EDOT-SO$_3^-$ (**35**), and a fullerene modified with two ammonium groups (**36** – Figure 22).[61] When the bilayers were deposited onto an ITO electrode (50 layers), uniform films were obtained as indicated by morphology studies.

In the LBL assemblies some interpenetration of the polymer chains within the layers was observed, leading to some kind of bulk heterojunction. After deposition, the active layer showed an absorbance of 0.5–0.7 units at the $\lambda_{max}$, allowing the capture of a 50% of visible light. When the device was completed with the vacuum deposition of LiF and aluminium layers and, irradiated under AM 1.5 conditions, it displayed a good photovoltaic response, although the overall efficiency resulted to be rather low (**34/36**: $\eta = 0.04\%$; **35/36**: $\eta = 0.01\%$). However, to date these values represent the best results obtained within the LBL approach.

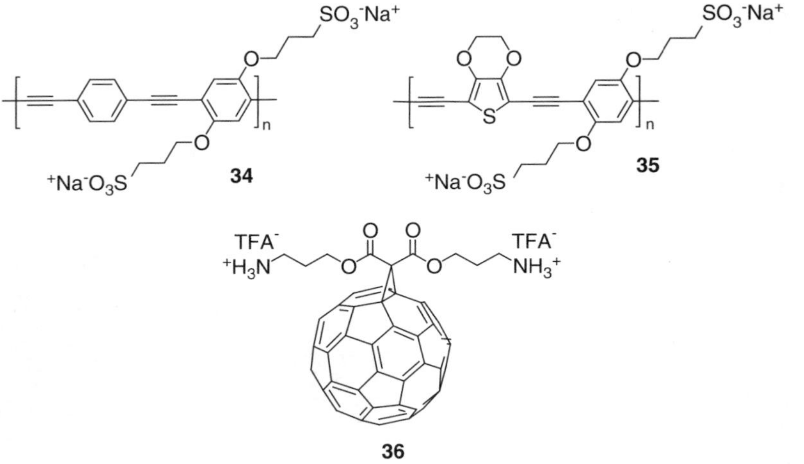

**Figure 22** π-conjugated polymers **34** and **35** and fullerene modified structure **36** used in the construction of photovoltaic devices by the LBL approach

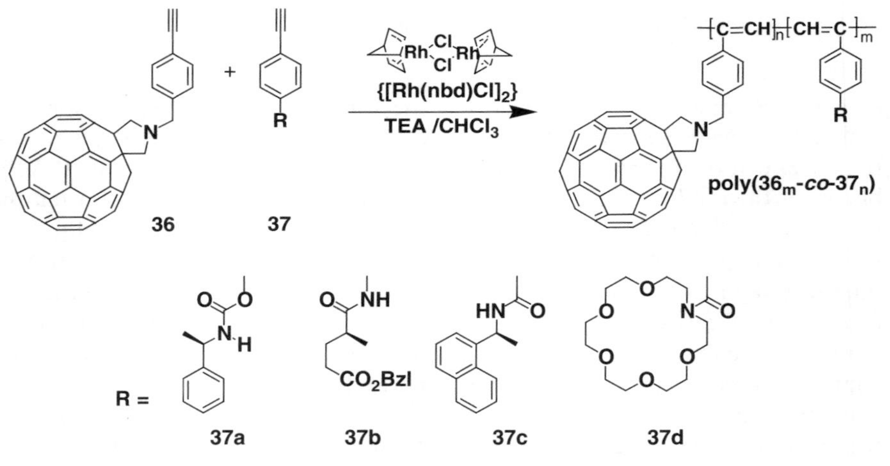

**Scheme 4** Synthesis of copolymers poly($36_m$-co-$37_n$)

The combination of pristine $C_{60}$ with macrocyclic hosts, such as crown ethers, cyclodextrines (CD), calixarenes or porphyrins results in the formation of spectacular inclusion polymers, where van der Waals dispersion interactions greatly add to H-bonding and donor–acceptor interactions in providing stability to the formed assemblies.[62] In the following paragraphs we will review the D–A supramolecular $C_{60}$-based polymers prepared considering this approach.

Yashima *el al.*, prepared a series of copolymers of fulleropyrrolidine **36** and phenyleneacetylenes (**37a-d**) and demonstrated that they form predominantly one-handed helical structures (Scheme 4). In the aggregates, the pendant

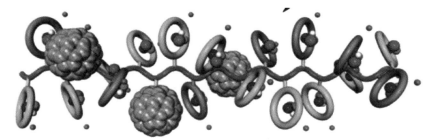

**Figure 23** *Cartoon representing poly(36$_{0.15}$-co-37d$_{0.85}$). The achiral lateral fullerene and crown ether moieties arrange in a helical array along the one-handed helical polymer backbone induced by non-covalent chiral interactions with L-alanine [64]*
(Reproduced by permission of the Royal Society of Chemistry)

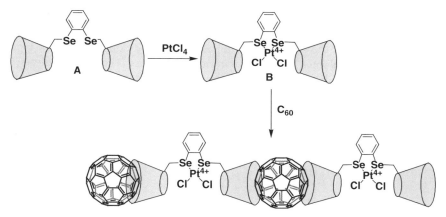

**Scheme 5** *Intermolecular inclusion complexation of metallobridged bis(β-CD) and C$_{60}$*

achiral fullerene groups are arranged in helical arrays with a predominant screw sense along the polymer backbone.[63]

More interestingly, when monomer **37d**, bearing a crown ether substituent, is incorporated in the final polymer, the desired predominantly one-handed helical conformation is reached by mixing together with L-or D-alanine perchlorate in acetonitrile. Upon complexation of the optically active amino acid by means of the pendant crown ether units, the achiral C$_{60}$ moieties arrange in a helical array with the desired helix-sense along the polymer backbone via chiral, non-covalent bonding interactions (Figure 23).[64]

Different CD-based polyrotaxane doubly end-capped with β-CD units have been employed to form aggregates in the presence of [60]fullerene.[65] In a remarkable example, Liu and co-workers prepared a water soluble assembly by the intermolecular inclusion complexation of metallobridged *bis*(β-CD) and C$_{60}$.[66]

The supramolecular fullerene polymer was prepared according to the procedures shown in Scheme 5. TEM micrographs display the presence of linear structures with length in the range of 150–250 nm, constituted by 60–80 units of the complex **B**:$C_{60}$. Moreover, this complex showed an effective DNA-cleavage ability under light irradiation, which has potential application in biological and medicinal chemistry.

Very recently, the complementary molecular affinity between $C_{60}$ and calix[5]arenes has been exploited as the driving force for the preparation of self-assembly networks. Thus, the mixing of the dumbbell [60]fullerene **38** with the ditopic calix[5]arene host **39**, resulted in its self-assembly directed by molecular recognition to afford the formal supramolecular copolymer **38**•**39** (Figure 24).[67]

At a low concentration, diffusion coefficients obtained by pulsed field gradient NMR studies ($2 \times 10^{-4}$ M), demonstrated that the supramolecular complexes mainly adopt a trimeric structure.

The capability that cyclic *bis*porphyrin systems have to create multicapsular structures has been elegantly considered by Shinkai and co-workers in the preparation of novel H-bonding supramolecular polymers.[68] For this purpose, the amide-appended porphyrin **40** (Figure 25) was designed as a system in which π–π stacking interactions among the porphyrins and hydrogen-bonding interactions among the amide moieties could operate cooperatively.[69] Upon addition of $C_{60}$ to **40**, on a gel phase, the formation of a 1:2 complex is observed and confirmed by SEM and TEM techniques (**40** forms a two-dimensional sheet like structure which in the presence of $C_{60}$ changes to a fibril one-dimensional multicapsular structure).

An interesting potential application of these assemblies could be in photo-induced electron transfer events, the preliminary experiments carried out by the authors evidenced a strong fluorescence quenching of porphyrin **40** in the one dimensional (**40**)$_2$•$C_{60}$ aggregate.

## 6.5 Non-Covalent Functionalization of Carbon Nanotubes (CNTs)

Carbon nanotubes (CNTs) have emerged as a new class of materials with extraordinary electrical and mechanical properties.[70] In particular, due to the unique ability of single-walled carbon nanotubes (SWNTs) to form both metallic and semiconducting species, they are very appealing candidates as key building blocks in the construction of photovoltaic devices.[71]

Major setbacks in the chemical processability of CNTs are the strong, mutual interactions through van der Waals forces that held together CNTs, yielding agglomerates of intimately associated long tubes. In this context, chemical functionalization is an especially powerful tool as it might lead to the improvement of solubility and processability of CNTs. In addition, chemical functionalization allows blending of the unique properties of CNTs with those of other functionalities. Thus, the development of chemical strategies aimed at solubilizing SWNTs has been an important motivation in driving the surface chemistry of SWNTs.[72]

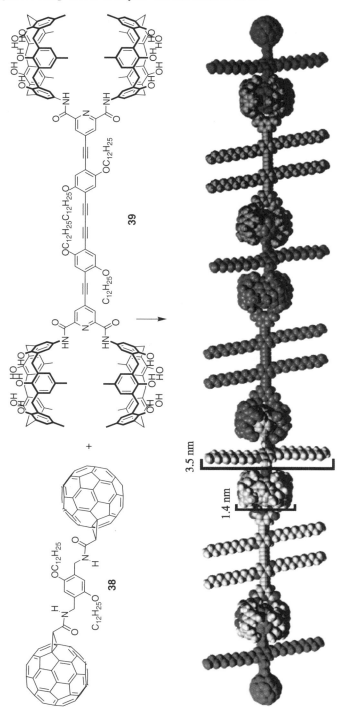

**Figure 24** *Representation of the self-assembling polymer formation by the host-guest complexation between fullerene **38** and calix[5]arene **39**.* (Reprinted with permission from ref. 67. Copyright (2005) American Chemical Society)

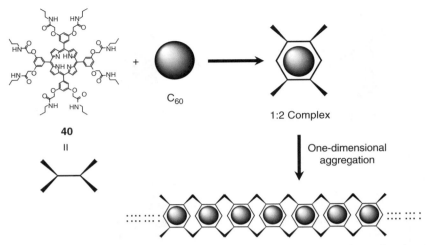

**Figure 25** *Synthesis of supramolecular polymers with programmed hydrogen bonding for the encapsulation of $C_{60}$*

Among the different chemical modification possibilities for CNTs, the non–covalent one is particularly attractive, because it offers the possibility of associating functional groups to the CNT surface without modifying the π system of the graphene sheets and, therefore, their electronic properties.[73]

As examples of non-covalent functionalization, SWNTs can be solubilized in water by "wrapping" them with linear polymers, through electrostatic interactions or using a combination of van der Waals and complementary electrostatic interactions. We will briefly review in this section some of the most representative examples of these approaches considering, in particular, the formation of D–A nanoconjugates.

### 6.5.1 Polymer Wrapping

Recent studies have demonstrated that, besides possibly improving the mechanical and electrical properties of polymers, the formation of polymer/CNT composites is considered a useful approach for incorporating CNTs into polymer-based devices.[74] In one of the first examples, SWNTs were added to a solution of poly-(*m*-phenylenevinylene) (PmPV) substituted with octyloxy alkyl chains, and a stable suspension of CNTs was obtained upon sonication due to the wrapping of the polymer around CNTs.[74a,75] The SWNT/PmPV complex exhibits conductivity eight-times higher than that of the pure polymer, without any restriction on its luminescence properties.

More recently, Stoddart and co-workers have synthesized new families of poly[*m*-phenylenevinylene)-co-(*p*-phenylenevinylene)]s, functionalized in the synthetically accessible C-5 position of the meta-disubstituted phenylene rings.[76] They are essentially poly{(5-alkoxy-*m*-phenylenevinylene)-co-[(2,5-dioctyloxy-*p*-phenylene)-vinylene]} (PA*m*PV) derivatives bearing tethers or rings that form pseudorotaxanes with matching rings or threads.

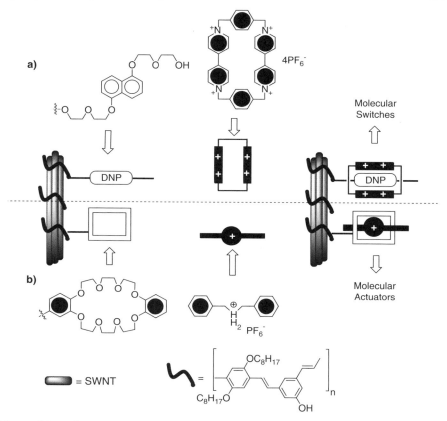

**Figure 26** *Schematic representation for the formation of polypseudorotaxanes grafted along the walls of SWNTs. (a) From a naphthalene-containing PAmPV polymer or (b) from a dibenzo[24]crown-8-containing PAmPV polymer*

The self-assembly of these pseudorotaxane-containing PA*m*PV polymers is based on two different recognition motifs, one involving hydrogen bonding interactions between secondary dialkylammonium centers (*i.e.*, dibenzylammonium ions) and suitable crown ethers (*i.e.*, benzo[24]crown-8) and, the other involving π–π stacking, [C-H•••O], and [C-H•••π] interactions between π-electron-deficient hosts [such as cyclobis(paraquat-*p*-phenylene)] and π-electron-rich guests [such as 1,5-*bis*(hydroxyethoxyethoxy)naphthalene]. Wrapping of these functional PA*m*PV polymers around SWNTs results in the grafting of pseudorotaxanes along the walls of the nanotubes in a periodic fashion (Figure 26). The results hold out the prospect of developing future arrays of molecular actuators and switches.[76]

The wrapping of SWNTs with polymers that bear polar side-chains, such as polyvinylpyrrolidine (PVP) or polystryrenesulfonate (PSS) in water has also been investigated.[77] CNTs were found to unwrap on changing the medium to less polar solvents. In similar approaches, it was shown that soluble nanotube-polymer composites can be prepared by in situ polymerization of phenylacetylene[78] or polycystine[79] and, the ability of short rigid conjugate polymers, such as, poly(aryleneethylene)s to solubilize CNTs was as well demonstrated.[80]

On the combination of CNT and conducting polymers, polyaniline (PANI) bears particularly great potential in synthesizing polymer/CNT composites due to its environmental stability, good processability and reversible control of conductivity both by protonation and charge-transfer doping. Several recent reports have focused on the design and the fabrication of PANI/CNT composites.[81] For example, Wu et al.[82] described the synthesis of doped polyaniline in its emeraldine salt form (PANI-ES) with multi-walled carbon nanotubes (MWNTs) fabricated by in situ polymerization. The as-prepared MWNTs were treated using a 3:1 mixture of concentrated $H_2SO_4:HNO_3$, which produced carboxylic acid groups at the defect sites (see Scheme 6). On the basis of the $\pi-\pi^*$ electron interaction between aniline monomers and MWNTs and hydrogen bonding interactions between the amino group of aniline monomers and the carboxylic acid groups of the modified MWNTs, aniline molecules were adsorbed and polymerized on the surface of MWNTs. The structural analysis of the composites formed by spectroscopic techniques showed the formation of tubular structures with diameters of several tens to hundreds of nanometers, depending on the PANI content. The electric conductivities at room temperature of PANI-ES/MWNTs composites are 50–70% higher than those of PANI without MWNTs.[82]

More recently, Guldi, Prato and co-workers succeeded in the supramolecular association of pristine SWNTs with linearly polymerized porphyrin polymers towards versatile electron–donor–acceptor nanohybrids.[83] The target SWNT nanohybrids, which are dispersable in organic media, were prepared employing soluble and redox-inert poly(methylmethacrylate) (PMMA) bearing surface immobilized porphyrins (i.e., $H_2P$-polymer). Absorption spectroscopy provided conclusive evidence for $H_2P$-polymer/SWNTs interactions: the fingerprints of SWNTs and $H_2P$-polymer are discernable throughout the UV, VIS, and NIR part of the solar spectrum. A similar conclusion was also derived from transmission electron microscopy (TEM) and atomic force microscopy (AFM). AFM images illustrate, for example, the debundling of individual SWNTs. An additional feature of $H_2P$-polymer/SWNTs is an intrahybrid charge separation, which has been shown to last for $2.1 \pm 0.1$ μs.

Other stable $H_2P$/SWNT composites were also synthesized by condensing tetraformylporphyrins and diaminopyrenes on SWNTs.[84] In these aggregates, UV-vis and fluorescence spectroscopy helped in determining the degree of interaction between SWNT and $H_2P$. In the electron–donor–acceptor nanohybrids, the Soret and Q-bands of $H_2P$ moieties were significantly broadened and their fluorescence was almost quenched. Apparent extinction coefficients at the Soret bands were decreased to ca. 20% (Scheme 7).

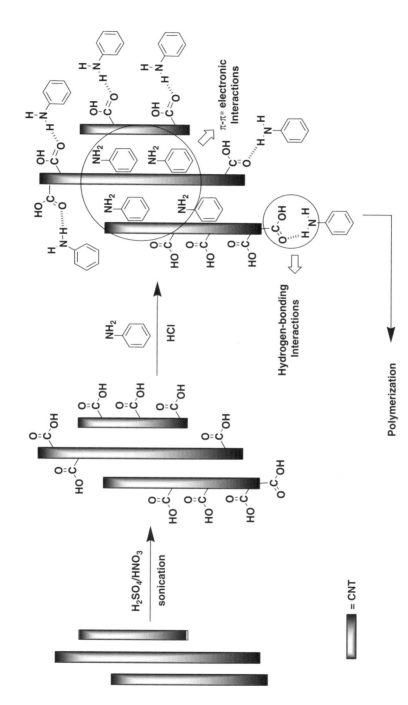

Scheme 6 Schematic representation of the mechanism governing the formation of PANI-ES/MWNTs composites

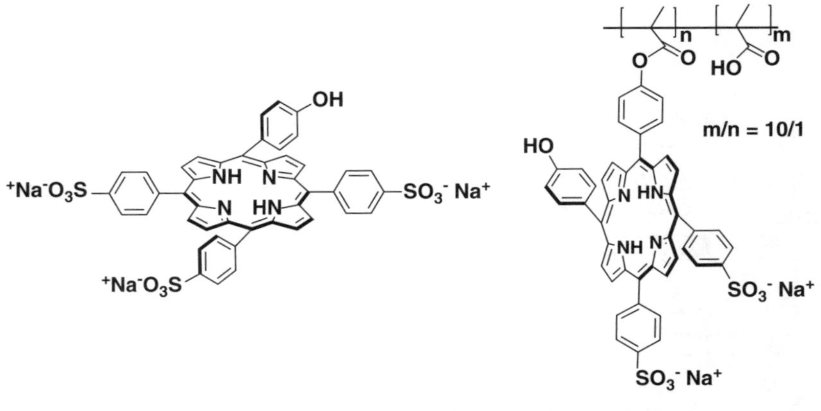

**Scheme 7** *Structures of $H_2P$ (left) and poly($H_2P$) (right)*

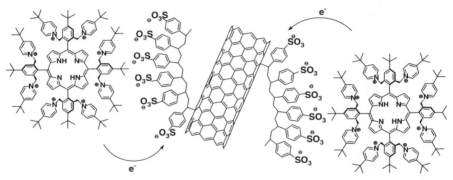

**Scheme 8** *SWNT-PSS$^{n-}$ electrostatic-$H_2P$ nanohybrid*

### 6.5.2 Electrostatic Interactions

A versatile approach towards maintaining the electronic structure of CNTs is the grafting of SWNTs with polymers such as poly(sodium 4-styrenesulfonate) (SWNT-PSS$^{n-}$).[85] Highly stable water dispersable SWNTs were prepared by using this methodology. Indeed, the attached PSS functionalities also assist in exfoliating individual SWNT-PSS$^{n-}$ from the larger bundles and stabilizing them. AFM and TEM analysis corroborated the presence of SWNTs with lengths reaching several micrometers and diameters around 1.2 nm.

In a more sophisticated approach, a coulomb complex formation was achieved with SWNT-PSS$^{n-}$ and an octapyridinium $H_2P$ salt (Scheme 8).[86] Photoexcitation of the octapyridinium $H_2P$ chromophore results in an efficient intrahybrid charge separation event (0.3 ns), which leads, subsequently, to a radical ion pair formation. The charge separation is governed by a large thermodynamic driving force of 0.81 eV. Importantly, the newly formed radical ion pair exhibits a remarkably long lifetime of 14 μs under anaerobic condition, which constitutes one of the longest reported for any CNT ensemble found so far.

### 6.5.3 van der Waals and Complementary Electrostatic Interactions

In the two supramolecular approaches that we have reviewed, carbon nanotubes are evidenced to be essential for the creation of functional device architectures. However, most common loadings of nanotubes in polymer matrix are within the 1–5 wt% due to problems of phase segregation, and more than 50% of the SWNT content is needed for materials with special mechanical performance without comprising the homogeneity of the composite at the nanometer level. The phase segregation between different materials can be avoided by applying the deposition technique often called LBL assembly.[87] This technique is based on the alternating adsorption of monolayer of individual components attracted to each other by electrostatic and van der Waals interactions.

The LBL approach has been successfully applied on CNTs[88] and a number of applications in future nanoscale biosensors[89] and photonic and electronic devices[90] have been highlighted.

The construction of donor–acceptor nanoensembles of soluble carbon nanotubes for photochemical energy conversion systems is probably one of the most interesting applications of the LBL technique. In this approach, developed in Guldi's and Prato's laboratories[91], water soluble SWNTs and MWNTs were obtained by treating them in aqueous solutions of 1-(trimethylammonium

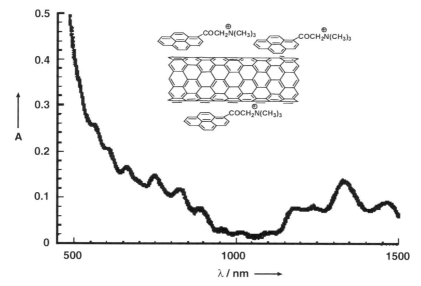

**Figure 27** *Absorption spectra of pyrene$^+$/SWNT in $D_2O$*
(Reprinted from ref. 91a with permission from Wiley-VCH)

acetyl)pyrene (pyrene$^+$), 1-pyreneacetic acid, 1-pyrenecarboxylic acid, 1-pyrenebutyric acid, or 8-hydroxy-1,3,6-pyrenetrisulfonic acid (pyrene$^-$). Excess free pyrene was removed from solution using vigorous centrifugation and the final CNT-($\pi$–$\pi$-interaction)-pyrene$^+$ or CNT-($\pi$–$\pi$-interaction)-pyrene- solids were resuspended in water. The solubility of CNTs in the resulting black suspensions is as high as 0.2 mg ml$^{-1}$ and the stability reaches several months under ambient conditions without showing any apparent precipitation.

The formation and the strength of the $\pi$–$\pi$ interactions is reflected in the instantaneous deactivation of the singlet excited state of pyrene. The Vis-NIR spectrum of the SWNT/pyrene$^+$ in D$_2$O is shown in Figure 27. The typical absorptions of the pristine tubes are easily discernable, with the classical transitions between van Hove singularities of both metallic and semiconducting nanotubes at 400–600 nm and 600–1500 nm respectively.

Once that the surface of CNTs is covered with positively or negatively charged ionic head groups, van der Waals and electrostatic interactions are utilized to complex oppositely charged electron donors. Water-soluble porphyrins (*i.e.*, octapyridinium ZnP/H$_2$P salts or octacarboxylate ZnP/H$_2$P salts)

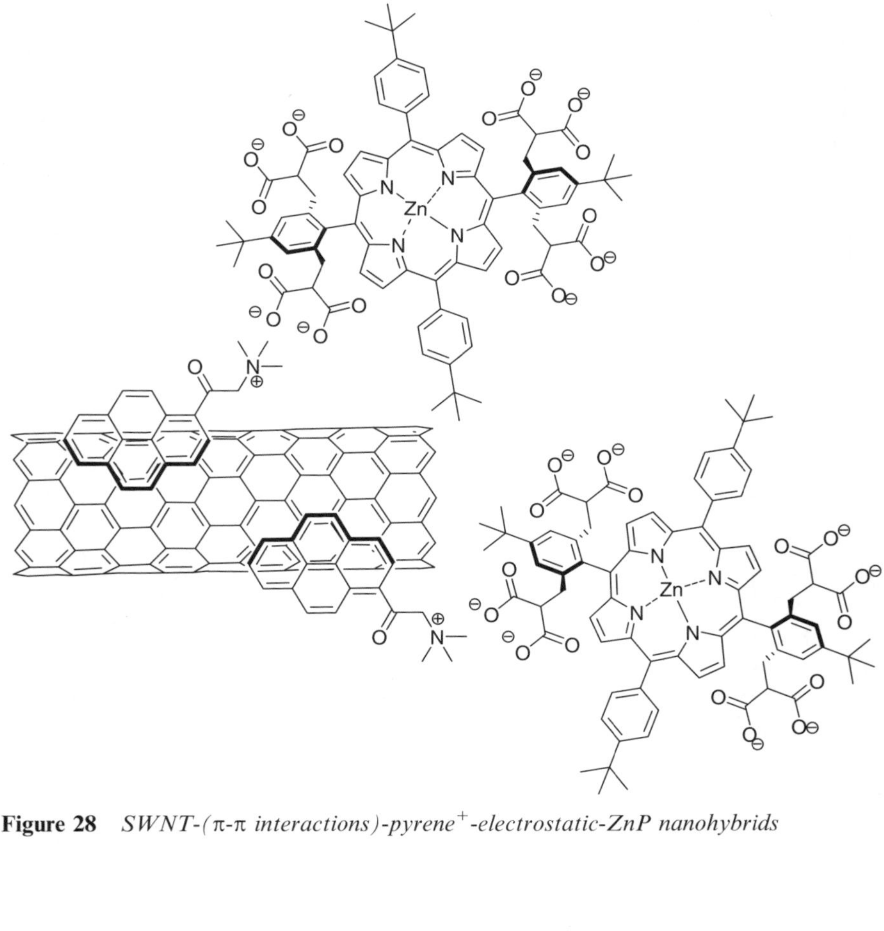

**Figure 28** *SWNT-($\pi$-$\pi$ interactions)-pyrene$^+$-electrostatic-ZnP nanohybrids*

emerged as ideal candidates to realize SWNT-(π–π-interaction)-pyrene$^+$-electrostatic-ZnP or SWNT-(π–π-interaction)-pyrene$^+$-electrostatic-H$_2$P electron-donor–acceptor nanohybrids (Figure 28).

These SWNT/pyrene$^+$ covered with positive charges and electrostatically associated to strong electron-donors that carry negative charges such as porphyrins and phtalocyanines (ZnP$^{8-}$ and H$_2$P$^{8-}$) have been tested in the construction of photoactive surfaces. Photoexcitation of the porphyrin chromophores is succeeded by the formation of radical pairs, namely, reduced SWNTs and oxidized ZnP$^{8-}$. The current generation process is achieved *via* the redox gradient from the reduced SWNTs to the ITO conduction band, while the oxidized donors are reduced at the solid-liquid interface by ascorbate.[91e] It is remarkable that a cell containing a single stack of SWNT/pyrene$^+$/ZnP$^{8-}$, namely, a single layer of SWNT/pyrene$^+$ and a single layer of ZnP$^{8-}$, leads to monochromatic incident-photon-to-photocurrent efficiency (IPCE) of up to 4.2%. Multilayered cells were also constructed and the repetitive growth was followed through absorption spectroscopy. Interestingly, for these 10 stacks a power conversion efficiency of up to 8.5% was observed.

## 6.6 Conclusions and Outlook

Electronic donor–acceptor interactions constitute one of the basic processes used by Nature to transform and store energy. Following biomimetic principles it is possible to carefully design new energy harvesting materials in which energy and electron transfer events are primary processes for which a better understanding is required in order to achieve a control on these phenomena for specific device applications.

Since the discovery of fullerenes it was realized that these new carbon allotropes offered new possibilities for the development of materials for the coming technologies. Although a lot of studies have been carried out on covalently bonded C$_{60}$-based donor–acceptor systems, this is not the case for H-bonding donor–acceptor dyads, which have only been studied during the last recent years. Thus, a wide variety of supramolecular architectures involving the highly directional and selective H-bond between [60]fullerene and a variety of electron-donor units have been described. Interestingly, the electron transfer through a H-bonding network results to be as efficient as that found in the covalent related systems. This finding prompted to the fast development of this new interdisciplinary field of supramolecular fullerenes, which present a closer resemblance to the natural photosynthetic process than previous covalently bonded supermolecules.

The imagination of chemists has produced a wide variety of H-bonding C$_{60}$-based donor–acceptor ensembles in which the induced supramolecular order results, for example, in improved photovoltaic properties.

H-bonded supramolecular C$_{60}$-based polymers is an emerging field with an enormous potential applicability. Because of the ease synthetic methodology, as well as, the intrinsic reversibility of the engaged weak forces, a wide variety of fullerene containing supramolecular polymers can be figured out. In this regard, although a few outstanding examples have recently been reported, more work is needed in order to fully exploit the potentiality of this novel approach for the fabrication of structurally organized optoelectronic devices.

The same basic principles of supramolecular organization can be also applied to carbon nanotubes. Although considerably less studied, the CNT-based donor–acceptor ensembles reveal that these new carbon allotropes are as efficient as the parent fullerenes in electron transfer events. In this regard, a remarkable advantage of the supramolecular functionalization, in contrast to the covalent functionalization, of CNTs is that they preserve the π-system structure and, therefore, the electronic properties of pristine CNTs.

In summary, H-bonding fullerene and carbon nanotube based supramolecular architectures emerge as a readily available and highly efficient approach to advance our understanding of the molecular principles that Nature deploys for energy harvesting, which offer new possibilities for their use in molecular electronics as well as in the emerging nanotechnologies.

## Acknowledgments

The authors gratefully acknowledge the outstanding research labour done by all the colleagues in this field and whose works appear cited in the references. A large contribution to the CNTs donor–acceptor nanohybrids referred to in this chapter derives from works done by Prof Dirk. M. Guldi. We specially thank him for his indirect but important contribution to this chapter.

This work was supported by the MCYT of Spain (CTQ2005-02609/BTQ), the Comunidad de Madrid (S-0505/PPQ/0225) and the Universidad Complutense de Madrid (PR1/05-13336).

## References

1. (a) T.J. Meyer, *Acc. Chem. Res.*, 1989, **22**, 163; (b) M.R. Wasielewski, *Chem. Rev.*, 1992, **92**, 435; (c) H. Kurreck and M. Huber, *Angew. Chem. Int. Ed. Engl.*, 1995, **34**, 849; (d) D. Gust, T.A. Moore and A.L. Moore, *Acc. Chem. Res.*, 2001, **34**, 40; (e) D. Holten, D.F. Bocian and J.S. Lindsey, *Acc. Chem. Res.*, 2002, **35**, 57; (f) D.M. Adams, L. Brus, E.D. Chidsey, S. Creager, C. Creutz, C.R. Kagan, P.V. Kamat, M. Lieberman, S. Lindsey, R.A. Marcus, R.M. Metzger, M.E. Michel-Beyerle, J.R. Miller, M.D. Newton, D.R. Rolison, O. Sankey, K.S. Schanze, J. Yardley and X. Zhu, *J. Phys. Chem. B.*, 2003, **107**, 6668; (g) N. Armaroli, *Photochem. Photobiol. Sci.*, 2003, **2**, 73; (h) H. Imahori, Y. Mori and Y. Matano, *J. Photochem. Photobiol.*, 2003, **C4**, 51; (i) *Energy Harvesting Materials*, D.I. Andrews, (ed), World Scientific, Singapore, 2005.

2. (*a*) A. Yazdani and C.M. Lieber, *Nature*, 1999, **401**, 1683; (*b*) H.L. Anderson, *Angew. Chem. Int. Ed.*, 2000, **39**, 2451; (*c*) Special issue on Nanotechnology, *Sci. Am.*, 2001, **285**, 32; (*d*) S. Bunk, *Nature*, 2001, **410**, 127; (*e*) V. Balzani, M. Venturi and A. Credi, *Molecular Devices and Machines: A Journey into the Nanoworld*, Wiley-VCH Verlag GmbH & Co., Weinheim, 2003.
3. (*a*) J. Barber, *Nature*, 1988, **333**, 114; (*b*) *Supramolecular Chemistry*, V. Balzani and L. de Cola (eds) NATO ASI Series, Kluwer Academic Publishers, Dordrech, 1992; (*c*) G. McDermott, S.M. Priece, A.A. Freer, A.M. Hawthornthwaite-Lawless, M.Z. Papiz, R.J. Cogdell and N.W. Isaacs, *Nature*, 1995, **374**, 517.
4. (*a*) L.F. Lindoy and I.M. Atkinson, *Self-Assembly*, in *Supramolecular Systems*, Royal Society of Chemistry, Cambridge, UK, 2000; (*b*) A.J. Goshe, I.M. Steele, C. Ceccarelli, A.L. Rheingold and B. Bosnich, *Proc. Natl. Acad. Sci.*, 2002, **99**, 4823; (*c*) S. Zhang, *Materials Today*, 2003, **6**, 20; (*d*) S. Zhang, *Nature Biotechnology*, 2003, **21**, 1171; (*e*) D.M. Guldi, F. Zerbetto, V. Georgakilas and M. Prato, *Acc. Chem. Res.*, 2005, **38**, 38; (*f*) M.W. Hosseini, *Chem. Commun.*, 2005, 5825; (*g*) M.D. Ward, *Chem. Commun.*, 2005, 5838.
5. (*a*) G.A. Jeffrey, *An introduction to hydrogen bonding*, Oxford University Press, Oxford, 1997; (*b*) R.P. Sijbesma and E.W. Meijer, *Chem. Commun.*, 2003, 5.
6. (*a*) C. Schmunk and W. Wienand, *Angew. Chem. Int. Ed.*, 2001, **40**, 4363; (*b*) B.P. Orner, X. Salvatella, J. Sánchez-Quesada, J. de Mendoza, E. Giralt and A.D. Hamilton, *Angew. Chem. Int. Ed.*, 2002, **41**, 117; (*c*) J. Otsuki, M. Takatsuki, M. Kaneko, H. Miwa, T. Takido, M. Seno, K. Okamoto, H. Imahori, M. Fujitsuka, Y. Araki, O. Ito and S. Fukuzumi, *J. Phys. Chem. A.*, 2003, **107**, 379.
7. (*a*) G.M. Whitesides, E.E. Simanek, J.P. Mathias, C.T. Seto, D.N. Chin, M. Mammen and D.M. Gordon, *Acc. Chem. Res.*, 1995, **28**, 37; (*b*) F.H. Beijer, R.P. Sijbesma, J.A.J.M. Vekemans, E.W. Meijer, H. Kooijman and A.L. Spek, *J. Org. Chem.*, 1996, **61**, 6371; (*c*) S.C. Zimmermann and P.S. Corbin, *Struct. Bonding*, Springer, Berlin, 2000, **96**, 63–94.
8. (*a*) F.H. Beijer, H. Kooijman, A.L. Spek, R.P. Sijbesma and E.W. Meijer, *Angew. Chem. Int. Ed.*, 1998, **37**, 75; (*b*) S.H.M. Söntjens, R.P. Sijbesma, M.H.P. van Genderen and E.W. Meijer, *J. Am. Chem. Soc.*, 2000, **122**, 7487.
9. (*a*) P. Harris, *Carbon Nanotubes and Related Structures: New Materials for the Twenty-First Century*, Cambridge University Press, Cambridge, 2001; (*b*) M.S. Dresselhaus, G. Dresselhaus and P. Avouris, *Carbon Nanotubes: Synthesis, Structure, Properties and Applications*, Springer, Berlin, 2001; (*c*) Special issue on *Functionalized Fullerene Materials*, M. Prato and N. Martín (eds), *J. Mater. Chem.* 2002, **12**, 1931; (*d*) D.M. Guldi and N. Martín, *From Synthesis to Optoelectronic Properties*, Kluwer Academic Publishers Dordrecht, The Netherlands, 2002; (*e*) Special issue on *Carbon Nanotubes*, R.C. Haddon, (ed), *Acc. Chem. Res.*, 2002, **35**, 997; (*f*) S. Reich,

C. Thomsen and J. Maultzsch, *Carbon Nanotubes Basic Concepts and Physical Properties*, VCH, Weinheim, 2004.
10. (*a*) Special issue on *Supramolecular Chemistry of Fullerenes*, N. Martín and J.F. Nierengarten (eds), *Tetrahedron Series-in-Print*, 2006, **62**(9), 1905–2132; (*b*) D.M. Guldi and N. Martín, *J. Mater. Chem.*, 2002, **12**, 1978.
11. For a recent review on H-bonded $C_{60}$ assemblies, see: L. Sánchez, N. Martín and D.M. Guldi, *Angew. Chem. Int. Ed.*, 2005, **44**, 5374.
12. For recent reviews on $C_{60}$-Donor compositions, see: (a) N. Martín, L. Sánchez, B.M. Illescas and I. Pérez, *Chem. Rev.*, 1998, **98**, 2527; (b) D.M. Guldi, *Chem. Commun.*, 2000, 321; (c) D. M. Guldi, *Chem. Soc. Rev.*, 2002, **31**, 22; (d) H. Imahori, *J. Phys. Chem B.*, 2004, **108**, 6130; (e) L. Sánchez, M.A. Herranz and N. Martín, *J. Mater. Chem.*, 2005, **15**, 1409; (f) J. Roncali, *Chem. Soc. Rev.*, 2005, **34**, 483.
13. (*a*) A.J. Myles and N.R. Branda, *J. Am. Chem. Soc.*, 2001, **123**, 177; (*b*) J.L. Sessler, M. Sathiosatham, C.T. Brown, T.A. Rhodes and G. Wiederrecht, *J. Am. Chem. Soc.*, 2001, **123**, 3655; (*c*) A.P.H.J. Schenning, J.V. Herrikhuyzen, P. Jonkheijm, Z. Chen, F. Würthner and E.W. Meijer, *J. Am. Chem. Soc.*, 2002, **124**, 10252; (*d*) N.H. Damrauer, J.M. Hodgkiss, J. Rosenthal and D.G. Nocera, *J. Phys. Chem. B.*, 2004, **108**, 6315.
14. (*a*) D.M. Guldi, J. Ramey, M.V. Martínez-Díaz, A. de la Escosura, T. Torres, T. da Ros and M. Prato, *Chem. Commun.*, 2002, 2774; (*b*) M.V. Martínez-Díaz, N.S. Fender, M.S. Rodríguez-Morgade, M. Gómez-López, F. Diederich, L. Echegoyen, J.F. Stoddart and T. Torres, *J. Mater. Chem.*, 2002, **12**, 2095.
15. D.M. Guldi, A. Gouloumis, P. Vázquez and T. Torres, *Chem. Commun.*, 2002, 2056.
16. (*a*) H. Yamada, H. Imahori, Y. Nishimura, I. Yamazaki, T.K. Ahn, S.K. Kim, D. Kim and S. Fukuzumi, *J. Am. Chem. Soc.*, 2003, **125**, 9129; (*b*) K. Ohkubo, H. Kotani, J. Shao, Z. Ou, K.M. Kadish, G. Li, R.K. Pandey, M. Fusitsuka, O. Ito, H. Imahori and S. Fukuzumi, *Angew. Chem., Int. Ed.*, 2004, **43**, 853 and references therein.
17. N. Watanabe, N. Kihara, Y. Forusho, T. Takata, Y. Araki and O. Ito, *Angew. Chem. Int. Ed.*, 2003, **42**, 681.
18. H. Imahori, M.E. El-Khouly, M. Fujitsuka, O. Ito, Y. Sakata and S. Fukuzumi, *J. Phys. Chem. A.*, 2001, **105**, 325.
19. (*a*) A.S. Sandanayaka, N. Watanabe, K.-I. Ikeshita, Y. Araki, N. Kihara, Y. Furusho, O. Ito and T. Tanaka, *J. Phys. Chem. B.*, 2005, **109**, 2516; (*b*) H. Sasabe, K.-I. Ikeshita, G.A. Rajkumar, N. Watanabe, N. Kihara, Y. Furusho, K. Mizumo, A. Ogawa and T. Tanaka, *Tetrahedron*, 2006, **62**, 1988.
20. H. Sasabe, N. Kihara, Y. Forusho, K. Mizuno, A. Ogawa and T. Takata, *Org. Lett.*, 2004, **6**, 3957.
21. F. D'Souza, G.R. Deviprasad, M.E. El-Khouly, M. Fujitsuka and O. Ito, *J. Am. Chem. Soc.*, 2001, **123**, 5277.

22. (a) F. D'Souza, G.R. Deviprasad, M.E. Zadler, M.E. El-Khouly, F. Fujitsuka and O. Ito, *J. Phys. Chem. A.*, 2003, **107**, 4801; (b) F. D'Souza, R. Chitta, S. Gadde, M.E. Zandler, A.S.D. Sandayanaka, Y. Araki and O. Ito, *Chem. Commun.*, 2005, 1279.
23. F. D'Souza, R. Chitta, S. Gadde, M.E. Zandler, A.L. McCarty, A.S.D. Sandanayaka, Y. Araki and O. Ito, *Chem. Eur. J.*, 2005, **11**, 4416.
24. J.L. Sessler, J. Jayawickramarajah, A. Gouloumis, T. Torres, D.M. Guldi, S. Maldonado and K.J. Stevenson, *Chem. Commun.*, 2005, 1892.
25. (a) N. Solladié, M.E. Walther, M. Gross, T.M. Figueira Duarte, C. Bourgogne and J.-F. Nierengarten, *Chem. Commun.*, 2003, 2412; (b) N. Solladié, M.E. Walther, H. Herschbach, E. Leize, A. Van Dorsselaer, T.M. Figueira Duarte and J.-F. Nierengarten, *Tetrahedron*, 2006, **62**, 1979.
26. M. Gutiérrez-Nava, H. Nierengarten, P. Masson, A. Van Dorsselaer and J.-F. Nierengarten, *Tetrahedron Lett.*, 2003, **44**, 3043.
27. (a) J.H. Bourroughes, D.D.C. Bradley, A.R. Brown, R.N. Marks, K. MacKay, R.H. Friend, P.L. Burns and A.B. Holmes, *Nature*, 1990, **347**, 539; (b) S.-H. Jin, M.-Y. Kim, J.Y. Kim, K. Lee and Y.-S. Gal, *J. Am. Chem. Soc.*, 2004, **126**, 2474.
28. (a) N.S. Sariciftci, L. Smilowitz, A.J. Heeger and F. Wudl, *Science*, 1992, **258**, 1474; (b) G. Yu, J. Gao, J.C. Hummelen, F. Wudl and A.J. Heeger, *Science*, 1995, **270**, 1789; (c) S.E. Shaheen, C.J. Brabec, N.S. Sariciftci, F. Padinger, T. Fromherz and J.C. Hummelen, *Appl. Phys. Lett.*, 2001, **78**, 841; (d) M.M. Wienk, J.M. Kroon, W.J.H. Verhees, J. Knol, J.C. Hummelen, P.A. van Hal and R.A.J. Janssen, *Angew. Chem. Int. Ed.*, 2003, **42**, 3371; (e) M. Svensson, F. Zhang, S.C. Veenstra, W.J.H. Verhees, J.C. Hummelen, J.M. Kroon, O. Inganäs and M.R. Andersson, *Adv. Mater.*, 2003, **15**, 988; (f) C. Winder and N.S. Sariciftci, *J. Mater. Chem.*, 2004, **14**, 1077; (g) I. Riedel, E. von Hauff, J. Parisi, N. Martín, F. Giacalone and V. Dyakonov, *Adv. Funct. Mater.*, 2005, **15**, 1979; (h) G. Li, V. Shrotriya, J. Huang, Y. Yao, T. Moriarty, K. Emery and Y. Yang, *Nature Mater.*, 2005, **4**, 864; (i) W. Ma, C. Yg, X. Gong, K. Lee and A.J. Heeger, *Adv. Funct. Mater.*, 2005, **15**, 1617.
29. (a) M.A. Fox, *Acc. Chem. Res.*, 1999, **32**, 201; (b) J.-L. Brédas, D. Beljonne, V. Coropceanu and J. Cornil, *Chem. Rev.*, 2004, **104**, 4971.
30. (a) E.H.A. Beckers, P.A. van Hal, A.P.H.J. Schenning, A. El-ghayoury, E. Peeters, M.T. Rispens, J.C. Hummelen, E.W. Meijer and R.A.J. Janssen, *J. Mater. Chem.*, 2002, **12**, 2054; (b) M.T. Rispens, L. Sánchez, E.H.A. Beckers, P.A. van Hal, A.P.H.J. Schenning, A. El-ghayoury, E. Peeters, E.W. Meijer, R.A.J. Janssen and J.C. Hummelen, *Synth. Met.*, 2003, **801–803**, 135.
31. (a) E. Peeters, P.A. van Hal, J. Knol, C.J. Brabec, N.S. Sariciftci, J.C. Hummelen and R.A.J. Janssen, *J. Phys. Chem. B.*, 2000, **104**, 10174; (b) P.A. van Hal, S.C.J. Meskers and R.A.J. Janssen, *Appl. Phys. Lett. A*, 2004, **79**, 41.

32. For a recent review on PV materials by using the $C_{60}$/oligomer approach, see: J.L. Segura, N. Martín, D.M. Guldi, *Chem. Soc. Rev.*, 2005, **34**, 31.
33. E.H.A. Beckers, A.P.H.J. Schenning, P.A. van Hal, A. El-ghayoury, L. Sánchez, J.C. Hummelen, E.W. Meijer and R.A.J. Janssen, *Chem. Commun.*, 2002, 2888.
34. M. Elhabiri, A. Trabolsi, F. Cardinali, U. Hahn, A.-M. Albrecht-Gary and J.-F. Nierengarten, *Chem. Eur. J.*, 2005, **11**, 4793.
35. (*a*) D. Philp and J.F. Stoddart, *Angew. Chem. Int. Ed. Engl.*, 1996, **35**, 1155; (*b*) H.R. Pouretedal and M. Shamsipur, *J. Chem. Eng. Data*, 1998, **43**, 742.
36. U. Hahn, M. Elhabiri, A. Trabolsi, H. Herschbach, E. Leize, A. Van Dorsselaer, A.-M. Albrecht-Gary and J.-F. Nierengarten, *Angew. Chem. Int. Ed.*, 2005, **44**, 5338.
37. H.C. Kolb, M.G. Finn and K.B. Sharpless, *Angew. Chem. Int. Ed.*, 2001, **40**, 2004.
38. N.D. McClenaghan, Z. Grote, K. Darriet, M. Zimine, R.M. Williams, L. De Cola and D.M. Bassani, *Org. Lett.*, 2005, **7**, 807.
39. C.-H. Huang, N.D. McClenaghan, A. Kuhn, J.W. Hofstraat and D.M. Bassani, *Org. Lett.*, 2005, **7**, 3409.
40. (*a*) J.-M. Lehn, Supramolecular Chemistry, Concepts and Perspectives, VCH, Weinheim (Germany) (1995); (*b*) *Comprehensive Supramolecular Chemistry*; Vols. 1–11; Pergamon: Oxford, UK, 1996; J.L. Atwood, J.E.D. Davies, D.D. MacNicol, F. Vögtle (eds); (*c*) J.L. Segura and N. Martín, *Angew. Chem. Int. Ed.*, 2001, **40**, 1372; (*d*) *TTF Chemistry: Fundamentals and Applications of Tetrathiafulvalene*, J. Yamada and T. Sugimoto (eds); Kodansha-Springer, (2004).
41. N. Martín, L. Sánchez, C. Seoane, E. Ortí, P.M. Viruela and R. Viruela, *J. Org. Chem.*, 1998, **63**, 1268.
42. (*a*) N. Martín, L. Sánchez, M.A. Herranz and D.M. Guldi, *J. Phys. Chem. A*, 2000, **104**, 4648; (*b*) D.M. Guldi, S. González, N. Martín, A. Antón, J. Garín and J. Orduna, *J. Org. Chem.*, 2000, **65**, 1978; (*c*) J.L. Segura, E.M. Priego, N. Martín, C. Luo and D.M. Guldi, *Org. Lett.*, 2000, **2**, 4021; (*d*) G. Kodis, P.A. Liddell, L. de la Garza, A.L. Moore, T.A. Moore and D. Gust, *J. Mater. Chem.*, 2002, **12**, 2100; (*e*) N. Martín, L. Sánchez and D.M. Guldi, *Chem. Commun.*, 2000, 113; (*f*) M.A. Herranz, N. Martín, J. Ramey and D.M. Guldi, *Chem. Commun.*, 2002, 2968; (*g*) S. González, N. Martín and D.M. Guldi, *J. Org. Chem.*, 2003, **68**, 779; (*h*) F. Giacalone, J.L. Segura, N. Martín and D.M. Guldi, *J. Am. Chem. Soc.*, 2004, **126**, 5340; (*i*) L. Sánchez, M. Sierra, N. Martín, D.M. Guldi, M.W. Wienk and R.A.J. Janssen, *Org. Lett.*, 2005, **7**, 1691.
43. M. Segura, L. Sánchez, J. de Mendoza, N. Martín and D.M. Guldi, *J. Am. Chem. Soc.*, 2003, **125**, 15093.
44. M.C. Díaz, B.M. Illescas, N. Martín, J.F. Stoddart, M.A. Canales, J. Jiménez-Barbero, G. Sarovad and D.M. Guldi, *Tetrahedron*, 2006, **62**, 1998.
45. (*a*) M. Rumi, J.E. Ehrlich, A.A. Heikal, J.W. Perry, S. Barlow, Z. Hu, D. McCord-Maughon, T.C. Parker, H. Röckel, S. Thayumanavan, S.R.

Marder, D. Beljonne and J.-L. Brédas, *J. Am. Chem. Soc.*, 2000, **122**, 9500; (*b*) M. Thelakkat, *Macromol. Mater. Eng.*, 2002, **287**, 442; (*c*) H.Y. Woo, J.W. Hong, B. Liu, A. Mikhailovsky, D. Korystov and G.C. Bazan, *J. Am. Chem. Soc.*, 2005, **127**, 820.
46. (*a*) Special Issue on Organic Electronics, *Chem. Mater.*, 2004, **16**, 4381; (*b*) R.A. Wassel and C.B. Gorman, *Angew. Chem. Int. Ed.*, 2004, **43**, 5120.
47. A.S.D. Sandanayaka, H. Sasabe, Y. Araki, Y. Forusho, O. Ito and T. Takata, *J. Phys. Chem. B.*, 2004, **108**, 5145.
48. L. Sánchez, M.T. Rispens and J.C. Hummelen, *Angew. Chem. Int. Ed.*, 2002, **41**, 838.
49. L. Brunsveld, B.J.B. Folmer, E.W. Meijer and R.P. Sijbesma, *Chem. Rev.*, 2001, **101**, 4071.
50. A.P.H.J. Schenning, P. Jonkheijm, E. Peeters and E.W. Meijer, *J. Am. Chem. Soc.*, 2001, **123**, 409.
51. (*a*) L. Dai, J. Lu, B. Matthews and A.W.H. Mau, *J. Phys. Chem. B.*, 1998, **102**, 4049; (*b*) L. Lu, L. Dai and A.W.H. Mau, *Acta Polym.*, 1998, **49**, 371.
52. (*a*) S. Morita, A.A. Zakhidov and K. Yoshino, *Solid State Commun.*, 1992, **82**, 249; (*b*) Y. Wei, J. Tian, A.G. MacDiarmid, J.G. Masters, A.L. Smith and D. Li, *J. Chem. Soc., Chem. Commun.*, 1993, 603; (*c*) S. Morita, A.A. Zakhidov, T. Kawai, H. Araki and K. Yoshino, *J. Phys. Condens. Mater.*, 1993, **5**, 2103; (*d*) A.A. Zakhidov, H. Araki, K. Tada and T.K. Yoshino, *Synth. Met.*, 1996, **77**, 127.
53. X.-D. Huang, S. H. Goh, S. Y. Lee, C. Hon and C.H.A. Huan, *Macromol. Chem. Phys.*, 2000, **201**, 281.
54. (*a*) H.W. Goh, S.H. Goh and G.Q. Xu, *J. Polym. Sci.: Part A: Polym. Chem.*, 2002, **40**, 1157; (*b*) H.W. Goh, S.H. Goh and G.Q. Xu, *J. Polym. Sci.: Part A: Polym. Chem.*, 2002, **40**, 4316; (*c*) H.L. Huang, S.H. Goh, J.W. Zheng, D.M.Y. Lai and C.H.A. Huan, *Langmuir*, 2003, **19**, 5332.
55. X.-D. Huang, S.H. Goh and S.Y. Lee, *Macromol. Chem. Phys.*, 2000, **201**, 2660.
56. X.-D. Huang and S. H. Goh, *Polymer*, 2002, **43**, 1417.
57. (*a*) X.-D. Huang, *Macromolecules*, 2000, **33**, 8894; (*b*) T. Song, S.H. Goh and S.Y. Lee, *Macromolecules*, 2002, **35**, 4133.
58. M. Wang, K.P. Pramoda and S.H. Goh, *Chem. Mater.*, 2004, **16**, 3452.
59. (*a*) Z. Shi, Y. Li, H. Gong, M. Liu, S. Xiao, H. Liu, H. Li, S. Xiao and D. Zhu, *Org. Lett.*, 2002, **4**, 1179; (*b*) H. Fang, S. Wang, S. Xiao, J. Yang, Y. Li, Z. Shi, H. Li, H. Liu, S. Xiao and D. Zhu, *Chem. Mater.*, 2003, **15**, 1593.
60. N. Fujita, T. Yamashita, M. Asai and S. Shinkai, *Angew. Chem. Int. Ed. Engl.*, 2005, **44**, 1257.
61. J.K. Mwaura, M.R. Pinto, D. Witker, N. Ananthakrishnan, K.S. Schanze and J.R. Reynolds, *Langmuir*, 2005, **21**, 10119.
62. F. Diederich and M. Gómez-López, *Chem. Soc. Rev.*, 1999, **28**, 263.

63. (a) T. Nishimura, K. Takatani, S. Sakurai, K. Maeda and E. Yashima, *Angew. Chem. Int. Ed.*, 2002, **41**, 3602; (b) T. Nishimura, K. Maeda, S. Ohsawa and H. Yashima, *Chem. Eur. J.*, 2005, **11**, 1181.
64. T. Nishimura, S. Ohsawa, K. Maeda and E. Yashima, *Chem. Commun.*, 2004, 646.
65. (a) S. Samal, B.-J. Choi and K.E. Geckeler, *Chem. Commun.*, 2000, 1373; (b) D. Nepal, S. Samal and K.E. Geckeler, *Macromolecules*, 2003, **36**, 3800; (c) Y. Liu, Y.-W. Chen and H.-X. Zou, *Macromolecules*, 2005, **38**, 5838.
66. Y. Liu, H. Wang, P. Liang and H.-Y. Zhang, *Angew. Chem. Int. Ed.*, 2004, **43**, 2690.
67. T. Haino, Y. Matsumoto and Y. Fukazawa, *Y. J. Am. Chem. Soc.*, 2005, **127**, 8936.
68. M. Shirakawa, N. Fujita and S. Shinkai, *J. Am. Chem. Soc.*, 2003, **125**, 9902.
69. M. Shirakawa, S.-I. Kawano, N. Fujita, K. Sada and S. Shinkai, *J. Org. Chem.*, 2003, **68**, 5037.
70. (a) A. Javea, J. Guo, Q. Wang, M. Ludstrom and H. Dai, *Nature*, 2003, **424**, 654; (b) M. Zheng, A. Jagota, E.D. Semke, B.A. Diner, R.S. Mclean, S.R. Lustig, R.E. Richardson and N.G. Tassi, *Nat. Mater.*, 2003, **2**, 338; (c) R. Krupke, F. Hennrich, H. Von Lohnenysen and M.M. Kappes, *Science*, 2003, **301**, 344; (d) J.A. Misewich, P. Avouris, R. Martel, J.C. Tsang, S. Heinz and J. Tersoff, *Science*, 2003, **300**, 783; (e) J. Li, Q. Ye, A. Cassell, H.T. Ng, R. Stevens and J. Han, *Appl. Phys. Lett.*, 2003, **82**, 2491.
71. (a) K.-I. Nakayama, Y. Asakura and M. Yokoyama, *Molec. Cryst. and Liq. Cryst.*, 2004, **424**, 217; (b) G.M.A. Rahman, D.M. Guldi, R. Cagnoli, A. Mucci, L. Schenetti, L. Vaccari and M. Prato, *J. Am. Chem. Soc.*, 2005, **127**, 10051; (c) B.J. Landi, S.L. Castro, H.J. Ruf, C.M. Evans, S.G. Bailey and R.P. Raffaelle, *Sol. En. Mat. Sol. Cells*, 2005, **87**, 733; (d) D.M. Guldi, G.M.A. Rahman, M. Prato, N. Jux, S. Qin and W. Ford, *Angew. Chem., Int. Ed.*, 2005, **44**, 2015; (e) B.J. Landi, R.P. Raffaelle, S.L. Castro and S.G. Bailey, *Progr. in Photovolt.*, 2005, **13**, 165; (f) J.A. Rud, L.S. Lovell, J.W. Senn, Q. Qiao and J.T. Mcleskey, *J. Mat. Science*, 2005, **40**, 1455.
72. For recent reviews, see: (a) J.L. Bahr and J.M. Tour, *J. Mater. Chem.*, 2002, **12**, 1952; (b) A. Hirsch, *Angew. Chem. Int. Ed.*, 2002, **41**, 1853; (c) S. Banerjee, M.G.C. Kahn, S.S. Wong, *Chem. Eur. J.*, 2003, **9**, 1898; (d) D. Tasis, N. Tagmatarchis, V. Georgakilas and M. Prato, *Chem. Eur. J.*, 2003, **9**, 4000; (e) Ch. A. Dyke and J.M. Tour, *Chem. Eur. J.*, 2004, **10**, 812; (f) D.M. Guldi, G.M.A. Rahman, F. Zerbetto and M. Prato, *Acc. Chem. Res.*, 2005, **38**, 871; (g) S. Banerjee, T. Hemraj-Benny and S.S. Wong, *Adv. Mater.*, 2005, **17**, 17.
73. M.A. Hamon, J. Chen, H. Hu, Y. Chen, A. Rao, P.C. Eklund and R.C. Haddon, *Adv. Mater.*, 1999, **11**, 834.

74. (a) S.A. Curran, P.M. Ajayan, W.J. Blau, D.L. Carroll, J.N. Coleman, A.B. Dalton, A.P. Davey, A. Drury, B. McCarthy, S. Maier and A. Stevens, *Adv. Mater.*, 1998, **10**, 1091; (b) H. Ago, K. Petritsch, M.S.P. Shaffer, A.H. Windle and R.H. Friend, *Adv. Mater.*, 1999, **11**, 1281; (c) E. Kymakis and G.A. Amaratunga, *J. Appl. Phys. Lett.*, 2002, **80**, 112.
75. (a) J.N. Coleman, A.B. Dalton, S. Curran, A. Rubio, A.P. Davey, A. Drury, B. McCarthy, B. Lahr, P.M. Ajayan, S. Roth, R.C. Barklie and W.J. Blau, *Adv. Mater.*, 2000, **12**, 213; (b) A. Star, J.F. Stoddart, D. Steuerman, M. Diehl, A. Boukai, E.W. Wong, X. Yang, S.-W. Chung, H. Choi and J.R. Heath, *Angew. Chem. Int. Ed.*, 2001, **40**, 1721.
76. A. Star, Y. Liu, K. Grant, L. Ridvan, J.F. Stoddart, D.W. Steuerman, M.R. Diehl, A. Boukai and J.R. Heath, *Macromolecules*, 2003, **36** 553.
77. M.J. O'Connell, P. Boul, L.M. Ericson, C. Huffman, Y. Wang, E. Haroz, C. Kuper, J. Tour, K.D. Ausman and R.E. Smalley, *Chem. Phys. Lett.*, 2001, **342**, 65.
78. Z. Tang and H. Xu, *Macromolecules*, 1999, **32**, 2569.
79. Wang, G.-C. Zhao and X.-W. Wei, *Chem. Lett.*, 2005, **34**, 518.
80. J. Chen, H. Liu, W.A. Weimer, M.D. Halls, D.H. Waldeck and G.C. Walker, *J. Am. Chem. Soc.*, 2002, **124**, 9034.
81. (a) Z.X. Wei, M.X. Wan, T. Lin and L.M. Dai, *Adv. Mater.*, 2003, **15**, 136; (b) G.B. Blanchet, C.R. Fincher and F. Gao, *Appl. Phys. Lett.*, 2003, **82**, 90.
82. T.-M. Wu, Y.-W. Lin and C.-S. Liao, *Carbon*, 2005, **43**, 734.
83. D.M. Guldi, H. Taieb, G.M.A. Rahman, N. Tagmatarchis and M. Prato, *Adv. Mater.*, 2005, **17**, 871.
84. A. Satake, Y. Miyajima and Y. Kobuke, *Chem. Mater.*, 2005, **17**, 716.
85. S. Qin, D. Qin, W.T. Ford, J.E. Herrera, D.E. Resasco, S.M. Bachilo and R.B. Weisman, *Macromolecules*, 2004, **37**, 3965.
86. D.M. Guldi, G.M.A. Rahman, J. Ramey, M. Marcaccio, D. Paolucci, F. Paolucci, S. Qin, W.T. Ford, D. Balbinot, N. Jux, N. Tagmatarchis and M. Prato, *Chem. Commun.*, 2004, 2034.
87. (a) G. Decher, *Science*, 1997, **277**, 1232; (b) A. Wu, D. Yoo, J.K. Lee and M.F. Rubner, *J. Am. Chem. Soc.*, 1999, **121**, 4883; (c) A.A. Mamedov, A. Belov, M. Giersig, M.M. Mamedova and N.A. Kotov, *J. Am. Chem. Soc.*, 2001, **123**, 7738.
88. (a) A.A. Mamedov, N.A. Kotov, M. Prato, D.M. Guldi, J.M. Wicksted and A. Hirsch, *Nature Mater.*, 2002, **1**, 190; (b) B. Kim, H. Park and W.M. Sigmund, *Langmuir*, 2003, **19**, 2525; (c) A.B. Artyukhin, O. Bakajin, P. Stroeve and A. Noy, *Langmuir*, 2004, **20**, 1442; (d) B. Kim and W.M. Sigmund, *Langmuir*, 2004, **20**, 8239; (e) Sh. Qin, D. Qin, W.T. Ford, J.E. Herrera and D.E. Resasco, *Macromolecules*, 2004, **37**, 9963.
89. (a) M. Guo, J. Chen, L. Nie and S. Yao, *Electrochimica Acta*, 2004, **49**, 2637; (b) L. Qian and X. Yang, *Electrochemistry Commun.*, 2005, **7**, 547; (c) M. Zhang, K. Gong, H. Zhang and L. Mao, *Biosensors and Bioelectronics*, 2005, **20**, 1270.

90. (*a*) H. Xin and A.T. Woolley, *J. Am. Chem. Soc.*, 2003, **125**, 8710; (*b*) A.-H. Bae, T. Hatano, N. Nakashima, H. Murakami and S. Shinkai, *Org. Biomol. Chem.*, 2004, **2**, 1139.
91. (*a*) D.M. Guldi, G.M.A. Rahman, N. Jux, N. Tagmatarchis and M. Prato, *Angew. Chem. Int. Ed.*, 2004, **43**, 5526; (*b*) D.M. Guldi and M. Prato, *Chem. Commun.*, 2004, 2517; (*c*) D.M. Guldi, G.M.A. Rahman, N. Jux, D. Balbinot, N. Tagmatarchis and M. Prato, *Chem. Commun.*, 2005, 2038; (*d*) D.M. Guldi, G.M.A. Rahman, N. Jux, D. Balbinot, U. Hartnagel, N. Tagmatarchis and M. Prato, *J. Am. Chem. Soc.*, 2005, **128**, 9830; (*e*) D.M. Guldi, *J. Phys. Chem. B.*, 2005, **109**, 11432.

CHAPTER 7
# Fullerenes for Material Science

STÉPHANE CAMPIDELLI, AURELIO MATEO-ALONSO AND MAURIZIO PRATO

Dipartimento di Scienze Farmaceutiche, Università degli Studi di Trieste, Piazzale Europa 1, I-34127, Trieste, Italy

## 7.1 Introduction

Fullerenes possess intriguing electrochemical,[1,2] photophysical,[3] optical,[4] semi-conducting,[5] superconducting,[6–8] magnetic[9–11] and aggregation[12] properties. A new research field has emerged in recent years, based on the functionalization and study of new fullerene derivatives, which intends to exploit the properties of these carbon hollow clusters in different technological fields. The high number of functionalization protocols together with the rich supramolecular chemistry of fullerenes allows the preparation of multiple derivatives with a wide diversity of architectures that are continuously explored as novel materials.

$C_{60}$ is the most studied and widely used of the fullerenes because of the availability, high symmetry and low price. Upon functionalization, the solubility of $C_{60}$ in organic solvents is increased, enhancing its proccessability. Some of the most promising fields of application of these novel materials, including artificial photosynthesis, non-linear optics and the preparation of photoactive well-organized films and nanostructures, will be reviewed in this chapter.

## 7.2 Donor–Acceptor Systems

The research on donor–acceptor systems involving fullerenes has attracted a lot of attention, as they can be used as artificial models of the photosynthetic reaction center to transform light into chemical energy. In the natural photosynthetic reaction center,[13] several photoactive units are coupled together, so that several photoinduced electron transfer (PET) events take place after irradiation, giving a long distance and long-lived charge separated state in which the energy is stored.

The easiest approach to an artificial model consists of two different units, an electron-donor and an electron-acceptor linked together (dyad).

Fullerenes and especially $C_{60}$ have been widely used electron-acceptors, since they posses a high electron affinity[1,14,15] and small reorganization energy.[16] The electrons are highly delocalized in the three-dimensional π-system.[17] Therefore fullerenes have shown to give very stable radical pairs.[3]

The lifetime of the radical pair is controlled by different factors including the nature of the electro/photoactive units and of the linker, along with the distance and orientation between the donor and the acceptor. These factors play an active role during the PET process and the rate of charge recombination in the charge separated state.

A wide range of electron donors and spacers has been explored. The following section has been organized by the chemical approach to combine the donor and the acceptor.

## 7.2.1 Covalently Linked Donor–Acceptor Systems

### 7.2.1.1 Photoactive Electron Donors

Electron transfer takes place easily between fullerenes and macrocycles that absorb light in the visible region, such as porphyrins, phthalocyanines, and their metallated homologous. The advantage of using such macrocycles lies in the strong absorption in the visible region and the minimal structural changes when electrons are released.

Porphyrin dyads have been the most investigated systems.[18–23] A wide range of architectures based on different functionalization protocols and on different spacers has been explored by several groups. Porphyrins have been attached covalently to fullerenes through 1,3-dipolar addition, Diels–Alder cycloaddition and Bingel cyclopropanation.

1,3-Dipolar addition was applied to prepare architectures bearing porphyrins in position 2 of fulleropyrrolidines.[24–26] These systems have shown the longest lifetimes reported in dyads to the moment (230 μs at 25°C) (**1** in Chart 1).[25] Multiple rigid spacers have been studied based on amides and diazo derivatives[27] and rigid chiral binaphthyl.[28]

Analogous monoadducts with rigid linkers were also prepared by Diels–Alder cycloaddition[29–31] and by Bingel cyclopropanation.[32]

Strapped and parachute-like fullerene–porphyrin architectures have been reported, allowing the porphyrin to get very close to the fullerene. These architectures were based on the synthesis of different Bingel mono and bisadducts (Chart 1).[33–37]

Efficient dyads containing free and metalated phthalocyanines have been prepared by 1,3-dipolar addition yielding fulleropyrrolidines substituted with phthalocyanines in position 2.[38–41] The lifetimes observed in solution

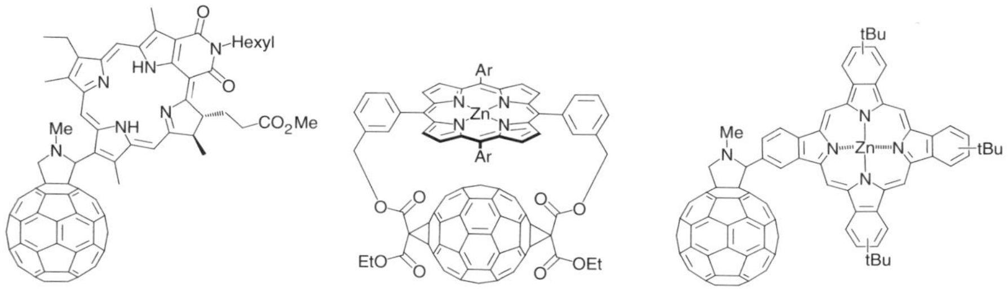

**Chart 1** *Fullerene-macrocycle dyads*

(nanosecond scale) were several orders of magnitude shorter than those observed for thin films deposited by spin coating (5 ms) (**3** in Chart 1).[42]

### 7.2.1.2 Non-Photoactive Electron–Donors

Donor–acceptor systems can be also prepared using non-photoactive electron donors such as tetrathiafulvalene (TTF),[43] π-extended-tetrathiafulvalenes (exTTF),[43] ferrocene (Fc), ruthenocene and conjugated oligomers. In this case, electron transfer is triggered by excitation of the fullerene while the donor remains in the ground state.[3]

The first synthesis of $C_{60}$-TTF and $C_{60}$-exTTF dyads was achieved though a 1,3 dipolar cycloaddition (Chart 2).[44–46] TTF forms only stable radical cations while exTTF can give stable dication species. In dyads where TTF was directly linked to the fulleropyrrolidine ring, no electronic interactions were observed.[47] Electron transfer events were detected when $C_{60}$ was functionalized by 1,3 dipolar addition with TTF[48,49] with vinyl spacers bearing TTF moieties[47,50] and exTTF,[51] along with azides bearing TTF[52] and π-extended-TTF units,[53] by Diels–Alder cycloaddition,[51,54–57] and Bingel cyclopropanation.[58,59]

Fullerene–ferrocene dyads have been synthesized using either variable spacing building blocks or rigid linkers, in order to tune the redox properties of the system.[45,60–64] Novel dyads that contain two covalently bound ferrocene units[65] or a ruthenocene[66] were recently synthesized and studied (Chart 2).

Oligomers have also shown to form long-lived radical-anions when coupled to fullerenes.[67,68] Oligophenylenevinylenes (OPV),[69–72] oligophenyleneethynylenes (OPE)[73] and oligothiophenes[74] are among the most studied.

### 7.2.2 Donor–Acceptor Systems Assembled by Supramolecular Interactions

### 7.2.2.1 Metal Complexation

*7.2.2.1.1 Axial Complexation.* If $C_{60}$ is functionalized with a pyridyl substituent, fullerene-porphyrin dyads can be assembled by axial coordination with metalloporphyrins.[75–77]

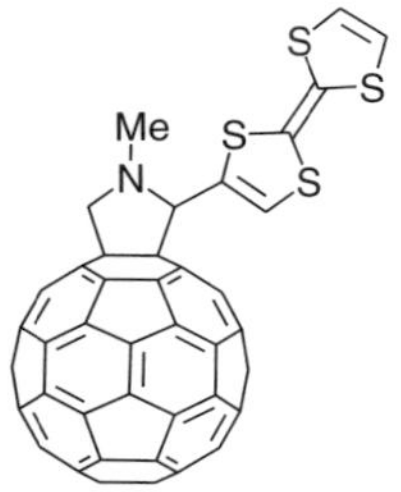

**Chart 2**  *Fullerenes functionalized with non-photoactive electron donors*

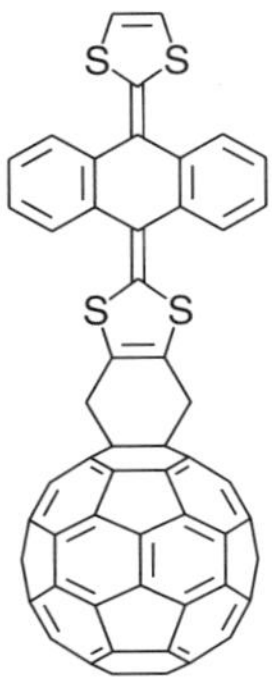

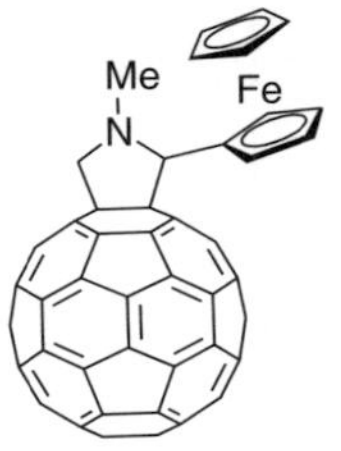

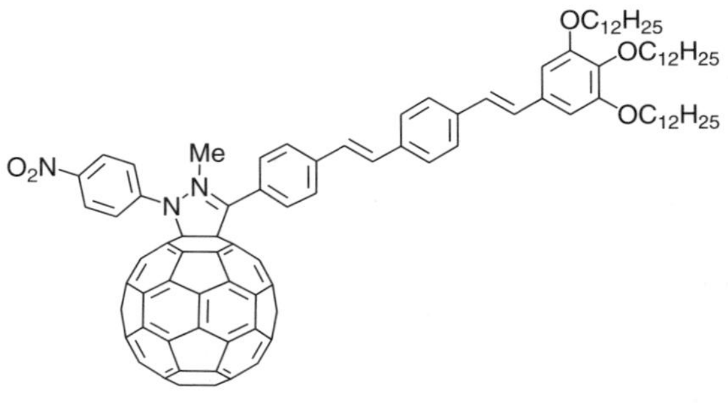

The self-assembly of $C_{60}$ substituted with a 4-pyridyl coordinating group[78,79] with zinc and ruthenium tetraphenyl porphyrins (ZnTPP and RuCOTPP) has been studied both in solution[80,81] and in the solid state (Chart 3).[82] The absorption and NMR studies with ZnTPP revealed the formation of the 1:1 complex by using equimolecular quantities of the ligand and the porphyrins. Photophysical studies showed a lifetime of 8.6 μs. Analogous systems with $C_{60}$ functionalized *via* Bingel reaction bearing a coordinating pyridyl group have been reported.[83]

Axial coordination can also be applied to porphyrins substituted with bulky groups. In a recent example a parachute-like fullerene-ZnTPP dyad was complexed to a pyridyl fullerene ligand.[84]

The versatility of axial coordination together with the different functionalization protocols for fullerenes allowed the preparation and study of multiple assemblies with different topologies, including $C_{60}$ functionalized with 3-pyridyl groups,[85–87] $C_{60}$- and azafullerene-based linear complexes,[88–91] coordination within covalently linked dyads[86,87] and fullerenes functionalized with two pyridyl groups that can bind either two porphyrins or porphyrin dimers (Chart 3).[92,93]

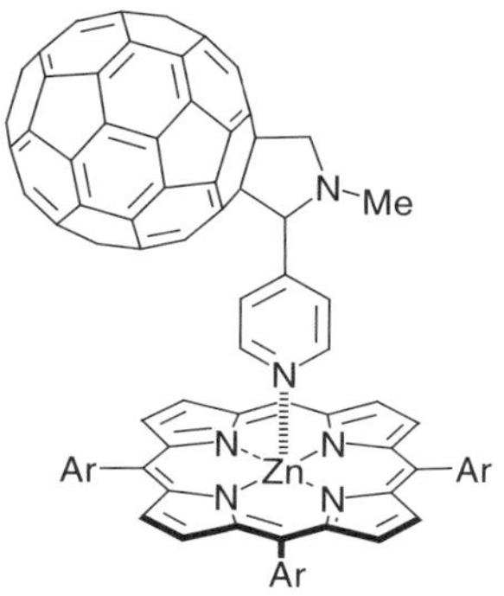

**Chart 3** *Supramolecular assemblies by axial coordination*

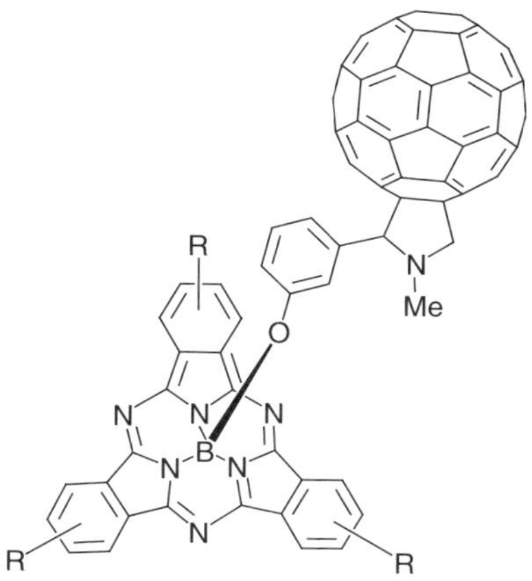

Metal complexation is not exclusive to porphyrins and pyridyl groups. Imidazolyl groups can also complex magnesium porphyrins[94] and titanium phthalocyanines can complex fullerenes functionalized with diphenoxy groups.[95]

$C_{60}$ substituted with phenoxy groups can also axially complex boron-subphthalocyanines (Chart 3). The peripheral substitution in the subphthalocyanine moiety, led to the tuning of the electron-donating properties of the dyads.[96,97] This strategy was also applied successfully to a subphthalocyanine dimer complexed with a fullerene bisadduct displaying two phenoxy groups.[98]

*7.2.2.1.2 Phenanthroline Based Complexes.* Other examples of donor–acceptor assemblies by metal complexation include the use of phenanthroline templates to produce Sauvage-like rotaxanes and catenanes.[99] Schuster and co-workers have successfully used this strategy to produce a rotaxane using fullerenes as stoppers with a porphyrin unit attached to the macrocycle (Chart 4). A lifetime of 32 μs was observed.[100] A similar value, of the order of microseconds, was observed in an analogous rotaxane with the opposite topology, in which the fullerene is placed in the macrocycle and the porphyrins are used as the stoppers.[101] The new structural distribution[102] allowed the preparation of novel catenane systems by complexation of the two terminal Zn-porphyrins when one equivalent of a bidentate ligands such as DABCO and 4,4′-bipyridine, were used. A subsequent addition of one extra-equivalent of the bidentate ligand converted the 1:1 catenane supramolecular complex into a 2:1 rotaxane complex (Chart 4).

## 7.2.2.2 Hydrogen Bonding

If the electon donor and the electron acceptor are functionalized with complementary hydrogen bond donors and acceptors the two units can be assembled by means of hydrogen bonding.[103]

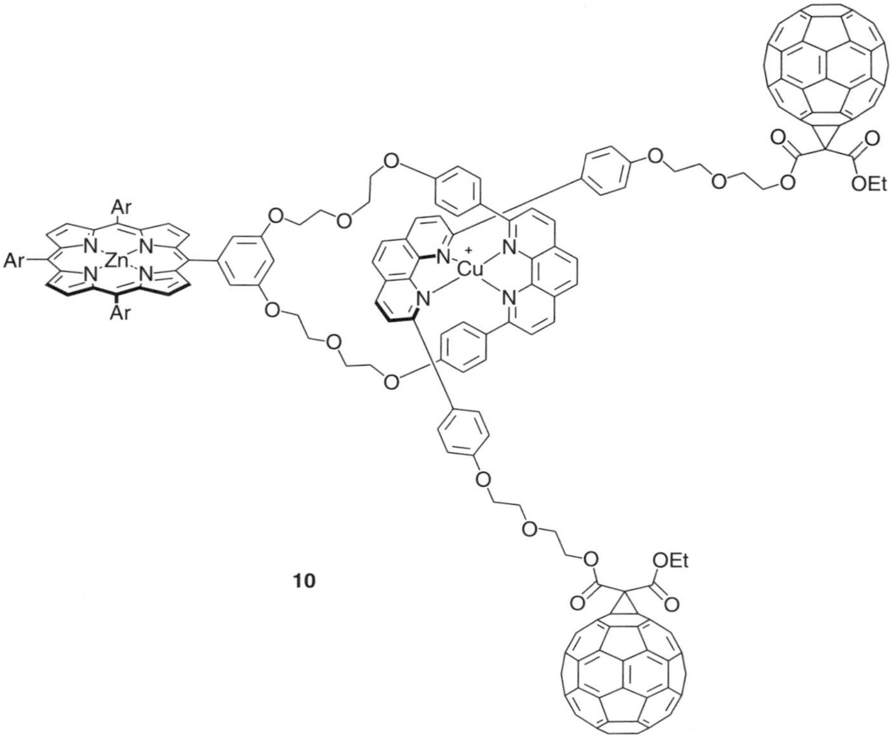

**Chart 4** *Rotaxane assembled by metal coordination*

*7.2.2.2.1 Crown Ether-Ammonium Salt Complexes.* Various pseudorotaxane architectures have been prepared with different electron donors by complexation of ammonium salts with crown ethers. The advantage of using such strategy relies on the possibility of switching "on" and "off" different modes of electron transfer or energy transfer by the addition of acid or base. Under acidic conditions the ammonium salt complexes the crown-ether placing the donor and the acceptor together. While under basic conditions, the two units are placed apart. Some examples of this structures have been studied with different electron donors including porphyrins,[104,105] phthalocyanines[106,107] and TTF.[108]

Rotaxanes containing porphyrin and ferrocene units as electron donor placed both in the macrocycle or as stoppers have been prepared and studied (Chart 5).[109,110]

*7.2.2.2.2 Complementary Hydrogen Bond Recognition Motifs.* A variety of supramolecular $C_{60}$-TTF dyads have been assembled through complementary recognition between guanidinium salts and carboxylic acids, based on donor–donor–acceptor-acceptor hydrogen bond motifs.[111] Eight different dyads were prepared and studied displaying different functional groups and spacers. Intramolecular electron transfer occurred between the two units but this was

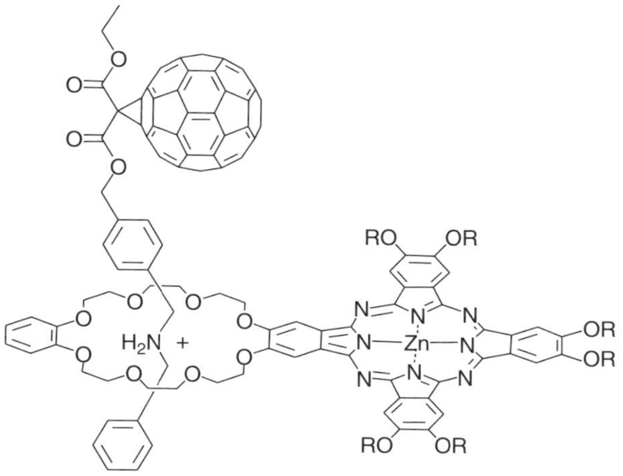

**11**

Chart 5  *Phthalocyanine–fullerene pseudorotaxane assembly*

not observed when solvents that disrupt hydrogen bonds were added. Alternatively intermolecular electron transfer was detected (Chart 6).

Rotaxanated dyads have been assembled by hydrogen bonding, the macrocycle was threaded by recognition of a hydrogen bond template contained in the thread and stoppered by two porphyrins[110,112,113] The macrocycle contained a butadiene sulfone moiety that was used to attach $C_{60}$ by Diels–Alder cycloaddition, as 1,3-butadiene was generated *in situ* from butadiene sulfone.

Recently, a novel Watson–Crick[114] guanosine–cytidine base-paired assembly was used to couple $C_{60}$ and porphyrin.

### 7.2.2.3  π-Stacking

Fullerene derivatives can form complexes with planar electron-rich macrocycles, such as porphyrins.[115] The association between fullerenes and porphyrins was first observed by X-ray crystallography in a covalently linked fullerene-porphyrin.[116,117] Since then, a variety of X-ray structures from co-crystals of different fullerenes and porphyrins have been reported.[118,119] The affinity of fullerenes and porphyrins was not only observed in the solid state, but also in solution as shown by upfield shifts in both $^1H$ and $^{13}C$ NMR experiments in toluene.[118]

Porphyrin hosts were designed and synthesized for binding fullerenes. A type of *bis*-porphyrin systems known as "porphyrin jaws", were prepared by Reed and co-workers.[120,121] The complexation of several "jaws" with fullerenes was studied both in solid state and in solution (**14** and **15** in Chart 7). The most successful binding was obtained with a palladium-based "jaw". Novel supramolecular peapod systems[122] have been developed using dendrimeric "jaws". The hexacarboxylic porphyrins containing the fullerenes polymerized unidirectionally through hydrogen bond dimerization of the side acid functionalization

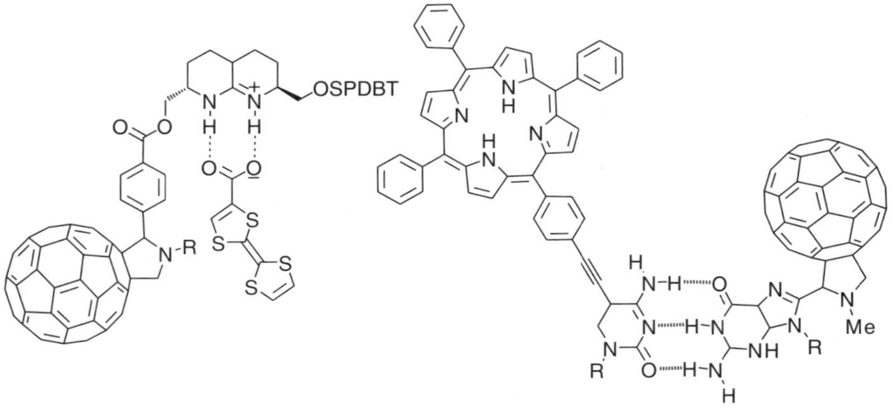

**Chart 6** *Dyads assembled by hydrogen bond recognition patterns*

affording the supramolecular peapods with the dendrimeric envelope. Recently novel jaws-like complexes have been published by Li (Chart 8).[123]

Cyclic porphyrin dimers showed high affinity for $C_{60}$.[124,125] Dendrimeric systems containing this cyclic *bis*-porphyrin dimers have been also prepared.[126] The high affinity between these dimers and fullerenes was observed even with the highly branched dendrimers. $C_{60}$ dimers were also hosted by porphyrin dimers[127] and tetramers.[128]

### 7.2.3 Polyads

Donor–acceptor systems with more than two electroactive units have been synthesized and studied.[20,22] In this case the different electroactive units can work both as the donor and the acceptor. The main goal of such systems is to produce a cascade of electron transfer events that should give a long-lived and long distance charge separated state.

These donor–acceptor systems are called triads, tetrads, pentads, hexads, *etc.* depending on the number of electroactive units.

A variety of triads and tetrads displaying combinations of fullerenes, porphyrins ($H_2P$), Zn-porphyrins (ZnP) and ferrocenes were synthesized and studied by Imahori and collaborators. A ZnP-$H_2$P-$C_{60}$ triad was synthesized and investigated.[129,130] After photostimulation energy transfer took place from the ZnP to the $H_2$P followed by an electron transfer from the excited $H_2$P to $C_{60}$ giving a long distance charge separated state. Analogous Fc-ZnP-$C_{60}$ and Fc-$H_2$P-$C_{60}$ triads that contain a terminal ferrocene unit[130,131] were prepared and examined. When irradiated, two consecutive electron transfers took place. First, an electron was transferred from the porphyrin to $C_{60}$ and then from the ferrocene to the porphyrin leading to the charge separated state. By combination of these two systems, a tetrad was reported,[132] which showed a very high

**14a** 

**15**

**14b** 

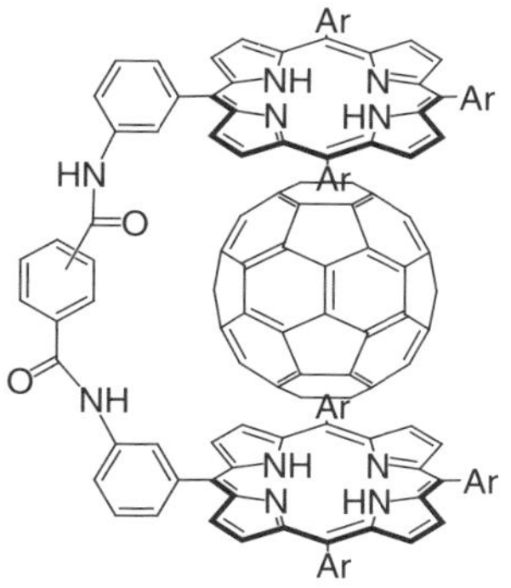

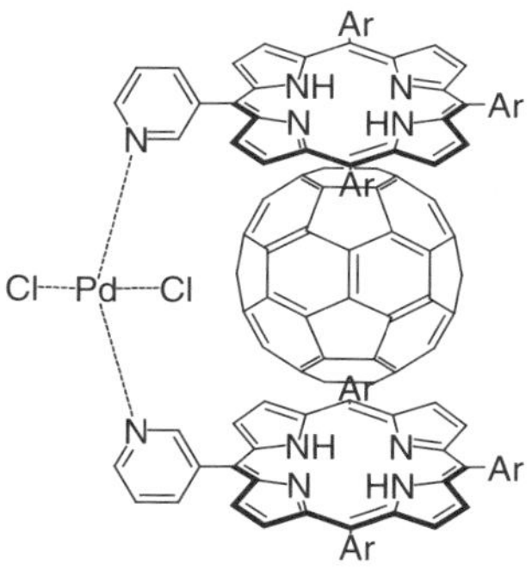

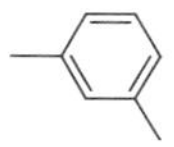

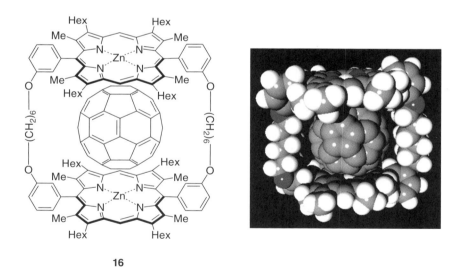

**Chart 7** *Fullerene complexed by "porphyrin jaws"*

**16**

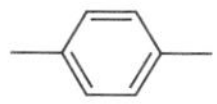

**Chart 8** *Fullerene complexed by a porphyrin dimer*

lifetime (380 μs), as the product of a sequence of energy and multistep electron transfer events (Chart 9).

Other multicompo‘nent polyads have been prepared and investigated, based on linear[133,134] and branched[135] multiporphyrin polyads, including a ferrocene-multiporphyrin pentad.[136] Multiple donor–acceptor systems containing TTF

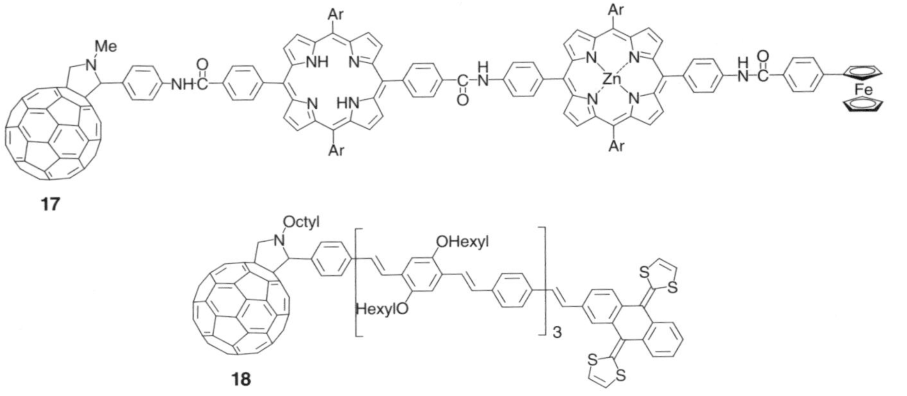

**Chart 9** *A Fc-ZnP-H₂P-C₆₀ tetrad and an ex-TTF-OPV-C₆₀ triad*

have been also explored. These include porphyrin-$C_{60}$ triads bearing terminal TTF[137] and π-extended-TTF[138] moieties.

Series of triads based on terminal π-extended-TTF[139,140] or porphyrin[141] linked to $C_{60}$ through oligophenylenevinylene (OPV) molecular wires of different lengths have been recently reported (Chart 9). After light irradiation, electron transfer occurred leading to up to 40 Å charge separated state. Analogous systems have been prepared with porphyrins connected to $C_{60}$ through oligothiophene,[142–145] OPE[146,147] and combination of oligothiophene and OPV.[148]

A recent example showed the assembly of the three electroactive units by a combination of covalent and supramolecular interactions.[149]

## 7.3 Fullerenes for Nonlinear Optical Applications

Several studies on fullerene derivatives have shown the great potential of this class of materials for optical limiting applications.[4,150,151] Optical limiting (OL) is an optical non-linear phenomenon in which the absorption of a material increases when the incident radiation intensity increases. This effect is due to the reverse saturable absorption (RSA) that is characteristic of materials in which the excited states absorb more strongly than the ground state. With the development of the laser in laboratories and in industries, such materials can find an important application for protecting optical sensors and also human eyes from damage caused by high-energy laser.

In the ground state, $C_{60}$ exhibits a broad absorption spectrum, characterized by a strong absorption in the UV region and weaker absorption extending over most of the visible region. Excited state properties of fullerenes have been studied[3] and it has been demonstrated that in the visible region, both the singlet and triplet excited states absorb more strongly than the ground state. These data, together with the values of the lifetimes of the excited states and the

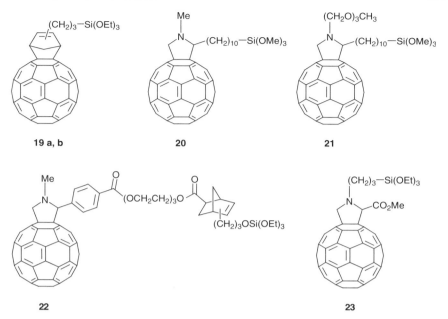

**Chart 10** *Fullerene derivatives containing siloxane functionality for incorporation into sol-gel matrices*

presence of a very efficient inter-system crossing, indicate that fullerenes are good candidates for OL in the visible range.

### 7.3.1 Functionalized Fullerenes for Nonlinear Optics

Even if solutions of fullerene derivatives could be efficient optical limiters, the use of solid devices is preferable for practical applications. For this purpose, several fullerene derivatives containing both solubilizing chains and siloxane groups have been synthesized (Chart 10).[152,153] Stable sol-gel glasses have been prepared and the optical-limiting properties of these derivatives have been investigated in solution and in sol-gel glasses. In solution, irradiation at 652 nm compounds **19–23** gave better results than unfunctionalized fullerene, while at 532 nm the opposite was observed. In the solid state, the measurements showed that the efficiency of the compounds as optical limiters was similar to that in solution.

Reduced fullerene derivatives have also been tested as optical limiters.[154] The neutral fulleropyrrolidine **24**[155] and its methylated counterpart **25**[156] as well as their reduced species **24$^{n-}$**, **25$^{n-}$** ($n = 1$–3) have been studied (Chart 11). The fulleropyrrolidines were reduced with rubidium and the reduced species showed a high non-linear response.

The amphiphilic methanofullerene **26** was synthesized by the group of Nierengarten (Chart 12).[157] Thanks to its high solubility in polar solvents, this compound could be easily included in sol-gel glasses and the optical limiting

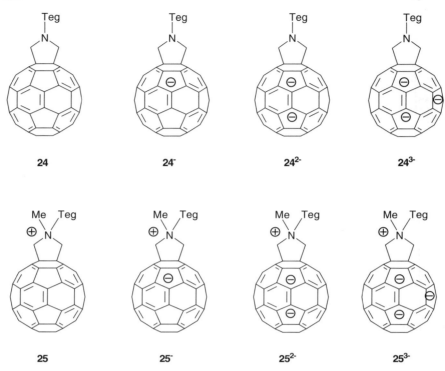

**Chart 11** *Fulleropyrrolidines and their reduced species*

property of the glass was investigated. The OL properties of **26** are similar to those of pristine $C_{60}$ in sol-gel this behaviour has been explained by the fact that the methanofullerene was not well dispersed but led to micellar aggregates in the solvent used for the preparation of the sol-gel.

The same group also synthesized and investigated the OL properties of dendritic fulleropyrrolidines **27–30** (Chart 13).[158,159] Encapsulation of fullerene in a dendrimer permitted to isolate the $C_{60}$ core from the external media leading to an enhancement of the lifetime of the triplet excited state of the fullerene. Compounds **27–30** were incorporated in sol-gel glasses and spectroscopic evidence showed that the molecules were dispersed and that there was no interaction between the $C_{60}$ units in the glasses. The sol-gel glasses containing fullerodendrimer exhibited a non-linear response; the threshold of the limiting action (for **30**) was about 5 mJ cm$^{-2}$. After this value the intensity of the absorption increases showing that the potential of these materials for optical limiting applications.

## 7.3.2 Donor–Acceptor Derivatives

Donor–acceptor derivatives containing $C_{60}$ were also tested as optical limiters. Two $C_{60}$-TTF dyads **31a,b** (Chart 14) were synthesized using the [4 + 2]

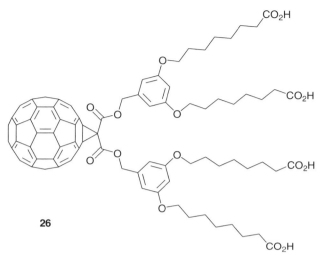

**Chart 12** *Amphiphilic methanofullerene containing acid groups*

Diels–Alder cycloaddition reactions[160] were studied for their non-linear optical and optical limiting applications.[161,162]

Compound **32** was synthesized from the $C_{60}^{2-}$ dianion,[163] its properties were investigated in solution and in a nematic liquid crystal matrix.[164] The $C_{60}$-TTF dyad **32** exhibited large third-order non-linear optical response.

These systems also exhibited dynamic holographic properties, *i.e.* holograms can be recorded and erased in real time without the need of any developing procedures. The diffraction efficiencies and speed of hologram formation could find application in many practical devices like optical memories, novel filters, multiplexers, optical processors or correlators.

Fullerene–ferrocene hybrids have been extensively studied for their electro- and photoelectrochemical properties[23,62] but they have also shown NLO properties. The group of Yamamoto synthesized three $C_{60}$-Fc dyads **33**, **34a,b** (Chart 15) and studied the properties of these compounds.[165] They have also studied the properties of donor–donor assemblies (*i.e.* fullerene-carborane dyads).[166,167]

Tour and co-workers[168] reported the synthesis of a series of OPE containing fullerene moieties at the periphery. A $C_{60}$-OPE hybrid presented an enhanced NLO performance relative to its OPE precursor; this behaviour is presumably due to the occurrence of periconjugation and/or charge transfer effects in the excited state.

The optical limiting behaviour of $C_{60}$-phthalocyanine **35** (Chart 16) has been investigated in solution and as a nanoparticle dispersion using nanosecond laser pulses at 532 nm. The nanoparticle dispersion was obtained by mixing a THF solution of **35** in water. An enhanced optical limiting performance of the nanoparticle sample as compared to that of the solution sample has been observed. The formation of ordered aggregates with a well-defined "face-to-face" packing

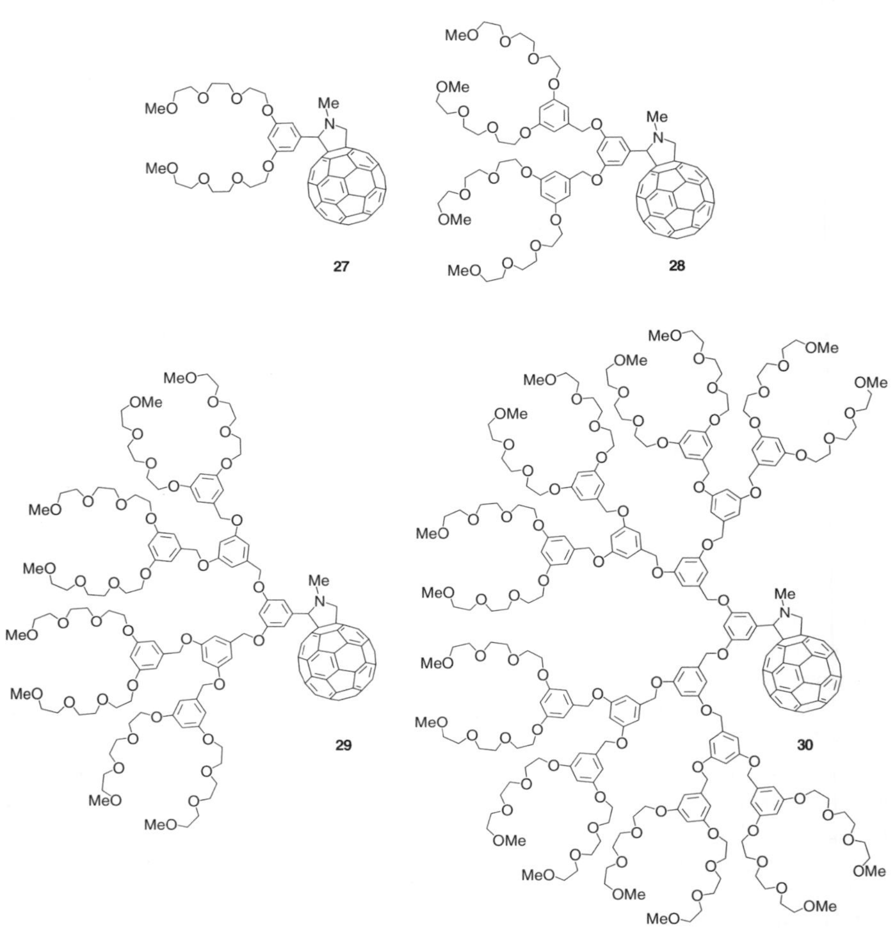

27 28 29 30

**Chart 13** *Dendritic fulleropyrrolidines from first to fourth generation*

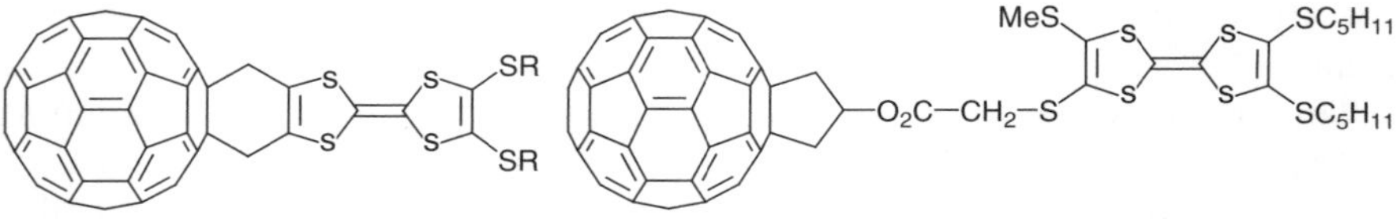

**31a** R = $CH_3$
**31b** R = $C_5H_{11}$

**32**

**Chart 14** *Fullerene–TTF dyads*

fashion is proposed to be responsible for the enhancement of the optical limiting performance of the nanoparticle sample.[169] Fullerene-silver nanocomposites have been prepared by reduction of $Ag^+$ in the presence of mono- and hexa-adducts methanofullerenes.[170,171] The nanocomposites possessed a reverse micelle-like

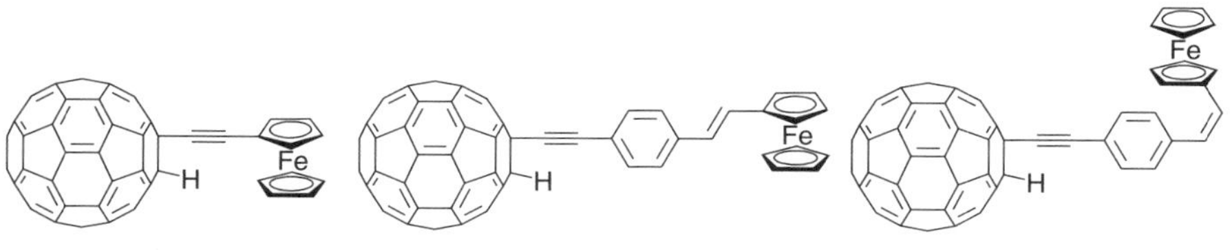

Chart 15  Fullerene–ferrocene dyads

structure with a size of about 10 nm and exhibited better optical limiting performance than fullerene and metal-free methanofullerenes.

## 7.4 Amphiphilic Fullerenes

Amphiphilic fullerenes result from the attachment of hydrophilic groups such as saccharides,[172,173] oligoethylene glycol chains[155,158] or ionic groups,[156,174] to the apolar $C_{60}$ moiety. These fullerenes are partially or totally soluble in water and can be used for the preparation of Langmuir films,[175] self assembling[176,177] or tested for biological applications.[178]

### 7.4.1 Langmuir Films

When some amphiphilic compounds are deposited on the surface of water, it is possible to observe the formation of a monolayer at the water–air interface. This film is called Langmuir film and can be transferred on solid substrates by Langmuir–Blodgett or Langmuir–Schäfer techniques. Many $C_{60}$ derivatives (dendrimers,[172,179–183] polyadducts,[184–186] supramolecular assemblies,[187] donor–acceptor assemblies[188–190]) have been tested for production of Langmuir films.

#### 7.4.1.1 Fullerenes in Langmuir Films

It has been shown that the very simple fulleropyrrolidine **24** forms stable monolayers at the water-air interface. This Langmuir film could be transferred onto a glass or quartz slide by Langmuir–Blodgett or Langmuir–Schäfer techniques.[155] The determined molecular area (105 Å$^2$) was in good agreement with the theoretical and experimental values of 93 and 96 Å$^2$ respectively for pure $C_{60}$.[191,192] When using fulleropyrrolidines, it is possible to introduce a positive charge on the nitrogen of the pyrrolidine ring either by protonation with a strong acid or by methylation using methyl iodide. This possibility becomes an advantage for the preparation of Langmuir films because the hydrophilicity can be tuned by the addition of a charge near the hydrophobic fullerene. A mixture of tricosanoic acid with the compounds **25**, **36** and **37**

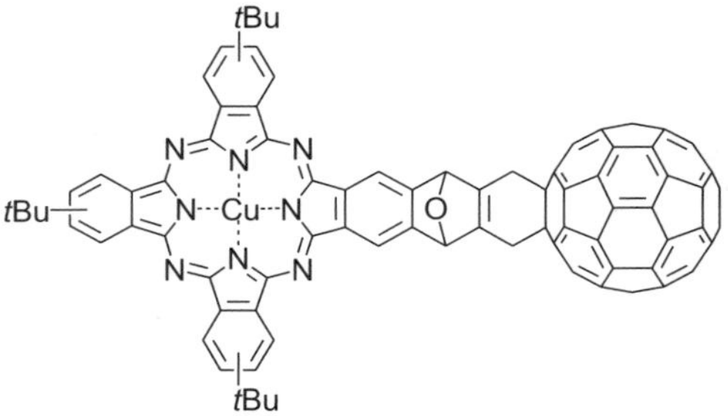

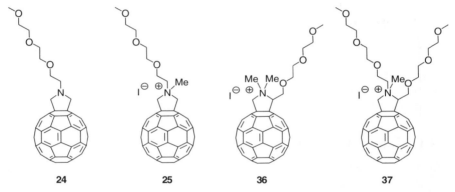

**Chart 16** *Fullerene containing phthalocyanine*

**Chart 17** *Example of amphiphilic fullerenes containing triethylene glycol chains*

(Chart 17) formed Langmuir films which were deposited on different substrates. The optical properties of the films were characterized using absorption spectroscopy and their structure was investigated by X-ray diffraction.[193]

### 7.4.1.2 Photoactive Langmuir Films

If a pyridyl group is introduced together with a triethylene glycol chain, the resulting fulleropyrrolidine can complex axially a metal porphyrin (compound **38**) (Chart 18).[82] The structure of the complex did not provide the adequate hydrophobic-hydrophilic balance to ensure the formation of stable Langmuir monolayers. In order to prepare stable films, a mixture of **38** and arachidic acid was deposited on water. The film was transferred to solid substrates by the Langmuir–Blodgett technique. The transfer was monitored by spectroscopy and the film was studied by transient absorption spectroscopy. Upon irradiation, compound **38** gave rise to an electron transfer and the lifetime of the charge-separated state was found to be 2.2 µs.

Supramolecular association of the positively charged fulleropyrrolidine **39** with a water-soluble porphyrin **40** was also used for the preparation a thin film

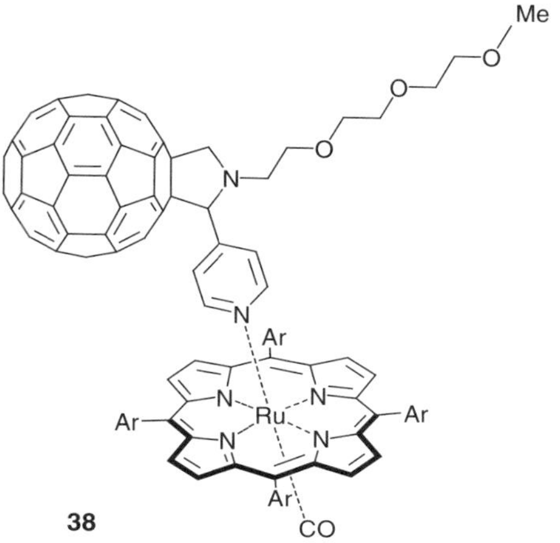

**Chart 18** Amphiphilc fullerenes containing photoactive moieties

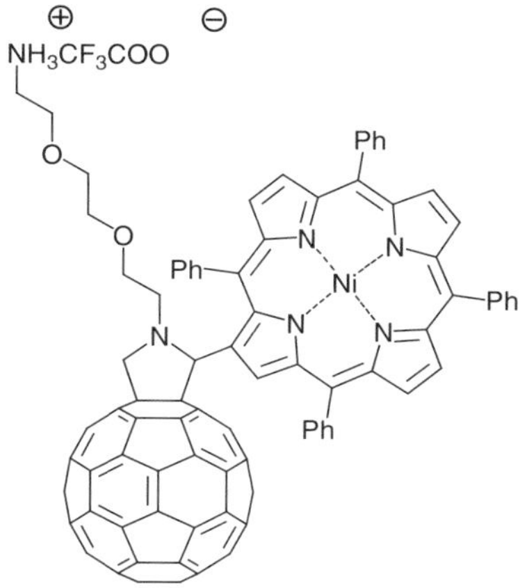

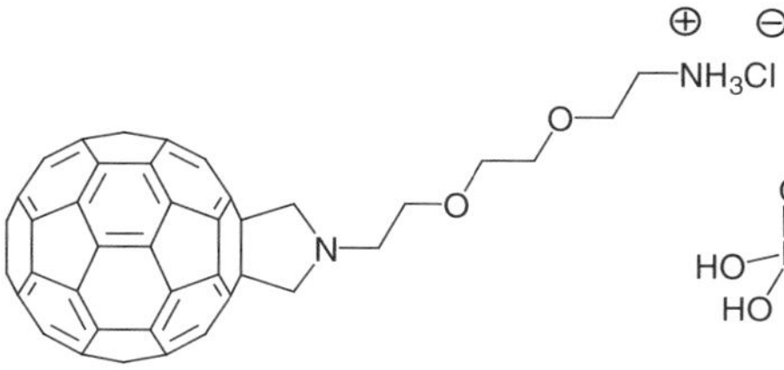

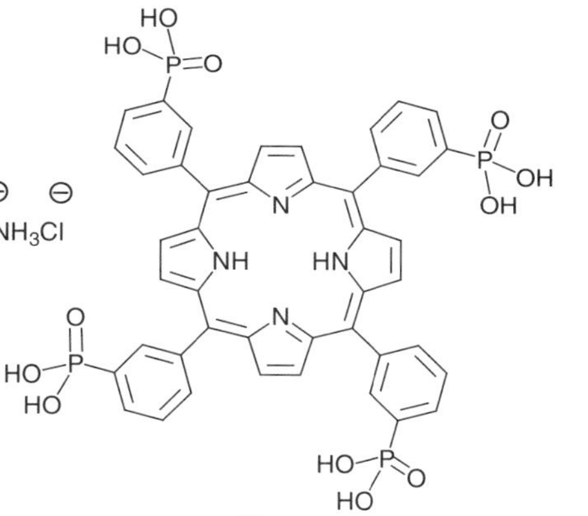

with the Langmuir–Shäfer technique.[194] The presence of the ethylene glycol chain and the $NH_3^+$ was found to be insufficient for the formation of stable monolayers at the water–air interface. The formation of stable Langmuir film was possible only when **39** was deposited on a solution of the porphyrin **40** in water. In this case, the coulombic interactions between the ammonium groups of the fullerenes and the phosphates of the porphyrins are the driving force for the formation of the Langmuir film. The films were characterized by Brewster angle microscopy and UV-Vis reflection spectroscopy while the Langmuir–Shäfer films were characterized by AFM, UV-Vis spectroscopy. The observed photocurrent increased notably with increasing transfer surface pressure.

Langmuir films have also been obtained from compound **41** (Chart 18). The films were transferred on ITO substrate by Langmuir–Shäfer technique while another electrode was prepared by layer by layer (LBL) deposition (deposition of **41** directly from a solution on ITO electrodes coated with a polyelectrolyte).[26] Modified ITO electrodes were probed in photocurrent experiments. The LBL-modified electrodes revealed smaller photon to current conversion efficiencies relative to the Langmuir–Shäfer-modified electrodes.

A series of covalently linked donor–acceptor dyads, composed of porphyrins and fullerene were synthesized by Vuorinien and co-workers[189] and were used for the fabrication of Langmuir films. The spectroscopic properties of the donor-acceptor films were studied with the steady-state absorption and fluorescence methods. All these examples show that the Langmuir technique is a powerful tool for the development of photoactive films based on fullerene and electron donor moieties like porphyrins and phthalocyanines.

### 7.4.2 Fullerene in Smectite Clays

Smectite clays are a class of layered aluminosilicate minerals with a unique combination of swelling, intercalation, and ion-exchange properties that make these nanostructures valuable in diverse fields.[195–197] The clays possess a cation exchange capacity, which allows doping with numerous cationic materials. The neutral fulleropyrrolidine **24** has been successfully incorporated into the interlayer space of the smectite clays by an ion exchange process between the charge balanced sodium ion of the clays and the fullerene derivative in the presence of trimethyl-hexadecylammonium chloride. The two positively charged fulleropyrrolidines **42** and **43** were also incorporated (Chart 19).[198] The fullerene-layered aluminosilicate composites were characterized by several techniques as powder X-ray diffraction, Raman, and $^{57}$Fe-Mössbauer spectroscopies and constitute new hybrid systems where the properties of the $C_{60}$ differ from its crystals or its solutions.

### 7.4.3 Self Organization

#### 7.4.3.1 Assemblies

Amphiphilic fullerenes are able to give rise to self-assembled nanostructures.[12] It is known that $C_{60}$ has a strong tendency to form clusters and aggregates. It is usually necessary to counter this tendency for the good development of the optic and electronic properties of $C_{60}$. Fine tuning of the balance between hydrophobicity of $C_{60}$ and hydrophilicity of the functional groups should permit to control the production of organized structures.

By treating a solution of **44** in DMSO/water with benzene or with sonication, Tour and co-workers[176] demonstrated that the fullerene derivative can self-assemble into either nanorods or vesicles depending on how the solution was treated. A pH-dependent organization has been also described recently.[199] At neutral pH, the amphiphilic fullerene organized into rod-shaped structure while at pH 9 globular micelles with a diameter of 85 ± 10 Å were predominant.

Amphiphilic fullerenes **45–48** (Chart 20) show different types of self-organization when dispersed in water by sonication.[177,200,201] The organization of the

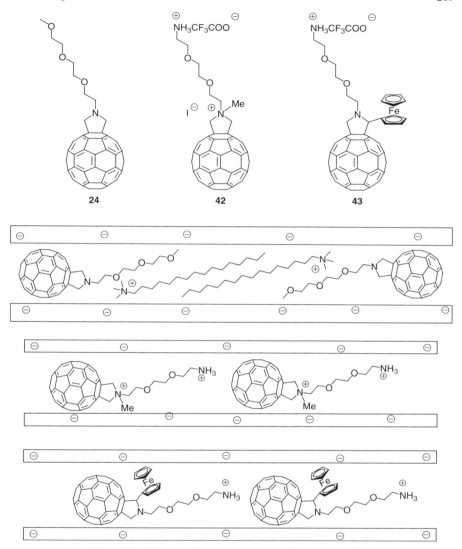

**Chart 19** *Fullerene derivatives incorporated in smectite clays*

amphiphilic compounds has been investigated mainly by transmission electron microscopy (TEM). The fullerene derivative **45a** reveals perfectly round shapes; the spheres have very similar sizes, with diameters ranging from 500 nm to 1.2 µm, and tend to aggregate with each other. Compounds **45b–c** and **46** form objects resembling rods or blocks with nanoscopic dimensions. In addition, the fullerene **45c** and **46** were studied by atomic force microscopy (AFM) and scanning electron microscopy (SEM). The images revealed well defined nanorods with typical length of 200 to 700 nm, from 40 to 100 nm in width, and from 40 to 60 nm in height. In contrast, no organization was observed in the case of

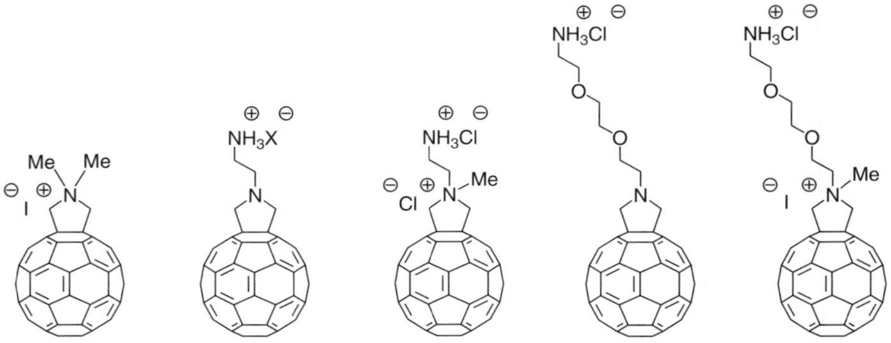


44

45a X = Cl

45b X = Br

45c X = $BF_4$

45d X = $CF_3COO$

46

47

48

**Chart 20** *Ionic fulleropyrrolidines*

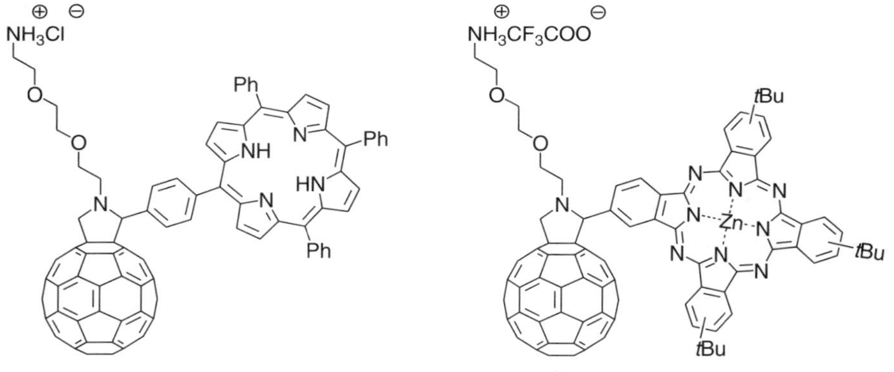


49

50

**Chart 21** *Ionic fulleropyrrolidines containing photoactive moieties*

compound **45d**, in which the counter ion is organic. Compounds **47** and **48** containing a triethylene glycol chain tend to aggregate in long uniform bundles with diameters of about 4 nm and lengths of several microns. The counter ion of the ammonium salt as well as the size and the nature of the substituent on the pyrrolidine play a crucial role in determining the morphology of the materials.

## *7.4.3.2 Assemblies Formed by Donor–Acceptor Dyads*

In order to improve the self-assembly process and to extend the applicability of the nanostructures, tetraphenylporphyrin[177] and (Zn)-phthalocyanine (ZnPc)[202] were linked to the skeleton of **47** leading to **49** and **50** (Chart 21).

Sonication of these compounds in water led to the formation of extremely well organized nanorods. It is known that porphyrins give rise to strong π-stacking interactions both among themselves and with fullerenes. The macrocycle-macrocycle, macrocycle-$C_{60}$, and $C_{60}$-$C_{60}$ interactions are the driving forces of the self-assembly of these systems. Compound **50** was studied by steady-state fluorescence and transient absorption measurements. The lifetime of the charge-separated state $ZnPc^{\bullet+}$-$C_{60}^{\bullet-}$ was found to be 1.4 ms in the aggregates compared to that of the monomeric ZnPc-$C_{60}$ **50** ($\tau \approx 3$ ns). These values implies a stabilization of the charge-separated state of six orders of magnitude; the lifetime found in nanotubules of **50** reaches into a time domain typically found in thin solid films of donor-acceptor composites.

The driving force for the formation of each superstructure is the result of a subtle balance between hydrophilic and hydrophobic interactions and specifically, directional interactions between the molecules.

## 7.5 Conclusion

Organic functionalization of $C_{60}$ has produced a wide range of derivatives, which combine the basic properties of the pristine fullerene with the properties of the organic subsitutents (as hydrophilicity, electro-photoactive behaviour, *etc*.). This continuous evolution of fullerene science and technology opens new horizons for applications.

By mixing fullerene and electron donating groups such as porphyrins, phthalocyanines, ferrocene, TTF, conjugated oligomers, it has been possible to observe photoinduced electron and/or transfers. These systems are still under heavy investigation as model of the photosynthetic reaction centers and to produce photovoltaic cells based on carbon.

Amphiphilic fullerenes have also demonstrated a great potential of organization by self-aggregation or in Langmuir films. The creation of efficient photoactive films is very promising for the future developments of the applications of $C_{60}$.

## References

1. L. Echegoyen and L.E. Echegoyen, *Acc. Chem. Res.*, 1998, **31**, 593–601.
2. C. Bruno, I. Doubitski, M. Marcaccio, F. Paolucci, D. Paolucci and A. Zaopo, *J. Am. Chem. Soc.*, 2003, **125**, 15738.
3. D.M. Guldi and M. Prato, *Acc. Chem. Res.*, 2000, **33**, 695–703.
4. L.W. Tutt and A. Kost, *Nature*, 1992, **356**, 225.
5. R.C. Haddon, A.S. Perel, R.C. Morris, T.T.M. Palstra, A.F. Hebard and R.M. Fleming, *Appl. Phys. Lett.*, 1995, **67**, 121.
6. E. Dagotto, *Science*, 2001, **293**, 2410.
7. P. Grant, *Nature*, 2001, **413**, 264.

8. A.F. Hebard, M.J. Rosseinsky, R.C. Haddon, D.W. Murphy, S.H. Glarum, T.T.M. Palstra, A.P. Ramirez and A.R. Kortan, *Nature*, 1991, **350**, 600.
9. P.M. Allemand, K.C. Khemani, A. Koch, F. Wudl, K. Holczer, G. Gruner and J.D. Thompson, *Science*, 1991, **253**, 301.
10. A. Lappas, K. Prassides, K. Vavekis, D. Arcon, R. Blinc, P. Cevc, A. Amato, R. Feyerherm, F.N. Gygax and A. Schenk, *Science*, 1995, **267**, 1799.
11. B. Narymbetov, A. Omerzu, V.V. Kabanov, M. Tokumoto, H. Kobayashi and D. Mihailovic, *Nature*, 2000, **407**, 883.
12. D.M. Guldi, F. Zerbetto, V. Georgakilas and M. Prato, *Acc. Chem. Res.*, 2005, **38**, 38.
13. *The Photosynthetic Reaction Center*, Academic Press, San Diego, 1993.
14. F. Arias, L. Echegoyen, S.R. Wilson and Q. Lu, *J. Am. Chem. Soc.*, 1995, **117**, 1422–1427.
15. L. Echegoyen, F. Diederich and L.E. Echegoyen, *Electrochemistry of Fullerenes*, in *Fullerenes: Chemistry, Physics, and Technology*, Wiley, New York, 2000, 1.
16. H. Imahori, K. Hagiwara, T. Akiyama, M. Aoki, S. Taniguchi, T. Okada, M. Shirakawa and Y. Sakata, *Chem. Phys. Lett.*, 1996, **263**, 545–550.
17. D.M. Guldi, *Chem. Commun.*, 2000, 321–327.
18. H. Imahori and Y. Sakata, *Adv. Mater.*, 1997, **9**, 537.
19. H. Imahori and Y. Sakata, *Eur. J. Org. Chem.*, 1999, 2445–2457.
20. H. Imahori, *Org. Biomol. Chem.*, 2004, **2**, 1425–1433.
21. D.M. Guldi, *Pure Appl. Chem.*, 2003, **75**, 1069–1075.
22. D.M. Guldi, *Chem. Soc. Rev.*, 2002, **31**, 22–36.
23. H. Imahori, Y. Mori and Y. Matano, *J. Photochem. Photobiol. C.*, 2003, **4**, 51–83.
24. Y. Kashiwagi, K. Ohkubo, J.A. McDonald, I.M. Blake, M.J. Crossley, Y. Araki, O. Ito, H. Imahori and S. Fukuzumi, *Org. Lett.*, 2003, **5**, 2719–2721.
25. K. Ohkubo, H. Kotani, J.G. Shao, Z.P. Ou, K.M. Kadish, G.L. Li, R.K. Pandey, M. Fujitsuka, O. Ito, H. Imahori and S. Fukuzumi, *Angew. Chem., Int.Ed.*, 2004, **43**, 853–856.
26. D.M. Guldi, I. Zilbermann, G.A. Anderson, K. Kordatos, M. Prato, R. Tafuro and L. Valli, *J. Mater. Chem.*, 2004, **14**, 303–309.
27. H. Imahori, H. Yamada, D.M. Guldi, Y. Endo, A. Shimomura, S. Kundu, K. Yamada, T. Okada, Y. Sakata and S. Fukuzumi, *Angew. Chem., Int. Ed.*, 2002, **41**, 2344–2347.
28. D.M. Guldi, F. Giacalone, G. de la Torre, J.L. Segura and N. Martín, *Chem. Eur. J.*, 2005, **11**, 7199–7210.
29. H. Imahori, K. Hagiwara, M. Aoki, T. Akiyama, S. Taniguchi, T. Okada, M. Shirakawa and Y. Sakata, *J. Am. Chem. Soc.*, 1996, **118**, 11711.
30. P.A. Liddell, J.P. Sumida, A.N. Macpherson, L. Noss, R. Seely, K.N. Clark, A.L. Moore, T.A. Moore and D. Gust, *Photochem. Photobiol.*, 1994, **60**, 537.

31. H. Imahori, K. Hagiwara, T. Akiyama, S. Taniguchi, T. Okada and Y. Sakata, *Chem. Lett.*, 1995, **24**, 265.
32. K. Tamaki, H. Imahori, Y. Nishimura, I. Yamazaki, A. Shimomura, T. Okada and Y. Sakata, *Chem. Lett.*, 1999, **28**, 227.
33. E. Dietel, A. Hirsch, E. Eichhorn, A. Rieker, S. Hackbarth and B. Roder, *Chem. Commun.*, 1998, 1981–1982.
34. P. Cheng, S.R. Wilson and D.I. Schuster, *Chem. Commun.*, 1999, 89–90.
35. D.I. Schuster, P. Cheng, S.R. Wilson, V. Prokhorenko, M. Katterle, A.R. Holzwarth, S. E. Braslavsky, G. Klihm, R.M. Williams and C.P. Luo, *J. Am. Chem. Soc.*, 1999, **121**, 11599–11600.
36. D.I. Schuster, P. Cheng, P.D. Jarowski, D.M. Guldi, C.P. Luo, L. Echegoyen, S. Pyo, A.R. Holzwarth, S.E. Braslavsky, R.M. Williams and G. Klihm, *J. Am. Chem. Soc.*, 2004, **126**, 7257–7270.
37. L.R. Sutton, M. Scheloske, K.S. Pirner, A. Hirsch, D.M. Guldi and J.P. Gisselbrecht, *J. Am. Chem. Soc.*, 2004, **126**, 10370–10381.
38. A. Sastre, A. Gouloumis, P. Vazquez, T. Torres, V. Doan, B. J. Schwartz, F. Wudl, L. Echegoyen and J. Rivera, *Org. Lett.*, 1999, **1**, 1807–1810.
39. D.M. Guldi, A. Gouloumis, P. Vazquez and T. Torres, *Chem. Commun.*, 2002, 2056–2057.
40. M.A. Loi, P. Denk, H. Hoppe, H. Neugebauer, C. Winder, D. Meissner, C. Brabec, N. S. Sariciftci, A. Gouloumis, P. Vazquez and T. Torres, *J. Mater. Chem.*, 2003, **13**, 700–704.
41. D.M. Guldi, I. Zilbermann, A. Gouloumis, P. Vázquez and T. Torres, *J. Phys. Chem. B.*, 2004, **108**, 18485–18494.
42. M.A. Loi, P. Denk, H. Hoppe, H. Neugebauer, D. Meissner, C. Winder, C.J. Brabec, N.S. Sariciftci, A. Gouloumis, P. Vázquez and T. Torres, *Synth. Met.*, 2003, **137**, 1491–1492.
43. J.L. Segura and N. Martín, *Angew. Chem., Int. Ed.*, 2001, **40** 1372–1409.
44. N. Martín, L. Sanchez, C. Seoane, R. Andreu, J. Garin and J. Orduna, *Tetrahedron Lett.*, 1996, **37**, 5979–5982.
45. M. Prato, M. Maggini, C. Giacometti, G. Scorrano, G. Sandona and G. Farnia, *Tetrahedron*, 1996, **52**, 5221–5234.
46. N. Martín, I. Perez, L. Sanchez and C. Seoane, *J. Org. Chem.*, 1997, **62**, 5690–5695.
47. N. Martín, L. Sanchez, M.A. Herranz and D.M. Guldi, *J. Phys. Chem. A.*, 2000, **104**, 4648–4657.
48. L. Sánchez, I. Pérez, N. Martín and D.M. Guldi, *Chem. Eur. J.*, 2003, **9**, 2457–2468.
49. L. Sánchez, M. Sierra, N. Martín, D.M. Guldi, M.W. Wienk and R.A.J. Janssen, *Org.Lett.*, 2005, **7**, 1691–1694.
50. N. Martín, L. Sanchez, B. Illescas, S. Gonzalez, M.A. Herranz and D.M. Guldi, *Carbon*, 2000, **38**, 1577–1585.
51. M.C. Díaz, M.A. Herranz, B.M. Illescas, N. Martín, N. Godbert, M.R. Bryce, C. Luo, A. Swartz, G. Anderson and D.M. Guldi, *J.Org.Chem.*, 2003, **68**, 7711–7721.

52. D.M. Guldi, S. Gonzalez, N. Martín, A. Anton, J. Garin and J. Orduna, *J. Org. Chem.*, 2000, **65**, 1978–1983.
53. S. Gonzalez, N. Martín, A. Swartz and D.M. Guldi, *Org.Lett.*, 2003, **5**, 557–560.
54. J. Llacay, M. Mas, E. Molins, J. Veciana, D. Powell and C. Rovira, *Chem.Commun.*, 1997, 659–660.
55. J. Llacay, J. Veciana, J. Vidal-Gancedo, J.L. Bourdelande, R. Gonzalez-Moreno and C. Rovira, *J. Org. Chem.*, 1998, **63**, 5201–5210.
56. M. Mas-Torrent, R.A. Rodriguez-Mias, M. Sola, M.A. Molins, M. Pons, J. Vidal-Gancedo, J. Veciana and C. Rovira, *J. Org. Chem.*, 2002, **67**, 566–575.
57. M.A. Herranz, N. Martín, J. Ramey and D.M. Guldi, *Chem. Commun.*, 2002, **8**, 2968–2969.
58. S. Gonzalez, N. Martín and D.M. Guldi, *J. Org. Chem.*, 2003, **68**, 779–791.
59. E. Allard, F. Oswald, B. Donnio, D. Guillon, J.L. Delgado, F. Langa and R. Deschenaux, *Org. Lett.*, 2005, **7**, 383–386.
60. T. Chuard and R. Deschenaux, *Helv. Chim. Acta*, 1996, **79**, 736–741.
61. M. Maggini, A. Karlsson, G. Scorrano, G. Sandona, G. Farnia and M. Prato, *J. Chem. Soc., Chem. Commun.*, 1994, 589–590.
62. D. M. Guldi, M. Maggini, G. Scorrano and M. Prato, *J. Am. Chem. Soc.*, 1997, **119**, 974–980.
63. D.M. Guldi, C.P. Luo, D. Koktysh, N.A. Kotov, T. Da Ros, S. Bosi and M. Prato, *Nano Lett.*, 2002, **2**, 775–780.
64. D.M. Guldi, C.P. Luo, T. Da Ros, S. Bozi and M. Prato, *Chem. Commun.*, 2002, 2320–2321.
65. K.Y. Kay, L.H. Kim and I.C. Oh, *Tetrahedron Lett.*, 2000, **41**, 1397–1400.
66. J.J. Oviedo, P. De La Cruz, J. Garín, J. Orduna and F. Langa, *Tetrahedron Lett.*, 2005, **46**, 4781–4784.
67. J.L. Segura, N. Martín and D.M. Guldi, *Chem. Soc. Rev.*, 2005, **34**, 31–47.
68. L. Sánchez, M.A. Herranz and N. Martín, *J. Mater. Chem.*, 2005, **15**, 1409–1421.
69. M.J. Gómez-Escalonilla, F. Langa, J.-M. Rueff, L. Oswald and J.-F. Nierengarten, *Tetrahedron Lett.*, 2002, **43**, 7507–7511.
70. N. Armaroli, G. Accorsi, Y. Rio, J.-F. Nierengarten, J.-F. Eckert, M.J. Gómez-Escalonilla and F. Langa, *Synth. Met.*, 2004, **147**, 19–28.
71. V. Parra, T. Del Caño, J.M. Gómez-Escalonilla, F. Langa, M.L. Rodríguez-Méndez and J.A. De Saja, *Synth. Met.*, 2005, **148**, 47–52.
72. F. Langa, M.J. Gómez-Escalonilla, J.-M. Rueff, T.M.F. Duarte, J.-F. Nierengarten, V. Palermo, P. Samori, Y. Rio, G. Accorsi and N. Armaroli, *Chem. Eur. J.*, 2005, **11**, 4405–4415.
73. C. Atienza, B. Insuasty, C. Seoane, N. Martín, J. Ramey, G.M.A. Rahman and D.M. Guldi, *J. Mater. Chem.*, 2005, **15**, 124–132.
74. T. Yamashiro, Y. Aso, T. Otsubo, H. Tang, Y. Harima and K. Yamashita, *Chem. Lett.*, 1999, 443–444.
75. F. D'Souza and O. Ito, *Coord. Chem. Rev.*, 2005, **249**, 1410–1422.

76. M.E. El-Khouly, O. Ito, P.M. Smith and F. D'Souza, *J. Photochem. Photobiol. C.*, 2004, **5**, 79–104.
77. A. Mateo-Alonso, C. Sooambar and M. Prato, *C. R. Chimie*, 2006, in Press.
78. M. Maggini, G. Scorrano and M. Prato, *J. Am. Chem. Soc.*, 1993, **115**, 9798–9799.
79. M. Prato and M. Maggini, *Acc. Chem. Res.*, 1998, **31**, 519–526.
80. T. Da Ros, M. Prato, D. Guldi, E. Alessio, M. Ruzzi and L. Pasimeni, *Chem. Commun.*, 1999, 635–636.
81. T. Da Ros, M. Prato, D. Guldi, M. Ruzzi and L. Pasimeni, *Chem. Eur. J.*, 2001, **7**, 816–827.
82. T. Da Ros, M. Prato, M. Carano, P. Ceroni, F. Paolucci, S. Roffia, L. Valli and D.M. Guldi, *J. Organomet. Chem.*, 2000, **599**, 62–68.
83. N. Armaroli, F. Diederich, L. Echegoyen, T. Habicher, L. Flamigni, G. Marconi and J.F. Nierengarten, *New J. Chem.*, 1999, **23**, 77–83.
84. D.M. Guldi, C.P. Luo, T. Da Ros, M. Prato, E. Dietel and A. Hirsch, *Chem. Commun.*, 2000, 375–376.
85. F. D'Souza, G.R. Deviprasad, M.S. Rahman and J.P. Choi, *Inorg. Chem.*, 1999, **38**, 2157–2160.
86. F. D'Souza, N.P. Rath, G.R. Deviprasad and M.E. Zandler, *Chem. Commun.*, 2001, 267–268.
87. F. D'Souza, G.R. Deviprasad, M.E. El-Khouly, M. Fujitsuka and O. Ito, *J. Am. Chem. Soc.*, 2001, **123**, 5277–5284.
88. S.R. Wilson, S. MacMahon, F.T. Tat, P.D. Jarowski and D.I. Schuster, *Chem. Commun.*, 2003, 226–227.
89. T. Galili, A. Regev, A. Berg, H. Levanon, D.I. Schuster, K. Möbius and A. Savitsky, *J. Phys. Chem. A.*, 2005, **109**, 8451–8458.
90. D.M. Guldi, T. Da Ros, P. Braiuca and M. Prato, *Photochem. Photobiol. Sci.*, 2003, **2**, 1067–1073.
91. F. Hauke, A. Swartz, D.M. Guldi and A. Hirsch, *J. Mater. Chem.*, 2002, **12**, 2088–2094.
92. F. D'Souza, S. Gadde, M.E. Zandler, M. Itou, Y. Araki and O. Ito, *Chem. Commun.*, 2004, **10**, 2276–2277.
93. A. Trabolsi, M. Elhabiri, M. Urbani, J.L. Delgado de la Cruz, F. Ajamaa, N. Solladie, A.M. Albrecht-Gary and J.F. Nierengarten, *Chem. Commun.*, 2005, 5736–5738.
94. F. D'Souza, M.E. El-Khouly, S. Gadde, A.L. McCarty, P.A. Karr, M.E. Zandler, Y. Araki and O. Ito, *J. Phys. Chem. B.*, 2005, **109**, 10107–10114.
95. B. Ballesteros, G. de la Torre, T. Torres, G.L. Hug, G.M.A. Rahman and D.M. Guldi, *Tetrahedron*, 2006, **62**, 2097–2101.
96. D. Gonzalez-Rodriguez, T. Torres, D.M. Guldi, J. Rivera, M.A. Herranz and L. Echegoyen, *J. Am. Chem. Soc.*, 2004, **126**, 6301–6313.
97. D. Gonzalez-Rodriguez, T. Torres, D.M. Guldi, J. Rivera and L. Echegoyen, *Org. Lett.*, 2002, **4**, 335–338.
98. R.S. Iglesias, C.G. Claessens, T. Torres, G.M.A. Rahman and D.M. Guldi, *Chem. Commun.*, 2005, 2113–2115.

99. J.-P. Sauvage, *Chem. Commun.*, 2005, 1507–1510.
100. K. Li, P.J. Bracher, D.M. Guldi, M.A. Herranz, L. Echegoyen and D.I. Schuster, *J. Am. Chem. Soc.*, 2004, **126**, 9156–9157.
101. K. Li, D.I. Schuster, D.M. Guldi, M.A. Herranz and L. Echegoyen, *J. Am. Chem. Soc.*, 2004, **126**, 3388–3389.
102. D.I. Schuster, K. Li, D.M. Guldi and J. Ramey, *Org. Lett.*, 2004, **6**, 1919–1922.
103. L. Sánchez, N. Martín and D.M. Guldi, *Angew. Chem., Int. Ed.*, 2005, **44**, 5374–5382.
104. N. Solladie, M.E. Walther, M. Gross, T.M.F. Duarte, C. Bourgogne and J.F. Nierengarten, *Chem. Commun.*, 2003, 2412–2413.
105. N. Solladié, M.E. Walther, H. Herschbach, E. Leize, A. Van Dorsselaer, T.M. Figueira Duarte and J.-F. Nierengarten, *Tetrahedron*, 2006, **62**, 1979–1987.
106. D.M. Guldi, J. Ramey, M.V. Martinez-Diaz, A. de la Escosura, T. Torres, T. Da Ros and M. Prato, *Chem. Commun.*, 2002 2774–2775.
107. M.V. Martinez-Diaz, N.S. Fender, M.S. Rodriguez-Morgade, M. Gomez-Lopez, F. Diederich, L. Echegoyen, J.F. Stoddart and T. Torres, *J. Mater. Chem.*, 2002, **12**, 2095–2099.
108. M.C. Díaz, B.M. Illescas, N. Martín, J.F. Stoddart, M.A. Canales, J. Jiménez-Barbero, G. Sarova and D.M. Guldi, *Tetrahedron*, 2006, **62**, 1998–2002.
109. H. Sasabe, N. Kihara, Y. Furusho, K. Mizuno, A. Ogawa and T. Takata, *Org. Lett.*, 2004, **6**, 3957–3960.
110. H. Sasabe, K.-I. Ikeshita, G.A. Rajkumar, N. Watanabe, K. Kihara, Y. Furusho, K. Mizuno, A. Ogawa and T. Takata, *Tetrahedron*, 2006, **62**, 1988–1997.
111. M. Segura, L. Sanchez, J. de Mendoza, N. Martín and D.M. Guldi, *J. Am. Chem. Soc.*, 2003, **125**, 15093–15100.
112. N. Watanabe, N. Kihara, Y. Furusho, T. Takata, Y. Araki and O. Ito, *Angew. Chem. Int. Ed. Engl.*, 2003, **42**, 681–683.
113. A.S.D. Sandanayaka, N. Watanabe, K.-I. Ikeshita, Y. Araki, N. Kihara, Y. Furusho, O. Ito and T. Takata, *J. Phys. Chem. B.*, 2005, **109**, 2516–2525.
114. J.L. Sessler, J. Jayawickramarajah, A. Gouloumis, T. Torres, D.M. Guldi, S. Maldonado and K.J. Stevenson, *Chem. Commun.*, 2005, 1892–1894.
115. P.D.W. Boyd and C.A. Reed, *Acc. Chem. Res.*, 2005, **38**, 235–242.
116. T. Drovetskaya, C.A. Reed and P. Boyd, *Tetrahedron Lett.*, 1995, **36**, 7971–7974.
117. Y.P. Sun, T. Drovetskaya, R.D. Bolskar, R. Bau, P.D.W. Boyd and C.A. Reed, *J. Org. Chem.*, 1997, **62**, 3642–3649.
118. P.D.W. Boyd, M.C. Hodgson, C.E.F. Rickard, A.G. Oliver, L. Chaker, P.J. Brothers, R.D. Bolskar, F.S. Tham and C.A. Reed, *J. Am. Chem. Soc.*, 1999, **121**, 10487–10495.

119. D.V. Konarev, I.S. Neretin, Y.L. Slovokhotov, E.I. Yudanova, N.V. Drichko, Y.M. Shul'ga, B.P. Tarasov, L.L. Gumanov, A.S. Batsanov, J.A.K. Howard and R.N. Lyubovskaya, *Chem. Eur. J.*, 2001, **7**, 2605–2616.
120. D. Sun, F.S. Tham, C.A. Reed and P.D.W. Boyd, *Proc. Natl. Acad. Sci. U.S.A.*, 2002, **99**, 5088–5092.
121. D.Y. Sun, F.S. Tham, C.A. Reed, L. Chaker and P.D.W. Boyd, *J. Am. Chem. Soc.*, 2002, **124**, 6604–6612.
122. T. Yamaguchi, N. Ishii, K. Tashiro and T. Aida, *J. Am. Chem. Soc.*, 2003, **125**, 13934–13935.
123. Z. Wu, X. Shao, C. Li, J. Hou, K. Wang, X. Jiang and Z. Li, *J. Am. Chem. Soc.*, 2005, **127**, 17460–17468.
124. K. Tashiro, T. Aida, J.Y. Zheng, K. Kinbara, K. Saigo, S. Sakamoto and K. Yamaguchi, *J. Am. Chem. Soc.*, 1999, **121**, 9477–9478.
125. Y. Shoji, K. Tashiro and T. Aida, *J. Am. Chem. Soc.*, 2004, **126**, 6570–6571.
126. T. Nishioka, K. Tashiro, T. Aida, J.Y. Zheng, K. Kinbara, K. Saigo, S. Sakamoto and K. Yamaguchi, *Macromolecules*, 2000, **33** 9182–9184.
127. K. Tashiro, Y. Hirabayashi, T. Aida, K. Saigo, K. Fujiwara, K. Komatsu, S. Sakamoto and K. Yamaguchi, *J. Am. Chem. Soc.*, 2002, **124**, 12086–12087.
128. H. Sato, K. Tashiro, H. Shinmori, A. Osuka, Y. Murata, K. Komatsu and T. Aida, *J. Am. Chem. Soc.*, 2005, **127**, 13086–13087.
129. C. Luo, D.M. Guldi, H. Imahori, K. Tamaki and K. Sakata, *J. Am. Chem. Soc.*, 2000, **122**, 6535–6551.
130. H. Imahori, K. Tamaki, D.M. Guldi, C.P. Luo, M. Fujitsuka, O. Ito, Y. Sakata and S. Fukuzumi, *J. Am. Chem. Soc.*, 2001, **123**, 2607–2617.
131. M. Fujitsuka, O. Ito, H. Imahori, K. Yamada, H. Yamada and Y. Sakata, *Chem. Lett.*, 1999, **28**, 721–722.
132. H. Imahori, D.M. Guldi, K. Tamaki, Y. Yoshida, C.P. Luo, Y. Sakata and S. Fukuzumi, *J. Am. Chem. Soc.*, 2001, **123**, 6617–6628.
133. D. Bonifazi and F. Diederich, *Chem. Commun.*, 2002, 2178–2179.
134. D. Bonifazi, M. Scholl, F.Y. Song, L. Echegoyen, G. Accorsi, N. Armaroli and F. Diederich, *Angew. Chem., Int. Ed.*, 2003, **42**, 4966–4970.
135. D. Kuciauskas, P.A. Liddell, S. Lin, T.E. Johnson, S.J. Weghorn, J.S. Lindsey, A.L. Moore, T.A. Moore and D. Gust, *J. Am. Chem. Soc.*, 1999, **121**, 8604–8614.
136. H. Imahori, Y. Sekiguchi, Y. Kashiwagi, T. Sato, Y. Araki, O. Ito, H. Yamada and S. Fukuzumi, *Chem. Eur. J.*, 2004, **10**, 3184–3196.
137. P.A. Liddell, G. Kodis, L. de la Garza, J.L. Bahr, A.L. Moore, T.A. Moore and D. Gust, *Helv. Chim. Acta*, 2001, **84**, 2765–2783.
138. G. Kodis, P.A. Liddell, L. de la Garza, A.L. Moore, T.A. Moore and D. Gust, *J. Mater. Chem.*, 2002, **12**, 2100–2108.
139. N. Martín, F. Giacalone, J.L. Segura and D.M. Guldi, *Synth. Met.*, 2004, **147**, 57–61.

140. F. Giacalone, J.L. Segura, N. Martín and D.M. Guldi, *J. Am. Chem. Soc.*, 2004, **126**, 5340–5341.
141. G. de la Torre, F. Giacalone, J.L. Segura, N. Martín and D.M. Guldi, *Chem. Eur. J.*, 2005, **11**, 1267–1280.
142. J. Ikemoto, K. Takimiya, Y. Aso, T. Otsubo, M. Fujitsuka and O. Ito, *Org. Lett.*, 2002, **4**, 309–311.
143. T. Nakamura, M. Fujitsuka, Y. Araki, O. Ito, J. Ikemoto, K. Takimiya, Y. Aso and T. Otsubo, *J. Phys. Chem. B.*, 2004, **108**, 10700–10710.
144. H. Kanato, K. Takimiya, T. Otsubo, Y. Aso, T. Nakamura, Y. Araki and O. Ito, *J. Org. Chem.*, 2004, **69**, 7183–7189.
145. T. Nakamura, J.-Y. Ikemoto, M. Fujitsuka, Y. Araki, O. Ito, K. Takimiya, Y. Aso and T. Otsubo, *J. Phys. Chem. B.*, 2005, **109**, 14365–14374.
146. S.A. Vail, J.P.C. Tomé, P.J. Krawczuk, A. Dourandin, V. Shafirovich, J.A.S. Cavaleiro and D.I. Schuster, *J. Phys. Org. Chem.*, 2004, **17**, 814–818.
147. S.A. Vail, P.J. Krawczuk, D.M. Guldi, A. Palkar, L. Echegoyen, J.P.C. Tomé, M.A. Fazio and D.I. Schuster, *Chem. Eur. J.*, 2005, **11**, 3375–3388.
148. S. Handa, F. Giacalone, S.A. Haque, E. Palomares, N. Martín and J.R. Durrant, *Chem. Eur. J.*, 2005, **11**, 7440–7447.
149. F. D'Souza, M.E. El-Khouly, S. Gadde, M.E. Zandler, A.L. McCarty, Y. Araki and O. Ito, *Tetrahedron*, 2006, **62**, 1967–1978.
150. D. Vincent and J. Cruickshank, *Appl. Opt.*, 1997, **36**, 7794.
151. Y.-P. Sun, G.E. Lawson, J.E. Riggs, B. Ma, N. Wang and D.K. Moton, *J. Phys. Chem. A.*, 1998, **102**, 5520.
152. M. Maggini, C. De Faveri, G. Scorrano, M. Prato, G. Brusatin, M. Guglielmi, M. Meneghetti, R. Signorini and R. Bozio, *Chem. Eur. J.*, 1999, **5**, 2501.
153. G. Brusatin and R. Signorini, *J. Mater. Chem.*, 2002, **12**, 1964.
154. E. Koudoumas, M. Konstantaki, A. Mavromanolakis, S. Couris, M. Fanti, F. Zerbetto, K. Kordatos and M. Prato, *Chem. Eur. J.*, 2003, **9**, 1529.
155. P. Wang, B. Chen, R.M. Metzger, T. Da Ros and M. Prato, *J. Mater. Chem.*, 1997, **7**, 2397.
156. T. Da Ros, M. Prato, M. Carano, P. Ceroni, F. Paolucci and S. Roffia, *J. Am. Chem. Soc.*, 1998, **120**, 11645.
157. D. Felder, D. Guillon, R. Lévy, A. Mathis, J.-F. Nicoud, J.-F. Nierengarten, J.-L. Rehspringer and J. Schell, *J. Mater. Chem.*, 2000, **10**, 887.
158. Y. Rio, G. Accorsi, H. Nierengarten, J.-L. Rehspringer, B. Hönerlage, G. Kopitkovas, A. Chugreev, A. Van Dorsselaer, N. Armaroli and J.-F. Nierengarten, *New J. Chem.*, 2002, **26**, 1146.
159. J.-F. Nierengarten, *C.R. Chimie*, 2003, **6**, 725.
160. C. Boulle, J.-M. Rabreau, P. Hudhomme, M. Cariou, M. Jubault, A. Gorgues, J. Orduna and J. Garín, *Tetrahedron Lett.*, 1997, **38**, 3909.
161. B. Sahraoui, I.V. Kityk, P. Hudhomme and A. Gorgues, *J. Phys. Chem. B.*, 2001, **105**, 6295.

162. D. Kreher, M. Cariou, S.-G. Liu, E. Levillain, J. Veciana, C. Rovira, A. Gorgues and P. Hudhomme, *J. Mater. Chem.*, 2002, **12**, 2137.
163. E. Allard, J. Delaunay, F. Cheng, J. Cousseau, J. Orduna and J. Garín, *Org. Lett.*, 2001, **3**, 3503.
164. B. Sahraoui, I. Fuks-Janczarek, S. Bartkiewicz, K. Matczyszyn, J. Mysliwiec, I.V. Kityk, J. Berdowski, E. Allard and J. Cousseau, *Chem. Phys. Lett.*, 2002, **365**, 327.
165. N. Tsuboya, R. Hamasaki, M. Ito, M. Mitsuishi, T. Miyashita and Y. Yamamoto, *J. Mater. Chem.*, 2003, **13**, 511.
166. M. Lamrani, R. Hamasaki, M. Mitsuishi, T. Miyashita and Y. Yamamoto, *Chem. Commun.*, 2000, 1595.
167. R. Hamasaki, M. Ito, M. Lamrani, M. Mitsuishi, T. Miyashita and Y. Yamamoto, *J. Mater. Chem.*, 2003, **13**, 21.
168. Y. Zhao, Y. Shirai, A.D. Slepkov, L. Cheng, L.B. Alemany, T. Sasaki, F.A. Hegmann and J.M. Tour, *Chem. Eur. J.*, 2005, **11**, 3643.
169. Z. Tian, C. He, C. Liu, W. Yang, J. Yao, Y. Nie, Q. Gong and Y. Liu, *Mater. Chem. Phys.*, 2005, **94**, 444.
170. N. Sun, Y. Wang, Y. Song, Z. Guo, L. Dai and D. Zhu, *Chem. Phys. Lett.*, 2001, **344**, 277.
171. N. Sun, Z.-X. Guo, L. Dai, D. Zhu, Y. Wang and Y. Song, *Chem. Phys. Lett.*, 2002, **356**, 175.
172. F. Cardullo, F. Diederich, L. Echegoyen, T. Habicher, N. Jayaraman, R.M. Leblanc, J.F. Stoddart and S. Wang, *Langmuir*, 1998, **14**, 1955.
173. H. Isobe, H. Mashima, H. Yorimitsu and E. Nakamura, *Org. Lett.*, 2003, **5**, 4461.
174. M. Brettreich and A. Hirsch, *Tetrahedron Lett.*, 1998, **39**, 2731.
175. J.-F. Nierengarten, *New J. Chem.*, 2004, **28**, 1177.
176. A.M. Cassell, C.L. Asplund and J.M. Tour, *Angew. Chem., Int. Ed.*, 1999, **38**, 2403.
177. V. Georgakilas, F. Pellarini, M. Prato, D.M. Guldi, M. Melle-Franco and F. Zerbetto, *Proc. Natl. Acad. Sci. U.S.A.*, 2002, **99**, 5075.
178. T. Da Ros and M. Prato, *Chem. Commun.*, 1999, 663.
179. J.-F. Nierengarten, C. Schall, J.-F. Nicoud, B. Heinrich and D. Guillon, *Tetrahedron Lett.*, 1998, **39**, 5747.
180. D. Felder, M. del Pilar Carreon, J.-L. Gallani, D. Guillon, J.-F. Nierengarten, T. Chuard and R. Deschenaux, *Helv. Chim. Acta*, 2001, **84**, 1119.
181. J.-F. Nierengarten, J.-F. Eckert, Y. Rio, M. del Pilar Carreon, J.-L. Gallani and D. Guillon, *J. Am. Chem. Soc.*, 2001, **123**, 9743.
182. D. Felder, M. Gutiérrez Nava, M. del Pilar Carreon, J.-F. Eckert, M. Luccisano, C. Schall, P. Masson, J.-L. Gallani, B. Heinrich, D. Guillon and J.-F. Nierengarten, *Helv. Chim. Acta*, 2002, **85**, 288.
183. C. Hirano, T. Imae, S. Fujima, Y. Yanagimoto and Y. Takaguchi, *Langmuir*, 2005, **21**, 272.
184. A.P. Maierhofer, M. Brettreich, S. Burghardt, O. Vostrowsky, A. Hirsch, S. Langridge and T.M. Bayerl, *Langmuir*, 2000, **16**, 8884.

185. Y. Gao, Z. Tang, E. Watkins, J. Majewski and H.-L. Wang, *Langmuir*, 2005, **21**, 1416.
186. D. Felder-Flesch, C. Bourgogne, J.-L. Gallani and D. Guillon, *Tetrahedron Lett.*, 2005, **46**, 6507.
187. E. Vuorimaa, T. Vuorinen, N. Tkachenko, O. Cramariuc, T. Hukka, S. Nummelin, A. Shivanyuk, K. Rissanen and H. Lemmetyinen, *Langmuir*, 2001, **17**, 7327.
188. D.M. Guldi, M. Maggini, S. Mondini, F. Guérin and J.H. Fendler, *Langmuir*, 2000, **16**, 1311.
189. T. Vuorinen, K. Kaunisto, N.V. Tkachenko, A. Efimov and H. Lemmetyinen, *Langmuir*, 2005, **21**, 5383.
190. F. Cardinali, J.-L. Gallani, S. Schergna, M. Maggini and J.-F. Nierengarten, *Tetrahedron Lett.*, 2005, **46**, 2969.
191. J. Milliken, D.D. Dominguez, H.H. Nelson and W.R. Barger, *Chem. Mater.*, 1992, **4**, 252.
192. N.C. Maliszewskyi, P.A. Heiney, D.H. Jones, R.M. Strongin, M.A. Cichy and A.B. Smith, *Langmuir*, 1993, **9**, 1439.
193. M. Carano, P. Ceroni, F. Paolucci, S. Roffia, T. Da Ros, M. Prato, M.I. Sluch, C. Pearson, M.C. Petty and M.R. Bryce, *J. Mater. Chem.*, 2000, **10**, 269.
194. S. Conoci, D.M. Guldi, S. Nardis, R. Paolesse, K. Kordatos, M. Prato, G. Ricciardi, M.G.H. Vicente, I. Zilbermann and L. Valli, *Chem. Eur. J.*, 2004, **10**, 6523.
195. T.J. Pinnavaia, *Science*, 1983, **220**, 365.
196. J. Konta, *J. Appl. Clay. Sci.*, 1995, **10**, 275.
197. G. Lagaly, *Solid State Ionics*, 1986, **22**, 43.
198. D. Gournis, V. Georgakilas, M.A. Karakassides, T. Bakas, K. Kordatos, M. Prato, M. Fanti and F. Zerbetto, *J. Am. Chem. Soc.*, 2004, **126**, 8561.
199. S. Burghardt, A. Hirsch, B. Schade, K. Ludwig and C. Böttcher, *Angew. Chem., Int. Ed.*, 2005, **44**, 2976.
200. P. Brough, D. Bonifazi and M. Prato, *Tetrahedron*, 2006, **62**, 2110–2114.
201. M. Mannsberger, A. Kukovecz, V. Georgakilas, J. Rechthalter, F. Hasi, G. Allmaier, M. Prato and H. Kuzmany, *Carbon*, 2004, **42**, 953.
202. D.M. Guldi, A. Gouloumis, P. Vázquez, T. Torres, V. Georgakilas and M. Prato, *J. Am. Chem. Soc.*, 2005, **127**, 5811.

CHAPTER 8

# Plastic Solar Cells Using Fullerene Derivatives in the Photoactive Layer

PIÉTRICK HUDHOMME AND JACK COUSSEAU

Groupe Synthèse Organique et Matériaux Fonctionnels (SOMaF), Laboratoire de Chimie et Ingénierie Moléculaire des Matériaux d'Angers (CIMMA), UMR CNRS 6200, Université d'Angers, 2 Bd Lavoisier, F-49045, Angers, France

## 8.1 Introduction

During the last 20 years following the discovery of $C_{60}$ fullerene[1] and its subsequent large scale production at the beginning of the 1990s,[2] the attention of scientists has continuously increased for its utilization in materials science. Among the different fullerenes, $C_{60}$ was rapidly considered as the most popular of these new carbon architectures. In particular, the interaction of $C_{60}$ with light has attracted considerable interest in the exploration of applications related to photophysical, photochemical and photoinduced charge transfer properties of [60]fullerene derivatives. Its unique electrochemical properties, with six reversible single-electron reduction waves,[3] combined with interesting photophysical properties,[4] make $C_{60}$ a fascinating chromophore to study photodriven redox phenomena. Photoinduced electron and energy transfer processes are of great significance since they govern natural photosynthesis, and considerable efforts have been devoted to the construction of $C_{60}$-based molecular structures as artificial photosynthetic systems.[5] Moreover, it was demonstrated that a π-conjugated polymer was able to efficiently transfer electrons to the $C_{60}$ core giving rise to long-lived charge-separated states.[6,7] Since then, intensive research programs are focused on the utilization of fullerene derivatives acting as electron acceptors in organic solar cells. The development of these devices was stimulated by the inherent advantages of organic materials, such as their light weight and their low cost, and by the possibility of fabricating large active surfaces, thanks to their processability. After only 10 years of studies, it is now clearly established that [60]fullerene-based materials are among the most important candidates for the expansion of plastic solar cells and for renewable sources of electrical energy.[8]

In this chapter, we will mention the current state of the art corresponding to the progressive development of organic [60]fullerene-based photovoltaic cells. Finally, we will briefly present the recent trends in terms of new approaches brought by organic chemists to improve the characteristics of photoactive materials.

## 8.2 From Inorganic to Fullerene-Based Organic Solar Cells

The discovery of the photovoltaic effect in 1839 is commonly ascribed to the french scientist Becquerel. While experimenting with an electrolytic cell made up of two metal electrodes exposed to light, he observed the generation of a photocurrent.[9] Smith[10] showed the photoconductivity in selenium in 1873 and Adams[11] demonstrated that illuminating a junction between selenium and platinum also presents a photovoltaic effect. These two discoveries were the basis of the first selenium solar cell, which was built in 1877. This phenomenon was not fully comprehensible until the theoretical explanation of the photoelectric effect published by Einstein[12] in 1905, this being experimentally proved by Millikan[13] in 1916. The development of monocrystalline silicon production by Czochralski in 1918 enabled the subsequent construction of monocrystalline solar cells and its production in 1941. Bell's Laboratories with Chapin, Fuller and Pearson[14] reported in 1954 on 4.5% efficient silicon solar cells, this efficiency raising to 6% only a few months later. Over the years the improvement reached progressively 24.7% for crystalline silicon solar cells,[15] which still represent the most efficient photovoltaic devices for the conversion of solar light into electrical energy.[16] On the contrary, the prediction for the development of amorphous silicon appears pessimistic at that time. Amorphous silicon, which is an alloy of silicon with hydrogen, was first incorporated in 1976,[17] but the expectancy in this material was curbed essentially by the light-induced degradation in this kind of solar cells. Today, stabilized cell efficiencies reach 13% but this performance is still difficult to achieve on a production level and consequently module efficiencies are in the 6–8% range.[16]

Since the observation in 1932 of the photovoltaic effect in cadmium–selenide (CdS), cadmium solar cells have attracted considerable interest among scientists. More recently, a wide range of novel inorganic semiconducting materials, including CdTe and GaAs, have been tested as photo- and electroactive building blocks, which nowadays belong to the most important materials for solar cells production with efficiency similar to that of silicon modules. These devices use multiple layers of semiconductor materials to absorb light and convert it into electricity more efficiently than single-junction cells. Recently, an exciting result was reported by an industrial company from a gallium indium phosphide/gallium arsenide/germanium (GaInP/GaAs/Ge) triple-junction solar cell approaching 35% efficiency under concentrated sunlight. Nevertheless, the instability of the corresponding cells and the environmental toxicity of these elements are the major drawbacks of this strategy.

Consequently the historical development of photovoltaic devices is traditionally attributed to inorganic semiconductors. Today, silicon-based solar cells are by far the most dominating materials industrially produced for photovoltaic applications.[18] Nevertheless, despite the recent advances in the development of silicon-based photovoltaic cells, the manufacturing of these materials remains rather expensive due to the high-temperature processing of semiconductors in a high-vacuum environment. This largely restricts the fabrication to batch processing onto glass substrates with associated costs. Moreover, traditional silicon solar cells convert only a small portion – about 25% – of the solar spectrum into electricity and waste the rest of the received energy as heat. Therefore, photovoltaic has been so far dominated by solid-state junction devices. But this dominance of silicon systems is now being challenged by the emergence of a new generation of photovoltaic cells based on organic materials. Considering that the solar energy market should grow exponentially in the beginning of the 21st century (the worldwide production of photovoltaic modules, which was 635 MW in 2003, has been growing since 2000 by 30% per year), scientists and private industry are racing to develop new photovoltaic materials and manufacturing processes that offer less expensive and more efficient generation of electricity. Thus there is a large tendency to move progressively from inorganic to organic-based materials for photovoltaic applications.[19]

A first alternative approach introduced by Grätzel[20] is the concept of nano-structured active materials with hybrid organic – inorganic dye-sensitized solar cells. The active junction of this solar cell is composed of a titanium dioxide ($TiO_2$) as the n-type inorganic semiconductor and a monolayer of a ruthenium *tris*(bipyridyl) complex acting as photon harvesting and electron-donor component.[21] The major advantages of these hybrid solar cells are the importance of the active area and the utilization of an appropriate sensitizer which absorbs quantitatively the incident light. Although the power conversion efficiency (PCE) of these devices can reach over 10%, the need for a liquid electrolyte for the photogenerated hole transport remains a major obstacle. Nevertheless, recent developments of dyes absorbing across the visible spectrum and also of all solid-state heterojunctions offer a great potentiality for further cost reduction, simplification of the manufacturing and also higher efficiencies for these dye solar cells.[22]

A second approach consists in using organic molecular semiconductors, which offer the prospect of cheap fabrication together with other attractive features such as mechanical flexibility. These materials are less expensive compared to inorganic semiconductors and can be processed over large areas at relatively low temperatures, either by vacuum sublimation or preferably by film-forming from solution through printing or coating techniques. Consequently, organic solar cells are expected to be low cost compared to traditional cells, thanks to easier and cheaper fabrication processes available at industrial scale. But these devices will have to fulfill simultaneously lifetime efficiency and cost requirements; otherwise they will be limited to a niche market.[23] However, reaching the performances of inorganic solar cells is not necessary to put such organic materials into the market. Moreover, the important point for the 21st century will be to consider ecological and economic aspects of renewable energy sources. Plastic-organic solar cells may contribute to this general context and may emerge as an energy source presenting these requirements.

The development of organic photovoltaic materials as potent alternatives to inorganic has drawn less attention in the past.[24,25] The efficiency remained low until the mid 1980s and the most significant advance was the contribution made by Tang in 1986 with the application of the p/n junction approach in organic photovoltaic cells.[26] A double-layer solar cell was fabricated by thermal evaporation of successive layers of copper phthalocyanine (CuPc) and 3,4,9,10-perylenetetracarboxylic *bis*-benzimidazole (PTCBI) sandwiched between silver and ITO-coated glass substrate. The AM2 PCE was 0.95%, a record value for an organic solar cell at that time.

At the present time, thanks to the invention of Sariciftci and Heeger,[27] one of the most used acceptors in heterojunction photovoltaic cells is $C_{60}$ fullerene. An impressive progress was accomplished in the last few years with an efficiency reaching 5% from polymer bulk-heterojunction devices[28] and 5.7% from low molecular-weight materials.[29] Owing to these results, the point of view concerning organic solar cells has deeply changed and there is undoubtedly an exciting and credible industrial future for fullerene-based plastic-organic solar cells.

## 8.3 Principle of Fullerene-Based Organic Solar Cells

Organic solar cells are based on the photosynthesis process in plants, in which the absorption of sunlight by the chlorophyll "dye" creates a charge separation, thus converting carbon dioxide, water and minerals into organic compounds and oxygen. Typically, solid-state heterojunctions are fabricated using p-type donor (D) and n-type acceptor (A) semiconductors. Experimentally, organic [60]fullerene-based solar cells are fabricated by inserting (or sandwiching) the p-type and n-type materials between two different electrodes. One of the electrode must be (semi-) transparent, often indium tin oxide (ITO), but a thin metal layer can also be used. The other electrode is very often aluminium (calcium, magnesium, gold are also used) (Figure 1).

From the theoretical point of view, the principal feature of the p–n heterojunction is the built-in potential at the interface between both materials presenting a difference of electronegativities.[27] In fact, the absorption of light induces the promotion of an electron from the highest occupied molecular orbital (HOMO) [or the valence band (VB)] of the donor to the lowest unoccupied molecular orbital (LUMO) [or the conducting band (CB)] of the acceptor, generating an exciton at the interface of the junction. Then the built-in potential and the associated difference in electronegativities of materials allow the exciton dissociation.[19] The charge separation occurs at D/A interfaces and free charge carriers are transported through semiconducting materials with the electron reaching the cathode (Al) and the hole reaching the anode (ITO) (Figure 2).

Practically, the photovoltaic characteristics of a solar cell can be evaluated from the $I$–$V$ curve which is shifted down compared to the curve obtained in the dark (Figure 3) and by the different parameters here after:[30]

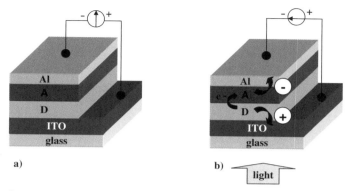

**Figure 1** *Representation of a donor–acceptor heterojunction: (a) structure of the solar cell; (b) under illumination, electron transfer from the donor to the acceptor and generation of excitons followed by charge separation and transport of carriers to the electrodes inducing a photocurrent.*

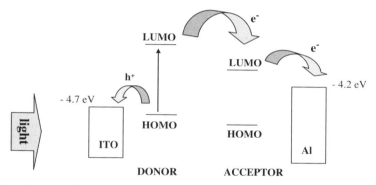

**Figure 2** *Theoretical principle of a donor–acceptor heterojunction. It is usually assumed that for a semiconductor the HOMO corresponds to the VB and the LUMO corresponds to the CB.*

- the open-circuit voltage ($V_{oc}$) is the maximum voltage difference attainable between the two electrodes, typically around 0.5–1.5 V;
- the short-circuit current ($I_{sc}$) or photocurrent is the maximum current that can run through the cell and is associated with an applied voltage equal to zero;
- the rectangle defined by the maximum power point (MPP) and the corresponding maximum power $P_{max}$ (mW cm$^{-2}$) delivered by the photovoltaic cell, with $P_{max} = I_{max} V_{max}$;
- the fill-factor FF (typically around 0.4–0.6) is given by the ratio $I_{max} \cdot V_{max}/V_{oc} \cdot I_{sc}$, where $V_{max}$ and $I_{max}$ are respectively voltage and current density at the MPP;
- the power conversion efficiency (PCE) corresponding to the photovoltaic yield $\eta$ (%) calculated by dividing the maximum power output by the

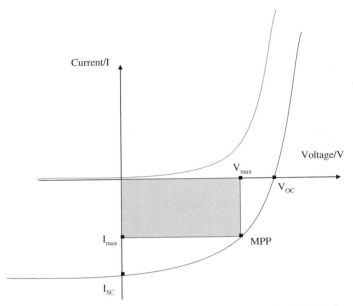

**Figure 3** *I–V characteristics of an organic photovoltaic cell under dark (broken line) and under illumination (solid line) conditions.*

incident light power: $\eta = P_{max}/P_{in}$ where $P_{in}$ is the intensity light power. This is standardized at 100 mW cm$^{-2}$ for solar cell testing with a spectral intensity distribution matching that on the sun on the earth's surface at an incident angle of 45°, which is called the AM1.5 spectrum;
- the internal photon to current efficiency (IPCE) is defined as IPCE = 1.24 $J_{sc}/G.\lambda$ in which $J_{sc}$ corresponds to the short-circuit current density (A cm$^{-2}$), $G$ the power of illumination (W cm$^{-2}$) and $\lambda$ the wavelength of the monochromatic lightning.

## 8.4 Polymer – C$_{60}$ Derivatives Heterojunctions

### 8.4.1 p/n Heterojunction Devices

The concept of heterojunction consists in incorporating two materials with different electron affinities and ionization potentials, thus favouring the exciton dissociation. Historically, this concept was the first approach used for the development of devices incorporating fullerene in the active layer. In their most basic configuration, these organic heterojunction solar cells are formed by the superposition of two layers of materials composed of an electron-donor molecule (p-type component) thin film and the fullerene derivative as the acceptor molecule (n-type component) thin film (Figure 4).

In 1991, preliminary results reported that solvent-cast films of [60]fullerene show photovoltaic response typical of n-type semiconductors.[31] Shortly after,

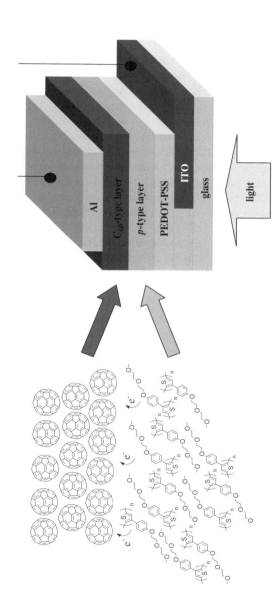

**Figure 4** Schematic representation of a bilayer heterojunction and photoinduced electron transfer occurring at the interface of the device ITO/PEDOT-PSS/PEOPT/C$_{60}$/Al in which PEDOT-PSS is commonly used in organic electronic devices since it lowers the electrode work function of ITO ($-4.7$ eV) to ca. $-5.1$ eV. In addition, it is known to reduce the surface roughness of the underlying ITO layer and thus the probability for electrical shorts.

Brabec et al.[32] demonstrated that, after photo-excitation of the poly[2-methoxy-5-(2′-ethylhexyloxy)-1,4-phenylenevinylene] (MEH-PPV) π-conjugated polymer, an ultra-fast photoinduced electron transfer towards $C_{60}$ occurred in approximately 45 fs. The quantum efficiency for the charge separation is close to unity, whereas the recombination is hindered and takes place in a microsecond regime.[6] Therefore, the charges live long enough to be collected at the electrodes. In 1992, Sariciftci et al. and Morita et al.[33] independently reported photophysical properties of blends composed of pristine $C_{60}$ and MEH-PPV[6] or poly(alkylthiophene) (PAT) respectively (Scheme 1).

The first heterojunction with a conjugated polymer and $C_{60}$ was reported in 1993.[34] In this work, the device ITO/MEH-PPV/$C_{60}$/Au (or Al) was fabricated by sublimation of fullerene onto a MEH-PPV layer spin-coated on ITO-covered glass. This solar cell showed a relative high FF of 0.48 and a PCE of 0.04% under monochromatic illumination. Moreover, it was demonstrated that the photocurrent increased by a factor of twenty when a bilayer photodiode polymer/$C_{60}$ was used instead of a single polymer layer device, indicating that [60]fullerene strongly assists the charge separation. Significant improvement was reported with the photodiode ITO/PEDOT-PSS/PEOPT/$C_{60}$/Al in which the aluminium electrode is used as the electron collector and the poly[3,4-ethylenedioxythiophene)-poly(styrenesulfonate) (PEDOT-PSS) (also called Baytron P)/ITO as the hole extracting collector. This device was characterized by a FF value of 0.78 and a PCE of 1.7% under monochromatic illumination (460 nm with 15 µW cm$^{-2}$ intensity).[35]

However, in such bilayer systems, the effective interaction between the electron donor and the electron acceptor materials is active at the interface, and is limited by the diffusion length of the exciton (near 20 nm maximum). As a consequence, rather low short-circuit photocurrent (Isc) values and

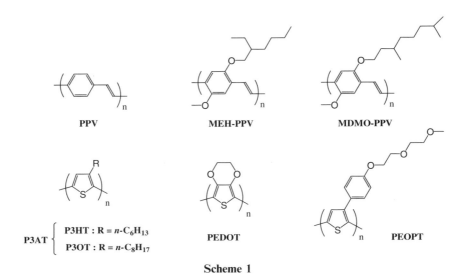

**Scheme 1**

conversion efficiency were obtained. For that reason, preliminary $C_{60}$-based devices yielded only low photovoltaic performances.

An interesting approach was proposed with the concept of interface, which consists in using a soluble $C_{60}$ derivative able to diffuse into a polymer film and to form an intermixed layer, similar in morphology to the interpenetrating donor–acceptor network.[36] Thus the 1-(3-methoxycarbonyl)propyl-1-phenyl[6,6]methanofullerene or [6,6]-phenyl $C_{61}$-butyric acid methyl ester ([60]PCBM) derivative, which is more soluble in organic solvents than pristine $C_{60}$, was associated with poly[2-methoxy-5-(3′,7′-dimethyloctyloxy)-1,4-phenylenevinylene] (MDMO-PPV) in the bilayer device ITO/PEDOT-PSS/MDMO-PPV/[60]PCBM:poly[3-(4-octylphenyl))thiophene] (POPT) (1:0.2)/Al for which the PCE reached 0.5% ($V_{oc}$ = 0.78 V, FF = 0.5). It was demonstrated that [60]PCBM molecules penetrate effectively onto the MDMO-PPV layer, and the POPT polymer added in a small amount was forming a continuous hole pathway in the [60]PCBM layer.

It should be noted that [60]PCBM was initially synthesized by Wudl et al.[37] with the aim of preparing methanofullerene derivatives useful for a large variety of applications. The straightforward synthesis started with the preparation of a stable diazo compound which could be generated in situ from the corresponding hydrazone (Scheme 2). The resulting diazoalcane reacted with $C_{60}$ to reach the [5,6]fulleroid ester which was isolated in 35% yield (83% based on converted $C_{60}$) and found to be essentially pure (95%) with the phenyl group over a former pentagon. The fulleroid was quantitatively isomerized to the [6,6]methanofullerene in refluxing o-dichlorobenzene solution or in the presence of trifluoroacetic acid.[38]

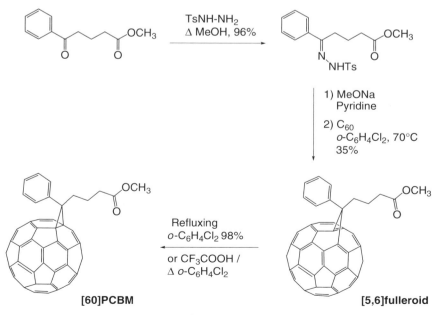

**Scheme 2**

The conversion efficiency of a bilayer heterojunction device is limited by several factors, *i.e.* the weak effective interfacial area available in the layer structure and the poor carrier collection efficiency, because excitons produced at the interface can recombine. In order to overcome these deficiencies, interpenetrating networks were developed as ideal photovoltaic materials for a high-efficiency photovoltaic conversion.

### 8.4.2 "bulk-heterojunction" Devices

#### 8.4.2.1 Principle and State of the Art

A crucial and major breakthrough towards efficient-organic devices was realized by Yu *et al.*[39] with the development of the "bulk-heterojunction" concept, that is an interpenetrating network of a (p-type) donor conjugated polymer and $C_{60}$ or another fullerene derivative as (n-type) acceptor material. Consequently, the photoactive layer of these solar cells consists of blending the conjugated polymer and the fullerene derivative (Figure 5). The effective interaction between the donor and the acceptor compounds within these so-called "bulk-heterojunction" solar cells can take place in the entire device's volume. Subsequently, the separated charge carriers are transported to the electrodes *via* an interpenetrating network. A major shortcoming of this kind of device is the tendency, especially for pristine $C_{60}$, to phase separate and then to crystallize. This aggregation phenomenon imposes important consequences on the solubility of $C_{60}$ within a conjugated polymer matrix. For that reason, intensive efforts on organic solar cells rapidly focused on interpenetrated networks of conjugated polymers with the soluble [60]PCBM derivative. The first example of blend between MEH-PPV and [60]PCBM exhibited a PCE of 2.9%,[8] but under monochromatic low intensity light.[39] Another major difference between $C_{60}$ and [60]PCBM concerns their relative-acceptor strength. Compared with $C_{60}$ ($E_{1/2}^{red} = -0.50$ V *vs.* Ag/AgCl), PCBM ($E_{1/2}^{red} = -0.58$ V *vs.* Ag/AgCl) appears to be a less effective-electron acceptor. The LUMO of $C_{60}$ ($-3.83$ eV) is consequently lower in energy than that of PCBM ($-3.75$ eV).[40] It was found that the $V_{oc}$ is directly related to the acceptor strength of the fullerene derivative incorporated in the bulk-heterojunction (Figure 6).[41]

So far, research programs have generated a continuous improvement of the performances of the cells under AM1.5 illumination using successively couples MDMO-PPV/[60]PCBM and poly(3-hexylthiophene) (P3HT)/[60]PCBM, the latter reaching now a PCE around 5%. In the recent years, regioregular polyalkylthiophenes (PAT) which combine the potential advantages of a better photostability and a smaller bandgap than *e.g.* PPV derivatives, were the most widely used π-donor polymers associated with in [60]PCBM bulk-heterojunction solar cells (Table 1).

It was rapidly evidenced that the PCE of organic-photovoltaic cells was dramatically affected by the morphology of the bulk-heterojunction.[42] First, the choice of the solvent for blending materials allowed to increase the efficiency from *ca.* 1–2.5% by using chlorobenzene instead of toluene. The influence of the

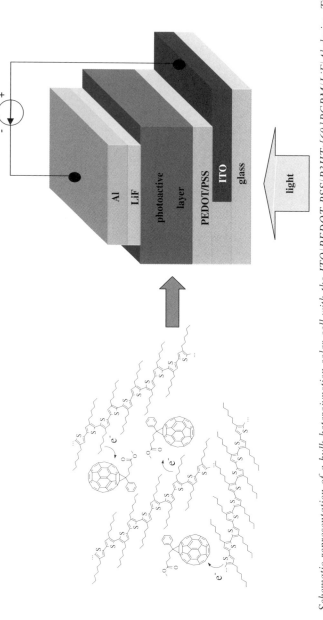

**Figure 5** Schematic representation of a bulk-heterojunction solar cell with the ITO/PEDOT-PSS/P3HT:[60]PCBM/LiF/Al device. The use of a LiF/Al electrode is now commonly adopted providing an ohmic contact between the metal and the organic layer.[122]

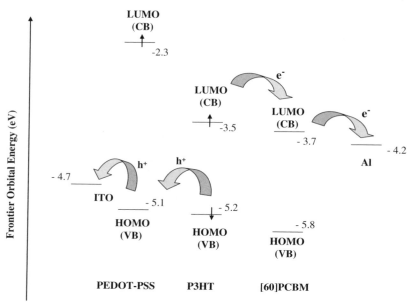

**Figure 6** *HOMO (or VB) and LUMO (or CB) of p-type and n-type semiconductors in the case of the ITO/PEDOT-PSS/P3HT:[60]PCBM/LiF/Al device. HOMO/ LUMO levels of PEDOT-PSS,[123] P3HT and PCBM were estimated from cyclic voltammetry data.[124] Comparatively, the values for MEH-PPV are: HOMO = −5.0 eV and LUMO = −2.8 eV and for MDMO-PPV: HOMO = −5.3 eV and LUMO: −2.8 eV.[125] HOMO value for PCBM is issued from literature.*

solvent on the crystal structure of PCBM was investigated and related to the photovoltaic efficiency.[43] This was also clearly evidenced with the comparison of composite films poly[3-octylthiophene] (P3OT):$C_{60}$, P3OT:[60]PCBM monoadduct and P3OT:[60]PCBM multiadduct (Scheme 3).[44] In this latter case, it was shown that [60]PCBM multiadduct is rather similar to [60]PCBM monoadduct from its electronic properties while the solubility, the film forming properties and the tendency against crystallization were enhanced compared to the monosubstituted [60]PCBM.

Secondly, a post-production treatment, that is annealing the devices and simultaneously applying an external voltage, was shown to improve the characteristics of solar cells.[117] It is suspected that annealing mainly helps in removing the residual solvent, in reducing the free volume and improving the interface with electrode. As a result of annealing, the number of trapping sites is reduced for carrier transport and extraction. Another possible mechanism is the improvement of cathode contact from morphology change. Moreover, it is known that at higher temperature, the hole mobility of the polymer increases dramatically and the applied external potential injects additional charges into the polymer bulk. Nevertheless, it is widely believed that the fundamental limitation of the photocurrent is due to the low mobility of the holes in donor

**Table 1** Evolution of the efficiency of polymer:[60]PCBM bulk-heterojunction solar cells

| π-donor polymer | Acceptor | Polymer/PCBM ratio | Solvent and condition of annealing | Illumination | $J_{sc}$ (mA cm$^{-2}$) | $V_{oc}$ (V) | FF | η | Year |
|---|---|---|---|---|---|---|---|---|---|
| MDMO-PPV | [60]PCBM | 1/3 | Toluene | 488 nm with 10 mW cm$^{-2}$ | | 0.53 | 0.35 | 1.2 | 1999[112] |
| MDMO-PPV | [60]PCBM | 1/4 | C$_6$H$_5$Cl | AM1.5 80 mW cm$^{-2}$ | 5.25 | 0.82 | 0.61 | 2.5 | 2001[113] |
| MDMO-PPV | [60]PCBM | 1/4 | C$_6$H$_5$Cl | AM1.5 100 mW cm$^{-2}$ | 4.9 | 0.87 | 0.6 | 2.55 | 2002[114] |
| MDMO-PPV | [60]PCBM | 1/4 | C$_6$H$_5$Cl | AM1.5 | 4.5 | 0.8 | 0.62 | 2.9 | 2002[115] |
| MDMO-PPV | [70]PCBM | 1/4.6 | C$_6$H$_5$Cl | AM1.5 100 mW cm$^{-2}$ | 7.6 | 0.77 | 0.51 | 3.0 | 2003[116] |
| P3HT | [60]PCBM | | $o$-C$_6$H$_4$Cl$_2$ 4 min at 75°C | AM1.5 80 mW cm$^{-2}$ | 8.5 | 0.55 | 0.6 | 3.5 | 2003[117] |
| P3HT | [60]PCBM | 1/4 | | AM1.5 100 mW cm$^{-2}$ | 15.9 | 0.56 | 0.47 | 3.8 | 2004[118] |
| P3HT | [60]PCBM | 1/1 | $o$-C$_6$H$_4$Cl$_2$ 10 min at 110°C | AM1.5 100 mW cm$^{-2}$ | 10.6 | 0.61 | 0.67 | 4.4 | 2005[119] |
| P3HT | [60]PCBM | 1/0.8 | 5 min at 155°C | AM1.5 80 mW cm$^{-2}$ | 11.1 | 0.6 | 0.54 | 4.9 | 2005[120] |
| P3HT | [60]PCBM | 1/0.8 | C$_6$H$_5$Cl 30 min at 150°C | AM1.5 80 mW cm$^{-2}$ | 9.5 | 0.63 | 0.68 | 5.1 | 2005[28] |
| P3HT | [60]PCBM | 1/1 | 3 min at 155°C | AM1.5 80 mW cm$^{-2}$ | 10.1 | 0.55 | 0.61 | 5.2 | 2005[121] |

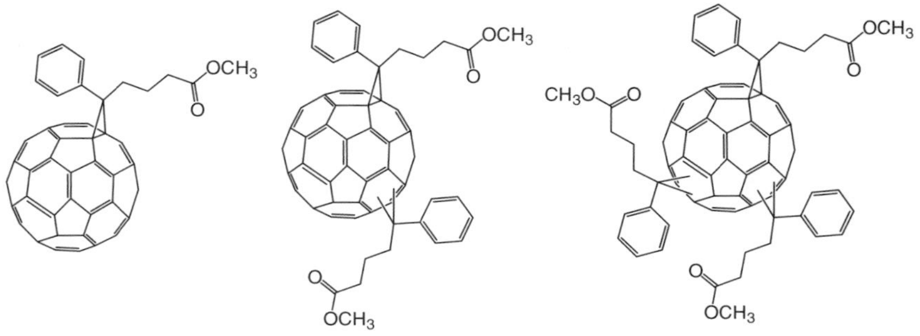

**[60]PCBM monoadduct**   **[60]PCBM bisadduct**   **[60]PCBM trisadduct**

**Scheme 3**

polymers. For example, at room temperature, the electron mobility of [60]PCBM is estimated as $\mu_e = 2.10^{-3}$ cm$^2$ V$^{-1}$ s$^{-1}$ (determined from *I–V* characteristics) which is 4000 greater than the hole mobility in MDMO-PPV due to the less energetic disorder in [60]PCBM.[45] Thirdly, the optimization of the composition weight ratio revealed the important role played by the nanomorphology for the transport properties in polymer:[60]PCBM interpenetrating networks. Concerning the PPV derivatives:[60]PCBM ratio, it was estimated for a rather long time that the 1:4 (w:w) ratio was an optimal value with MDMO-PPV,[113] whereas some recent studies showed that the 1:1 ratio should be superior in the case of P3HT.[46]

However, the performance of bulk-heterojunction solar cells is still affected by some critical factors. Whereas the morphology of the blends plays a crucial role, phase separation and aggregation phenomenon of fullerene can occur, thus reducing the effective donor–acceptor interfacial area as compared to the ideal value. Moreover, in this device architecture, a balanced transport of both photogenerated electrons and holes would be highly important. Although the percolation phenomenon inside donor and acceptor materials is necessary for carrying holes and electrons respectively, clustering effects dramatically affect the transport of charges. A further improvement of the performance of polymer:fullerene solar cells is expected to be attained by controlling the device photoactive layer structural organization at the nanoscale precision. This would avoid the phase separation of donor and acceptor into discrete domains, this phenomenon being responsible of decreasing the charge-carrier generation, the mobility and the collection at the external electrodes.

Some other aspects should be taken into account, such as the solar light absorption coefficients of the materials. Effectively, any further improvement seems limited by an incomplete absorption of the incident light because of a poor match between the absorption spectrum of both polymer and fullerene materials and the solar emission spectrum.[47,116] Therefore, the absorption for the active layer maximizing the absorption of solar radiation in thin films of 100–200 nm is a crucial issue to pursue. However, polymers with bandgap above 2 eV absorb radiation only in the ultraviolet and green part of the visible range. Consequently, chemists are making efforts to synthesize new low bandgap polymers to harvest more photons in the red and near infrared part of the solar spectrum in order to increase the photocurrent.[48] One example was presented with the preparation of hybrid systems, such as fused tetrathiafulvalene (TTF) incorporated into polythiophene architectures (Scheme 4). Thus the polymer PTV-TTF exhibited a low bandgap (1.44 eV) and a photovoltaic effect was observed in the device PTV-TTF: [60]PCBM (1:2) with a PCE of 0.13% ($V_{oc}$ = 0.52 V, $J_{sc}$ = 0.68 mA cm$^{-2}$, FF = 0.3, AM1.5 under 80 mW cm$^{-2}$ illumination).[49] In order to get more efficient devices, it is indeed desirable that the polymer absorbs up to 800 nm since most solar photons are found in this region. In this context, preliminary results were reported for the oligomer PTPTB which exhibited an optical bandgap of 1.60 eV.[50] Incorporating this oligomer in a bulk heterojunction with [60]PCBM in a 1:1 ratio, the device gave rise to a PCE of 0.34% ($V_{oc}$ = 0.67 V, $J_{sc}$ = 0.8 mA cm$^{-2}$, FF = 0.35 under 55 mW cm$^{-2}$ white light illumination). More recently, the fluorene moiety was introduced inside the copolymer backbone thus leading to an alternating polyfluorene copolymer with a green colour (APFO-Green 2).[51] A bulk-heterojunction of this polymer with [60]PCBM was described with a PCE of

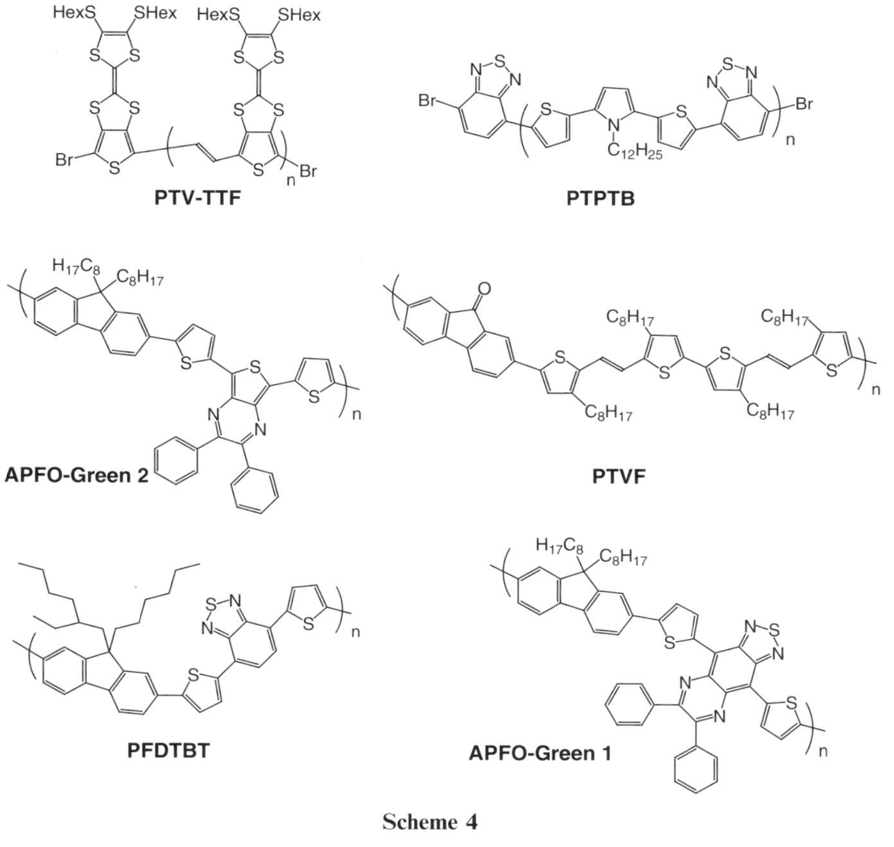

**Scheme 4**

0.9% ($V_{oc}$ = 0.78 V, $J_{sc}$ = 3.0 mA cm$^{-2}$, AM1.5 under 100 mW cm$^{-2}$). Regioregular alternative copolymers containing fluorenone units (PTVF) were also used as donor components in the active layer of bulk-heterojunction solar cells. It was demonstrated that fluorenone-based donors are efficient in converting photons to electrons. These devices exhibit efficiencies up to 1.1% ($V_{oc}$ = 0.5 V, $J_{sc}$ = 2.97 mA cm$^{-2}$, FF = 0.58, AM1.5, 80 mW cm$^{-2}$). These results have to be considered as preliminary because the device fabrication conditions (*i.e.* acceptor:donor ratio, thicknesses, post-annealing treatment) have not been yet optimized.[52] Up to now, better results were obtained with the polymer PFDTBT based on conjugated segments with internal donor–acceptor functions between substituted fluorene. A photodiode was prepared by spin-coating a solution of PFDTBT with [60]PCBM. The device ITO/PEDOT-PSS/PFDTBT:PCBM (1:4)/Al showed a PCE of 2.2% with a remarkable high open-circuit voltage ($V_{oc}$ = 1.04 V, $J_{sc}$ = 4.66 mA cm$^{-2}$, FF = 0.46, $P_{in}$ = 100 mW cm$^{-2}$).[53]

### 8.4.2.2 Newer Fullerene Derivatives in the Photoactive Layer

The absorption spectrum of [60]PCBM is quite similar to that of $C_{60}$ with the main absorption peak at 340 nm and smaller peaks located at 420–500 nm and 550–720 nm. However, the improvement of the light absorption of fullerene derivatives and its relationship with the efficiency of photovoltaic cells was recently demonstrated by Janssen *et al.* with the replacement of [60]PCBM by [70]PCBM.[116] When the acceptor material [70]PCBM was used instead of [60]PCBM in a bulk-heterojunction with MDMO-PPV, 50% higher current densities were obtained and an overall PCE of 3% instead of 2.5% with [60]PCBM. The synthesis of [70]PCBM was performed according to the procedure described for [60]PCBM. However, the 1,3-dipolar addition to $C_{70}$ afforded a mixture of three isomers: the major one (85%) was the chiral α isomer in which the addend is fixed on the C(8)–C(25) bond and two minor achiral β isomers (15%) in which the addend is bound to the C(9)–C(10) bond (Scheme 5).

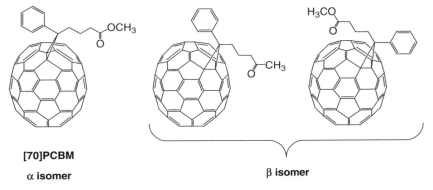

[70]PCBM

α isomer

β isomer

**Scheme 5**

Very recently, a novel methanofullerene derivative called 1,1-*bis*(4,4′-dodecyloxyphenyl)-(5,6)C61 diphenylmethanofullerene (DPM-12) was investigated as a possible electron acceptor in photovoltaic devices (Scheme 6).[54] The synthesis of DPM-12 started from 4,4′-dialkoxysubstituted benzophenone which reacted with tosylhydrazine under acidic catalysis. The diazo compound was generated *in situ* in the presence of sodium methoxide and subsequent 1,3-dipolar cycloaddition with $C_{60}$ occurred in refluxing *o*-dichlorobenzene.[55] Similar 1,3-dipolar cycloaddition to $C_{70}$ led to a mixture of α and β regioisomers of [70]DPM-12.[56] The presence of two alkyl chains in DPM-12 provides the required solubility for blending this material with the polymer. The field-effect electron mobility in DPM-12 ($\mu_e = 2.10^{-4}$ cm$^2$ V$^{-1}$ s$^{-1}$) was found to be forty times lower than the one measured in [60]PCBM ($\mu_e = 8.10^{-3}$ cm$^2$ V$^{-1}$ s$^{-1}$). Nevertheless, interesting photovoltaic characteristics were evidenced in the bulk-heterojunction device ITO/PEDOT-PSS/P3HT:DPM-12 (1:2)/Al with a PCE of 2.3% ($V_{oc} = 0.65$ V, $J_{sc} = 4.7$ mA cm$^{-2}$, FF = 0.58 under white-light illumination 80 mW cm$^{-2}$).

An interesting efficiency/fullerene structure relationship was presented with plastic solar cells using a low bandgap copolymer of fluorene (APFO-Green 1, Scheme 4) blended with [60]PCBM or either the pyrazolino[60] or [70]fullerene derivatives, called 60BTPF and 70BTPF respectively (Scheme 6). It should be noted that the 70BTPF was isolated as a mixture of three regioisomers in relative proportions 4:3:3, deduced from $^1$H and $^{19}$F NMR data. Whereas a PCE of 0.17% was obtained for APFO-Green1:60PCBM (1:4) based devices ($V_{oc} = 0.69$ V, $J_{sc} = 0.53$ mA.cm$^{-2}$, FF = 0.47 under AM 1.5 illumination 100mW.cm$^{-2}$), a significant increase of the PCE (0.3%) was observed for solar cells using 60BTPF as the acceptor ($V_{oc} = 0.54$ V, $J_{sc} = 1.76$ mA.cm$^{-2}$, FF = 0.32).[126] As a consequence of improved photocurrent spectral response, an enhanced photocurrent density ($J_{sc} = 3.4$ mA.cm$^{-2}$) was obtained when 70BTPF was used in the place of 60BTPF. Consequently, a PCE of 0.7% was determined for APFO-Green1:70BTPF devices ($V_{oc} = 0.58$ V, FF = 0.35 under AM 1.5 illumination 100mW.cm$^{-2}$).[127]

In parallel to the development of the polymer:fullerene derivative bulk-heterojunction, $C_{60}$ functionalized macromolecules were investigated for the preparation of all-polymer type solar cells (Scheme 7). The controlled incorporation of fullerenes into well-defined linear polymers was achieved by polycondensation of bifunctional fullerene derivative affording a $C_{60}$-based polymer with an average polymerization degree of 25.[57] Photovoltaic cells were prepared by blending this soluble $C_{60}$-polymer with MDMO-PPV. The device ITO/PEDOT-PSS/MDMO-PPV:$C_{60}$-polymer(1:5)/LiF/Al showed clear photovoltaic behaviour ($V_{oc} = 0.36$ V, $J_{sc} = 0.7$ μA cm$^{-2}$, FF = 0.36) but these weak values of short- and open-circuit currents might be ascribed to the low conductivity of the fullerene polymer. The presence of large solubilizing groups may inhibit interactions between fullerenes, thus decreasing the charge transport.[58]

Solar cells described in this chapter are still far from being optimized and, in future years, it will be necessary to synthesize new materials for improving significantly the performances of organic solar cells. Considering all these new

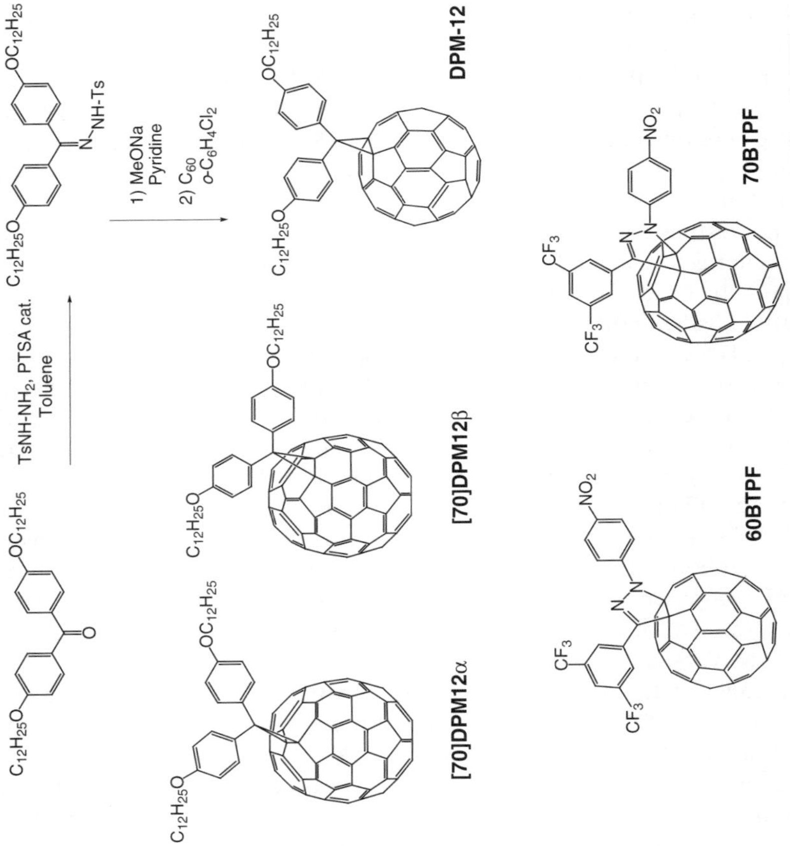

238

Scheme 6

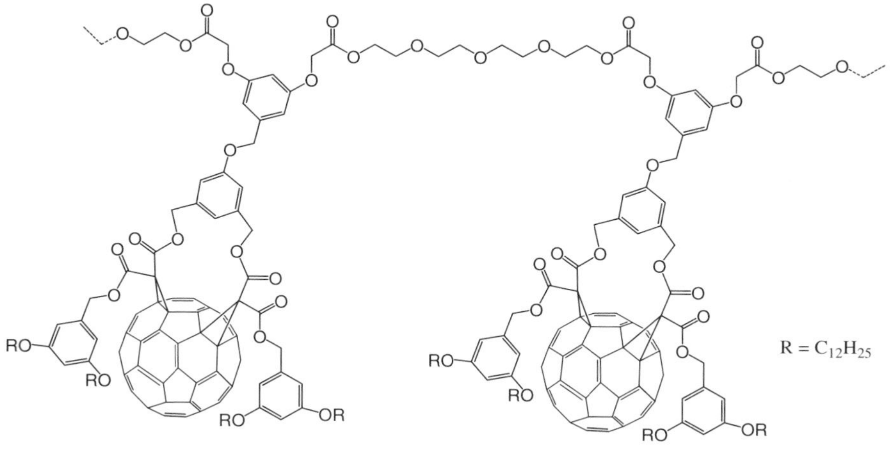

**Scheme 7**

$C_{60}$ derivatives, of course, it will be necessary to convince physicists that other fullerene derivatives represent a possible alternative to [60]PCBM and may allow an improvement of the solar cells efficiency.

At the present time, some derivatives have been recently presented as able to compete with [60]PCBM (Scheme 8). The motivation behind the choice of these designed structures **1–5** was their good solubility in common organic solvents and/or the possibility to be stronger acceptors than [60]PCBM.[59] Effectively, the choice of soluble fullerenes should help the homogeneous mixing with the polymer component and lead to good-quality films.[60] The performance of photovoltaic cells using PCBM analogues **1** with different alkyloxy chains was evaluated and it was concluded that, when the chain does not exceed C-12 substitution, the cell performances are not particularly affected.[61] A structure/activity relationship was also proposed when comparing efficiencies of solar cells using MEH-PPV and two fullerene derivatives $TC_{60}$ and $TDC_{60}$. It was concluded that the more flexible $TDC_{60}$ compared to $TC_{60}$ strongly affects the interpenetrating network morphology leading to better results due to the better compatibility of $TDC_{60}$ with MEH-PPV.[62]

One of the biggest challenges is the access to materials which present the optimal absorption range, matching as well as possible the solar irradiation spectrum. This can be achieved with the development of low bandgap polymers, which may improve the efficiency of organic photovoltaic devices by increasing the absorption in the visible and near infrared region of the solar spectrum. An alternative approach was recently presented with the synthesis of dyad **6**[63] or triad **7**[64] molecular systems associating the perylenediimide dye to $C_{60}$. In this case, the dye could act as an antenna by absorption of sunlight with the aim of inducing an intramolecular energy transfer to the fullerene (Scheme 9). Dendrimer-based light-harvesting structures have attracted attention, the peripheral chromophores being able to transfer the collected energy to the central core of

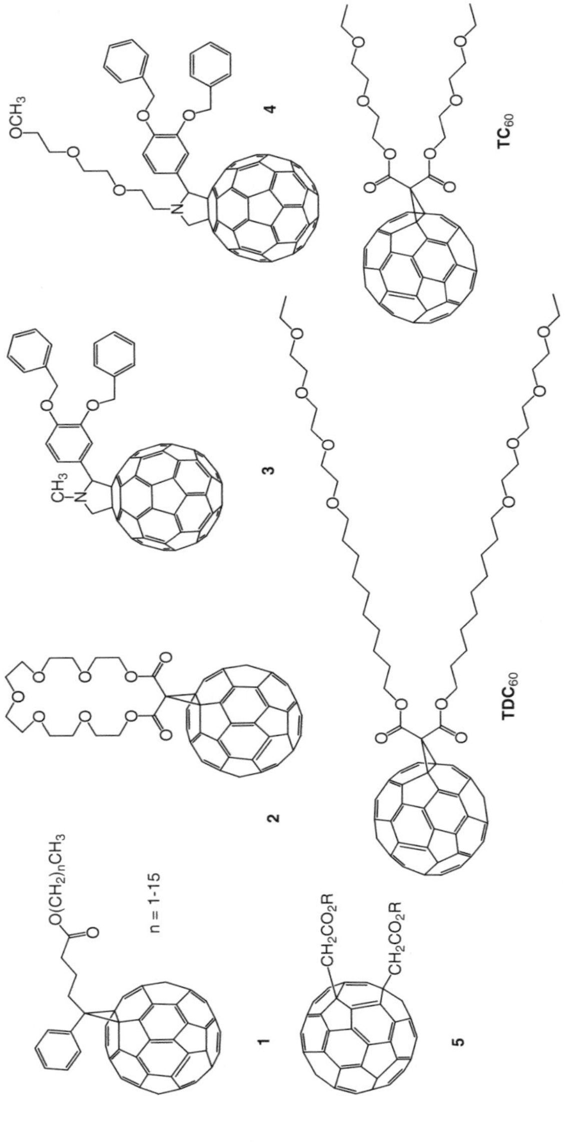

Scheme 8

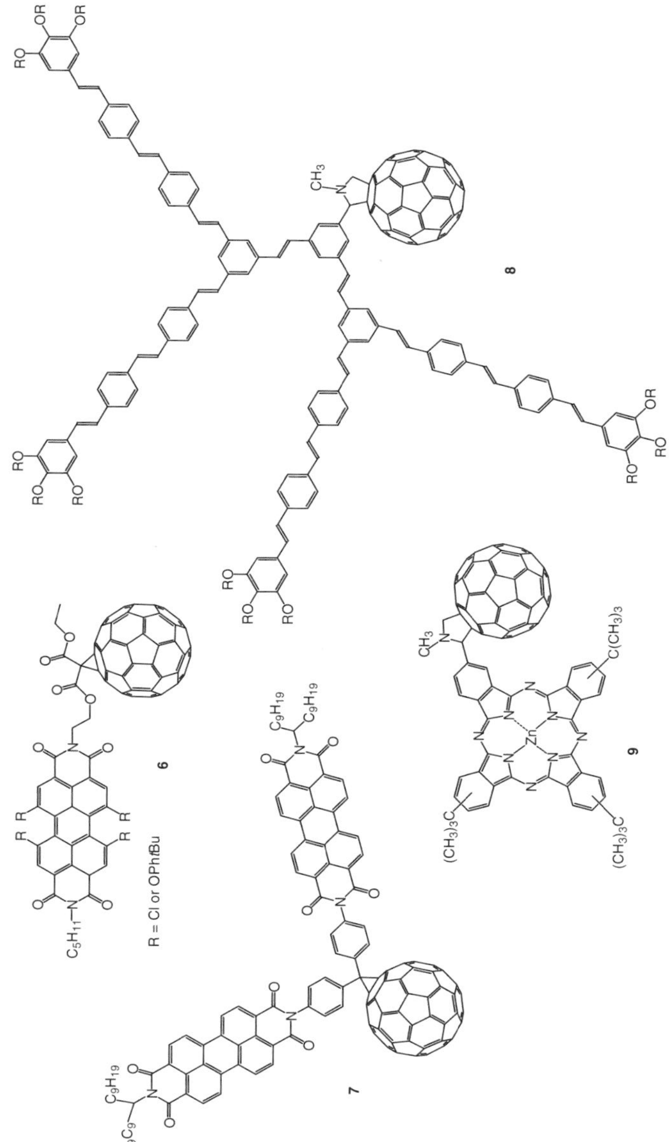
RO
OR
CH3
8
C5H11
R
R = Cl or OPhtBu
6
C9H19
H19C9
7
C(CH3)3
(CH3)3C
Zn
N
9

**Scheme 9**

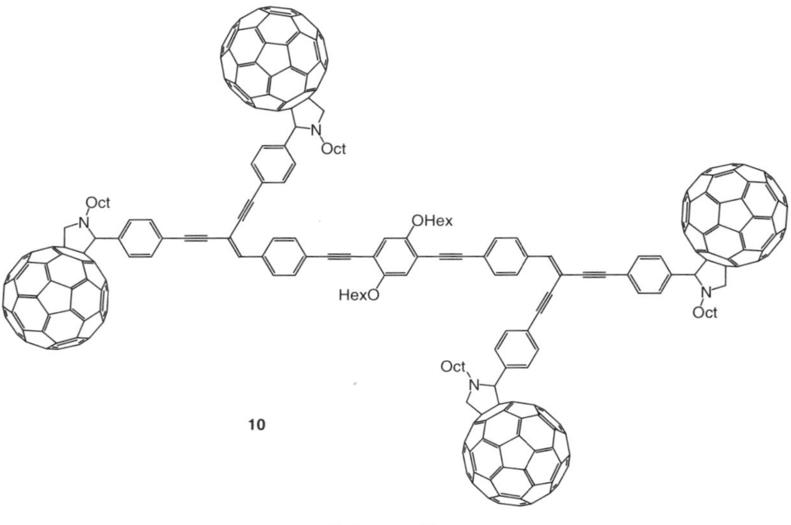

**Scheme 10**

the dendrimer. In this topic, dendrimers **8** comprising a fullerene core and peripheral oligophenylenevinylene (OPV) subunits appear as potentially interesting systems with light-harvesting properties.[65] The extension of the spectral range was also presented with the addition of phthalocyanine (Pc)-$C_{60}$ dyad **9** as a third component to a blend of MDMO-PPV:[60]PCBM, thus increasing the photocurrent in the range of 700 nm. Another strategy consists in the utilization of MDMO-PPV as an antenna system in photovoltaic devices using Pc-$C_{60}$ as donor–acceptor system for charge separation.[66]

Very recently, a visible light-absorbing nanoarray integrating four $C_{60}$ moieties and a single π-conjugated oligomer was synthesized (Scheme 10). It was demonstrated that pure oligomer-tetrafullerene **10** gave no significant photovoltaic effect. On the contrary, such an effect was noted when compound **10** was incorporated in combination with P3HT ($V_{oc} = 0.65$ V, $J_{sc} = 1.2$ mA cm$^{-2}$, FF = 0.65 under white-light illumination 75 mW cm$^{-2}$), this phenomenon being explained by an energy transfer occurring between $C_{60}$ units and the π-conjugated oligomer.[67]

## 8.4.3 "Double-cable" Polymer-$C_{60}$ Derivatives

In the bulk-heterojunction device, the miscibility of both donor and acceptor materials appears to be limited, especially because clusters of fullerenes can be formed within the photoactive film, so that the transport of electrons is located in separated domains. In order to overcome phase segregation and to reach a homogeneous electron transport pathway, an elegant solution has been proposed. It consists in the preparation of the so-called double-cable molecular systems, in which fullerene-based substituents are grafted onto the conjugated polymer chain. In this way, the effective donor–acceptor interfacial area is

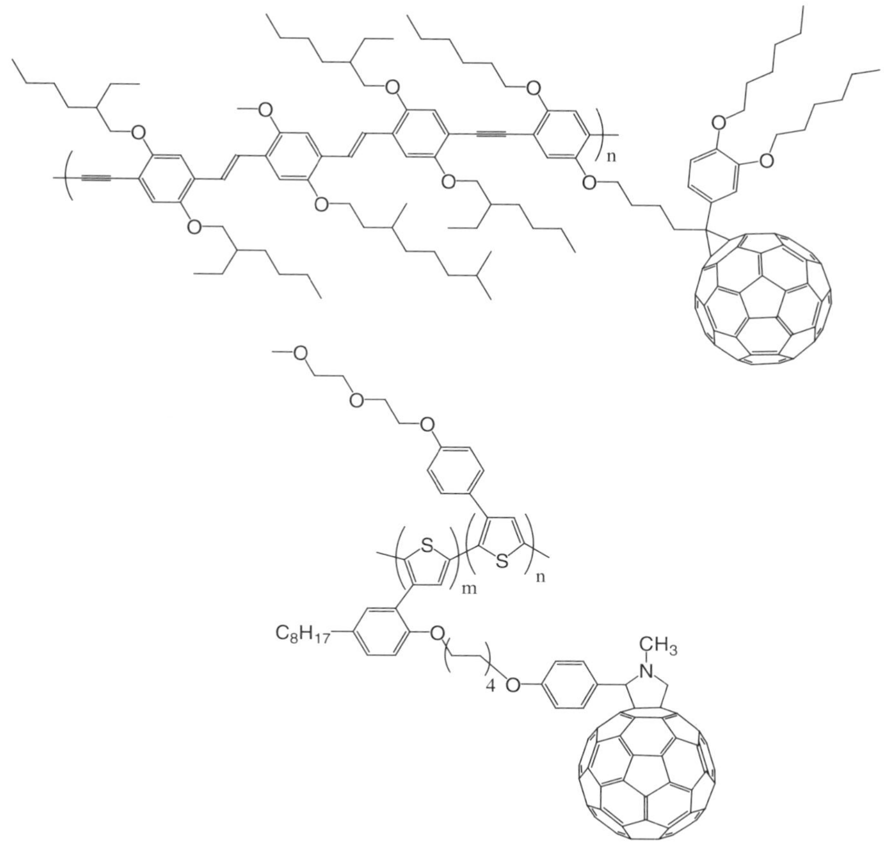

## Scheme 11

expected to be maximized, and the phase separation and clustering phenomena should be prevented as well. Basically, the realisation of effective double-cable polymers will bring the p/n heterojunction at the molecular level.[68] Double-cable materials can be consequently seen at the frontier with the molecular heterojunction.[69]

The first example of double-cable-based solar cell was using an hybrid of poly(p-phenylenevinylene) and poly(p-phenyleneethynylene) covalently linked to methanofullerene moieties. The photovoltaic cell was prepared from a spin-coated film of this double-cable polymer-$C_{60}$ and this device was competitive with those resulting from bilayer or bulk-heterojunction ($V_{oc} = 0.83$ V, $J_{sc} = 0.42$ mA cm$^{-2}$, FF $= 0.29$ under white light irradiation (AM1.5 100 mW cm$^{-2}$) (Scheme 11).[70]

Soluble double-cable copolythiophenes carrying fulleropyrrolidine moieties and solubilizing polyether chains were synthesized.[71] The photodiode composed of this double-cable polymer-$C_{60}$ presented promising photovoltaic characteristics. A PCE of 0.60% ($V_{oc} = 0.75$, FF $= 0.25$) was obtained under monochromatic irradiation at $\lambda = 505$ nm at low irradiation intensity (0.10 mW cm$^{-2}$).[35] These characteristics clearly confirm the potentialities of $C_{60}$-derivatized conjugated polymers for the realization of organic solar cells.

## 8.5 Molecular Scale $C_{60}$-Based Heterojunctions

### 8.5.1 Low-Molecular-Weight Materials

The PCE of photovoltaic cells using low-molecular-weight organic materials has increased steadily over the past decade and this progress is attributable to the introduction of the donor-acceptor (p/n) heterojunction and the possibility of vacuum-deposition for these "small molecules".[72] Since the 1% efficient thin-film photovoltaic cell described in 1986 using copper phthalocyanine (CuPc) as the donor and 3,4,9,10-perylenetetracarboxylic *bis*-benzimidazole (PTCBI) as the acceptor,[26] a higher efficiency of $2.4 \pm 0.3\%$ was obtained from these materials with the device (ITO/PEDOT/CuPc/PTCBI/BcP/Al) thanks to the incorporation of the exciton-blocking layer BcP† inserted between the photoactive organic layers and the metal electrode.[73] The major improvement related to the utilization of "small molecules" in p/n heterojunction concerned the substitution of PTCBI by the fullerene $C_{60}$ and the design of new device architectures. The improved results with $C_{60}$ replacing PTCBI are essentially the consequence of a longer exciton diffusion length [(77±10) Å for $C_{60}$ compared to (30±3) Å for PTCBI]. As an aside, it should be noted that the first example can be attributed to the observation of a photocurrent in a heterojunction using the electron-donor tetrathiafulvalene (TTF) and $C_{60}$ sandwiched between ITO and Au electrodes (ITO/$C_{60}$/TTF/Au), with the evidence of a photoinduced charge generation at the interface (Scheme 12).[74]

Forrest *et al.* reported a double-heterostructure solar cell ITO/PEDOT-PSS/CuPc/$C_{60}$/BcP/Al exhibiting a PCE of $3.6 \pm 0.2\%$ ($V_{oc} = 0.58$ V, $J_{sc} = 18.8$ mA cm$^{-2}$, FF = 0.52) under AM1.5 with a spectral illumination of 150 mW cm$^{-2}$ (1.5 suns) (Figure 7a).[75] The efficiency increased with the incident optical power density reaching $4.2 \pm 0.2\%$ with a FF superior to 0.6 under 4–12 suns simulated AM1.5 illumination for the device ITO/CuPc/$C_{60}$/BcP/Ag.[76] Stacking a number of ultrathin organic photovoltaic cells in series has proven to be another means to increase device efficiency.[77] Applying the concept of bulk-heterojunction to small molecules, the advantage of a mixed donor–acceptor molecular heterojunction was demonstrated with a film mixture of donor and acceptor molecules inserted between two layers of pure donor and acceptor composition (Figure 7b). Using the donor CuPc and the acceptor $C_{60}$, this hybrid planar-mixed heterojunction device showed a PCE of $5.0 \pm 0.3\%$ under 1–4 suns

---

†Bathocuproine or 2,9-dimethyl-4,7-diphenyl-1,10-phenanthroline has been found to better transport electrons to the cathode from the adjoining acceptor layer.

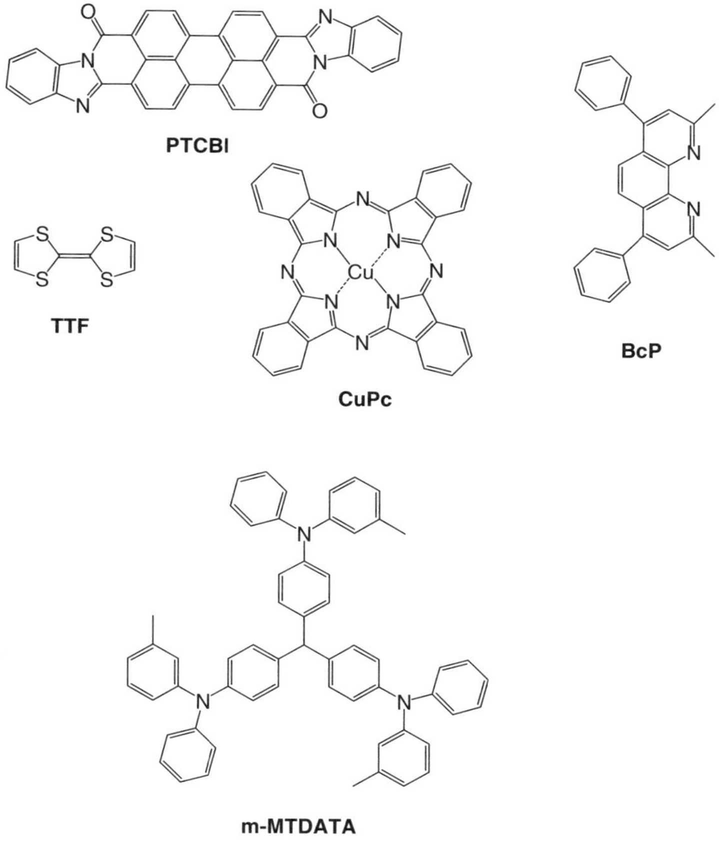

**Scheme 12**

Structures shown: PTCBI, TTF, CuPc, BcP, m-MTDATA

simulated AM1.5 solar illumination.[78] The considerable potential of the tandem geometry consisting of two hybrid planar-mixed molecular junction cells stacked in series was proved with the increase of the efficiency up to 5.7% on an active device area of 0.8 mm$^2$ ($V_{oc}$ = 1.03 V, $J_{sc}$ = 9.7 mA cm$^{-2}$, FF = 0.59 under 1 sun simulated AM1.5) (Figure 7c).[29]

## 8.5.2 Molecular π-Donor – C$_{60}$ Derivatives Solar Cells

Intensive efforts have been recently focused on the development of a molecular strategy especially to limiting the tendency of C$_{60}$ to phase separate from the polymer in the polymer:C$_{60}$ blend. A first molecular strategy was investigated with the preparation by co-evaporation of blends using conjugated oligomers and C$_{60}$ as model systems for understanding optoelectronic properties of polymers and the morphology of the devices. This concept was developed using medium-sized and well-defined π-conjugated oligomers, such as sexithiophene (6T)[79] or oligo(phenylenevinylene) derivatives with the major objective to study

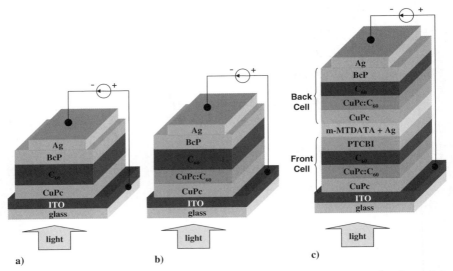

**Figure 7** *Device structures (a) using the donor CuPc, the acceptor $C_{60}$ molecules and the exciton-blocking layer BcP; (b) organic hybrid planar-mixed heterojunction solar cell; (c) asymmetric tandem organic photovoltaic cells with planar-mixed molecular heterojunctions (the two subcells are connected in series by a charge recombination zone for electrons generated in the front cell and holes generated in the back cell, this recombination centres being Ag clusters buried in a layer of p-doped m-MTDATA.*

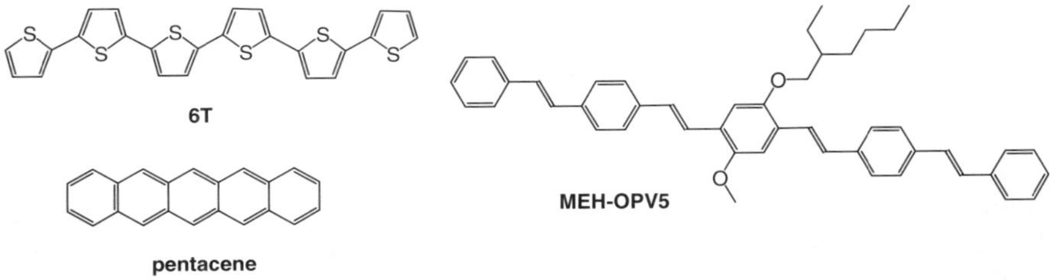

**Scheme 13**

the morphology of the heterojunction in relation with the π-conjugation of the donor (Scheme 13).[80] A complete study was performed with the MEH-OPV5 donor and photovoltaic devices with $C_{60}$ were prepared by co-evaporation of both materials forming films presenting a good percolation of donor and acceptor. A high PCE of 2.2% ($V_{oc}$ = 0.85 V, $J_{sc}$ = 6.3 mA cm$^{-2}$, FF = 0.41) was reached for the solar cell ITO/PEDOT/MEH-OPV5/$C_{60}$ (1:4)/Al, this study being completed with the relation of the morphology of the device films with photovoltaic characteristics.[81] An efficient solar cell based on a junction between polycrystalline pentacene as p-type semiconductor and $C_{60}$ was recently reported. The device ITO/pentacene/$C_{60}$/BcP/Al exhibited a PCE of 2.7 ± 0.4% [$V_{oc}$ =

0.36 V, $J_{sc}$ = 15 mA cm$^{-2}$, FF = 0.50 under illumination of the broadband light (100 mW cm$^{-2}$, 350–900 nm)].[82]

Donor and acceptor molecules are usually incompatible and tend to undergo uncontrolled macrophase separation with in particular the phase separation and clustering of fullerene units, thus reducing the effective donor–acceptor interfacial area and also the efficiency of the devices. In order to prevent such undesirable effects, a second molecular strategy was proposed by simply chemically linking the hole-conducting moiety to the electron-conducting fullerene subunit. Consequently, thanks to the considerable development of fullerene chemistry during the past decade, the direct covalent bonding of different π-electron donors[83] and π-conjugated oligomeric systems[84] has emerged as a very active field of research to reach new organic photovoltaic devices. This has given rise to the synthesis of a huge number of dyads or triads involving donor groups covalently attached onto $C_{60}$. Such molecular assemblies are of particular interest in that they can exhibit characteristic electronic and excited-state properties, which make them promising candidates for the investigation of photoinduced electron transfer processes and long-lived charge-separated states $D^{\cdot+}$-$C_{60}^{\cdot-}$. Effectively, when $C_{60}$ is involved in electron transfer processes, it gives rise to a rapid photoinduced charge separation and a further slow charge recombination in the dark, this phenomenon being ascribed to the small reorganization energy of [60]fullerene. That could explain the acceleration effect of photoinduced charge separation as well as charge shift and the deceleration effect of charge recombination in donor-linked fullerenes. This has been perfectly demonstrated in the case of the electroactive triad $C_{60}$-exTTF$_1$-exTTF$_2$ which presents a remarkable lifetime of 111 μs for the charge-separated state $C_{60}^{\cdot-}$-exTTF$_1^{+\cdot}$-exTTF$_2$.[85] This value is among the highest lifetimes found for electroactive $C_{60}$-based triads and this particular stabilization is provided by the gain of aromaticity and planarity of the π-extended TTF unit upon oxidation (Scheme 14).

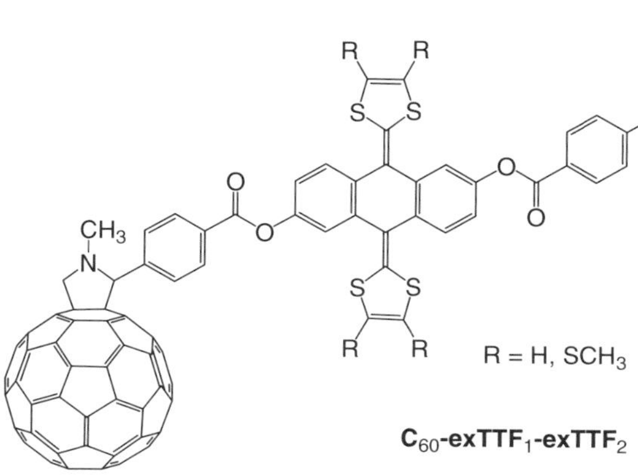

R = H, SCH$_3$

**C$_{60}$-exTTF$_1$-exTTF$_2$**

**Scheme 14**

Considering the covalent fixation of a π-conjugated oligomer onto $C_{60}$, oligomers present some important drawbacks compared to polymers. Soluble conjugated polymers have in general better film-forming properties than oligomers, this being important for the device fabrication and for the mechanical properties of the solar cell. Furthermore, polymers are generally prepared by more straightforward synthetic methods than the tedious multi-step syntheses required for the preparation of well-defined π-conjugated oligomers. Finally, because of the decrease of the HOMO–LUMO gap with the chain extension, conjugated polymers should present an optimized absorption of the solar emission spectrum.

However, several advantages are inherent in oligomers. Because of their well-defined chemical structure, $C_{60}$-derivatized π-conjugated oligomers make possible the establishment of structure–properties relationships in series of conjugated structures with increasing chain length and thus decreasing HOMO–LUMO gap. In particular, this allows detailed investigations of the effects of chain extension on the elemental photophysical processes of light absorption, photoinduced energy and/or electron transfer and performances of corresponding solar cells. Using well-defined conjugated oligomers should avoid the statistical distribution of chain lengths resulting from the non-regioselective synthesis of conjugated polymers. Furthermore, covalent donor-$C_{60}$ systems can help to solve the problem of phase segregation and clustering phenomena. The molecular approach could provide new synthetic tools for a better control of the interface between the π-conjugated donor and the fullerene acceptor, thus allowing a fine tuning of relevant parameters, such as the ratio, the distance, the relative orientation and the mode of connection of the donor and acceptor groups. Besides, $C_{60}$-derivatized π-conjugated systems can also open interesting opportunities to develop nanoscale or molecular photovoltaic devices.

The utilization of a fulleropyrrolidine derivative attached to an oligophenylenevinylene can be considered as the first example of a molecular heterojunction specifically designed for photovoltaic conversion.[86] In 1999, Nierengarten et al.[87] reported the synthesis of a trimeric oligophenylenevinylene (OPV) moiety covalently linked to $C_{60}$. This fulleropyrrolidine OPV-$C_{60}$ **11** was incorporated in a photovoltaic device by spin-casting between aluminium and ITO electrodes (Figure 8). The device ITO/OPV-$C_{60}$ **11**/Al delivered a low PCE of 0.01% ($V_{oc}$ = 0.46 V, $J_{sc}$ = 10 µA cm$^{-2}$, FF = 0.3) under monochromatic irradiation (400 nm, 12 mW cm$^{-2}$). This study showed that plastic solar cells can be obtained by chemically linking the hole-conducting and the electron-conducting units. The length of the OPV was increased from three to four units and the performances of the ITO/OPV-$C_{60}$/**12**/Al were significantly improved with a monochromatic PCE of 0.03%. The limited efficiency of these devices was attributed to the competition between energy transfer and electron transfer.

The OPV-$C_{60}$ hybrid **13** was tested as an active material in photovoltaic cells and the device ITO/PEDOT-PSS/**13**/Al presented enhanced I–V characteristics ($V_{oc}$ = 0.65 V, $J_{sc}$ = 235 µA cm$^{-2}$ under white light illumination at 65 mW cm$^{-2}$) (Scheme 15). Even if these solar cells are not prepared under similar conditions

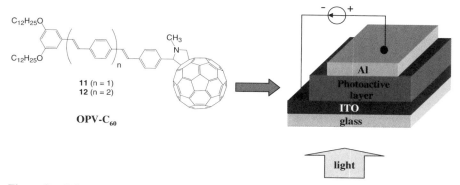

**Figure 8** *Schematic structure of the ITO/11/Al device.*

used for OPV-$C_{60}$ derivatives **11** and **12**, the increased donating ability of the OPV moiety is an important argument for the improvement of the device performances.[88] Photovoltaic devices were prepared from oligophenyleneethynylene – $C_{60}$ (OPE-$C_{60}$) oligomers **14** and **15** and it is to be noted that the performances were significantly improved from the *N,N*-dialkylaniline derivative **15** when compared to the analogue **14**.[89] This clearly demonstrated the interest of the molecular approach, which allows to establish a structure/activity relationship.

Oligothiophene-$C_{60}$ systems were also investigated in this molecular area as interesting opportunities to improve the conversion efficiency. Photovoltaic cells based on oligomer-$C_{60}$ dyad **16** were described displaying an efficiency of about 0.2% under white light illumination at 80 mW cm$^{-2}$.[90] Moreover, the highly efficient photoinduced electron transfer occurring in oligothiophene-$C_{60}$ dyads (nT-$C_{60}$) was exploited to investigate the potential of these dyads as photovoltaic materials. Thus the sandwich device structure Al/nT-$C_{60}$/Au was fabricated by spin-coating the organic layer on a vacuum-deposited Al film and the best results were obtained for the **16T-$C_{60}$** with a reasonably high monochromatic ($\lambda = 456$ nm) PCE of 0.4% ($V_{oc} = 0.65$ V, $J_{sc} = 148$ nA cm$^{-2}$, FF $= 0.34$).[91] The role of the absorption parameter was evidenced with the fullerene-azothiophene dyad **17**. This photoactive material exhibits a wide absorption spectrum in the visible range centred at about 570 nm. The device showed a relatively high performance with an efficiency of 0.37% under white-light irradiation of intensity 80 mW cm$^{-2}$ ($V_{oc} = 0.66$ V, $J_{sc} = 1.6$ mA cm$^{-2}$, FF $= 0.28$).[92]

## 8.6 Supramolecular Nanostructured $C_{60}$-Based Devices

It will be necessary in the future to consider a chemical self-organization of materials inside a physical nanostructured device, able to improve efficiently both charge separation and charge transport. For such purposes, the supramolecular design of photovoltaic devices present inherent advantages due to

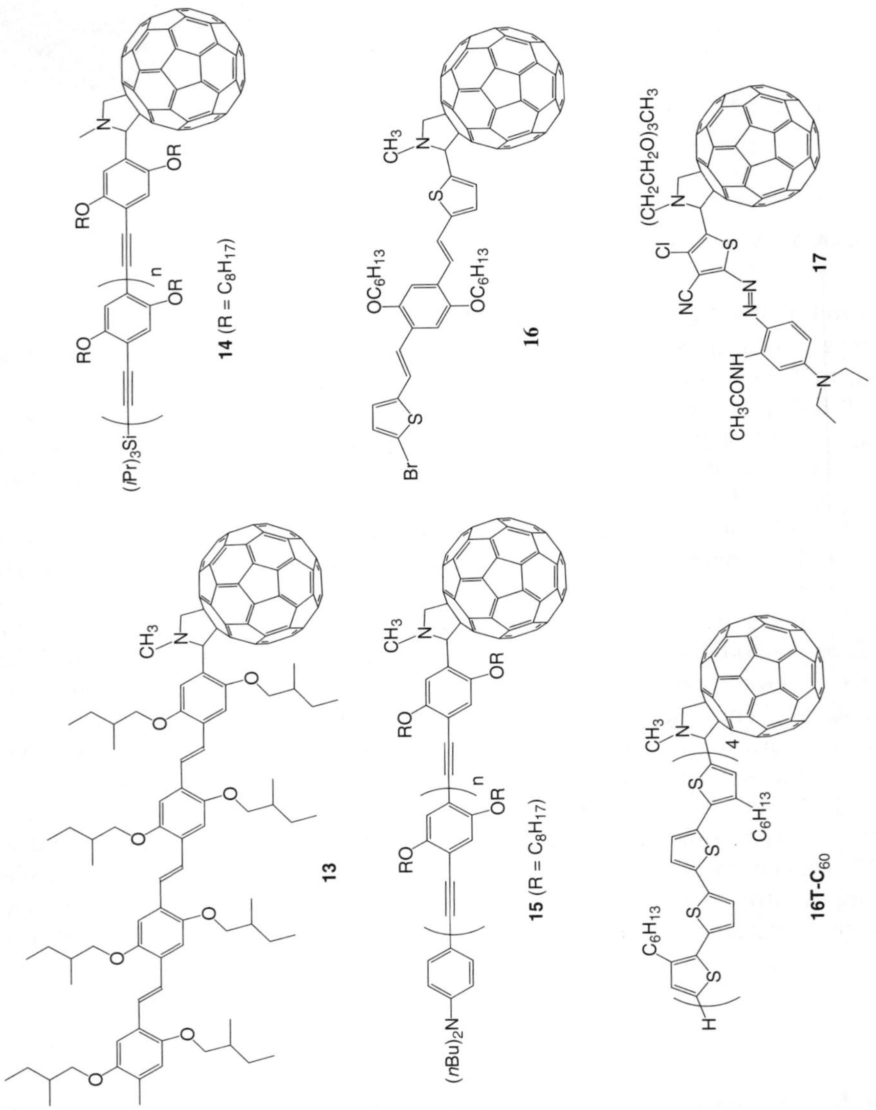

Scheme 15

the possibility to anchor the molecule onto the electrode, but also to control the spatial organization of molecular assemblies.[93]

## 8.6.1 Self-Assembled Monolayers

Significant efforts have been made in recent years to explore properties of donor–acceptor systems. Since covalently linked donor-$C_{60}$ arrays can produce a long-lived charge-separated state with a high quantum yield, the challenge was to employ them in organized structures. Self-assembled monolayers (SAMs) of donor-$C_{60}$ linked systems constitute a highly promising methodology for the construction of nanostructured molecular photovoltaic devices, because donor-$C_{60}$ dyads can be well-packed on electrode surfaces to reveal unidirectional orientation.[94]

The first example was the utilization of oligothiophene-$C_{60}$ dyad **18** linked to a gold surface with a disulfide anchoring group (Scheme 16). Photo-electrochemical experiments were carried out on a monolayer of this dyad immersed in a solution containing methylviologen ($MV^{2+}$) as electron carrier. A photocurrent of several hundreds of nA cm$^{-2}$ was observed under 440 nm monochromatic irradiation at 22 mW cm$^{-2}$.[95] Another class of SAMs was proposed using a three-armed anchor of oligothiophene-$C_{60}$ dyad **19** onto the surface, this tripodal mode of fixation leading to a more dense surface coverage than the previous single-point attachment. Under monochromatic light irradiation (0.65 mW cm$^{-2}$ at $\lambda = 400$ nm) the device delivered a current density approaching 1 µA cm$^{-2}$ in the presence of methylviologen as electron carrier.[96] Very recently, the rigidified and higher nanostructured triad system **20** was presented in which the fixation of the anchoring site onto the nitrogen atom of the pyrrole group allowed the conjugated chain to adopt an orientation parallel to the surface. Photo-electrochemical experiments under 400 nm irradiation with 0.85 mW cm$^{-2}$ illumination power showed that the device Au/$C_{60}$-triad **20**/$MV^{2+}$/Pt delivered a current density of 3200 nA cm$^{-2}$ with a high quantum yield (51%). This current density, considerably larger than those reported for related systems, was attributed to the possible orientation of the conjugated system perpendicularly to the direction of the incident light.[97]

Molecular photovoltaic devices were built from porphyrin-$C_{60}$ linked systems self-assembled onto an ITO electrode (Scheme 17). In this case, a $C_{60}$ diacid was directly attached on the ITO surface to enhance electrical communication between the $C_{60}$ unit and the ITO electrode. The high quantum yield (99%) and the long lifetime (770 ns) of the charge-separated state of the zinc porphyrin-$C_{60}$ dyad play essential roles for the generation of photocurrent in these systems attached to ITO surface.[98]

Another representative example was presented with the construction of a photovoltaic device based on SAMs of covalently linked $C_{60}$-porphyrin-ferrocene triads perpendicularly oriented on the gold electrode thanks to the thioalkyl group as anchor (Scheme 18). Through irradiation of the triad-modified gold electrode and in the presence of methylviologen, photocurrent

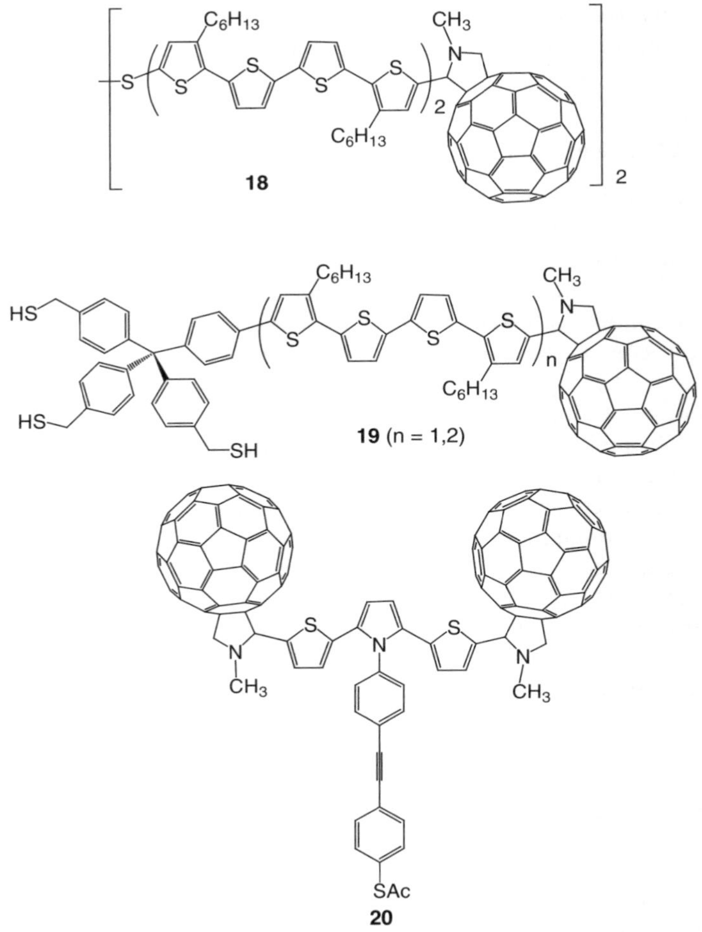

**Scheme 16**

generation was observed with the demonstration of a vectorial electron transfer process inside covalently linked molecules.[99]

### 8.6.2 Langmuir and Layer by Layer (LBL) Films

Methodologies able to organize fullerenes into well-defined two- and three-dimensional films are important to design new fullerene-based materials. One of the most widely pursued approaches has been the preparation of Langmuir monolayers at the air–water interface and their subsequent transfer onto solid substrates. However, monolayers of fullerene derivatives are difficult to prepare because of the strong aggregation of fullerene. In addition, the resulting Langmuir films are usually rigid and, as a result, their transfer onto solid substrates is generally complicated.[100] An alternative to Langmuir films has emerged with

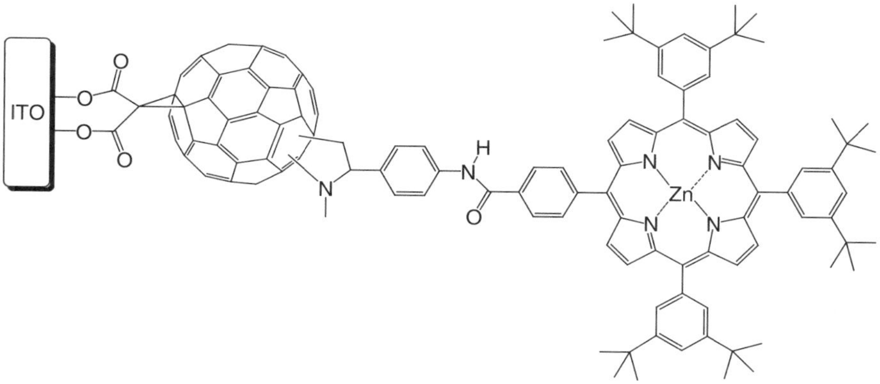

**Scheme 17**

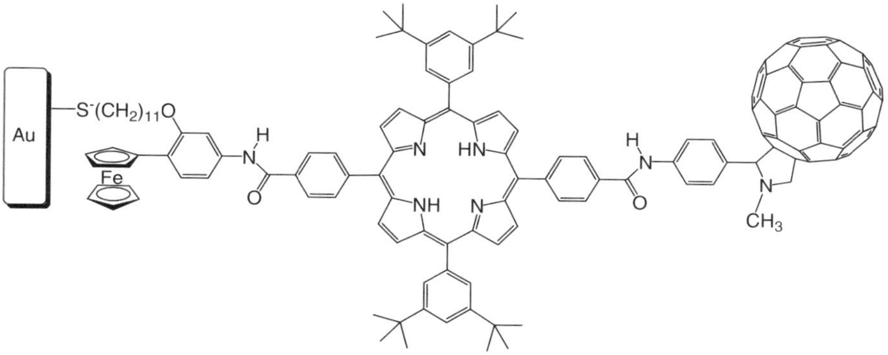

**Scheme 18**

the development of the LBL technique as a powerful possibility towards the modification of photoactive electrodes. For instance, a fullerene-nickel porphyrin dyad $C_{60}$-NiP was designed for both Langmuir and LBL deposition experiments (Scheme 19). Modified ITO electrodes were probed in photocurrent experiments, in which the LBL-modified electrodes reveal smaller photon to current conversion efficiencies compared to the Langmuir-modified electrodes.[100] In this field, a promising IPCE of 5.4% was reported from LBL assembled photovoltaic thin films due to electrostatic interactions between CdTe nanoparticules and water-soluble $C_{60}$ derivative **21**.[101] Moreover, multilayered assemblies obtained thanks to the LBL deposition of positively charged PPV derivative and negatively charged $C_{60}$ derivative (sodium salt of hexa(sulfobutyl)fullerene) were obtained. This approach provides a new avenue for producing materials for photovoltaic conversion and an alternative for the fabrication of a molecular array for efficient energy and electron transfer.[102] The water-based self-assembly procedure, as opposed to the more conventional spin-coating and/or vacuum evaporation techniques, is emerging as a new means to create organized organic photovoltaic

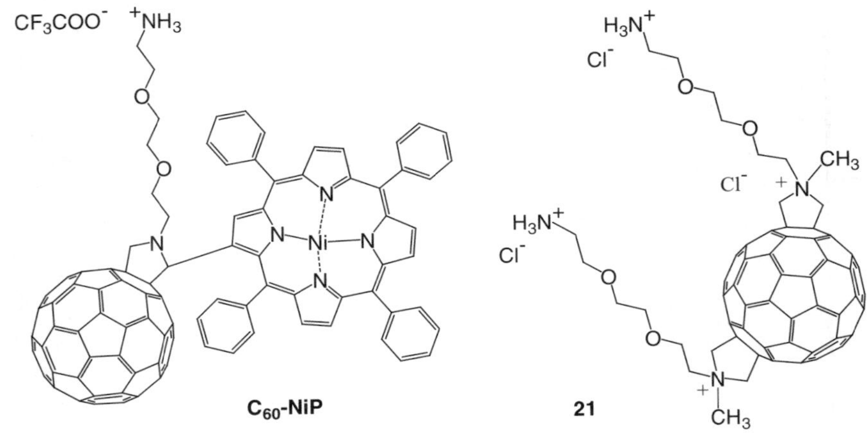

**Scheme 19**

devices. In this process, layers of oppositely charged species are sequentially adsorbed onto a substrate from an aqueous solution and a film is built due to electrostatic attraction between the layers.[103]

A supramolecular LBL system was reported with the adsorption of a cationic-charged $C_{60}$-*tris*(2,2'-bipyridine)ruthenium(II) complex dyad on an anion-covered ITO surface thanks to coulombic interaction using the cationic charge of the ruthenium complex moiety (Scheme 20). Through irradiation of the dyad-modified electrode, progressive an intensification of the photocurrent was measured while increasing the number of layers.[104]

A conjugated polyelectrolyte, which features anionic sulfonate solubilizing groups, and a water-soluble cationic methanofullerene were recently used as electron donor and acceptor, respectively (Figure 9). Although the overall PCE of 0.04% for this solar cell under AM1.5 (100 mW cm$^{-2}$) was relatively low ($V_{oc}$ = 0.26 V, $J_{sc}$ = 0.5 mA cm$^{-2}$, FF = 0.31), this multi-layer deposition strategy could be promising when fundamental aspects of photoinduced electron transfer and exciton transport inside these nanostructured films will be understood.[105]

## 8.6.3 Hydrogen-Bonding Supramolecular Devices

Supramolecular auto-organization can be also reached with the introduction of hydrogen-bonding molecular recognition motifs for the preparation of ordered supramolecular devices. A methodology was reported with the formation of homogeneous films by codeposition from a solution of an aminopyrimidine-terminated oligothiophene-bearing hydrogen-bonding units and a covalently linked barbiturate fullerene derivative (Scheme 21). The incorporation in photovoltaic devices gave a 2.5-fold enhancement in energy conversion when compared to analogous systems using the non-hydrogen-bonding parent $C_{60}$. The improvement of the photovoltaic response was attributed to the presence of complementary hydrogen-bonding between the active components. This

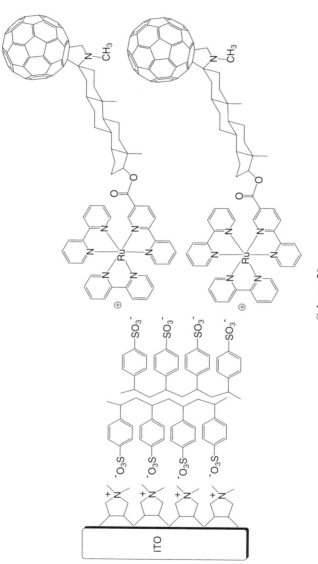

**Scheme 20**

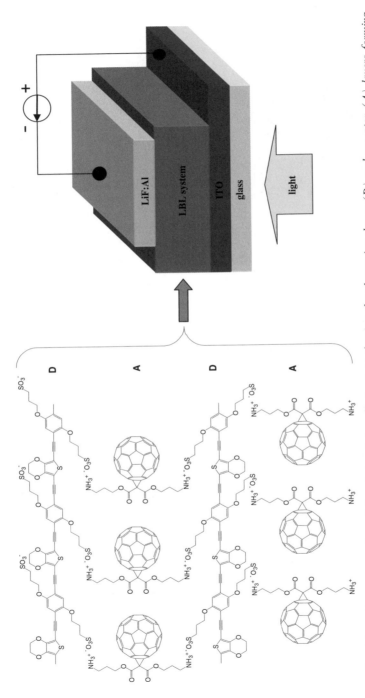

**Figure 9** *Schematic representation of the photovoltaic cell structure showing the alternating donor (D) and acceptor (A) layers forming the photoactive layer*

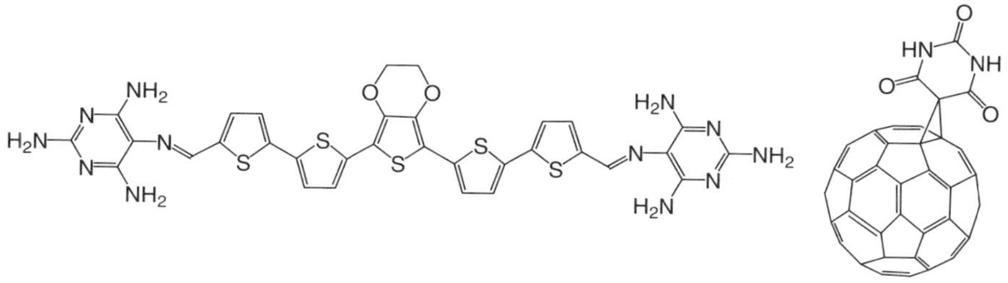

**Scheme 21**

strategy is expected to provide, in the medium term, an increased compatibility between poorly miscible materials.[106]

## 8.7 Conclusion and Outlook

The advent of [60]fullerene and its production in multigram quantities at the beginning of the 1990s launched a major effort to explore its outstanding electron-acceptor features. Among the wide perspectives offered by this structural fascinating molecule, the possibility to develop efficient organic solar cells seems increasingly attractive. Some of the important advantages of the so-called plastic organic solar cells as an alternative to silicon-based solar cells include low cost of fabrication for large-area modules, ease of processing, low weight of plastic materials, mechanical flexibility and versatility of chemical structure from advances in organic chemistry. The synthetic approach offers great potentiality for tailoring the design of the photovoltaic system. For example, the ability to add solubilizing groups to organic molecules can be used for new and inexpensive techniques in the manufacturing process.

Comparatively, there are only 10 years that [60]fullerene derivatives are involved in the development of organic solar cells and, effectively, the progress obtained over these last years in organic photovoltaics is impressive. Concerning $C_{60}$-based photovoltaics using polymeric materials in bulk-heterojunction,[107] the efficiency was improved from only 0.04% in 1993[34] to 1.2% in 1999 on large area flexible plastic solar cells with MDMO-PPV and [60]PCBM, reaching 2.5% in 2001, 3.5% in 2003 using P3HT and rapidly 5% in 2005 thanks to technological improvements.[28]

In the case of the utilization of "small molecules", the 1% efficient CuPc/PTCBI based solar cell was rapidly out of date due to the substitution of PTCBI by [60]fullerene and the design of new device architectures. The efficiency progressed rapidly to 3.6%,[75] then reached 4.2%[76] to finally approaching 5.7% using the couple CuPc/$C_{60}$ in a tandem geometry.[29] Considering this evolution, a PCE of 10% seems accessible as a research target for the next few years and this might be attainable at the latest in 2010 (Figure 10).

Beside the tremendous progress made through molecular engineering around the $C_{60}$ scaffold, it is clear that a lot of work still needs to be accomplished in particular on the optimization of the device characteristics:

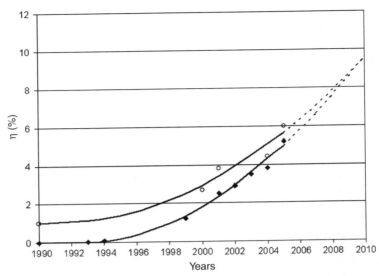

**Figure 10** *Evolution of solar cells efficiencies using low molecular-weight materials ( ○ ) and polymer-$C_{60}$ based heterojunctions ( ■ ).*

- optimization of the choice of metallic electrodes to achieve good ohmic contacts of both sides for the collection of charge carriers;
- optimization of the choice of the donor–acceptor pair, the energy levels of the HOMO and LUMO of materials influencing directly the open-circuit voltage;
- optimization of the network morphology to maximize the mobility of the charge carriers within the different materials of the heterojunction. Improved mobilities will increase the solar cell fill factor and, thereby, will lead to higher solar cell efficiencies.

A further improvement of the performance of polymer-fullerene solar cells is expected to be attained by controlling the device photoactive layer structural organization with a nanoscale precision. For instance, it was recently demonstrated that the nanoscale organization in the photoactive layer of P3HT thin fibrillar crystals and homogeneous nanocrystalline [60]PCBM improves charge transport towards a morphology in which both materials present a large interfacial area.[108]

Besides, from the chemical point of view, the improvement in the efficiency of these heterojunction cells can be considered through the synthesis of new p-type donors and n-type $C_{60}$ derivatives organic semiconductors specifically designed for photovoltaic conversion. The molecule [60]PCBM is now considered by physicists as the reference accepting material with ten years of development in corresponding plastic solar cells. Nevertheless, there is no valuable reason justifying that another $C_{60}$ derivative should not be able to attain then pass beyond the [60]PCBM properties. Progress in this field lies in the development of an intensive research effort of creative synthetic chemists to offer other $C_{60}$

derivatives as potential electronic acceptors able to improve organic solar cells. The major parameters required for these new derivatives concern particularly the solubility, the stability towards degradation induced by light or oxygen and the optimal absorption in the visible spectrum. Solar cells have to operate in intense sunlight without degradation for many years. It is not yet clear whether organic based systems can easily achieve this goal, and the degradation mechanisms of organic solar cells need to be thoroughly explored. However, the promise of cheap energy that is also good for environment will propel future research in this direction.[109]

The design of fullerene derivatives containing an electron donor moiety covalently attached to the $C_{60}$ core constitute also a promising field due to the interesting electronic properties they can display. Consequently, the molecular approach to the photovoltaic conversion is not only an interesting model for a better understanding of the devices using bulk-heterojunction materials but it also appears as an excellent alternative to the polymeric approach. Devices prepared with fullerene-($\pi$-donor) dyads are quite promising, but a major problem results from the relative incompatibility in the dimensionality of the transport process. Whereas $C_{60}$ is recognized as an efficient 3D electron carrier, unidirectional $\pi$-stacking interactions are generally required for the hole transport in $\pi$-donor systems.

Consequently, it could be advantageous to improve the two-component systems by different strategies such as the realization of nano-structured materials. The richness of the $C_{60}$ chemistry will allow developments of materials prone to be self-organized in a nanostructured device. The supramolecular design of molecular assemblies constitutes for the future the possibility to reach new efficient photovoltaic devices in a perfect control of the spatial organization of materials in the photoactive layer.

Just as fullerenes have boosted the efficiency of organic solar cells, carbon nanotubes constitute also for the future an exciting challenge to improve the device performance. Promising results were recently reported suggesting that single-wall carbon nanotubes (SWNTs) thin films can be used as transparent conducting current collectors in organic solar cells. A preliminary encouraging device was obtained with the construction of the device ITO/P3OT:SWNT/Al by spin-casting of polymer and SWNTs. In this configuration of nano heterojunctions with P3OT for the exciton dissociation, SWNTs play the role of electron acceptors acting also for the electron transport. This device exhibited a photovoltaic behaviour close to 1% ($V_{oc} = 0.7$–$0.9$) and the increase of the short-circuit current by two orders of magnitude compared with the pristine polymer devices was clearly attributed to the excellent electronic properties of carbon nanotubes and their large surface areas.[110] However, it was recently demonstrated another exciting behaviour of SWNTs which were simply used as transparent conducting current collectors by replacing the ITO transparent electrode for hole collection in a bulk-heterojunction P3HT:[60]PCBM device.[111]

There are certainly obstacles, chemical and technological challenges which will have to be overcome, but there is undoubtedly an exciting future for fullerene-based plastic solar cells.

# References

1. H.W. Kroto, J.R. Heath, S.C. O'Brien, R.F. Curl and R.E. Smalley, *Nature*, 1985, **318**, 162.
2. (*a*) W. Krätschmer, L. Lamb, K. Fostiropoulos and D.R. Huffman, *Nature*, 1990, **357**, 354; (*b*) W. Krätschmer, K. Fostiropoulos and D.R. Huffman, *Chem. Phys. Lett.*, 1990, **170**, 167; (*c*) H.W. Kroto, A.W. Allaf and S.P. Balm, *Chem. Rev.*, 1991, **91**, 1213.
3. Q. Xie, E. Pèrez-Cordero and L. Echegoyen, *J. Am. Chem. Soc.*, 1992, **114**, 3978.
4. (*a*) C.S. Foote, in *Topics in Current Chemistry;* Photophysical and Photochemical Properties of Fullerenes, J. Matty, (eds), Springer Berlin, 1994, **169**, 347; (*b*) Y.-P. Sun, in *Molecular and Supermolecular Photochemistry; Photophysics and Photochemistry of Fullerene Materials*, V. Ramamurthy and K.S. Schanze (eds), Marcel Dekker, New York, 1997, **1**, 325; (*c*) D.M. Guldi and P.V. Kamat, in *Fullerenes, Chemistry, Physics, and Technology*, K.M. Kadish and R.S. Ruoff (eds), Wiley, New York, 2000, **5**, 225.
5. (*a*) H. Imahori and Y. Sakata, *Adv. Mater.*, 1997, **9**, 537; (*b*) H. Imahori, *Org. Biomol. Chem.*, 2004, **2**, 1425.
6. N.S. Sariciftci, L. Smilowitz, A.J. Heeger and F. Wudl, *Science*, 1992, **258**, 1474.
7. N.S. Sariciftci and A.J. Heeger, in *Handbook of Organic Conductive Molecules and Polymers*, Vol 1 H.S. Nalwa (ed), Wiley, New York, 1997, **8**, 413.
8. C.J. Brabec, N.S. Sariciftci and J.C. Hummelen, *Adv. Funct. Mater.*, 2001, **11**, 15.
9. (*a*) A.E. Becquerel, *C. R. Acad. Sc.*, 1839, **9**, 145; (*b*) E. Becquerel, *C. R. Acad. Sc.*, 1839, **9**, 561.
10. W. Smith, *Nature*, 1873, **7**, 303.
11. W.G. Adams and R.E. Day, *Proc. R. Soc. London*, 1876, **25**, 113.
12. A. Einstein, *Annalen der Physik*, 1905, **17**, 132.
13. R.A. Millikan and W.H. Souder, *Proc. Nat. Acad. Sci.*, 1916, **2**, 19.
14. D.M. Chapin, C.S. Fuller and G.L. Pearson, *J. Appl. Phys.*, 1954, **25**, 676.
15. M.A. Green, K. Emery, D.L. King, S. Igari and W. Warta, *Prog. Photovolt.:Res. Appl.*, 2003, **11**, 347.
16. For a comprehensive review on silicon and other types of solar cells, see: A. Goetzberger, C. Hebling and H.-W. Schock, *Mater. Sci. Eng. R40*, 2003, 1.
17. D. Carlson and C. Wronski, *Appl. Phys. Lett.*, 1976, **28**, 671.
18. M.D. Archer and R. Hill (eds), *Clean Electricity from Photovoltaics, Series on Photoconversion of Solar Energy*, Vol 1, Imperial College Press, London, 2001.
19. H. Spanggaard and F.C. Krebs, *Solar Energy Mater. Solar Cells*, 2004, **83**, 125.
20. B. O'Regan and M. Grätzel, *Nature*, 1991, **353**, 737.
21. M. Grätzel, *Nature*, 2001, **414**, 338.

22. M. Grätzel, *Prog. Photovolt. Res. Appl.*, 2000, **8**, 171.
23. C.J. Brabec, *Solar Energy Mater. Solar Cells*, 2004, **83**, 273.
24. G.A. Chamberlain, *Solar Cells*, 1983, **8**, 47.
25. D. Wöhrle and D. Meissner, *Adv. Mater.*, 1991, **3**, 129.
26. C.W. Tang, *Appl. Phys. Lett.*, 1986, **48**, 183.
27. N.S. Sariciftci and A.J. Heeger, *U.S. Patent 005331183A*, 1994.
28. W. Ma, C. Yang, X. Gong, K. Lee and A.J. Heeger, *Adv. Funct. Mater.*, 2005, **15**, 1617.
29. J. Xue, S. Uchida, B.P. Rand and S.R. Forrest, *Appl. Phys. Lett.*, 2004, **85**, 5757.
30. J.-M. Nunzi, *C.R. Physique*, 2002, **3**, 523.
31. B. Miller, J.M. Rosamilia, G. Dabbagh, R. Tycko, R.C. Haddon, A.J. Muller, W. Wilson, D.W. Murphy and A.F. Hebard, *J. Am. Chem. Soc.*, 1991, **113**, 6291.
32. C.J. Brabec, G. Zerza, G. Cerullo, S. De Silvestri, S. Luzzati, J.C. Hummelen and N.S. Sariciftci, *Chem. Phys. Lett.*, 2001, **340**, 232.
33. S. Morita, A.A. Zakhidov and K. Yoshino, *Solid State Commun.*, 1992, **82**, 249.
34. N.S. Sariciftci, D. Braun, C. Zhang, V.I. Srdanov, A.J. Heeger, G. Stucky and F. Wudl, *Appl. Phys. Lett.*, 1993, **62**, 585.
35. L.S. Roman, W. Mammo, L.A.A. Pettersson, M.R. Andersson and O. Inganäs, *Adv. Mater.*, 1998, **10**, 774.
36. L. Chen, D. Godovsky, O. Inganäs, J.C. Hummelen, R.A.J. Janssen, M. Svensson and M.R. Andersson, *Adv. Mater.*, 2000, **12**, 1367.
37. J.C. Hummelen, B.W. Knight, F. LePeq and F. Wudl, *J. Org. Chem.*, 1995, **60**, 532.
38. R. González, J.C. Hummelen and F. Wudl, *J. Org. Chem.*, 1995, **60**, 2618.
39. G. Yu, J. Gao, J.C. Hummelen, F. Wudl and A.J. Heeger, *Science*, 1995, **270**, 1789.
40. S. Sensfuss and M. Al-Ibrahim, in *Organic Photovoltaics*, S. Sun and N.S. Sariciftci (eds), 2004, **23**, 529.
41. C.J. Brabec, A. Cravino, D. Meissner, N.S. Sariciftci, T. Fromherz, M.T. Rispens, L. Sanchez and J.C. Hummelen, *Adv. Funct. Mater.*, 2001, **1**, 374.
42. H. Hoppe and N.S. Sariciftci, *J. Mater. Chem.*, 2006, **16**, 45.
43. M.T. Rispens, A. Meetsma, R. Rittberger, C.J. Brabec, N.S. Sariciftci and J.C. Hummelen, *Chem. Commun.*, 2003, **17**, 2116.
44. D. Gebeyehu, C.J. Brabec, F. Padinger, T. Fromherz, J.C. Hummelen, D. Badt, H. Schindler and N.S. Sariciftci, *Synth. Met.*, 2001, **118**, 1.
45. V.D. Mihailetchi, J.K.J. van Duren, P.W.M. Blom, J.C. Hummelen, R.A.J. Janssen, J.M. Kroon, M.T. Rispens, W.J.H. Verhees and M.M. Wienk, *Adv. Funct. Mater.*, 2003, **13**, 43.
46. D. Chirvase, J. Parisi, J.C. Hummelen and V. Dyakonov, *Nanotechnology*, 2004, **15**, 1317.
47. V. Shrotriya, J. Ouyang, R.J. Tseng, G. Li and Y. Yang, *Chem. Phys. Lett.*, 2005, **411**, 138.

48. C. Winder and N.S. Sariciftci, *J. Mater. Chem.*, 2004, **14**, 1077.
49. R. Berridge, P.J. Skabara, C. Pozo-Gonzalo, A. Kanibolotsky, J. Lohr, J.J.W. McDouall, E.J.L. McInnes, J. Wolowska, C. Winder, N.S. Sariciftci, R.W. Harrington and W. Clegg, *J. Phys. Chem. B.*, 2006, **110**, 3140.
50. A. Dhanabalan, J.K.J. van Duren, P.A. van Hal, J.L.J. van Dongen and R.A.J. Janssen, *Adv. Funct. Mater.*, 2001, **11**, 255.
51. F. Zhang, E. Perzon, X. Wang, W. Mammo, M.R. Andersson and O. Inganäs, *Adv. Funct. Mater.*, 2005, **15**, 745.
52. R. Demadrille, M. Firon, J. Leroy, P. Rannou and A. Pron, *Adv. Funct. Mater.*, 2005, **15**, 1547.
53. M. Svensson, F. Zhang, S.C. Veenstra, W.J.H. Verhees, J.C. Hummelen, J.M. Kroon, O. Inganäs and M.R. Andersson, *Adv. Mater.*, 2003, **15**, 988.
54. I. Riedel, E. von Hauff, J. Parisi, N. Martín, F. Giacalone and V. Dyakonov, *Adv. Funct. Mater.*, 2005, **15**, 1979.
55. M.-P. Hernández, F. Monroy, F. Ortega, R.G. Rubio, A. Martín Domenech, E.V. Priego, L. Sánchez and N. Martín, *Langmuir*, 2001, **17**, 3317.
56. J.L. Segura, F. Giacalone, R. Gómez, N. Martín, D.M. Guldi, C. Luo, A. Swartz, I. Riedel, D. Chirvase, J. Parisi, V. Dyakonov, N.S. Sariciftci and F. Padinger, *Mater. Sci. Eng.*, 2005, **C 25**, 835.
57. (a) N. Gutiérrez-Nava, S. Setayesh, A. Rameau, P. Masson and J.-F. Nierengarten, *New J. Chem.*, 2002, **26**, 1584; (b) N. Gutiérrez-Nava, P. Masson and J.-F. Nierengarten, *Tetrahedron Lett.*, 2003, **44**, 4487.
58. J.-F. Nierengarten, *New. J. Chem.*, 2004, **28**, 1177.
59. S. Alem, R. De Bettignies, M. Cariou, E. Allard, S. Chopin, J. Cousseau, S. Dabos-Seignon and J.-M. Nunzi, *Proc. SPIE – The International Society of Optical Engineering*, 2004, **5351**, 284.
60. M. Camaioni, G. Ridolfi, G. Casalbore-Miceli, G. Possamai, L. Garlaschelli and M. Maggini, *Solar Energy Mater. Solar Cells*, 2003, **76**, 107.
61. Q. Zhou, L. Zheng, D. Sun, X. Deng, G. Yu and Y. Cao, *Synth. Met.*, 2003, **135–136**, 825.
62. J. Li, N. Sun, Z.-X. Guo, C. Li, Y. Li, L. Dai, D. Zhu, D. Sun, Y. Cao and L. Fan, *J. Phys. Chem. B.*, 2002, **106**, 11509.
63. J. Baffreau, L. Perrin, S. Leroy-Lhez and P. Hudhomme, *Tetrahedron Lett.*, 2005, **46**, 4599.
64. R. Gómez, J.L. Segura and N. Martín, *Org. Lett.*, 2005, **7**, 717.
65. G. Accorsi, N. Armaroli, J.-F. Eckert and J.-F. Nierengarten, *Tetrahedron Lett.*, 2002, **43**, 65.
66. H. Neugebauer, M.A. Loi, C. Winder, N.S. Sariciftci, G. Cerullo, A. Gouloumis, P. Vásquez and T. Torres, *Solar Energy Mater. Solar Cells*, 2004, **83**, 201.
67. C.M. Atienza, J. Fernández, L. Sánchez, N. Martín, I. Sá Dantas, M.M. Wienk, R.A.J. Janssen, G.M.A. Rahman and D.M. Guldi, *Chem. Commun.*, 2006, **5**, 514.
68. A. Cravino and S. Sariciftci, *J. Mater. Chem.*, 2002, **12**, 1931.
69. A. Cravino and S. Sariciftci, *Nature Mater.*, 2003, **2**, 360.

70. A. Marcos Ramos, M.T. Rispens, J.K.J. van Duren, J.C. Hummelen and R.A.J. Janssen, *J. Am. Chem. Soc.*, 2001, **123**, 6714.
71. F. Zhang, M. Svensson, M.R. Andersson, M. Maggini, S. Bucella, E. Menna and O. Inganäs, *Adv. Mater.*, 2001, **13**, 1871.
72. P. Peumans, S. Uchida and S.R. Forrest, *Nature*, 2003, **425**, 158.
73. P. Peumans, V. Bulović and S.R. Forrest, *Appl. Phys. Lett.*, 2000, **76**, 2650.
74. Y. Yamashita, W. Takashima and K. Kaneto, *Jpn J. Appl. Phys.*, 1993, **32**, 1017.
75. P. Peumans and S.R. Forrest, *Appl. Phys. Lett.*, 2001, **79**, 126.
76. J. Xue, S. Uchida, B.P. Rand and S.R. Forrest, *Appl. Phys. Lett.*, 2004, **84**, 3013.
77. (a) A. Yakimov and S.R. Forrest, *Appl. Phys. Lett.*, 2002, **80**, 1667; (b) B.P. Rand, P. Peumans and S.R. Forrest, *J. Appl. Phys.*, 2004, **96**, 7519.
78. (a) B.R. Rand, J. Xue, S. Uchida and S.R. Forrest, *J. Appl. Phys. Part I*, 2005, **98**, 124902; (b) J. Xue, B.R. Rand, S. Uchida and S.R. Forrest, *J. Appl. Phys. Part II*, 2005, **98**, 124903.
79. S.C. Veenstra, G.G. Malliaras, H.J. Brouwer, F.J. Esselink, V.V. Krasnikov, P.F. van Hutten, J. Wildeman, H.T. Jonkman, G.A. Sawatzky and G. Hadziioannou, *Synth. Met.*, 1997, **84**, 971.
80. L. Ouali, V.V. Krasnikov, U. Stalmach and G. Hadziioannou, *Adv. Mater.*, 1999, **11**, 1515.
81. W. Geens, T. Aernouts, J. Poortmans and G. Hadziioannou, *Thin Solid Films*, 2002, **403–404**, 438.
82. S. Yoo, B. Domercq and P. Kippelen, *Appl. Phys. Lett*, 2004, **85**, 5427.
83. N. Martín, L. Sánchez, B. Illescas and I. Pérez, *Chem. Rev.*, 1998, **98**, 2527.
84. (a) J.-F. Nierengarten, *Solar Energy Mater. Solar Cells*, 2004, **83**, 187; (b) J.L. Segura, N. Martín and D.M. Guldi, *Chem. Soc. Rev.*, 2005, **34**, 31; (c) J. Roncali, *Chem. Soc. Rev.*, 2005, **34**, 483.
85. L. Sánchez, I. Pérez, N. Martín and D.M. Guldi, *Chem. Eur. J.*, 2003, **9**, 2457.
86. J.-F. Nierengarten, J.-F. Eckert, J.-F. Nicoud, L. Ouali, V. Krasnikov and G. Hadziioannou, *Chem. Commun.*, 1999, **7**, 617.
87. J.-F. Eckert, J.-F. Nicoud, J.-F. Nierengarten, S.-G. Liu, L. Echegoyen, F. Barigelletti, N. Armaroli, L. Ouali, V. Krasnikov and G. Hadziioannou, *J. Am. Chem. Soc.*, 2000, **122**, 7467.
88. E. Peeters, P.A. van Hal, J. Knol, C.J. Brabec, N.S. Sariciftci, J.C. Hummelen and R.A.J. Janssen, *J. Phys. Chem. B.*, 2000, **104**, 10174.
89. T. Gu, D. Tsamouras, C. Melzer, V. Krasnikov, J.-P. Gisselbrecht, M. Gross, G. Hadziioannou and J.-F. Nierengarten, *Chem. Phys. Chem.*, 2002, **1**, 124.
90. D.M. Guldi, C. Luo, A. Swartz, R. Gómez, J.L. Segura, N. Martín, C.J. Brabec and N.S. Sariciftci, *J. Org. Chem.*, 2002, **67**, 1141.
91. N. Negishi, K. Yamada, K. Takimiya, Y. Aso, T. Otsubo and Y. Harima, *Chem. Lett.*, 2003, **32**, 404.

92. (a) M. Maggini, G. Possamai, E. Menna, G. Scorrano, N. Camaioni, G. Ridolfi, G. Casalbore-Miceli, L. Franco, M. Ruzzi and C. Corvaja, *Chem. Commun.*, 2002, **18**, 2028; (b) G. Possamai, M. Maggini, E. Menna, G. Scorrano, L. Franco, M. Ruzzi, C. Corvaja, G. Ridolfi, P. Samori, A. Geri and N. Camaioni, *Appl. Phys. A.*, 2004, **79**, 51.
93. T. Konishi, A. Ikeda and S. Shinkai, *Tetrahedron*, 2005, **61**, 4881.
94. T. Otsubo, Y. Aso and K. Takimiya, *J. Mater. Chem.*, 2002, **12**, 2565.
95. D. Hirayama, T. Yamashiro, K. Takimiya, Y. Aso, T. Otsubo, H. Norieda, H. Imahori and Y. Sakata, *Chem. Lett.*, 2000, **5**, 570.
96. D. Hirayama, K. Takimiya, Y. Aso, T. Otsubo, T. Hasobe, H. Yamada, H. Imahori, S. Fukuzumi and Y. Sakata, *J. Am. Chem. Soc.*, 2002, **124**, 532.
97. K.-S. Kim, M.-S. Hang, H. Ma and A.K.-Y. Jen, *Chem. Mater.*, 2004, **16**, 5058.
98. H. Imahori, M. Kimura, K. Hosomizu and S. Fukuzumi, *J. Photochem. Photobiol. A: Chem.*, 2004, **166**, 57.
99. (a) H. Imahori, H. Yamada, S. Ozawa, K. Ushida and Y. Sakata, *Chem. Commun.*, 1999, **13**, 1165; (b) H. Imahori, H. Yamada, Y. Nishimura, I. Yamazaki and Y. Sakata, *J. Phys. Chem. B.*, 2000, **104**, 2099; (c) H. Imahori, H. Norieda, H. Yamada, Y. Nishimura, I. Yamazaki, Y. Sakata and S. Fukuzumi, *J. Am. Chem. Soc.*, 2001, **123**, 100; (d) H. Yamada, H. Imahori, Y. Nishimura, I. Yamazaki, T.K. Ahn, S.K. Kim and S. Fukuzumi, *J. Am. Chem. Soc.*, 2003, **125**, 9129.
100. D.M. Guldi, I. Zilbermann, G.A. Anderson, K. Kordatos, M. Prato, R. Tafuro and L. Valli, *J. Mater. Chem.*, 2004, **14**, 303.
101. D.M. Guldi, I. Zilbermann, G. Anderson, N.A. Kotov, N. Tagmatarchis and M. Prato, *J. Mater. Chem.*, 2005, **15**, 114.
102. H. Li, Y. Li, J. Zhai, G. Cui, H. Liu, S. Xiao, Y. Liu, F. Lu, L. Jiang and D. Zhu, *Chem. Eur. J.*, 2003, **9**, 6031.
103. (a) M.F. Durstock, B.E. Taylor, R.J. Spry, L. Chiang, S. Reulbach, K. Heitfeld and J.W. Baur, *Synth. Met.*, 2001, **116**, 373; (b) M.F. Durstock, R.J. Spry, J.W. Baur, B.E. Taylor and L. Chiang, *J. Appl. Phys.*, 2003, **94**, 3253.
104. C. Luo, D.M. Guldi, M. Maggini, E. Menna, S. Mondini, N.A. Kotov and M. Prato, *Angew. Chem. Int. Ed.*, 2000, **39**, 3905.
105. J.K. Mwaura, M.R. Pinto, D. Witker, N. Ananthakrishnan, K.S. Schanze and J.R. Reynolds, *Langmuir*, 2005, **21**, 10119.
106. (a) C.-H. Huang, N.D. McClenaghan, A. Kuhn, J.W. Hofstraat and D.M. Bassani, *Org. Lett.*, 2005, **7**, 3409; (b) C.-H. Huang, N.D. McClenaghan, A. Kuhn, G. Bravic and D.M. Bassani, *Tetrahedron*, 2006, **62**, 2050.
107. N.S. Sariciftci, *Curr. Opinion Solid State Mater. Sci.*, 1999, **4**, 373.
108. X. Yang, J. Loos, S.C. Veenstra, W.J.H. Verhees, M.M. Wienk, J.M. Kroon, M.A.J. Michels and R.A.J. Janssen, *Nano Lett.*, 2005, **5**, 579.
109. Z. Valy Vardeny, A.J. Heeger and A. Dodabalapur, *Synth. Met.*, 2005, **148**, 1.

110. (a) E. Kymakis and G.A.J. Amaratunga, *Appl. Phys. Lett.*, 2002, **80**, 112; (b) E. Kymakis, I. Alexandrou and G.A.J. Amaratunga, *Appl. Phys. Lett.*, 2003, **93**, 1764.
111. A. Du Pasquier, H.E. Unalan, A. Kanwal, S. Miller and M. Chhowalla, *Appl. Phys. Lett.*, 2005, **87**, 203511.
112. C.J. Brabec, F. Padinger, J.C. Hummelen, R.A.J. Janssen and N.S. Sariciftci, *Synth. Met.*, 1999, **102**, 861.
113. (a) S.E. Shaheen, C.J. Brabec, N.S. Sariciftci, F. Padinger, T. Fromherz and J.C. Hummelen, *Appl. Phys. Lett.*, 2001, **78**, 841; (b) C.J. Brabec, S.E. Shaheen, T. Fromherz, F. Padinger, J.C. Hummelen, A. Dhanabalan, R.A.J. Janssen and N.S. Sariciftci, *Synth. Met.*, 2001, **121**, 1517.
114. J.M. Kroon, M.M. Wienk, W.J.H. Verhees and J.C. Hummelen, *Thin Solid Films*, 2002, **403–404**, 223.
115. T. Munters, T. Martens, L. Goris, V. Vrindts, J. Manca, L. Lutsen, W. De Ceuninck, D. Vanderzande, L. De Schepper, J. Gelan, N.S. Sariciftci and C.J. Brabec, *Thin Solid Films*, 2002, **403–404**, 247.
116. M.M. Wienk, J.M. Kroon, W.J.H. Verhees, J. Knol, J.C. Hummelen, P.A. van Hal and R.A.J. Janssen, *Angew. Chem. Int. Ed.*, 2003, **42**, 3371.
117. F. Padinger, R.S. Rittberger and N.S. Sariciftci, *Adv. Funct. Mater.*, 2003, **13**, 85.
118. C. Waldauf, P. Schilinsky, J. Hauch and C.J. Brabec, *Thin Solid Films*, 2004, **451–452**, 503.
119. G. Li, V. Shrotriya, J. Huang, Y. Yao, T. Moriarty, K. Emery and Y. Yang, *Nature Mater.*, 2005, **4**, 864.
120. M. ReyesReyes, K. Kim and D.L. Carroll, *Appl. Phys. Lett.*, 2005, **87**, 83506.
121. M. Reyes-Reyes, K. Kim, J. Dewald, R. López-Sandoval, A. Avadhanula, S. Curran and D.L. Carroll, *Org. Lett.*, 2005, **7**, 5749.
122. (a) G.E. Jabbour, B. Kippelen, N.R. Armstrong and N. Peyghambarian, *Appl. Phys. Lett.*, 1998, **73**, 1185; (b) L.S. Hung, C.W. Tang and M.G. Mason, *Appl. Phys. Lett.*, 1997, **70**, 152.
123. A. Petr, F. Zhang, H. Peisert, M. Knupfer and L. Dunsch, *Chem. Phys. Lett.*, 2004, **385**, 140.
124. M. Al-Ibrahim, H.-K. Roth, U. Zhokhavets, G. Gobsch and S. Sensfuss, *Solar Energy Mater Solar Cells*, 2005, **85**, 13.
125. M. Al-Ibrahim, A. Konkin, H.-K. Roth, D.A.M. Egbe, E. Klemm, U. Zhokhavets, G. Gobsch and S. Sensfuss, *Thin Solid Films*, 2005, **474**, 201.
126. X. Wang, E. Perzon, J.L. Delgado, P. de la Cruz, F. Zhang, F. Langa, M. Andersson and O. Inganäs, *Appl. Phys. Lett.*, 2004, **85**, 5081.
127. X. Wang, E. Perzon, F. Oswald, F. Langa, S. Admassie, M.R. Andersson and O. Inganäs, *Adv. Funct. Mater.*, 2005, **15**, 1665.

CHAPTER 9

# Fullerene Modified Electrodes and Solar Cells

HIROSHI IMAHORI AND TOMOKAZU UMEYAMA

Department of Molecular Engineering, Graduate School of Engineering, Kyoto University, Nishikyo-ku, Kyoto 615-8510, Japan

## 9.1 Introduction

Photochemical and photophysical properties of fullerenes and their derivatives have been extensively studied because of their intriguing optical and redox properties.[1–16] In particular, fullerenes have small reorganization energies of electron transfer, which result in remarkable acceleration of photoinduced charge separation and of charge shift as well as deceleration of charge recombination.[17–24] Thus, they have been frequently employed as an electron acceptor in donor–acceptor-linked systems to yield a long-lived charge-separated state with a high quantum yield.[1–24] Extensive efforts have also been made in recent years to explore the photovoltaic and photoelectrochemical properties of fullerenes and their derivatives.[25–112] To optimize the device performance, it is of importance to (i) collect the visible light extensively, (ii) produce a charge-separated state efficiently, and (iii) inject resultant electron and hole into respective electrodes, minimizing undesirable charge recombination. Accordingly, the fabrication of fullerenes and their derivatives onto electrode surfaces is a vital step for controlling the morphology of the fullerene assemblies on the electrode surface in molecular scale. Versatile methods such as Langmuir–Blodgett (LB) films,[25–32] self-assembled monolayers (SAMs),[33–57] layer-by-layer deposition,[58–66] vacuum deposition,[67–71] electrophoretic deposition,[72–92] chemical adsorption and spin coating,[93–112] have been adopted to construct photoelectrochemical devices and solar cells.

In this chapter, we focus on fullerene-modified electrodes which exhibit photoinduced energy and electron transfer processes. Specifically, we emphasize the utilization of $C_{60}$ as an electron acceptor for photoelectrochemical devices and solar cells.

# 9.2 Langmuir–Blodgett Films

LB technique has been proven to be a powerful, convenient method to construct organic thin films controlled in the order of the molecular level.[113] It has been frequently used to organize fullerenes and fullerene derivatives on electrode surfaces.[25–32]

Jin *et al.*[25] investigated LB films of $C_{60}$ derivatives bearing methoxycarbonyl groups **1** and carboxylic groups **2** (Figure 1). The photovoltaic properties of the

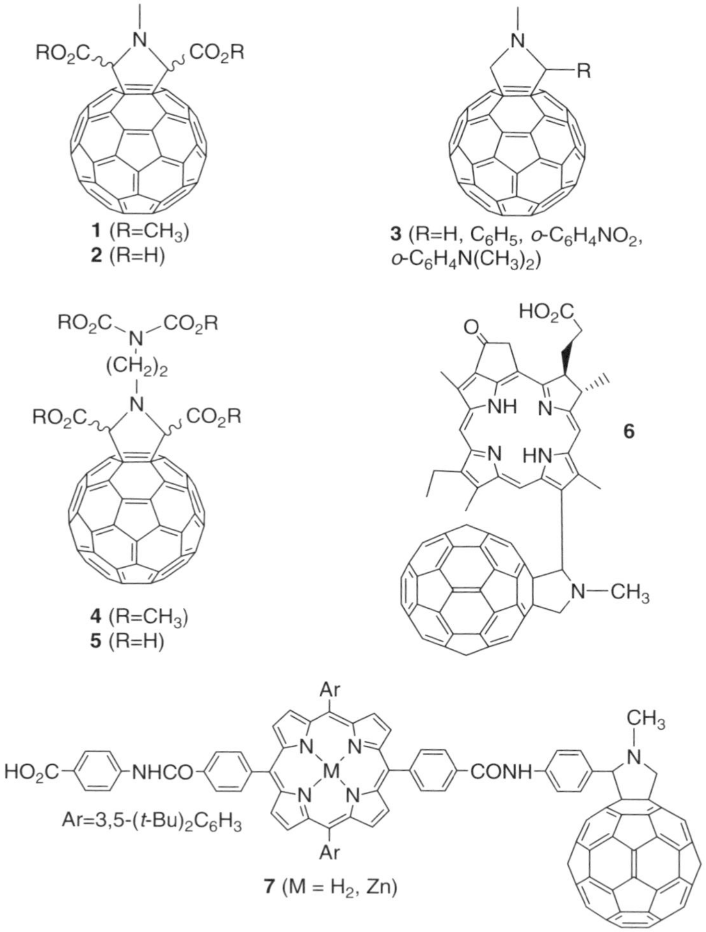


**Figure 1** *Fullerene derivatives 1–7 incorporated into LB films*

LB films on $n$-type Si substrates and their powders were studied by using surface photovoltage spectroscopy. The reverse response values of surface photovoltage spectra are obtained when the powders of **1** and **2** contact indium-tin oxide (ITO) substrates. The authors proposed that such an effect results from the opposite surface band bending arising from the difference in the molecular orbital levels in **1** and **2** with different electron-withdrawing substituents. When they are deposited onto n-Si substrates, the LB films reveal the same increments on the direct transition of n-Si.[25]

Luo et al.[26] reported photoelectrochemical properties of fullerene derivatives **3** on a $SnO_2$ electrode (Figure 1). For instance, a mixed monolayer of arachitic acid and the $C_{60}$-pyrrolidine derivatives **3** (1:1) is deposited onto the $SnO_2$ electrode with a transfer ratio of 0.9 at 30 mN m$^{-1}$ by LB technique. The photoelectrochemical measurements were performed using the modified $SnO_2$ electrode as a working electrode, a platinum wire as a counter electrode, and an Ag/AgCl as a reference electrode in the presence of ascorbic acid (AsA) (Au/**3**/AsA/Pt device). The internal quantum yield of photocurrent generation (1.2–8.9%) in Au/**3**/AsA/Pt device at an applied potential of 0.1 V vs. SCE increases with introducing more electron-donating substituents into the $C_{60}$ moiety.[26]

Zhang et al. and Zhou et al. [27,28] prepared the monolayer and multilayer LB films of fullerene derivatives including an amphiphilic $C_{60}$-tetramethylester **4** and its tetraacid **5** on an ITO electrode (Figure 1). The photoelectrochemical measurements were carried out using the modified ITO electrode as a working electrode, a platinum wire as a counter electrode, and an Ag/AgCl as a reference electrode in the presence of hydroquinone ($H_2Q$) (ITO/**4** or **5**/$H_2Q$/Pt device). The internal quantum yields of photocurrent generation in ITO/**4** or **5**/$H_2Q$/Pt device at an applied potential of 0.2 V vs. SCE are 4.8 and 3.8%, respectively. It is noteworthy that the direction of electron flow can be modulated by the pH in the electrolyte solution.[27]

Tkachenko et al. and Vuorinen et al.[30–32] examined photoinduced electron transfer and photocurrent generation in LB films of phytochlorin–fullerene dyad and porphyrin–fullerene dyads. For instance, they investigated photoinduced electron transfer in the LB films of **6** (Figure 1) possessing a hydrophilic propionic acid residue.[30] A mixed monolayer of octadecylamine and **6** can be transferred onto a solid substrate which is characterized by uniform orientation of dyad **6**. From time-resolved Maxwell displacement charge measurements, the LB films containing **6** are found to exhibit vectorial photoinduced electron transfer with a poor quantum yield of 0.2%. The lifetime of the charge-separated state in the LB films is $ca.$ 30 ns, being almost independent of the concentration of **6** (2–50 mol%).[30]

Tkachenko et al.[31] also investigated photophysical properties of porphyrin–fullerene dyad **7** (M=$H_2$) in LB films as well as in solutions (Figure 1). Being incorporated into an LB film, dyad **7** (M=$H_2$) exhibits fast photoinduced electron transfer with a time constant of $8 \times 10^9$ s$^{-1}$ compared to the value in benzonitrile ($5 \times 10^8$ s$^{-1}$). The fast photoinduced electron transfer of **7** (M=$H_2$) in the LB film can be rationalized by the shorter donor–acceptor separation arising from the compact conformer in the LB film.[31]

## 9.3 Self-Assembled Monolayers

### 9.3.1 Self-Assembled Monolayers of Fullerene-Containing Systems on Gold Electrodes

SAMs have recently attracted much attention as a new methodology for molecular assembly.[114] They enable the molecules of interest to be bound covalently on the surface such as metals, semiconductors, and insulators in a highly organized manner. The well-ordered structure in SAMs is in sharp contrast with those in conventional LB films and lipid bilayer membranes in terms of stability, uniformity, and manipulation. Therefore, they make it possible to arrange functional molecules unidirectionally at the molecular level on electrodes when substituents that would self-assemble covalently on the substrates are attached to a terminal of the molecules. A wide variety of examples have been reported to date involving fullerenes on gold,[33–47,115,116] ITO,[48–56,115] and other surfaces.[57] In the following section self-assembled monolayer systems are described for which experimental data on photoelectrochemical properties are available.

#### 9.3.1.1 Fullerene Single Components

So far the largest number of SAMs of fullerenes studied to date utilize fullerenes as a photosensitizer as well as an electron acceptor. Photoinduced electron transfer at the gold electrodes modified with fullerenes have been extensively studied.[33–47]

Imahori et al.[33,34] presented SAMs of fullerenes, prepared from $C_{60}$-tethered alkanethiols **8** (Figure 2), on a gold electrode (Au/**8**). To explore the relationship between the structure and the photoelectrochemical properties of the monolayers, they employ three different kinds of $C_{60}$ alkanethiols **8** where the linking position is varied systematically from *ortho* to *para* at the phenyl group on the pyrrolidine ring fused to the $C_{60}$ moiety. Photoelectrochemical measurements were performed using Au/**8** as a working electrode, a platinum wire as a counter electrode, and an Ag/AgCl as a reference electrode in the presence of ascorbic acid (AsA) (Au/**8**/AsA/Pt device). The internal quantum yield of anodic photocurrent generation in Au/**8**/AsA/Pt device varies from 7.5 to 9.8%, depending on the linking positions.[33,34] The internal quantum yields are comparable to those reported for similar $C_{60}$ devices using LB membranes.[26,27] The highly ordered monolayer as well as the small reorganization energy of $C_{60}$ in electron transfer improves the performance of the photoelectrochemical device greatly.[33,34]

Enger et al.[35] prepared two photoelectrochemical devices based on self-assembled monolayer of $C_{60}$-alkanethiol **9** (Figure 2) on a gold electrode (Au/**9**) and the $C_{60}$-modified gold electrode on which an ion-selective polyurethane membrane (Mb) is cast (Au/**9**/Mb). The high internal quantum yields of anodic photocurrent generation in Au/**9**/Pt (31%) and Au/**9**/Mb/Pt (25%) devices are obtained without an external electron sacrificer. However, the photocurrent generation mechanism is unclear.[35]

## 9.3.1.2 Fullerene-Containing Dyads

Akiyama *et al.* and Imahori *et al.*[36,37] reported the first multistep electron transfer in SAMs of porphyrin–fullerene dyads **10** (Figure 2). A combination of porphyrin–fullerene is chosen to exhibit a long-lived charge-separated state with a high quantum yield owing to the small reorganization energies of fullerenes.[17–24] The porphyrin-$C_{60}$ molecules are tilted and nearly parallel onto the gold surface, leading to the formation of the loosely packed structures ($(1.6–3.7) \times 10^{-11}$ mol cm$^{-2}$).[37] Photoelectrochemical measurements were carried out in argon-saturated $Na_2SO_4$ aqueous solution containing methylviologen ($MV^{2+}$) as an electron carrier using the modified gold electrode as a working electrode and a platinum counter electrode (Au/**10**/$MV^{2+}$/Pt device). Under the short-circuit conditions cathodic photocurrent is observed for the photoelectrochemical device; the internal quantum yield is *ca.* 0.5%. The photocurrent intensity in Au/**10**(M=H$_2$)/$MV^{2+}$/Pt device is larger by five fold, relative to the corresponding free-base porphyrin monomer device, demonstrating that $C_{60}$ is an effective

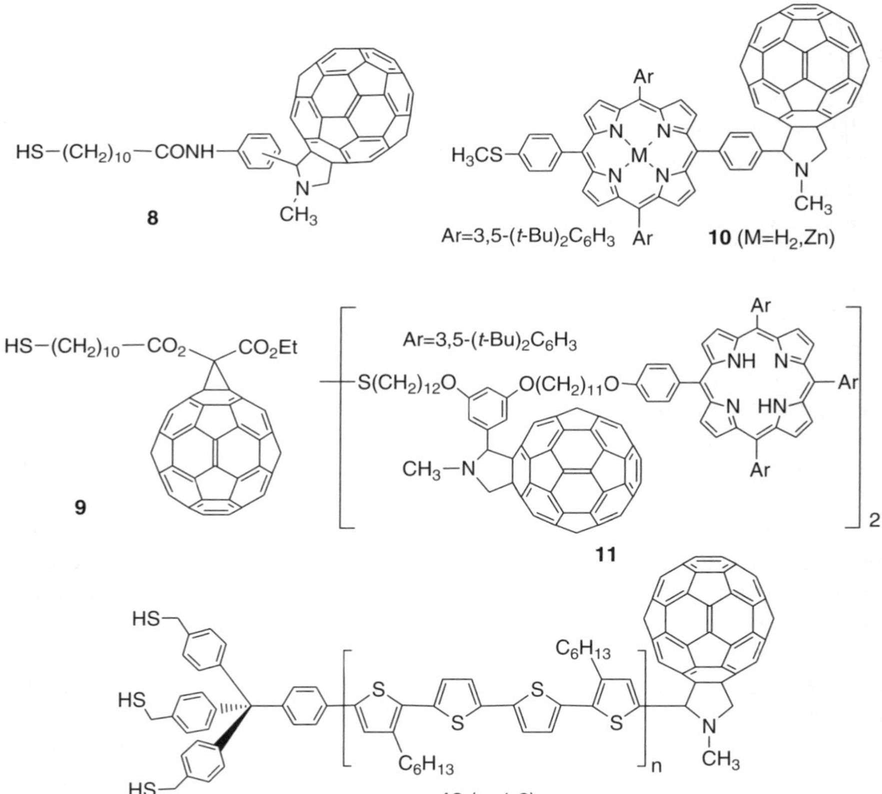

**Figure 2** *Sulfur-containing fullerene derivatives **8–12** which are linked to a gold electrode*

mediator in multistep electron transfer processes. The photocurrent intensity in Au/**10**(M=H$_2$)/MV$^{2+}$/Pt device is about one order of magnitude larger than that in the reference device without C$_{60}$ moiety. The shorter lifetime in the charge-separated state of the zinc porphyrin-C$_{60}$ device leads to a poor generation of the photocurrent, whereas the longer lifetime in the exciplex state of the freebase porphyrin-C$_{60}$ device results in a pronounced increase of the photocurrent.[36,37]

Yamada et al.[38] also investigated a self-assembled monolayer from fullerene–porphyrin linked dyad dimer **11** (Figure 2, Au/**11**) on a gold electrode (Figure 2) which exhibits anodic photocurrent generation instead of cathodic photocurrent generation. The spectroscopic and electrochemical studies reveal densely packed structure of Au/**11**. The photoelectrochemical properties of Au/**11** in the presence of electron donor (triethanolamine (TEA), ascorbic acid (AsA), and EDTA) were examined to compare with those of the porphyrin reference device without C$_{60}$ otherwise under the same conditions (+0.70 vs. Ag/AgCl (sat. KCl) in a standard three-electrode arrangement; Au/**11**/electron donor/Pt device). The internal quantum yield (2.1%) of photocurrent generation in Au/**11**/TEA/Pt device is 20 times as large as that in the porphyrin reference device (0.09%) without C$_{60}$ moiety. These results clearly demonstrate that C$_{60}$ acts as a good electron mediator between the porphyrin moiety and the gold electrode.[38]

Aso, Hirayama et al.[39,40] designed gold electrodes modified with oligothiophene–fullerene dyads **12** (Figure 2) bearing a tripodal rigid anchor, which is a tetraphenylmethane core with three mercaptomethyl arms, allowing such molecules to be well-organized on the electrodes (Au/**12**). Photoelectrochemical measurements were performed in an argon-saturated 0.1 M Na$_2$SO$_4$ solution containing 5 × 10$^{-3}$ M of methyl viologen as an electron acceptor using the modified gold electrode as a working electrode and a platinum counter electrode at –0.1 V bias against an Ag/AgCl reference electrode under illumination of 400 nm light for Au/**12** ($n = 1$) or 440 nm for Au/**12** ($n = 2$), where the oligothiophene absorbs the light mainly (Au/**12**/MV$^{2+}$/Pt device). The internal quantum yields of photocurrent generation are estimated to be 17% for Au/**12** ($n = 1$)/MV$^{2+}$/Pt and 35% for Au/**12** ($n = 2$)/MV$^{2+}$/Pt devices, respectively.[40]

### 9.3.1.3 Fullerene-Containing Triads

To our knowledge, only three examples on a self-assembled monolayer of fullerene-containing triad appeared. Imahori et al.[41,42] reported the first alkanethiol-attached triads **13** (M=H$_2$, Zn) with a linear array of ferrocene (Fc), porphyrin (P), and C$_{60}$ (Figure 3). The triad molecules are densely packed with an almost perpendicular orientation on the gold surface. Photoelectrochemical experiments using the triad modified gold, a platinum wire, and an Ag/AgCl electrodes were carried out in the presence of electron carriers such as oxygen and/or methylviologen (Au/**13**/MV$^{2+}$/Pt device). Under the optimal conditions using the oxygen-saturated electrolyte solution in the presence of methylviologen, the internal quantum yields of photocurrent generation in Au/**13**/MV$^{2+}$/Pt device are 25% (M=H$_2$) and 20% (M=Zn), respectively,[41,42] which are

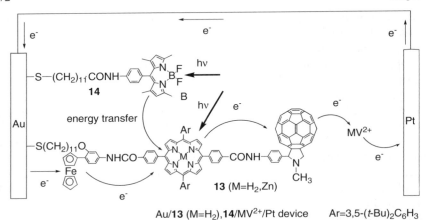

**Figure 3** *Schematic diagram for photocurrent generation in Au/**13,14**/MV$^{2+}$/Pt device*

comparable to the highest values among the previously reported photoinduced multistep electron transfer at monolayer-modified metal electrodes using donor–acceptor linked molecules.[33–47] Utilization of $C_{60}$ with the small reorganization energy allows the device to accelerate the forward-electron transfer processes and decelerate the undesirable back-electron transfer ones, thus leading to the high quantum yields.

Imahori et al.[43] further incorporated ferrocene (Fc)-freebase porphyrin ($H_2P$)-$C_{60}$ triad alkanethiol (**13** in Figure 3) into the boron dipyrrin-alkanethiol **14** to mimic light-harvesting in antenna complexes and multistep electron transfer in reaction centers (Figure 3). The photoelectrochemical measurements were performed using the mixed SAM of **13** and **14** on the gold electrodes in a three electrodes system under the optimized conditions described above (electrolyte solution: $O_2$-saturated 0.1 M $Na_2SO_4$ solution containing 30 mM methyl viologen ($MV^{2+}$) as an electron carrier; applied potential: –0.20 V vs. Ag/AgCl) (Au/**13,14**/$MV^{2+}$/Pt device). The internal quantum yield in Au/**13,14**/$MV^{2+}$/Pt device at 430 and 510 nm increases with increasing the content of **13** in the SAM. The maximum values of 21 ± 3% (430 nm) and 50 ± 8% (510 nm) are obtained at the ratio of **13**:**14** = 63:37 (Au/**13,14**/$MV^{2+}$/Pt device).[43] The internal quantum yield (50 ± 8%) at 510 nm is the highest value ever reported for photocurrent generation at monolayer-modified metal electrodes using donor–acceptor linked molecules.[33–47]

Saha et al.[44,45] designed a photoactive molecular triad **15** as a nanoscale power supply for a supramolecular machine (Figure 4). The fullerene-based molecular triad, bearing a porphyrin as the sensitizer unit and tetrathiafulvalene (TTF) as the donor unit, constitutes the basis of their design. Photoelectrochemical experiments were carried out in a standard three-electrode electrochemical device by using a gold-foil electrode functionalized with **15** as the working electrode, a platinum gauze as the counter electrode, Ag/AgCl as the reference electrode, and $CH_3CN$ containing 0.1 M $TBAPF_6$ as

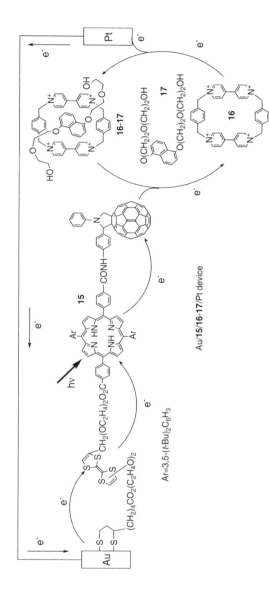

**Figure 4** *Schematic diagram for photoactive triad as a nanoscale power supply for supramolecular machine in Au/15/16-17/Pt device*

the electrolyte solution. Cathodic photocurrent generation (internal quantum yield = 1%) is observed upon irradiation with a 413 nm laser light (26 mW cm$^{-2}$), while the device is held with a bias potential at 0 V vs. Ag/AgCl.[44,45] Next, this nanoscale source of electrical energy is utilized to drive the dethreading of a pseudorotaxane **16–17** comprised of cyclobis(paraquat-*p*-phenylene) cyclophane **16** complexed with 1,5-bis-[(2-hydroxyethoxy)ethoxy]naphthalene **17** (Au/**15**/**16–17**/Pt device). The pseudorotaxane can be considered as a reasonably stable charge-transfer complex between thread **17** and cyclophane **16**. However, reduction of **16** to the corresponding radical cation weakens the charge-transfer interaction, which in turn, results in the decomplexation of the radical cation from thread **17**. From the experiments, the authors concluded that the decomplexation of the pseudorotaxane by the photosensitization of the light-harvesting molecular triad **15** at the applied potential of 0 V vs. Ag/AgCl is a direct consequence of the photoreduction of the electron-acceptor $C_{60}$ component, which passes its electron onto the pseudorotaxanes present in the adjacent electrolyte solution, causing the electrochemical reduction of cyclophane **16**, resulting in its dethreading from **16** in a chain process that can be controlled by a specific wavelength of laser light.[44,45]

Kim et al.[46] reported a self-assembled monolayer of $C_{60}$-tethered 2,5-dithienylpyrrole triad on a gold electrode. Photoelectrochemical measurements were performed using the modified gold electrode as a working electrode, a platinum wire as a counter electrode, and an Ag/AgCl as a reference electrode in the presence of $MV^{2+}$. The internal quantum yield of cathodic photocurrent generation in the photoelectrochemical device is as high as 51%, although the absorbance of the adsorbed chromophores on the gold electrode is assumed to be the same as that in solution.[46]

### 9.3.1.4 Fullerene-Containing Multiporphyrins

Morisue et al.[47] developed a self-assembled monolayer of multiporphyrin arrays terminated with $C_{60}$ on a gold electrode. The basic strategy employs the mutual coordination of two imidazolylporphyrinatozinc(II) units to form a cofacial dimer. Namely, *meso,meso*-linked bis(imidazolylporphyrinatozinc) $(Zn_2(ImP)_2)$ is organized onto the gold by forming a complex with imidazolylporphyrinatozinc preorganized on the gold electrode as a self-assembled monolayer. The organized $Zn_2(ImP)_2$ bearing allyl side chains is covalently linked by ring-closing olefin methathesis catalyzed with Grubbs catalyst. Alternating coordination/metathesis reactions allow the stepwise accumulation of multiporphyrin arrays on the gold electrode. Finally, the gold electrode modified with multiporphyrin arrays is further modified with imidazolylporphyrinatozinc-$C_{60}$ dyad, followed by treatment with Grubbs catalyst to yield the $C_{60}$-terminated antenna on the gold electrode. The maximum IPCE (1.2% at 480 nm) in the photoelectrochemical device is three times as large as that in the reference device without $C_{60}$ moiety.[47]

## 9.3.2 Self-Assembled Monolayers of Fullerene-Containing Systems on ITO Electrodes

Some of SAM devices on gold electrodes have exhibited efficient photocurrent generation.[33–47] However, strong quenching of the excited singlet state of the adsorbed dyes by gold surfaces has precluded achievement of a high internal quantum yield for charge separation on the surfaces as attained in photosynthesis ($\sim 100\%$). To surmount such a quenching problem, ITO which has high optical transparency and electrical conductivity seems to be the most promising candidate as an electrode, since semiconductors such as ITO can suppress the quenching of the excited states of adsorbed dyes on the surfaces.

### *9.3.2.1 Fullerene-Containing Dyads*

Yamada *et al.*[48,49] prepared ITO electrodes modified chemically with SAMs of free base porphyrin-$C_{60}$ linked dyad and zinc porphyrin-$C_{60}$ linked dyad. The photoelectrochemical measurements were performed in an oxygen-saturated 0.1 M $Na_2SO_4$ aqueous solution containing 5 mM 1,1′-dihexyl-4,4′-dipyridinium diperchlorate ($HV^{2+}$) as an electron carrier using the modified ITO electrode as a working electrode, a Pt counter electrode, Ag/AgCl (sat. KCl) as a reference electrode ($\lambda_{ex} = 430$ nm at a bias potential of –0.2 V *vs.* Ag/AgCl). The IPCE values of the freebase porphyrin-$C_{60}$ linked dyad and zinc porphyrin-$C_{60}$ linked dyad devices are 6.4 and 3.9%, which are 10–30 times as large as those of the porphyrin reference devices (0.21–0.38%) without $C_{60}$ moiety.[48,49] These results clearly show that remarkable enhancement of the photocurrent generation in the present systems results from an incorporation of the $C_{60}$ moiety as an acceptor into the porphyrin SAM on the ITO. This is the first example of the remarkable enhancement of photocurrent generation using SAMs of donor–acceptor linked molecules on ITO electrodes, compared to the corresponding single chromophore systems.[48,49]

Imahori *et al.*[51] examined the molecular structures, photophysical and photoelectrochemical properties of SAMs of porphyrin-containing dyads **18** and **19** on ITO to obtain the relationship between the photodynamics of the dyads in solutions and the photocurrent generation (Figure 5). In this case, the

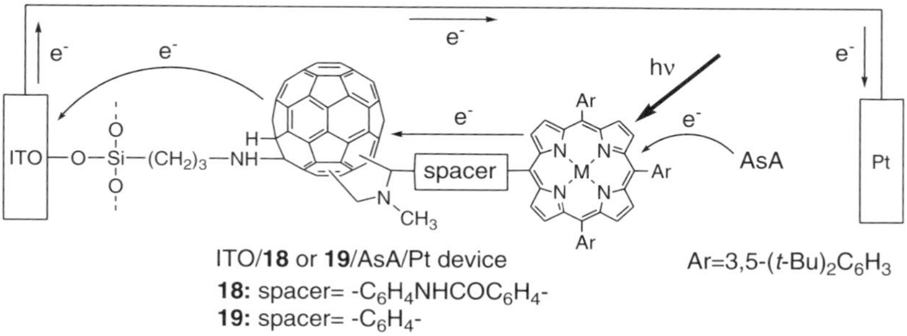


ITO/**18** or **19**/AsA/Pt device

**18:** spacer= $-C_6H_4NHCOC_6H_4-$

**19:** spacer= $-C_6H_4-$

Ar=3,5-($t$-Bu)$_2C_6H_3$

**Figure 5** *Schematic diagram for photocurrent generation in ITO/**18** or **19**/AsA/Pt device*

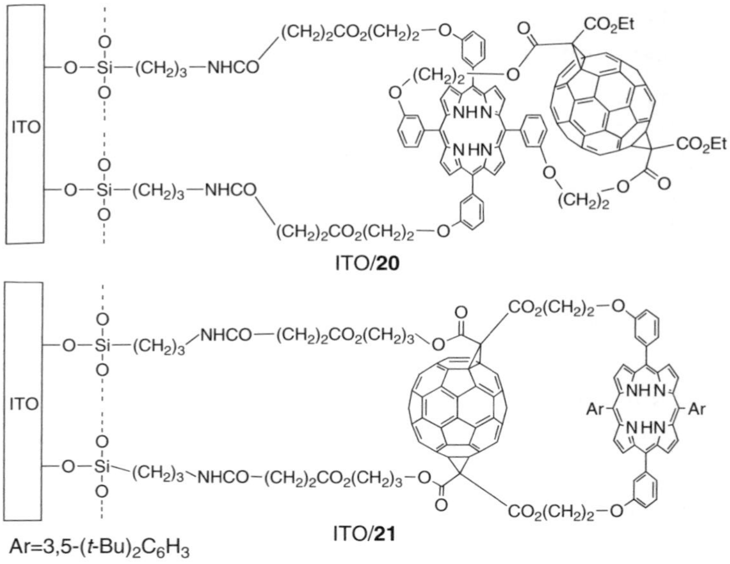

**Figure 6** *Schematic diagram for molecular arrangements in ITO/20 and ITO/21*

chromophores are arranged in reverse direction on ITO [ITO/**18** (M=H$_2$,Zn) and ITO/**19** (M=H$_2$,Zn)] relative to the aforementioned dyad devices. Photoelectrochemical measurements were carried out in an argon-saturated 0.1 M Na$_2$SO$_4$ aqueous solution containing 50 mM AsA acting as an electron sacrificer using the modified ITO electrode as a working electrode, a platinum counter electrode, and an Ag/AgCl (sat. KCl) reference electrode (ITO/**18** or **19**/AsA/Pt device). The internal quantum yields of photocurrent generation are compared for ITO/**18**/AsA/Pt and ITO/**19**/AsA/Pt devices at an applied potential of +0.15 V vs. Ag/AgCl (sat. KCl) with $\lambda_{ex}$ = 419 nm for freebase porphyrins and $\lambda_{ex}$ = 430 nm for zinc porphyrins. The value increases in the order of ITO/**19** (M=Zn)/AsA/Pt (1.7%) < ITO/**19** (M=H$_2$)/AsA/Pt (3.4%) < ITO/**18** (M=H$_2$)/AsA/Pt (5.1%) < ITO/**18** (M=Zn)/ITO (8.0%) devices. The lifetime of a charge-separated state as well as charge separation efficiency is an important controlling factor for achieving the high quantum yield of photocurrent generation in donor–acceptor-linked systems, which are covalently attached to ITO surface.[51]

The collaborative efforts of Tkachenko, Imahori, and Chukharev et al.[52] have continued to prepare two porphyrin–fullerene dyads to form SAMs on ITO electrodes with ITO-porphyrin-C$_{60}$ (ITO/**20**) and ITO-C$_{60}$-porphyrin (ITO/**21**) orientations, respectively (Figure 6). Specifically, the dyads contain two linkers for connecting the porphyrin and C$_{60}$ moieties and enforcing them essentially to similar geometries of the donor–acceptor pair, and other two linkers for linking the ITO surface and the porphyrin or C$_{60}$ moiety. The transient photovoltage responses were measured for the dyad films covered by insulating LB films, thus ensuring that the dyads interact only with the ITO

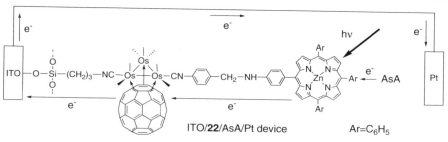

**Figure 7** *Schematic diagram for photocurrent generation in ITO/22/AsA/Pt device*

electrode. The direction of the electron transfer from the photoexcited dyad to the ITO is independent of the dyad orientation. The response amplitude for ITO/**21**, where the primary intramolecular electron transfer direction coincides with that of the final electron transfer from the dyad to ITO, is 25 times as large as the value for ITO/**20**. Static photocurrent measurements under the standard three electrode arrangement conditions (0.1 M $Na_2SO_4$ aqueous electrolyte solution containing 50 mM TEA (0.5 V vs. Ag/AgCl) or $HV^{2+}$ (0 V vs. Ag/AgCl)), however, show only a minor orientation effect (ITO/**20** or **21**/TEA or $HV^{2+}$/Pt device). The internal quantum yields of cathodic and anodic photocurrent generation are 0.36 and 0.45% for ITO/**20**/$HV^{2+}$/Pt and ITO/**21**/$HV^{2+}$/Pt devices, and 2.1 and 2.8% for ITO/**20**/TEA/Pt and ITO/**21**/TEA/Pt devices, respectively.[52] These results indicate that the photocurrent generation is controlled by the processes at the SAM-electrolyte solution interface.[52]

Cho et al.[54] designed ITO electrodes covalently modified with $C_{60}$-metal cluster moiety which was further tethered with zinc porphyrin unit (ITO/**22**, Figure 7). Photocurrent measurements for SAMs were carried out using AsA as a sacrificial electron donor at a bias potential of 0.1 V vs. Ag/AgCl with $\lambda_{ex}$ = 435 nm (ITO/**22**/AsA/Pt device). The internal quantum yield of anodic photocurrent generation in ITO/**22**/AsA/Pt device is estimated to be 10.4%.[54] Surprisingly, an addition of DABCO (1,4-diazabicyclo[2.2.2]octane) into the device results in the improvement of the quantum yield (19.5%), which is one of the highest quantum yields ever reported for molecular photoelectrochemical devices based on the covalently linked donor–acceptor molecules on ITO.[48–56] From the fluorescence lifetime measurements together with femtosecond transient absorption studies, the authors draw the conclusion that the complexation of DABCO between the two zincporphyrins precludes aggregation with adjacent porphyrins and increase of the donor–acceptor separation, leading to the high performance in photocurrent generation.[54]

### 9.3.2.2 Fullerene-Containing Triads

Imahori et al.[51] reported the first SAMs of fullerene-containing triads on ITO. An additional donor moiety (Fc) is attached to the porphyrin moiety of ITO/**23** to facilitate the electron relay between the porphyrin and AsA in the electrolyte solution for the improvement of the internal quantum yield of photocurrent

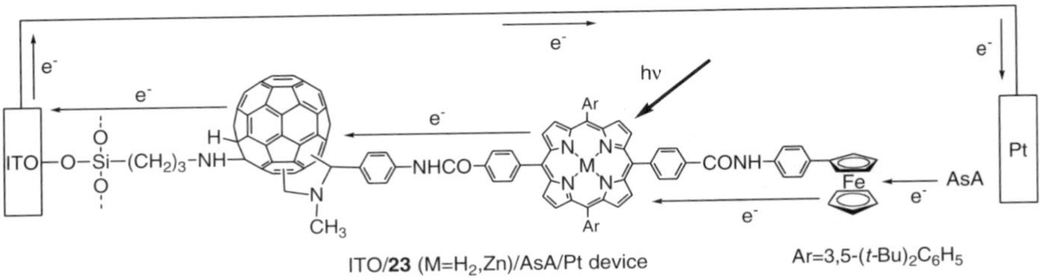

**Figure 8** *Schematic diagram for photocurrent generation in ITO/23 (M=H$_2$,Zn)/AsA/ Pt device*

generation (Figure 8). Photoelectrochemical measurements were carried out in an argon-saturated 0.1 M Na$_2$SO$_4$ aqueous solution containing 50 mM AsA acting as an electron sacrificer using ITO/23 (M=H$_2$,Zn) as a working electrode, a platinum counter electrode, and an Ag/AgCl (sat. KCl) reference electrode (ITO/23/AsA/Pt device). The internal quantum yields are determined at an applied potential of +0.15 V vs. Ag/AgCl (sat. KCl) at $\lambda_{ex}$ = 419 nm for the freebase porphyrin and $\lambda_{ex}$ = 430 nm for the zinc porphyrin. The quantum yield of ITO/23(M=Zn)/AsA/Pt device (11%) is larger than that of ITO/18(M=Zn)/AsA/Pt device (8.0%), whereas the values of ITO/23(M=H$_2$)/AsA/Pt (4.5%) and ITO/18(M=H$_2$)/AsA/Pt (5.1%) devices are comparable. The quantum yield of ITO/23(M=Zn)/AsA/Pt device (11%) is one of the highest values ever reported for photocurrent generation using donor–acceptor linked molecules which are covalently tethered to ITO surface.[48–56] It is noteworthy that the highest internal quantum yield (11%) of photocurrent generation for the ITO electrodes is rather lower than the corresponding values (20–25%) for the gold electrodes when similar ferrocene–porphyrin–fullerene triad molecules are attached chemically to the electrodes. The complex surface structure together with the reduced electron accepting ability of the C$_{60}$ arising from the bisaddition may rationalize the low photocurrent generation efficiency in the ITO electrode system.[51]

### 9.3.2.3 *Fullerene-Containing Multilayers*

Ikeda *et al.* and Konishi *et al.*[55,56] prepared ITO electrodes which were self-assembled with sodium 3-mercaptoethanesulfonate (first layer), hexacationic homooxacalix[3]arene 24-C$_{60}$ 2:1 complex (24-C$_{60}$) (second layer), and anionic porphyrin polymers 25 (M=H$_2$,Zn) (third layer) sequentially (ITO/24-C$_{60}$/25; Figure 9). Photocurrent measurements were carried out for ITO/24-C$_{60}$, ITO/24-C$_{60}$/25 (M=H$_2$), and ITO/24-C$_{60}$/25 (M=Zn) electrodes in 0.1 M Na$_2$SO$_4$ electrolyte solution (pH 3.5) containing 50 mM AsA as an electron sacrificer under a standard three electrode conditions (ITO/24-C$_{60}$/AsA/Pt or ITO/24-C$_{60}$/25/AsA/Pt device). The internal quantum yields of anodic photocurrent generation are estimated as 10% ($\lambda_{ex}$ = 400 nm) for ITO/24-C$_{60}$/AsA/Pt, 14% ($\lambda_{ex}$ = 420 nm) for ITO/24-C$_{60}$/25 (M = H$_2$)/AsA/Pt, and 21% ($\lambda_{ex}$ = 430 nm) for ITO/24-C$_{60}$/25 (M = Zn)/AsA/Pt devices, respectively. To reduce self-quenching of the porphyrin excited singlet state in the polymers, the porphyrin unit in the

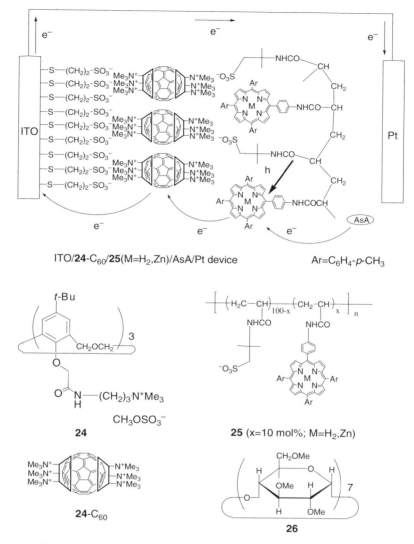

**Figure 9** *Schematic diagram for photocurrent generation in ITO/**24**-C$_{60}$/**25** (M=H$_2$,Zn)/AsA/Pt device*

porphyrin polymers **25** is isolated from the surrounding porphyrins by adding cyclodextrin **26**. The internal quantum yields of anodic photocurrent generation are estimated as 15% ($\lambda_{ex} = 420$ nm) for ITO/**24**-C$_{60}$/**25** (M=H$_2$)/AsA/Pt device in the absence of **26** and 20% ($\lambda_{ex} = 420$ nm) for ITO/**24**-C$_{60}$/**25** (M=H$_2$)/AsA/Pt devices in the presence of **26**. The authors propose that the porphyrin unit is insulated by **26** to suppress the undesirable self-quenching of the porphyrin excited singlet state.[55,56]

Bustos et al.[57] described the preparation and photoelectrochemical properties of a multilayered C$_{60}$-PAMAM G0.0 (a commercially available poly(amidoamine) dendrimer) film prepared on the top of a silanized nanocrystalline semiconductor $TiO_2$ electrode. Photocurrent measurements were carried out using the modified $TiO_2$ electrode as a working electrode, Ti substrate covered with a film of colloidal graphite as a counter electrode, and a deoxygenated 0.3 M KI and 0.015 M $I_2$ aqueous electrolyte solution. The maximum IPCE value at 430 nm increases dramatically with increasing the number of $C_{60}$ layers, reaching value as high as 82% (three layers).[57]

The examples given demonstrate that SAMs of fullerenes on gold and ITO electrodes are excellent systems for the realization of efficient photocurrent generation on electrode surfaces. Self-assembled monolayer will open a door for the development of photoactive molecular devices in which the highly ordered, well-designed architecture acts as efficient photocatalysts, photosensors, and photodiodes.

## 9.4 Layer-by-Layer Deposition

Guldi et al.[59,60] have extensively explored the use of electrostatic and *van der Waals* interactions to fabricate photoactive ITO electrodes deposited with porphyrin–fullerene dyad. Initially, a base layer of poly(diallyldimethylammonium) (PDDA) is deposited onto the hydrophobic surfaces *via* simple hydrophobic–hydrophobic interactions (step 1). The resulting hydrophilic and positively charged surface promotes the electrostatically driven deposition of poly(sodium 4-styrenesulfonate) (PSS) (step 2), bearing sulfonic acid functionalities, to yield the PDDA/PSS templates. With the surface sufficiently overlaid with negative charges, the porphyrin–fullerene dyads can be deposited by immersing the substrates into *o*-dichlorobenzene/DMSO (1:1, v/v) solution of the porphyrin–fullerene dyad (step 3). Sequential repetition of step 2, and 3, deposition of PSS and the porphyrin–fullerene dyads for up to 12 times (absorbance$_{435\ nm}$ = 1.3), leads to the systematic stacking of PSS and the porphyrin–fullerene dyad sandwich layers. Photoelectrochemical experiments were performed in an $O_2$-saturated 0.1 M KCl aqueous solution containing 50 mM AsA as an electron sacrificer using the modified ITO working electrode, a Pt gauze counter electrode, and an Ag/AgCl reference electrode. The IPCE value of anodic photocurrent generation in the one-layer device is as low as 0.6% at 440 nm with a bias potential of 0 V *vs.* Ag/AgCl.[59,60] The poor IPCE value may be caused by the flexible cationic tail of the porphyrin–fullerene dyad, leading to the ill-defined arrangement of the porphyrin–fullerene dyad on ITO. In addition, a relatively short lifetime of the exciplex state may be responsible for the low IPCE value.

Zilbermann et al.[61] described the photoelectrochemical properties of ITO electrodes deposited with porphyrin–fullerene dyads **7** (M=$H_2$, Zn) using electrostatic and *van der Waals* interactions (Figure 1). The PDDA-modified, hydrophilic and positively charged surface promotes the electrostatically driven deposition of **7**, to yield the PDDA/**7** templates. With the surface sufficiently overlaid with negative charges, PSS can be deposited. Sequential repetition of these steps, deposition of **7** and PSS for up to seven times (absorbance$_{435\,nm}$ = 0.3), results in the alternative stacking. Photoelectrochemical experiments were performed in an $O_2$-saturated 0.1 M $NaH_2PO_4$ aqueous solution using the modified ITO working electrode, a Pt gauze counter electrode, and an Ag/AgCl reference electrode (ITO/PDDA/**7**/$O_2$/Pt device). The IPCE values of cathodic photocurrent generation are 0.3 and 0.1% for ITO/PDDA/(**7**(M=$H_2$))/$O_2$/Pt and ITO/PDDA/(**7**(M=Zn))/$O_2$/Pt devices, respectively, at 440 nm with a bias potential of –0.20 V vs. Ag/AgCl. Guldi[58] also reported stepwise-assembled photoelectrochemical devices containing ruthenium dye-$C_{60}$ dyads on ITO electrodes. However, the maximum IPCE value remains low ($\sim$1.1% at 375 nm).

Guldi[62] tailored the arrangement of photoactive ITO electrodes at the molecular level (Figure 10). $C_{60}$ molecules with negatively charged moieties **27** are adsorbed at the PDDA-modified, positively charged ITO surface (ITO/PDDA/**27**). Strong van der Waals interactions between $C_{60}$ cores facilitate ITO/PDDA/**27** device layer formation, leaving the anionic dendrimer branches on the surface. In the next step octacationic porphyrins **28** (M=$H_2$, Zn) are deposited via electrostatic interactions with the anionic dendrimer branches in a monolayer fashion (ITO/PDDA/**27**/**28**). Subsequent layers are built up analogously utilizing cationic/anionic contacts to yield ITO/PDDA/**27**/**28**/**29** and ITO/PDDA/**27**/**28**/**29**/**30** electrodes, respectively. Photoelectrochemical experiments were performed in a $N_2$-saturated 0.1 M $NaH_2PO_4$ aqueous solution containing 1 mM AsA as an electron sacrificer using the modified ITO working electrode, a Pt gauze counter electrode, and an Ag/AgCl reference electrode (ITO/PDDA/**27**/**28**/**29**/**30**/AsA/Pt, ITO/PDDA/**27**/**28**/**29**/AsA/Pt, ITO/PDDA/**27**/**28**/AsA/Pt, and ITO/PDDA/**27**/AsA/Pt devices). The IPCE values of anodic photocurrent generation are 1.6, 1.0, 0.08, and 0.01% for ITO/PDDA/**27**/**28**/**29**/**30**, and ITO/PDDA/**27**/**28**/**29**, ITO/PDDA/**27**/**28** and ITO/PDDA/**27** devices, respectively, at 440 nm with a bias potential of 0 V vs. Ag/AgCl. The approach differs from the layer-by-layer strategy, where sandwich layers of covalently linked donor–acceptor dyads are integrated between layers of polyelectrolytes, affording smaller IPCE values.[62]

Guldi et al.[63] developed a molecular-level switch, a negatively charged dendritic fullerene ($C_{60}$DF)/cytochrome c (FeCytc) modified ITO electrode, that reversibly transmits and processes solar energy. The stepwise assembly of $C_{60}$DF and FeCytc onto the PDDA-modified, positively charged ITO surface (ITO/PDDA/$C_{60}$DF/FeCytc) is carried out using electrostatic interaction. The photocurrent generation in the ITO/PDDA/$C_{60}$DF/Fe$^{III}$Cytc device is very small (0.2 µA, OFF state), whereas that in the ITO/PDDA/$C_{60}$DF/Fe$^{II}$Cytc device is enhanced by a factor of 10 (1.8 µA, ON state). Thus, the reversible switching between the two defined states is attained by altering the redox state of the iron center as a result of applying different electrochemical potentials.[63]

## 9.5 Vacuum Deposition

Vacuum deposition technique has been frequently employed to fabricate fullerenes on electrode surfaces.[67–71] Typically, bulk heterojunction solar cells have

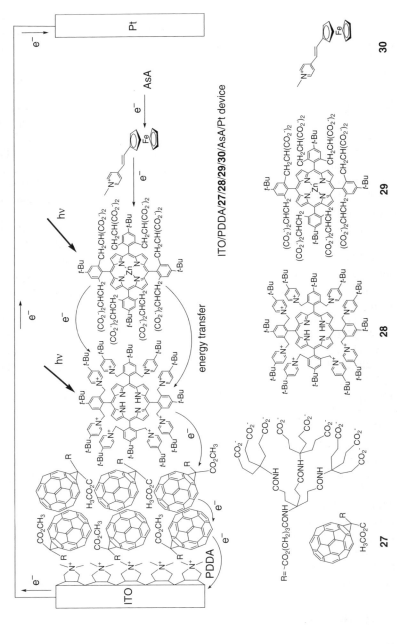

**Figure 10** *Schematic diagram for photocurrent generation in ITO/PDDA/27/28/29/30/AsA/Pt device*

been prepared using vacuum codeposition of donor (phthalocyanines) and fullerenes onto electrode surfaces. So far the power conversion efficiency has reached up to 5.6%.[67–71]

## 9.6 Electrochemical Deposition

### 9.6.1 Fullerenes and Their Derivatives

Although SAMs of donor-fullerene linked systems on gold or ITO electrodes have improved the internal quantum yields of photocurrent generation, such assemblies possess poor light-harvesting capability, leading to low IPCE. Kamat et al.[72] reported a novel approach of enhancing the light-harvesting efficiency by electrodeposited $C_{60}$ films from a cluster (aggregate) solution of acetonitrile/toluene (3:1, v/v), which absorb visible light intensively and exhibit much improved photoelectrochemical response. Specifically, a toluene solution of $C_{60}$ molecules is rapidly injected into acetonitrile to form $C_{60}$ clusters due to the lyophobic nature of $C_{60}$ in the mixed solvent. The clusters of $C_{60}$ are deposited as thin films on nanostructured $SnO_2$ electrodes under the influence of an electric field. A maximum IPCE of $\sim 4\%$ is noted at 420 nm. Direct electron transfer between the excited $C_{60}$ clusters deposited on $SnO_2$ nanocrystallites and the redox couple ($I_3^-/I^-$) in solution is responsible for the enhanced photocurrent generation.[72]

Hotta et al.[73] designed two fullerene derivatives with a phenylpyrrolidine moiety to examine the effect of the fullerene substituents on the structure and photoelectrochemical properties of fullerene clusters electrophoretically deposited on nanostructured $SnO_2$ electrodes. The cluster sizes increase and the incident photon-to-current efficiencies decrease with introduction of larger substituents into $C_{60}$. The trend for photocurrent generation efficiency as well as surface morphology on the electrode can be explained by the steric bulkiness around the $C_{60}$ molecules. A $C_{60}$ molecule with two alkoxy chains is suggested to give a bilayer vesicle structure, irrespective of the hydrophobic nature of both the $C_{60}$ and alkoxy chain moieties.[73] Such information will be valuable for the design of photoactive molecules, which are fabricated onto electrode surfaces to exhibit high energy conversion efficiency.

Sudeep et al.[74] assembled alkanethiol-tethered $C_{60}$ onto a gold nanoparticle for photoelectrochemical device. By incorporating alkanethiol-tethered $C_{60}$ into dodecanethiol-capped gold nanoparticles during the gold reduction step, it is possible to bind ca. 90 $C_{60}$ moieties per gold nanocore. The $C_{60}$-modifed gold nanoparticles in toluene are deposited electrophoretically onto a nanostructured $SnO_2$ electrode. A maximum IPCE value of 0.8% is obtained at 450 nm.[74]

The interaction between the excited sensitizer and the redox couple in dye-sensitized solar cells is an important factor that can decrease the power conversion efficiency. Kamat et al.[77] employed $C_{60}$ clusters to separate ruthenium dye and $I_3^-/I^-$ couple to minimize the sensitizer-redox couple interaction. The $ITO/TiO_2/Ru(II)/C_{60}/LiI + I_2/Pt$ device exhibits a maximum IPCE of

64%, which is larger by 15% than that in ITO/TiO$_2$/Ru(II)/LiI + I$_2$/Pt device without the C$_{60}$ layer under similar experimental conditons.[77]

### 9.6.2 Donor–Acceptor Linked Systems Involving Fullerene

Kamat et al.[78] deposited clusters of C$_{60}$-aniline linked dyad **31** as thin films on a nanostructured SnO$_2$ electrode under the influence of an electric field (Figure 11). Photoelectrochemical measurements were carried out under the similar conditions they have employed. A maximum IPCE of $\sim 4\%$ is noted at 450 nm,[78] which is similar to that ($\sim 4\%$) of similar photoelectrochemical device with C$_{60}$ itself.[72]

Imahori et al. and Hasobe et al.[79,86] applied electrophoretic deposition method to porphyrin–fullerene linked dyads **32** (M=H$_2$, Zn) to assemble the dyad clusters on a nanostructured SnO$_2$ electrode using an electrophoretic deposition technique (Figure 11). The porphyrin–fullerene dyad cluster films when electrophoretically deposited on nanostructured SnO$_2$ films are photoactive and generate anodic photocurrent under standard three electrode arrangement containing the modified SnO$_2$ working electrode, a Pt wire counter electrode, and an Ag/AgCl reference electrode (electrolyte: 0.5 M LiI and 1 mM I$_2$ in acetonitrile; 0 V vs. Ag/AgCl; ITO/SnO$_2$/**32**/LiI + I$_2$/Pt device). Although, the light-harvesting properties (maximal absorbance=1.6) are improved by a factor of 11 compared to the monolayer system ($\sim 0.04$), the maximum IPCE values are found to be still low (0.36% for ITO/SnO$_2$/**32** (M=H$_2$)/LiI + I$_2$/Pt device and 0.42% for ITO/SnO$_2$/**32** (M=Zn)/LiI + I$_2$/Pt device).[79,86] The maximum IPCE values are smaller than those of photoelectrochemical device ($\sim 9\%$) in which similar porphyrin-C$_{60}$ dyad is adsorbed onto SnO$_2$ electrode.[94] The poor IPCE values may result from macroscopic cancellation of the charge-separated state in the

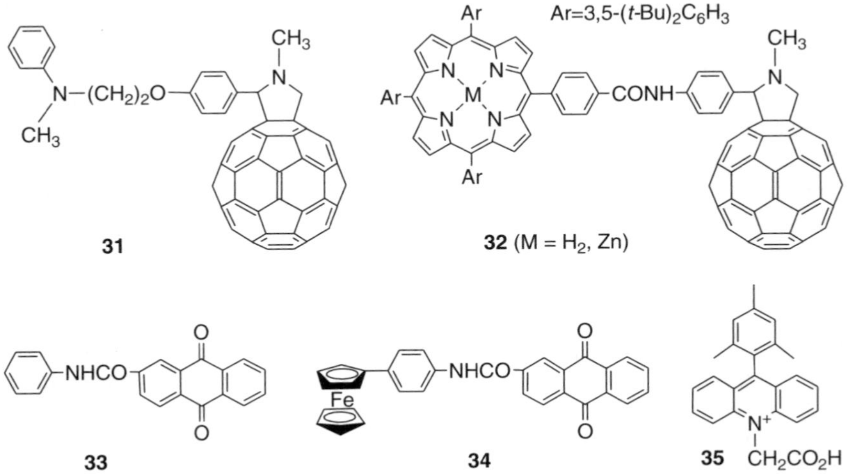

**Figure 11** *Donor–acceptor linked molecules **31–35** for photoelectrochemical devices*

clusters of **32** due to the rather random orientation of the dyad molecules in the clusters.

Okamoto et al.[80] employed formanilide–anthraquinone dyad **33** and ferrocene–formanilide–anthraquinone dyad **34** (Figure 11) as components of photoelectrochemical devices where composite molecular nanoclusters of the formanilide–anthraquinone dyad or ferrocene–formanilide–anthraquinone triad with fullerene are assembled onto a $SnO_2$ electrode using an electrophoretic deposition method. Photocurrent measurements were performed in acetonitrile containing 0.5 M NaI and 0.01 M $I_2$ as redox electrolyte and a Pt gauge counter electrode (ITO/$SnO_2$/(**33** + $C_{60}$)$_m$ or (**34** + $C_{60}$)$_m$/NaI + $I_2$/Pt device). The maximum IPCE value (9.7% at 470 nm) in ITO/$SnO_2$/(**33** + $C_{60}$)$_m$/NaI + $I_2$/Pt device is 10 times as large as that (0.98% at 465 nm) in ITO/$SnO_2$/(**34** + $C_{60}$)$_m$/NaI + $I_2$/Pt device under the same conditions (0 V vs. SCE). Such enhancement in photocurrent generation can be ascribed to the dramatic difference in the lifetime of the charge-separated state of **33** (900 μs) vs. **34** (20 ps) in DMSO.[80]

Hasobe et al.[81] reported photovoltaic cell comprising of 9-mesityl-10-carboxymethylacridinium ion **35** and fullerene clusters (Figure 11). 9-Mesityl-10-carboxymethylacridinium ion **35** are adsorbed chemically onto a $SnO_2$ electrode, followed by electrophoretic deposition of the fullerene cluster onto the $SnO_2$ electrode. Photoelectrochemical measurements were performed using a standard three electrode system consisting of a working electrode and a Pt gauge electrode in air-saturated acetonitrile containing 0.5 M NaI and 0.01 M $I_2$ (ITO/$SnO_2$/**35** + ($C_{60}$)$_m$/NaI + $I_2$/Pt device). The IPCE value in ITO/$SnO_2$/**35** + ($C_{60}$)$_m$/NaI + $I_2$/Pt device reaches 25% at 480 nm at an applied potential of 0.2 V vs. SCE, whereas the maximum IPCE value of ITO/$SnO_2$/**35**/NaI + $I_2$/Pt device is 2% at 460 nm.[81]

Hasobe et al.[82] deposited electrophoretically the composite clusters of $C_{60}$ and $TiO_2$ nanoparticles modified with **35** ($TiO_2$-**35**) in acetonitrile-toluene (3:1, v/v) onto a $SnO_2$ electrode. The $TiO_2$ nanoparticles act as scafford to organize **35** and $C_{60}$. The IPCE value in ITO/$SnO_2$/($TiO_2$-**35** + $C_{60}$)$_m$/NaI + $I_2$/Pt device reaches 37% at 480 nm at an applied potential of 0.2 V vs. SCE, whereas the maximum IPCE value of ITO/$SnO_2$/($TiO_2$-**35**)$_m$/NaI + $I_2$/Pt device is 5% at 480 nm.[82] Although **35** is reported to exhibit an extremely long-lived charge-separated state with a high quantum yield,[117–119] $C_{60}$ as an electron mediator between the charge-separated state and the $SnO_2$ electrode is necessary to achieve efficient photocurrent generation.

## 9.6.3 Donor and Fullerene Composite Systems

Composites of donor and fullerene moieties in the form of clusters have been assembled as three-dimensional network on a conducting surface for attaining efficient photocurrent generation. Biju et al.[83] prepared composite clusters of various donors (N,N-dimethylaniline, N,N-dimethyl-p-toluidine, ferrocene, N-methylphenothiazine, and N,N-dimethyl-p-anisidine) and 1,2,5-triphenylpyrrolidinofullerene which are deposited electrophoretically onto a nanostructured $SnO_2$ electrode. A maximum IPCE of ~0.05% is obtained at 420 nm,[83] which is much smaller than those of similar photoelectrochemical device with $C_{60}$ (1.6–4% at 420 nm)[72] and fullerene derivatives (0.2–0.8% at 400 nm).[73]

Imahori and Hasobe *et al.*[15,16,84] applied step-by-step self-assembly to porphyrin and fullerene single components and the both composites to construct a novel organic solar cell (dye-sensitized bulk heterojunction solar cell), possessing both the dye-sensitized and bulk heterojunction characters. Namely, first, porphyrin and fullerene make a complex in nonpolar solvent due to the π–π interaction[15,16] (step 1). Then, the supramolecular complex of porphyrin and fullerene is self-assembled into larger clusters in a mixture of polar and nonpolar solvents (step 2). Finally, the larger clusters can be further associated onto a nanostructured $SnO_2$ electrode using an electrophoretic deposition technique (step 3).[15,16,84]

When a concentrated solution of $C_{60}$ and/or porphyrin **36** (Figure 12) in toluene is mixed with acetonitrile by fast injection method, the molecules aggregate and form stable clusters $[(C_{60})_m$ or $(36)_m$ or $(36 + C_{60})_m]$.[84] Then, electrophoretic deposition method is adopted to prepare film of $(36 + C_{60})_m$ on an ITO/$SnO_2$ electrode. Photocurrent measurements were performed in acetonitrile containing NaI (0.5 M) and $I_2$ (0.01 M) as redox electrolyte and Pt gauge counter electrode (ITO/$SnO_2$/$(36 + C_{60})_m$/NaI + $I_2$/Pt device). The short circuit photocurrent density ($I_{SC}$) of 0.18 μA cm$^{-2}$, open-circuit voltage ($V_{OC}$) of 0.21 V, the fill factor (FF) of 0.35, and the power conversion efficiency of 0.012% (input power=110 mW cm$^{-2}$) are obtained for ITO/$SnO_2$/$(36 + C_{60})_m$/NaI + $I_2$/Pt device. The maximum IPCE value of 4.0% at 430 nm for ITO/$SnO_2$/$(36 + C_{60})_m$/NaI + $I_2$/Pt device is larger than the sum (~2%) of 0.6% for ITO/$SnO_2$/$(36)_m$/NaI + $I_2$/Pt and 1.6% for ITO/$SnO_2$/$(C_{60})_m$/NaI + $I_2$/Pt devices, respectively. Such an enhancement in the photocurrent generation of the composite

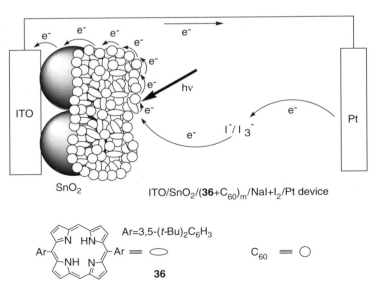

**Figure 12** *Schematic diagram for photocurrent generation in ITO/$SnO_2$/$(36 + C_{60})_m$/ NaI + $I_2$/Pt device*

cluster devices from **36** and $C_{60}$ demonstrates that charge separation between the excited porphyrin and $C_{60}$ is a dominating factor for efficient photocurrent generation. Photoinduced electron transfer occurs from the porphyrin singlet excited state ($^1H_2P^*/H_2P^{\bullet+} = -0.7$ V vs. NHE) to $C_{60}$ ($C_{60}/C_{60}^{\bullet-} = -0.2$ V vs. NHE). $C_{60}^{\bullet-}$ transfers electron to the conduction band of $SnO_2$ nanocrystallites ($E_{CB} = 0$ V vs. NHE), to produce the current in the circuit. The regeneration of $H_2P$ clusters ($H_2P/H_2P^{\bullet+} = 1.2$ V vs. NHE) is achieved by the iodide/triiodide couple ($I^-/I_3^- = 0.5$ V vs. NHE) present in the electrolyte system (Figure 12).[84]

The present organic solar cell (dye-sensitized bulk heterojunction solar cell) is unique in that it possesses both the dye-sensitized and bulk heterojunction characteristics. Moreover, the blend films exhibit the multilayer structure on ITO/$SnO_2$, which presents a striking contrast to monolayer structure of adsorbed dyes on $TiO_2$ electrodes of dye-sensitized solar cells. Therefore, we can expect improvement of the photovoltaic properties by modulating both the structures of electrode surfaces and D–A multilayers.

### 9.6.4 Pre-Organized Multi-Donor Systems

Supramolecular assembly of D–A molecules is a potential approach to create a desirable phase-separated, interpenetrating network involving molecular-based nanostructured electron and hole highways. However, different, complex hierarchies of self-organization going from simple molecules to devices have limited improvement of the device performance. To construct such complex hierarchies comprising of donor and acceptor molecules on electrode surfaces pre-organized molecular systems are excellent candidates for achieving the molecular architectures. In particular, porphyrins have been three-dimensionally organized using dendrimers,[85,86] oligomers,[87] and nanoparticles[88–93] to combine with fullerenes for organic solar cells.

#### 9.6.4.1 Multiporphyrin Dendrimers

The collaborative efforts of Hasobe et al.[85,86] have continued to develop the novel light energy conversion system using supramolecular complexes of multiporphyrin dendrimers with fullerene by clusterization in mixed solvent and electrophoretic deposition onto nanostructured $SnO_2$ electrodes. Porphyrin dendrimers (**37**(n); n = number of the porphyrins in **37** (Figure 13), step 1), form supramolecular complexes with fullerene molecules in toluene (step 2) and they are clusterized in an acetonitrile/toluene mixed solvent system (step 3). Then, the clusters are deposited on nanostructured $SnO_2$ electrodes by an electrophoretic deposition method according to Kamat's method[72] (step 4). Upon subjecting the resultant cluster suspension to a high electric DC field, mixed **37** and $C_{60}$ clusters [(**37** + $C_{60}$)$_m$] and reference clusters [(**38** + $C_{60}$)$_m$] are deposited onto an ITO/$SnO_2$ to give modified electrodes [ITO/$SnO_2$/(**37**(n) + $C_{60}$)$_m$ (n = 4, 8, 16) and ITO/$SnO_2$/(**38** + $C_{60}$)$_m$], respectively.

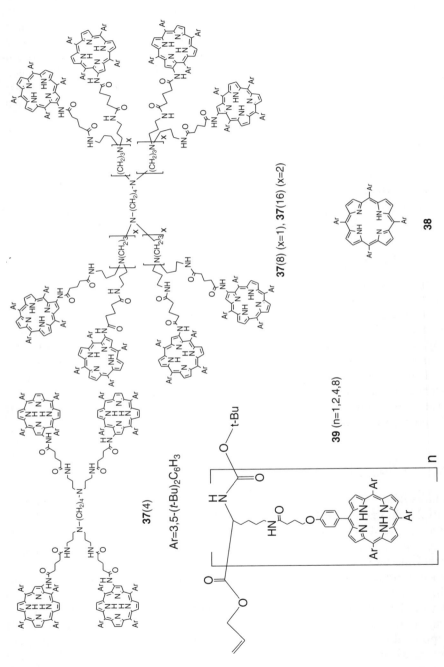

**Figure 13** *Molecular structures of multi-porphyrin dendrimers **37**, porphyrin monomer **38**, and porphyrin oligomers **39** for photoelectrochemical devices*

Photoelectrochemical measurements were performed with a standard two electrodes system consisting of the modified ITO working electrode and a Pt wire gauze electrode in 0.5 M NaI and 0.01 M $I_2$ in air-saturated acetonitrile (ITO/$SnO_2$/(**37**(n) + $C_{60}$)$_m$/NaI + $I_2$/Pt device). It should be noted here that the IPCE value of ITO/$SnO_2$/(**37**(n) + $C_{60}$)$_m$/NaI + $I_2$/Pt (n = 4, 8, 16) device decreases with increasing the dendritic generation. Such a trend may result from the steric effect of porphyrin moieties of dendrimers in the higher generation, which inhibits the π–π interaction with $C_{60}$. The ITO/$SnO_2$/(**37**(4) + $C_{60}$)$_m$/NaI + $I_2$/Pt device has fill factor (FF) of 0.31, open circuit voltage ($V_{OC}$) of 0.22 V, short circuit current density ($I_{SC}$) of 0.29 mA cm$^{-2}$, and power conversion efficiency ($\eta$) of 0.32% at input power ($W_{IN}$) of 6.2 mW cm$^{-2}$. The $\eta$ value of ITO/$SnO_2$/(**37**(4) + $C_{60}$)$_m$/NaI + $I_2$/Pt device is 10 times as large as that of ITO/$SnO_2$/(**38** + $C_{60}$)$_m$/NaI + $I_2$/Pt device ($\eta$ = 0.035%) under the same experimental conditions.[85,86]

### 9.6.4.2 Porphyrin Oligomers

Hasobe et al.[87] presented organic photovoltaic cells using supramolecular complexes of porphyrin-peptide oligomers **39** (n = 1, 2, 4, 8) with fullerene (Figure 13). The $SnO_2$ electrodes modified with the composite clusters of **39** with $C_{60}$ are prepared as in the case of the porphyrin dendrimer and fullerene composite system. Photoelectrochemical measurements were performed with a standard two electrode system consisting of the modified $SnO_2$ electrode as a working electrode and a Pt wire gauze electrode in 0.5 M NaI and 0.01 M $I_2$ in air-saturated acetonitrile (ITO/$SnO_2$/(**39** + $C_{60}$)$_m$/NaI + $I_2$/Pt device). A high power conversion efficiency ($\eta$) of 1.3% and a maximum incident photon-to-photocurrent efficiency (IPCE) of 42% at 600 nm (FF = 0.47, $V_{OC}$ = 0.30 V, $W_{IN}$ = 3.4 mW cm$^{-2}$) are attained using composite clusters of porphyrin peptide octamer and fullerene (ITO/$SnO_2$/(**39**(n = 8) + $C_{60}$)$_m$/NaI + $I_2$/Pt device). The $\eta$ value (1.3%) of ITO/$SnO_2$/(**39**(n = 8) + $C_{60}$)$_m$/NaI + $I_2$/Pt device is 30 times as large as that (0.043%) of ITO/$SnO_2$/(**39**(n = 1) + $C_{60}$)$_m$/NaI + $I_2$/Pt device. These results clearly show that the formation of a molecular assembly between fullerene and multi-porphyrin arrays with a polypeptide backbone controls the electron transfer efficiency in the supramolecular complex, which is essential for the efficient light-energy conversion.[87]

### 9.6.4.3 Multiporphyrin-Modified Gold Nanoparticles

Hasobe et al.[88,89] extend the ceoncept of dye-sensitized bulk heterojunction into more sophisticated organic solar cell prepared using quaternary self-organization of porphyrin (donor) and fullerene (acceptor) dye units by clusterization with gold nanoparticles on $SnO_2$ electrodes as shown in Figure 14. First, porphyrin-modified gold nanoparticles **40** with well-defined size (8–9 nm) and spherical shape [**40** (n = 5, 11, 15)] are prepared starting from porphyrin-alkanethiol **41**[120,121] (step 1). These nanoparticles form supramolecular complexes with fullerene molecules, which are incorporated between the porphyrin

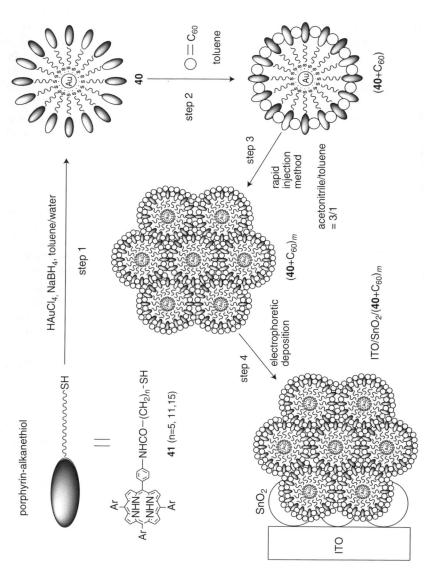

**Figure 14** *Schematic diagram for step-by-step organization of porphyrin-modified gold nanoparticle **40** and $C_{60}$ onto a nanostructured $ITO/SnO_2$ electrode*

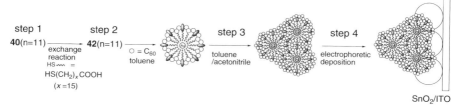

**Figure 15** *Schematic diagram for step-by-step organization of porphyrin-modified gold nanoparticle **42** with large, bucket-shaped holes on the surface and $C_{60}$ onto a nanostructured $ITO/SnO_2$ electrode*

moieties (step 2) and the supramolecular complexes are grown into larger clusters in an acetonitrile/toluene mixed solvent (step 3). Then, the large clusters are deposited electrophoretically onto $SnO_2$ electrodes (step 4). Photoelectrochemical measurements were performed with a standard two-electrode system consisting of a working electrode and a Pt wire gauze electrode in 0.5 M NaI and 0.01 M $I_2$ in acetonitrile (ITO/$SnO_2$/(**40** + $C_{60}$)$_m$/NaI + $I_2$/Pt device). The IPCE value of ITO/$SnO_2$/(**40** + $C_{60}$)$_m$/NaI + $I_2$/Pt device (up to 54% in the case of $n = 15$) increases with increasing the chain length between the porphyrin and the gold nanoparticle. The longer methylene spacer of **40** allows suitable space for $C_{60}$ molecules to insert themselves between the neighboring two porphyrin rings effectively compared to the clusters with a shorter methylene spacer, leading to more efficient photocurrent generation. Unfortunately, however, replacement of $C_{60}$ with $C_{70}$ or freebase porphyrin with zinc porphyrin results in a decrease of the photoelectrochemical response. This trend may be explained by the difference in the complexation abilities between the porphyrin and fullerene molecules as well as in the electron or hole hopping efficiency in the composite clusters. The ITO/$SnO_2$/(**40**($n = 15$) + $C_{60}$)$_m$/NaI + $I_2$/Pt device has fill factor ($FF$) of 0.43, open circuit voltage ($V_{OC}$) of 0.38 V, short circuit current density ($I_{SC}$) of 1.0 mA cm$^{-2}$, and overall power conversion efficiency ($\eta$) of 1.5% at input power ($W_{IN}$) of 11.2 mW cm$^{-2}$. The $I$–$V$ characteristics of ITO/$SnO_2$/(**40**($n = 15$) + $C_{60}$)$_m$/NaI + $I_2$/Pt device is also remarkably enhanced by a factor of 45 in comparison with ITO/$SnO_2$/(**38** + $C_{60}$)$_m$/NaI + $I_2$/Pt device under the same experimental conditions. These results clearly show that the large improvement of photoelectrochemical properties results from three-dimensional structure of interpenetrating network of the porphyrin and $C_{60}$ molecules on the nanostructured $SnO_2$ electrode which facilitates the injection of the separated electron into the conduction band.[88,89]

Imahori et al.[91,92] further extend this novel approach to construct light energy conversion system using supramolecular incorporation of $C_{60}$ molecules into tailored, large and bucket-shaped surface holes on porphyrin-modified gold nanoparticles **42** in mixed solvent, followed by electrophoretic deposition on nanostructured $SnO_2$ electrodes (Figure 15). Specifically, gold nanoparticles **42** modified with a mixed self-assembled monolayer of porphyrin alkanethiol **41** ($n = 11$) and short alkanethiol are prepared to examine the size and shape effects of surface holes (host) on the surface of the porphyrin-modified gold nanoparticles. The difference in the porphyrin–$C_{60}$ ratio is found to affect the structures and photoelectrochemical properties of the composite clusters in the mixed solvents as well as on the $SnO_2$ electrodes. Photoelectrochemical measurements were performed with a standard three electrodes system consisting of the working electrode, a Pt wire electrode, and an $Ag/AgNO_3$ reference electrode in 0.5 M LiI and 0.01 M $I_2$ in air-saturated acetonitrile (ITO/$SnO_2$/ (**42**+$C_{60}$)$_m$/LiI+$I_2$/Pt device). The authors compared the photoelectrochemical properties of the ITO/$SnO_2$/(**42** + $C_{60}$)$_m$/LiI + $I_2$/Pt and ITO/$SnO_2$/(**40**($n$ = 11) + $C_{60}$)$_m$/LiI + $I_2$/Pt devices at an applied potential of +0.06 V vs. SCE. The maximum IPCE value (42%) of ITO/$SnO_2$/(**42** + $C_{60}$)$_m$/LiI + $I_2$/Pt device with large, bucket-shaped holes is larger by a factor of 3 than that (16%) of ITO/ $SnO_2$/(**40**($n$ = 11) + $C_{60}$)$_m$/LiI + $I_2$/Pt device with small, wedge-shaped surface holes on **40** ($n$ = 11). This clearly demonstrates that the shape and size of the host holes on the three-dimensional porphyrin-modified gold nanoparticles have large impact on the photoelectrochemical properties. Remarkable enhancement in the photoelectrochemical performance as well as broader photoresponse in the visible and infrared relative to the reference systems demonstrates that the bottom-up, step-by-step organization approach provides novel perspective for the development of efficient organic solar cells.[91,92]

## 9.7 Chemical Adsorption and Spin Coating Deposition

For typical bulk heterojunction solar cells involving blend films of conjugated polymers and fullerenes, spin-coating method has been used for the fabrication on electrode surfaces.[99–112] Such examples will be described in other chapters and we will focus on other examples involving non-covalently bonded fullerenes.

Imahori et al.[95,96] presented the first mixed films of porphyrin and fullerene with hydrogen bonding on ITO electrode to reveal efficient photocurrent generation. To evaluate the hydrogen bonding effect on the photoelectrochemical properties of the donor–acceptor systems, carboxylic acid and ethoxycarbonyl groups are introduced to $C_{60}$ to yield **43** (R=$CO_2H$) and **43** (R=$CO_2Et$) as an acceptor, respectively (Figure 16).[95] Porphyrin **44** without hydrogen bonding unit is employed as a donor. An equimolar THF or benzonitrile solution of porphyrin **44** and/or fullerene **43** is cast or spin-coated on ITO electrode. Ultraviolet-visible absorption and infrared spectra for the modified ITO electrodes reveal the direct evidence for the formation of hydrogen bonding in the mixed films where the $C_{60}$ moiety is located in close proximity to the porphyrin. Photoelectrochemical measurements were performed in an air-saturated 0.1 M $Na_2SO_4$ aqueous solution containing 5 mM methylviologen ($MV^{2+}$) as an electron acceptor using ITO/**44** + **43**(R=$CO_2H$) as the working electrode, a platinum counter electrode, and an Ag/AgCl (sat. KCl) reference electrode (ITO/ **44** + **43**(R=$CO_2H$)/$MV^{2+}$/Pt device). The internal quantum yields of cathodic photocurrent generation are compared for ITO/**44** + **43** (R=$CO_2H$)/$MV^{2+}$/Pt, ITO/**44** + **43**(R=$CO_2Et$)/$MV^{2+}$/Pt, and ITO/**44**/$MV^{2+}$/Pt devices under the

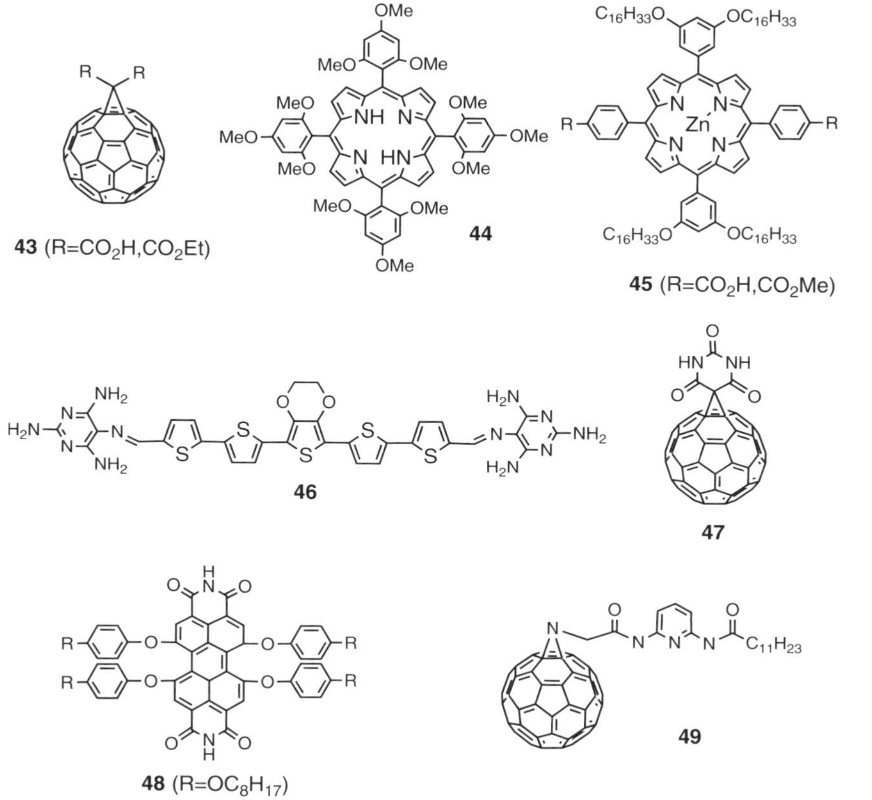

**Figure 16** *Donor and acceptor molecules **43–49** for hydrogen-bonded photoelectrochemical devices*

same conditions (applied potential: $-0.10$ V vs. Ag/AgCl (sat. KCl)). The quantum yield ($4.0 \pm 0.6\%$) of ITO/**44** + **43**(R=CO$_2$H)/MV$^{2+}$/Pt device is larger by a factor of 5–7 than those of ITO/**44** + **43**(R=CO$_2$Et)/MV$^{2+}$/Pt ($0.60 \pm 0.10\%$) and ITO/**44**/MV$^{2+}$/Pt devices ($0.76 \pm 0.03\%$). These results reveal that photocurrent generation efficiency is much improved in the mixed system with hydrogen bonding compared with the reference systems.[95]

Imahori et al.[96] further introduced carboxylic acid and alkoxycarbonyl groups into both the porphyrin and C$_{60}$ to yield zincporphyrin (ZnP) acid **45** (R=CO$_2$H) and ester **45** (R=CO$_2$Me) as a donor and C$_{60}$ acid **43** (R=CO$_2$H) and C$_{60}$ ester **43** (R=CO$_2$Et) as an acceptor, respectively. Modification of electrodes is carried out according to two different methods. A known amount of ZnP, C$_{60}$ or the mixed solution in THF is spin-coated onto ITO or ITO/SnO$_2$ or gold electrodes (spin-coating method), while ITO or ITO/SnO$_2$ electrode is immersed into the THF solution containing the compounds, followed by washing the substrate with THF repeatedly (dipping method). Infrared reflection absorption (IRRA) spectra and AFM measurements support the formation of the favorable hydrogen bonding in the mixed films (see below). Photoelectrochemical measurements were performed in a nitrogen-saturated acetonitrile solution containing 0.5 M LiI and 0.01 M $I_2$ using the modified $ITO/SnO_2$ as the working electrode, a platinum counter electrode, and a $I^-/I_3^-$ reference electrode under the same conditions (applied potential: 0.15 V vs. SCE, $\lambda_{ex}$ = 435 nm (absorbance = 1.00), $ITO/SnO_2$/ **45**($R=CO_2H$) + **43**($R=CO_2H$)/LiI + $I_2$/Pt device). The IPCE values obtained by the dipping method are in the order of

$ITO/SnO_2$/**45**($R=CO_2H$) + **43**($R=CO_2H$)/LiI + $I_2$/Pt (36%)
> $ITO/SnO_2$/**45**($R=CO_2H$) + **43**($R=CO_2Et$)/LiI + $I_2$/Pt (28%)
> $ITO/SnO_2$/**45**($R=CO_2H$)/LiI + $I_2$/Pt (26%)
> $ITO/SnO_2$/**45**($R=CO_2Me$) + **43**($R=CO_2H$)/LiI + $I_2$/Pt (15%)
> $ITO/SnO_2$/**45**($R=CO_2Me$) + **43**($R=CO_2Et$)/LiI + $I_2$/Pt (7%)
> $ITO/SnO_2$/**45**($R=CO_2Me$)/LiI + $I_2$/Pt (6%) devices.

The IPCE values are larger by 1–2 orders of magnitudes than those of similar porphyrin–hydrogen-bonded $C_{60}$ system on ITO electrode.[95] The IPCE values from the ZnP-acid systems are much larger than those from the ZnP-ester systems. This difference may result from direct adsorption of the ZnP-acid onto the $SnO_2$ surface where an electron is directly injected into the conduction band of the $SnO_2$ surface from the excited singlet state of the porphyrin to generate photocurrent. In both the porphyrin carboxylic acid system and the porphyrin ester system, IPCE values increase significantly when $C_{60}$ acid rather than $C_{60}$ ester is employed. This suggests that not only the direct electron injection takes place, but also a competitive electron transfer occurs from the porphyrin excited singlet state to $C_{60}$ acid, followed by the electron injection from the reduced $C_{60}$ to the conduction band of the $SnO_2$ surface (vide infra). Besides, electron and hole relay may occur in the arrays of hydrogen-bonded $C_{60}$-acid and ZnP-acid, respectively, as indicated from the dominant hydrogen bonding interaction between the identical ZnP-acid molecules or $C_{60}$-acid molecules rather than between different types of molecules (vide supra). It is noteworthy that the IPCE value is improved in the mixed system with hydrogen bonding compared to the reference systems. These results demonstrate that photocurrent generation is much enhanced in hydrogen-bonded porphyrin–fullerene system compared with the reference system without hydrogen bonding.[96]

Huang et al.[97] reported supramolecular photovoltaic devices comprising of oligothiophene **46** and fullerene derivative **47** (Figure 16). Codeposition from a solution of symmetric melamine-terminated electron-donor oligomers with a complementary barbiturate-labeled electron-acceptor fullerene resulted in homogeneous films. Incorporation into photovoltaic devices gave a 2.5-fold enhancement in photocurrent generation efficiency compared to the reference device with the non-hydrogen-bonding parent $C_{60}$, although the maximum internal and external quantum yields of the photocurrent generation are moderate (25, 9.3%).[97]

Liu et al.[98] developed a novel hydrogen-bonded supramolecular device of perylene bisimide **48** and fullerene derivative **49** (Figure 16). The poor photocurrent response of the mixed film of **48** and **49** (2:1) on a ITO electrode is observed,[98] although photocurrent generation mechanism is ambiguous.

## 9.8 Summary

This review has focused on recent advances in fullerene-modified electrodes and solar cells. In particular, photoinduced electron transfer reactions between donors and fullerenes have been extensively examined for the last decade and they are now well understood on the basis of fundamental parameters including reorganization energies and redox properties in the light of Marcus theory of electron transfer. In the next stage, chemists should design donor-fullerene ensembles, which reveal desirable photofunctions. Recent advances in SAMs on gold and ITO electrodes have given a deep insight for developing molecular machines including photodiodes and sensors. More importantly, utilization of weak intermolecular interactions for fabricating donor–acceptor molecules is an excellent methodology for the construction of organic molecular electronics such as organic transistors, solar cells, and light-emitting devices. Preprogrammed molecules will allow us to achieve such functions without difficulty in terms of synthesis and fabrication. In each case the search for novel and better systems involving other promising components such as carbon nanotubes and nanoparticles continues to be inspired by elegantly constructed natural systems.

## References

1. N. Martín, L. Sánchez, B. Illescas and I. Pérez, *Chem. Rev.*, 1998, **98**, 2527.
2. M. Prato, *J. Mater. Chem.*, 1997, **7**, 1097.
3. F. Diederich and M. Gómez-López, *Chem. Soc. Rev.*, 1999, **28**, 263.
4. Y.-P. Sun, J.E. Riggs, Z. Guo and H.W. Rollins, in *Optical and Electronic Properties of Fullerenes and Fullerene-Based Materials*, J. Shinar, Z.V. Vardeny and Z.H. Kafafi (eds), Marcel Dekker, New York, 2000, 43.
5. D.M. Guldi and P.V. Kamat, in *Fullerenes*, K.M. Kadish and R.S. Ruoff (eds), Wiley, New York, 2000, Chapter 5, 225.
6. S. Fukuzumi and D.M. Guldi, in *Electron Transfer in Chemistry*, Vol. 2, V. Balzani (ed), Wiley-VCH, Weinheim, 2001, 270.
7. D. Gust, T.A. Moore and A.L. Moore, *Acc. Chem. Res.*, 2001, **34**, 40.
8. N.-F. Nierengarten, *New J. Chem.*, 2004, **28**, 1177.
9. N. Armaroli, *Photochem. Photobiol. Sci.*, 2003, **2**, 73.
10. H. Imahori and Y. Sakata, *Adv. Mater.*, 1997, **9**, 537.
11. H. Imahori and Y. Sakata, *Eur. J. Org. Chem.*, 1999, 2445.
12. H. Imahori and S. Fukuzumi, *Adv. Mater.*, 2001, **13**, 1197.
13. H. Imahori, Y. Mori and Y. Matano, *J. Photochem. Photobiol. C*, 2003, **4**, 51.
14. H. Imahori, *Org. Biomol. Chem.*, 2004, **2**, 1425.
15. H. Imahori, *J. Phys. Chem. B*, 2004, **108**, 6130.
16. H. Imahori and S. Fukuzumi, *Adv. Funct. Mater.*, 2004, **14**, 525.

17. H. Imahori, K. Hagiwara, T. Akiyama, M. Aoki, S. Taniguchi, T. Okada, M. Shirakawa and Y. Sakata, *Chem. Phys. Lett.*, 1996, **263**, 545.
18. N.V. Tkachenko, C. Guenther, H. Imahori, K. Tamaki, Y. Sakata, H. Lemmetyinen and S. Fukuzumi, *Chem. Phys. Lett.*, 2000, **326**, 344.
19. H. Imahori, K. Tamaki, D.M. Guldi, C. Luo, M. Fujitsuka, O. Ito, Y. Sakata and S. Fukuzumi, *J. Am. Chem. Soc.*, 2001, **123**, 2607.
20. H. Imahori, D.M. Guldi, K. Tamaki, Y. Yoshida, C. Luo, Y. Sakata and S. Fukuzumi, *J. Am. Chem. Soc.*, 2001, **123**, 6617.
21. H. Imahori, N.V. Tkachenko, V. Vehmanen, K. Tamaki, H. Lemmetyinen, Y. Sakata and S. Fukuzumi, *J. Phys. Chem. A*, 2001, **105**, 1750.
22. H. Imahori, H. Yamada, D.M. Guldi, Y. Endo, A. Shimomura, S. Kundu, K. Yamada, T. Okada, Y. Sakata and S. Fukuzumi, *Angew. Chem. Int. Ed.*, 2002, **41**, 2344.
23. S. Fukuzumi, K. Ohkubo, H. Imahori and D.M. Guldi, *Chem. Eur. J.*, 2003, **9**, 1585.
24. H. Imahori, Y. Sekiguchi, Y. Kashiwagi, T. Sato, Y. Araki, O. Ito, H. Yamada and S. Fukuzumi, *Chem. Eur. J.*, 2004, **10**, 3184.
25. J. Jin, L.S. Li, Y. Li, Y.J. Zhang, X. Chen, D. Wang, S. Jiang and T.J. Li, *Langmuir*, 1999, **15**, 4565.
26. C. Luo, C. Huang, L. Gan, D. Zhou, W. Xia, Q. Zhuang, Y. Zhao and Y. Huang, *J. Phys. Chem.*, 1996, **100**, 16685.
27. W. Zhang, Y. Shi, L. Gan, C. Huang, H. Luo, D. Wu and N. Li, *J. Phys. Chem. B*, 1999, **103**, 675.
28. D. Zhou, L. Gan, C. Luo, H. Tan, C. Huang, G. Yao, X. Zhao, Z. Liu, X. Xia and B. Zhang, *J. Phys. Chem.*, 1996, **100**, 3150.
29. W. Zhang, L. Gan and C. Huang, *J. Mater. Chem.*, 1998, **8**, 1731.
30. N.V. Tkachenko, E. Vuorimaa, T. Kesti, A.S. Alekseev, A.Y. Tauber, P.H. Hynninen and H. Lemmetyinen, *J. Phys. Chem. B*, 2000, **104**, 6371.
31. N.V. Tkachenko, V. Vehmanen, J.-P. Nikkanen, H. Yamada, H. Imahori, S. Fukuzumi and H. Lemmetyinen, *Chem. Phys. Lett.*, 2002, **366**, 245.
32. T. Vuorinen, K. Kaunisto, N.V. Tkachenko, A. Efimov and H. Lemmetyinen, *Langmuir*, 2005, **21**, 5383.
33. H. Imahori, T. Azuma, S. Ozawa, H. Yamada, K. Ushida, A. Ajavakom, H. Norieda and Y. Sakata, *Chem. Commun.*, 1999, 557.
34. H. Imahori, T. Azuma, A. Ajavakom, H. Norieda, H. Yamada and Y. Sakata, *J. Phys. Chem. B*, 1999, **103**, 7233.
35. O. Enger, F. Nuesch, M. Fibbioli, L. Echegoyen, E. Pretsch and F. Diederich, *J. Mater. Chem.*, 2000, **10**, 2231.
36. T. Akiyama, H. Imahori, A. Ajavakom and Y. Sakata, *Chem. Lett.*, 1996, **25**, 907.
37. H. Imahori, S. Ozawa, K. Ushida, M. Takahashi, T. Azuma, A. Ajavakom, T. Akiyama, M. Hasegawa, S. Taniguchi, T. Okada and Y. Sakata, *Bull. Chem. Soc. Jpn.*, 1999, **72**, 485.
38. H. Yamada, H. Imahori and S. Fukuzumi, *J. Mater. Chem.*, 2002, **12**, 2034.

39. D. Hirayama, T. Yamashiro, K. Takimiya, Y. Aso, T. Otsubo, H. Norieda, H. Imahori and Y. Sakata, *Chem. Lett.*, 2000, **29**, 570.
40. D. Hirayama, K. Takimiya, Y. Aso, T. Otsubo, T. Hasobe, H. Yamada, H. Imahori, S. Fukuzumi and Y. Sakata, *J. Am. Chem. Soc.*, 2002, **124**, 532.
41. H. Imahori, H. Yamada, S. Ozawa, K. Ushida and Y. Sakata, *Chem. Commun.*, 1999, 1165.
42. H. Imahori, H. Yamada, Y. Nishimura, I. Yamazaki and Y. Sakata, *J. Phys. Chem. B*, 2000, **104**, 2099.
43. H. Imahori, H. Norieda, H. Yamada, Y. Nishimura, I. Yamazaki, Y. Sakata and S. Fukuzumi, *J. Am. Chem. Soc.*, 2001, **123**, 100.
44. S. Saha, L.E. Johansson, A.H. Flood, H.-R. Tseng, J.I. Zink and J.F. Stoddart, *Small*, 2005, **1**, 87.
45. S. Saha, E. Johansson, A.H. Flood, H.-R. Tseng, J.I. Zink and J.F. Stoddart, *Chem. Eur. J.*, 2005, **11**, 6846.
46. K.-S. Kim, M.-S. Kang, H. Ma and A.K.-Y. Jen, *Chem. Mater.*, 2004, **16**, 5058.
47. M. Morisue, S. Yamatsu, N. Haruta and Y. Kobuke, *Chem. Eur. J.*, 2005, **11**, 5563.
48. H. Yamada, H. Imahori, Y. Nishimura, I. Yamazaki and S. Fukuzumi, *Adv. Mater.*, 2002, **14**, 892.
49. H. Yamada, H. Imahori, Y. Nishimura, I. Yamazaki, T.K. Ahn, S.K. Kim, D. Kim and S. Fukuzumi, *J. Am. Chem. Soc.*, 2003, **125**, 9129.
50. H. Imahori, M. Kimura, K. Hosomizu and S. Fukuzumi, *J. Photochem. Photobiol. A*, 2004, **166**, 57.
51. H. Imahori, M. Kimura, K. Hosomizu, T. Sato, T.K. Ahn, S.K. Kim, D. Kim, Y. Nishimura, I. Yamazaki, Y. Araki, O. Ito and S. Fukuzumi, *Chem. Eur. J.*, 2004, **10**, 5111.
52. V. Chukharev, T. Vuorinen, A. Efimov, N.V. Tkachenko, M. Kimura, S. Fukuzumi, H. Imahori and H. Lemmetyinen, *Langmuir*, 2005, **21**, 6385.
53. M. Isosomppi, N.V. Tkachenko, A. Efimov, K. Kaunisto, K. Hosomizu, H. Imahori and H. Lemmetyinen, *J. Mater. Chem.*, 2005, **15**, 4546.
54. Y.-J. Cho, T.K. Ahn, H. Song, K.S. Kim, C.Y. Lee, W.S. Seo, K. Lee, S.K. Kim, D. Kim and J.T. Park, *J. Am. Chem. Soc.*, 2005, **127**, 2380.
55. A. Ikeda, T. Hatano, S. Shinkai, T. Akiyama and S. Yamada, *J. Am. Chem. Soc.*, 2001, **123**, 4855.
56. T. Konishi, A. Ikeda and S. Shinkai, *Tetrahedron*, 2005, **61**, 4881.
57. E. Bustos, J. Manríquez, L. Echegoyen and L.A. Godínez, *Chem. Commun.*, 2005, 1613.
58. C. Luo, D.M. Guldi, M. Maggini, E. Menna, S. Mondini, N.A. Kotov and M. Prato, *Angew. Chem. Int. Ed.*, 2000, **39**, 3905.
59. D.M. Guldi, F. Pellarini, M. Prato, C. Granito and L. Troisi, *Nano Lett.*, 2002, **2**, 965.
60. D.M. Guldi, I. Zilbermann, G.A. Anderson, K. Kordatos, M. Prato, R. Tafuro and L. Valli, *J. Mater. Chem.*, 2004, **14**, 303.

61. I. Zilbermann, G.A. Anderson, D.M. Guldi, H. Yamada, H. Imahori and S. Fukuzumi, *J. Porphyrins Phthalocyanines*, 2003, **7**, 357.
62. D.M. Guldi, I. Zilbermann, G.A. Anderson, A. Li, D. Balbinot, N. Jux, M. Hatzimarinaki, A. Hirsch and M. Prato, *Chem. Commun.*, 2004, 726.
63. D.M. Guldi, I. Zilbermann, A. Lin, M. Braun and A. Hirsch, *Chem. Commun.*, 2004, 96.
64. S. Conoci, D.M. Guldi, S. Nardis, R. Paolesse, K. Kordatos, M. Prato, G. Ricciardi, M.G.H. Vicente, I. Zilbermann and L. Valli, *Chem. Eur. J.*, 2004, **10**, 6523.
65. D.M. Guldi and M. Prato, *Chem. Commun.*, 2004, 2517.
66. D.M. Guldi, *J. Phys. Chem. B*, 2005, **109**, 11432.
67. M. Hiramoto, H. Fujiwara and M. Yokoyama, *Appl. Phys. Lett.*, 1991, **58**, 1062.
68. M. Hiramoto, H. Fujiwara and M. Yokoyama, *J. Appl. Phys.*, 1992, **72**, 3781.
69. T. Tsuzuki, Y. Shirota, J. Rostalski and D. Meissner, *Solar Energy Mater. Solar Cells*, 2000, **61**, 1.
70. J. Xue, S. Uchida, B.P. Rand and S.R. Forrest, *Appl. Phys. Lett.*, 2004, **84**, 3013.
71. J. Xue, B.P. Rand, S. Uchida and S.R. Forrest, *Adv. Mater.*, 2005 **17**, 66.
72. P.V. Kamat, S. Barazzouk, K.G. Thomas and S. Hotchandani, *J. Phys. Chem. B*, 2000, **104**, 4014.
73. H. Hotta, S. Kang, T. Umeyama, Y. Matano, K. Yoshida, S. Isoda and H. Imahori, *J. Phys. Chem. B*, 2005, **109**, 5700.
74. P.K. Sudeep, B.I. Ipe, K.G. Thomas, M.V. George, S. Barazzouk, S. Hotchandani and P.V. Kamat, *Nano Lett.*, 2002, **2**, 29.
75. P.V. Kamat, *J. Phys. Chem. B*, 2002, **106**, 7729.
76. K.G. Thomas and P.V. Kamat, *Acc. Chem. Res.*, 2003, **36**, 888.
77. P.V. Kamat, M. Haria and S. Hotchandani, *J. Phys. Chem.*, 2004, **108**, 5166.
78. P.V. Kamat, S. Barazzouk, S. Hotchandani and K.G. Thomas, *Chem. Eur. J.*, 2000, **6**, 3914.
79. H. Imahori, T. Hasobe, H. Yamada, P.V. Kamat, S. Barazzouk, M. Fujitsuka, O. Ito and S. Fukuzumi, *Chem. Lett.*, 2001, **30**, 784.
80. K. Okamoto, T. Hasobe, N.V. Tkachenko, H. Lemmetyinen, P.V. Kamat and S. Fukuzumi, *J. Phys. Chem. A*, 2005, **109**, 4662.
81. T. Hasobe, S. Hattori, H. Kotani, K. Ohkubo, K. Hosomizu, H. Imahori, P.V. Kamat and S. Fukuzumi, *Org. Lett.*, 2004, **6**, 3103.
82. T. Hasobe, S. Hattori, P.V. Kamat, Y. Wada and S. Fukuzumi, *J. Mater. Chem.*, 2005, **15**, 372.
83. V. Biju, S. Barazzouk, K.G. Thomas, M.V. George and P.V. Kamat, *Langmuir*, 2001, **17**, 2930.
84. T. Hasobe, H. Imahori, S. Fukuzumi and P.V. Kamat, *J. Phys. Chem. B*, 2003, **107**, 12105.

85. T. Hasobe, Y. Kashiwagi, M. Absalom, K. Hosomizu, M.J. Crossley, H. Imahori, P.V. Kamat and S. Fukuzumi, *Adv. Mater.*, 2004, **16**, 975.
86. T. Hasobe, P.V. Kamat, M.A. Absalom, Y. Kashiwagi, J. Sly, M.J. Crossley, K. Hosomizu, H. Imahori and S. Fukuzumi, *J. Phys. Chem. B*, 2004, **108**, 12865.
87. T. Hasobe, P.V. Kamat, V. Troiani, N. Solladié, T.K. Ahn, S.K. Kim, D. Kim, A. Kongkanand, S. Kuwabata and S. Fukuzumi, *J. Phys. Chem. B*, 2005, **109**, 19.
88. T. Hasobe, H. Imahori, S. Fukuzumi and P.V. Kamat, *J. Am. Chem. Soc.*, 2003, **125**, 14962.
89. T. Hasobe, H. Imahori, P.V. Kamat, T.K. Ahn, S.K. Kim, D. Kim, A. Fujimoto, T. Hirakawa and S. Fukuzumi, *J. Am. Chem. Soc.*, 2005, **127**, 1216.
90. H. Imahori, A. Fujimoto, S. Kang, H. Hotta, K. Yoshida, T. Umeyama, Y. Matano and S. Isoda, *Tetrahedron*, 2006, **62**, 1955.
91. H. Imahori, A. Fujimoto, S. Kang, H. Hotta, K. Yoshida, T. Umeyama, Y. Matano and S. Isoda, *Adv. Mater.*, 2005, **17**, 1727.
92. H. Imahori, A. Fujimoto, S. Kang, H. Hotta, K. Yoshida, T. Umeyama, Y. Matano, S. Isoda, M. Isosomppi, N.V. Tkachenko and H. Lemmetyinen, *Chem. Eur. J.*, 2005, **11**, 7265.
93. H. Imahori, K. Mitamura, T. Umeyama, K. Hosomizu, Y. Matano, K. Yoshida and S. Isoda, *Chem. Commun.*, 2006, 406.
94. F. Fungo, L. Otero, C.D. Borsarelli, E.N. Durantini, J.J. Silber and L. Sereno, *J. Phys. Chem. B*, 2002, **106**, 4070.
95. H. Imahori, J.-C. Liu, K. Hosomizu, T. Sato, Y. Mori, H. Hotta, Y. Matano, Y. Araki, O. Ito, N. Maruyama and S. Fujita, *Chem. Commun.*, 2004, 2066.
96. H. Imahori, J.-C. Liu, H. Hotta, A. Kira, T. Umeyama, Y. Matano, G. Li, S. Ye, M. Isosomppi, N.V. Tkachenko and H. Lemmetyinen, *J. Phys. Chem. B*, 2005, **109**, 18465.
97. C.-H. Huang, N.D. McCenaghan, A. Kuhn, J.W. Hofstraat and D.M. Bassani, *Org. Lett.*, 2005, **7**, 3409.
98. Y. Liu, S. Xiao, H. Li, Y. Li, H. Liu, F. Lu, J. Zhuang and D. Zhu, *J. Phys. Chem. B*, 2004, **108**, 6256.
99. G. Yu, J. Gao, J.C. Hummelen, F. Wudl and A.J. Heeger, *Science*, 1995, **270**, 1789.
100. E. Peeters, P.A. van Hal, J. Knol, C.J. Brabec, N.S. Sariciftci, J.C. Hummelen and R.A.J. Janssen, *J. Phys. Chem. B*, 2000, **104**, 10174.
101. J.-F. Eckert, J.-F. Nicoud, J.-F. Nierengarten, S.-G. Liu, L. Echegoyen, F. Barigelletti, N. Armaroli, L. Ouali, V. Krasnikov and G. Hadziioannou, *J. Am. Chem. Soc.*, 2000, **122**, 7467.
102. S.E. Shaheen, C.J. Brabec, N.S. Sariciftci, F. Padinger, T. Fromherz and J.C. Hummelen, *Appl. Phys. Lett.*, 2001, **78**, 841.
103. A.M. Ramos, M.T. Rispens, J.K.J. van Duren, J.C. Hummelen and R.A.J. Janssen, *J. Am. Chem. Soc.*, 2001, **123**, 6714.

104. C.J. Brabec, A. Cravino, G. Zerza, N.S. Sariciftci, R. Kiebooms, D. Vanderzande and J.C. Hummelen, *J. Phys. Chem. B*, 2001, **105**, 1528.
105. D.L. Vangeneugden, D.J.M. Vanderzande, J. Salbeck, P.A. van Hal, R.A.J. Janssen, J.C. Hummelen, C.J. Brabec, S.E. Shaheen and N.S. Sariciftci, *J. Phys. Chem. B*, 2001, **105**, 11106.
106. F. Zhang, M. Svensson, M.R. Andersson, M. Maggini, S. Bucella, E. Menna and O. Inganäs, *Adv. Mater.*, 2001, **13**, 1871.
107. W.U. Huynh, J.J. Dittmer and A.P. Alivisatos, *Science*, 2002, **295**, 2425.
108. D.M. Guldi, C. Luo, A. Swartz, R. Gómez, J.L. Segura, N. Martín, C. Brabec and N.S. Sariciftci, *J. Org. Chem.*, 2002, **67**, 1141.
109. A. Cravino and N.S. Sariciftci, *J. Mater. Chem.*, 2002, **12**, 1931.
110. F. Padinger, R.S. Rittberger and N.S. Sariciftci, *Adv. Funct. Mater.*, 2003, **13**, 85.
111. M.M. Wienk, J.M. Kroon, W.J.H. Verhees, J. Knol, J.C. Hummelen, P.A. Van Hal and R.A.J. Janssen, *Angew. Chem. Ind. Ed.*, 2003, **42**, 3371.
112. J.-F. Nierengarten, *New J. Chem.*, 2004, **28**, 1177.
113. G. Robert, *Langmuir–Blodgett Films*, Plenum Press, New York, 1990.
114. A. Ullman, *An Introduction to Ultrathin Organic Films*, Academic Press, San Diego, 1991.
115. C.A. Mirkin and W.B. Caldwell, *Tetrahedron*, 1996, **52**, 5113.
116. L. Echegoyen and L.E. Echegoyen, *Acc. Chem. Res.*, 1998, **31**, 593.
117. S. Fukuzumi, H. Kotani, K. Ohkubo, S. Ogo, N.V. Tkachenko and H. Lemmetyinen, *J. Am. Chem. Soc.*, 2004, **126**, 1600.
118. A.C. Benniston, A. Harriman, P.Y. Li, J.P. Rostron, H.J. van Ramesdonk, M.M. Groeneveld, H. Zhang and J.W. Verhoeven, *J. Am. Chem. Soc.*, 2005, **127**, 16054.
119. K. Ohkubo, H. Kotani and S. Fukuzumi, *Chem. Commun.*, 2005, 4520.
120. H. Imahori, M. Arimura, T. Hanada, Y. Nishimura, I. Yamazaki, Y. Sakata and S. Fukuzumi, *J. Am. Chem. Soc.*, 2001, **123**, 335.
121. H. Imahori, Y. Kashiwagi, Y. Endo, T. Hanada, Y. Nishimura, I. Yamazaki, Y. Araki, O. Ito and S. Fukuzumi, *Langmuir*, 2004, **20**, 73.

CHAPTER 10
# Biological Applications of Fullerenes

ALBERTO BIANCO[1] AND TATIANA DA ROS[2]

[1] Institut de Biologie Moléculaire et Cellulaire – UPR 9021, CNRS, Immunologie et Chimie Thérapeutiques – 15, Rue René Descartes – 67084 Strasbourg, France
[2] Dipartimento di Science Farmaceutiche – Piazzale Europa 1 – 34127 Trieste, Italy

In the present chapter we will report the most recent progress on applications of fullerenes in biology and medicine. Unless necessary, we will review only papers published in the last 3 years considering that for the previous period many reviews are available.[1–7]

## 10.1 Methodologies for Fullerene Solubilisation

Many different strategies have been explored to render the fullerene biocompatible. This was considered the starting point to integrate this material into living systems. In addition, functionalization of fullerenes seems of fundamental importance to reduce their toxic effects (see below).

The design of fullerene derivatives containing amino acids is particularly interesting because the resulting building blocks can be further derivatised and eventually inserted, for example, into peptide sequences. Different addition reactions to $C_{60}$ have been already proposed,[8] but the most recent are based on natural amino acids directly linked to $C_{60}$ through the amino group (Figure 1). In this case, the fullerene derivative retains the stereochemistry of the amino acid residue.[9]

Both enantiomers of different $C_{60}$-amino acids have been prepared to measure their selectivity in transmembrane diffusion, their lipid peroxidation rate and their activity *vs.* monoamine oxidase A in mitochondria of rat brain cells. Stereospecificity was observed in all cases.[9] These findings are of special interest in the evaluation of the membranotropic properties of fullerene derivatives and

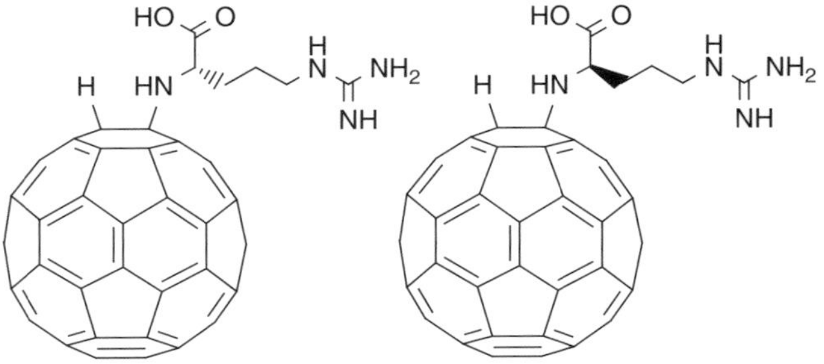

**Figure 1** *Molecular structures of $C_{60}$-L-Arg (1) and $C_{60}$-D-Arg (2)*

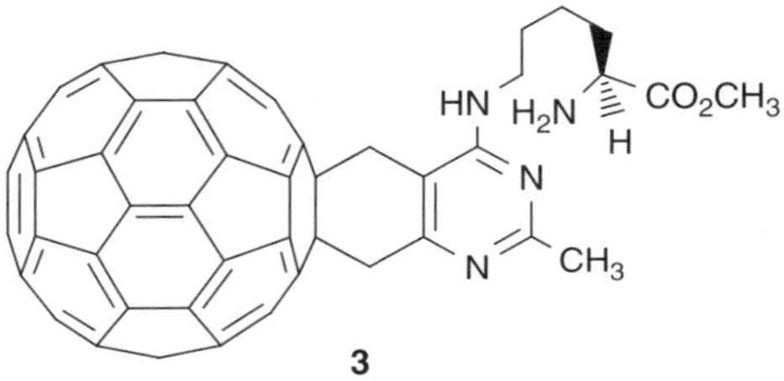

**Figure 2** *Molecular structure of a N-(2-methyl-5,6,7,8-tetrahydro[60]fullero[1,2-g]quinazolin-4-yl) α-amino acid*

will have important implications in the development of fullerenes as drug delivery systems.

Using a different approach, $C_{60}$-amino acid derivatives have been obtained by Diels–Alder addition of substituted pyrimidine ortho-quinodimethanes to fullerene. For this purpose, pyrimidine 3-sulfolenes functionalised with α-amino acids were prepared.[10] These precursors were subsequently added to $C_{60}$ and the obtained fullero-amino acids (Figure 2) displayed a reduced aggregation in contact with phosphatidyl choline liposomes and can be potentially incorporated into peptides.

Alternatively, fullerene substituted phenylalanine derivatives have been synthesised by condensing a N- and C-terminal protected (4-amino)phenylalanine with 1,2-(4′-oxocyclohexano) fullerene (Figure 3). The imine was converted to amine using acid catalysed hydroboration, which did not affect the fullerene cage.[11]

The fullero-amino acid **4** was used to form a dipeptide by coupling the carboxylic function to a glycine residue. Multifullerene peptides were prepared using the protected fulleropyrrolidino-L-glutamic acid **5** (Figure 4).[12,13] This fullero-amino acid was easily protected at the nitrogen with Boc (*tert*-butyloxycarbonyl) or Fmoc (9-fluorenylmethyloxycarbonyl), which render it particularly useful for the solid-phase peptide synthesis using Boc/Bzl and Fmoc/*t*Bu strategies, respectively.[12,14]

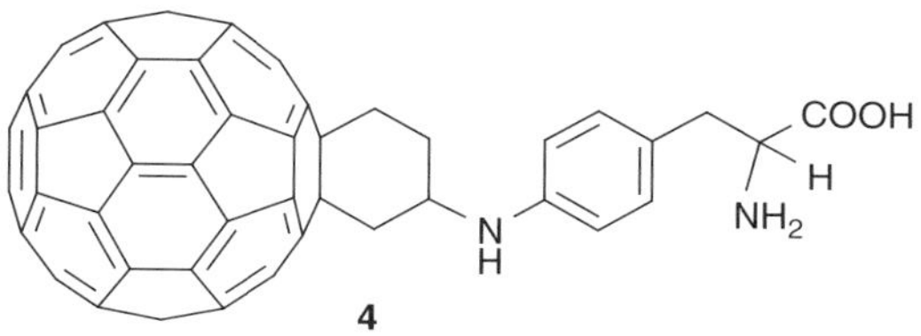

*Biological Applications of Fullerenes* 303

**Figure 3** *Molecular structure of a phenyalanine-$C_{60}$-amino acid derivative*

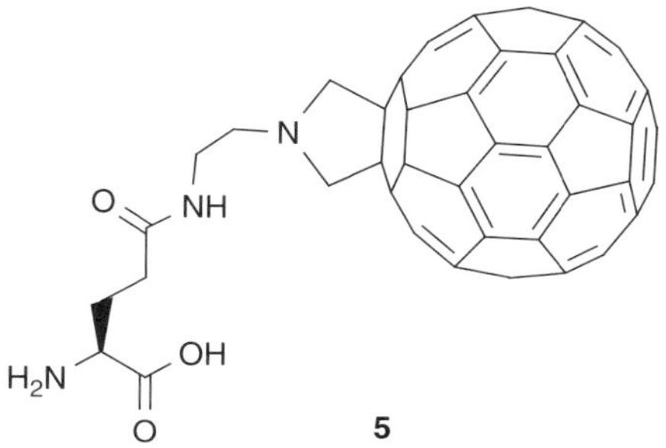

**Figure 4** *Molecular structure of the fulleropyrrolidino-L-glutamic acid*

Another possibility to make $C_{60}$ biocompatible is to modify its surface by conjugation with sugar moieties. It has been shown that fullerene glyco derivatives had an activity similar to lectins and participate in molecular recognition between cells.[15] Coupling of 2-azidoethyl tetra-*O*-acetyl-α-D-mannopyranoside to $C_{60}$ *via* thermal addition afforded two isomers which were separated and fully characterised.[16] Using the same cycloaddition reaction a bis derivative has been also prepared (Figure 5).

The amphiphilic compound was able to form supramolecular structure of nanometric size.[17] The self-assembly process in water-generated bilayer vesicles or liposomes with an internal cavity, which could encapsulate ions or fluorescent probes. This property might be useful for the development of a new slow-release drug delivery system. In addition, the carbohydrate moiety strongly interacted with lectin Con A (concanavalin A) by fitting into the sugar-binding domain of the protein. Oligosaccharides can be also covalently linked to fullerene by selective chemical ligation.[18] The strategy was based on the preparation of functionalised $C_{60}$ with an oxylamine, which subsequently underwent the reaction with the hemiacetal terminal part of a carbohydrate. The selective reaction formed a stable oxime linkage (Figure 6). The method generated *cis* and *trans* isomers but has the advantage of using deprotected sugars.

Cyclodextrins (CDs) are composed of hydrophilic sugar-building blocks and they are able to highly increase the water solubility of $C_{60}$.[19] $C_{60}$-cyclodextrin conjugates have been prepared by linking a methanofullerene to the secondary rim of β-CD.[20] This 2:1 cyclodextrin–fullerene derivative displayed remarkable

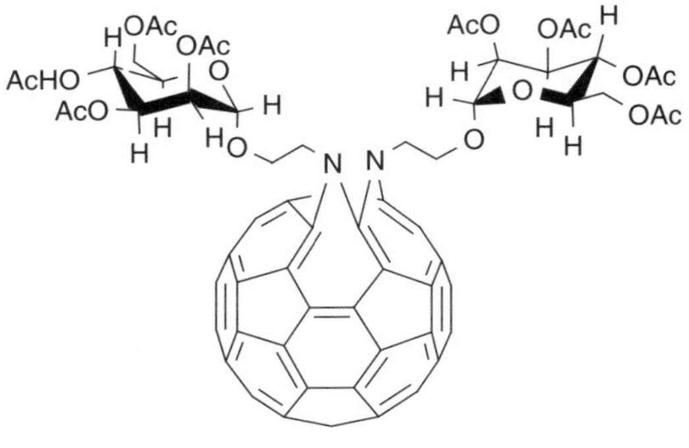

**Figure 5** *Molecular structure of a bis-(α-D-mannopyranosyl)-$C_{60}$ conjugate*

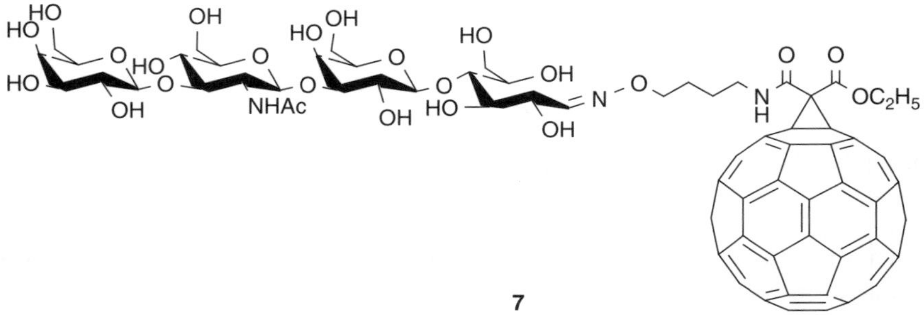

**Figure 6** *Molecular structure of a lacto-N-tetraose-$C_{60}$ conjugate*

solubility in water. However, this compound formed aggregates in aqueous solutions because the affinity of β-CD for fullerene is probably not sufficient to induce an efficient complexation.

All these fullerene derivatives can be relatively easily prepared and certainly improve the solubility of pristine $C_{60}$. Their use for biological scopes will need a careful evaluation of the biocompatibility profile and the impact in living systems.

## 10.2 Health and Environment Impact of Fullerenes

The applications of carbon nanomaterials are expanding very fast and the word nanotechnology is becoming commonly used in the daily life. Together with the popularity, the nano-objects are also growing some concerns about their environmental and health effects.[21–23] Potential toxicity of nanoparticles based on carbon materials, particularly fullerenes, is under investigation, but the studies are still scarce. A general feature is emerging from different studies: chemically modified fullerenes are less harmful for humans, animals and environment. This behaviour is strictly related to the solubility of this type of

# Biological Applications of Fullerenes

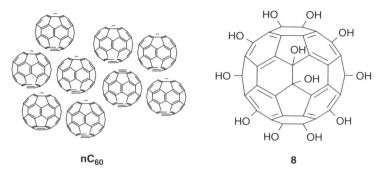

**Figure 7** *Molecular structures of insoluble fullerene cluster and soluble fullerenol*

material. Fullerenes are practically insoluble in aqueous solutions and once in contact with water they form toxic colloidal aggregates ($nC_{60}$) (Figure 7). It has been found that clusters of $C_{60}$ can cause oxidative damage to lipids in the brains of fish.[21] Indeed, colloidal $nC_{60}$ can translocate into the brain *via* the olfactory bulb in mammals and fishes similarly to other nanosized particles. $C_{60}$ is highly hydrophobic and redox active, therefore, it can potentially cause oxidative damage in aquatic species if dispersed into water sources. In fact, it has been found that water dispersed $C_{60}$ provokes oxidative stress in the brain of largemouth bass and depletion of glutathione in the gills of fish. In addition, the fullerene clusters increased water clarity, probably due to antibacterial activity. This *in vivo* study, showing the adverse effects of fullerene in aquatic species, may predict potential effect in humans. Manufactured nanomaterials must be carefully tested before they are used for human and industrial applications in order to define their benefits and risks.

$C_{60}$ nanocrystals generated in water at low concentration showed bacterial inhibition as evidenced by lack of growth and decreased aerobic respiration rates.[24] In the case of hydrated fullerenes no effect was instead observed, confirming that the surface modification of $C_{60}$ influences its biological activity. In addition, $nC_{60}$ was found cytotoxic to human dermal fibroblasts, human liver cancer cells and neuronal human astrocytes.[25] The damage of cell membrane was induced by reactive oxygen radicals produced by $nC_{60}$ in water, and it was completely prevented by addition of the antioxidant ascorbic acid. In a different approach, it has been shown that hydroxy fullerenes reacted rapidly and irreversibly with metal salts containing $Fe^{3+}$, $Al^{3+}$, $Ca^{2+}$, $Co^{2+}$, $Cu^{2+}$, $K^+$, $Ag^+$ and $Zn^{2+}$, and produced insoluble cross-linked polymers, identified as metal-hydroxy-fullerene.[26] In aqueous solutions this behaviour limited the mobility of soluble fullerenols, providing a possible solution for their accumulation in the environment by reducing their ease of transport. However, immobilisation might be not very convenient due to the high toxicity of the fullerene derivatives less mobile in water. Cytoxicity of fullerenes on different types of human cell lines was attributed to the degree of their surface functionalisation.[27] In particular, the toxicity of $C_{60}$ forming colloidal aggregates was several orders of magnitude higher than modified and soluble $C_{60}$

derivatives. Toxic effects were observed at the level of cellular membrane as oxidative damages. Indeed, fullerenes generate superoxide anions and therefore these oxygen radicals are responsible for membrane disruption and consequent cell death. Hydroxylation of $C_{60}$ cage can be considered a possible remediation for undesired biological effect of pristine fullerene (Figure 7). Similarly, the toxic effect of water-soluble malonic $C_{60}$·tris-adduct **9a** was compared to dendritic fullerene mono-adduct **10** (Figure 8) on Jurkat cells.[28] Dendrofullerene inhibited cell proliferation while the carboxyfullerene was ineffective. The growth was reversible since replacement of the fullerene solution with the cell culture medium restimulated again the cells. In addition, the fullerene derivatives became toxic when irradiated with UV light. Again the cell death was mainly due to membrane damages. In this case, phototoxicity was higher for compound **9a** in comparison to dendrofullerene **10**. Although the singlet oxygen generation by tris-adducts was less efficient, the dendrofullerene seemed to form clusters that probably interact with the cell membrane in a way that the singlet oxygen is inactivated before targeting and damaging phospholipids and membrane proteins.

It has been also shown that, although surface modifications of fullerene remarkably reduced the toxic effect, the type of functional groups may play an

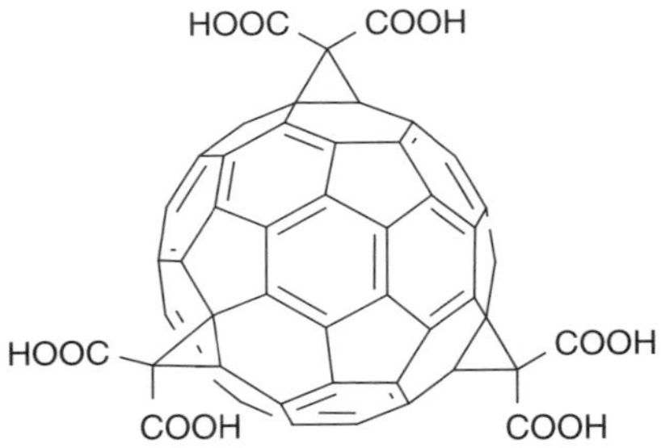

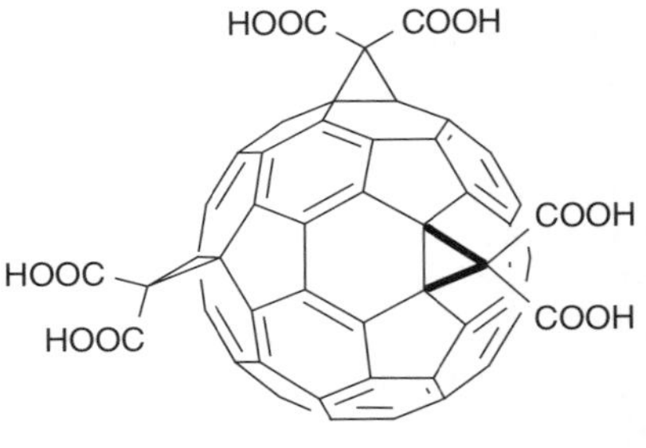

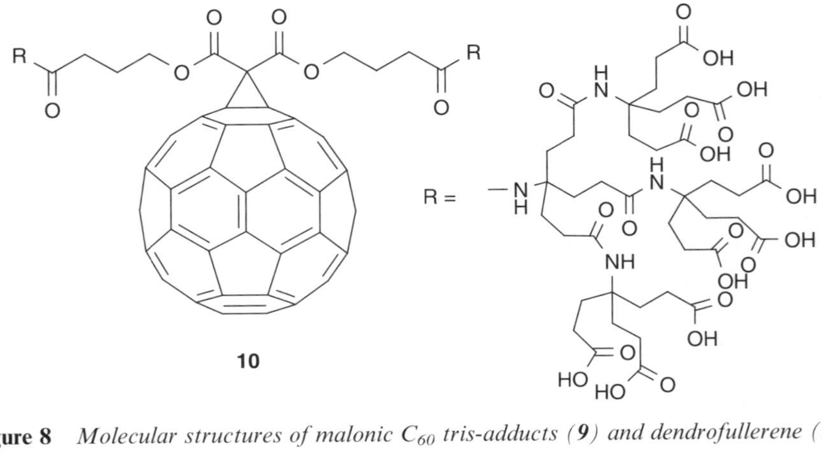

**Figure 8** *Molecular structures of malonic $C_{60}$ tris-adducts (**9**) and dendrofullerene (**10**)*

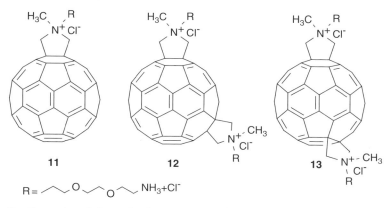

**Figure 9** *Examples of the molecular structures of polycationic fullerenes*

important role on this issue.[29] Indeed, the introduction of solubilising cationic chains-induced significant toxicity. Cytotoxic and haemolytic properties of a series of cationic fullerene derivatives were evaluated and correlated to their hydrophilic and hydrophobic surface area (Figure 9). This preliminary study highlighted that red cells tolerated better polycationic fullerenes with a relative high ratio of hydrophobic/hydrophilic surface area as established by experimental evidence and theoretical calculations.

Part of the results concerning the toxicity effects of pristine fullerene described above are in contrast with those found by the group of Andrievsky.[30] According to the data of the Russian group, it seems that nano-$C_{60}$ dispersions cannot be easily characterised in their chemical composition. The organic solvents often used to facilitate the suspension of pristine $C_{60}$ in water might influence the toxic profile of this material. Hydrated fullerenes did not display any toxic effect in various *in vitro* and *in vivo* experiments.[30] To support these observations, it has been found that $C_{60}$ did not induce observable alveolar macrophage cytotoxicity up to a dose of 226 µg cm$^{-2}$ (the dose was reported as available $C_{60}$ surface for the interaction with the cells).[31] In addition, aqueous suspensions of $C_{60}$ prepared without any organic solvent did not display acute or subacute toxic effects in rats and they protected the liver of the animals against free-radical attack in a dose-dependent manner.[32] Therefore, pristine fullerene can be considered a powerful protective agent towards intoxication of liver for example by carbon tetrachloride, a well-known *in vivo* radical initiator.

Although a certain number of studies are emerging and disclosing the toxic profile of fullerenes, the fundamental issue of health and environmental impact of pristine $C_{60}$ still need to find convergent opinions.

## 10.3  Fullerenes for Drug Delivery

Fullerenes can be considered potentially interesting systems for drug delivery because they can be multifunctionalised, they can act as drug adsorbents and they can form particles in the nanorange scale.[33] However, up to now, only very few examples on the utilisation of fullerenes for the delivery of therapeutic agents have been reported in the literature. This can be likely due to the fact that they are more difficult to integrate into the living systems in comparison to their higher family member carbon nanotubes.[34]

The demonstration that water-soluble fullerene derivatives were localised closely to mitochondria has opened new perspectives on the use of this material as a new drug delivery system.[35] A methanofullerene has been conjugated to paclitaxel and exploited for slow-released drug delivery system (Figure 10). The paclitaxel–fullerene conjugate was included into a liposome aerosol formulation and demonstrated to possess significant anticancer activity in tissue cultures.[36]

In a different application, fullerene has been used in the presence of adsorption enhancers for the development of new types of nanoparticles.[37] Liquid-filled nano- and microparticles have been prepared in the presence of solid adsorbents including $C_{60}$. These new delivery tools were engineered for the administration of erythropoietin (EPO) to the small intestine. The use of fullerene as a component of liquid-filled nanoparticles improved the bioavailability of EPO to 5.7% following intestinal administration. Although promising, better results were however obtained when fullerene was replaced by carbon nanotubes. Since EPO is highly sensitive to the strong intestinal enzymes and rapidly degraded, the new nanoparticulate system containing carbon materials might become particularly useful for the delivery of this drug provided that toxicity issues will be addressed and elucidated.

In a different approach, bis-acid methanofullerene **15** (Figure 11) has been used as counter-ions to modulate the cellular uptake and the anion carrier activity of polyarginines.[38,39] Counter-ions can act as activators of the carriers at the level of membrane interactions. Efficiency and selectivity of cellular internalisation can therefore be modulated using different types of polyaromatic hydrophobic molecules. The efficiency of carboxylic acid derivative **15** strongly depends on the composition of the cell membrane as shown using different membrane models.[39] The possibility of complex formation between

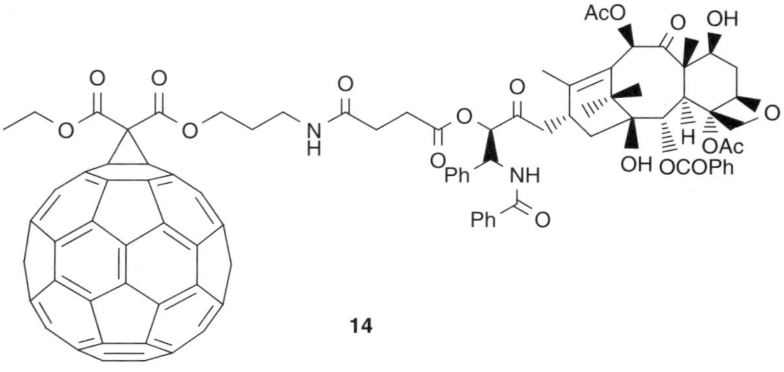

**14**

**Figure 10** *Molecular structure of the paclitaxel-fullerene conjugate*

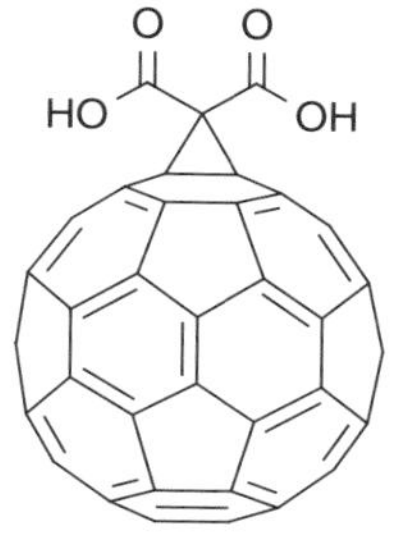

# Biological Applications of Fullerenes

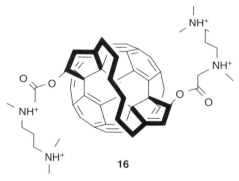

**Figure 11** *Molecular structure of the bis-acid methanofullerene or carboxyfullerene*

**Figure 12** *Molecular structure of cationic fullerene 16 for pDNA complexation and transfection*

the guanidinium groups of polyarginines and the carboxylates at the fullerene surface is at the origin of the anion activation performance in the translocation process.

In the field of nucleic acid delivery, including DNA, RNA, siRNA, LNA and plasmid DNA (pDNA), only one example has been reported until now. Nakamura and co-workers[40] designed a fullerene derivative that has been used to condense pDNA (Figure 12). This cationic fullerene represented the first $C_{60}$-based artificial vector for gene transfer. Transfection of COS-1 cells was then achieved by optimising various parameters (fullerene/base pair ratio, transfection time, amount of pDNA). However, the low solubility of the fullerene derivative in aqueous solution required an organic solvent to completely solubilise the molecule before forming the complex with pDNA. It was shown that the presence of the organic solvent increased significantly cell mortality. This somehow limits the use of this complex for gene transfer. It cannot be also excluded that the complexes between $C_{60}$ and pDNA, which appeared as big black clusters, influenced the cell viability and as a consequence the transfection efficiency. Certainly, this can be considered the first generation

## 10.4 Neuroprotection and Antioxidant Activity

Fullerene has been demonstrated to be able to uptake up to six electrons in solution.[41] This property renders $C_{60}$ very appealing as radical scavenger also for biomedical purposes.

An important class of fullerene derivatives studied into details for their biological properties, mainly as neuroprotective agents and radical scavengers, comprises carboxyfullerenes, including in particular two tris-adducts (*trans-3, trans-3, trans-3* or $D_3$, **9a**, and *equatorial, equatorial, equatorial* or $C_3$, **9b**, Figure 8).[42–44] Recently, the differential effect of carboxyfullerene **15** (Figure 11) on the neurotoxicity induced by 1-methyl-4-phenyl-pyridinium (MMP$^+$) and 1-methyl-4-phenyl-1,2,3,6-tetrahydropyridine (MPTP) has been reported.[45] The intracranial infusion of carboxyfullerene attenuates dopamine release and suppresses lipid peroxidation and cytochrome *c* depletion into cytosol, clearly indicating neuroprotection. A completely different behaviour was reported for systemic fullerene infusion, with increasing of dopamine depletion induced by MPTP and death of the animals. The results are in contrast with Dugan's findings, who described the decrease of mortality in mice affected by familial amyotrophic lateral sclerosis after systemic administration of carboxyfullerene.[43]

A very intriguing work on carboxyfullerene $C_3$ has been performed by Ali *et al.*[46] They investigated the mechanism of action of fullerene derivative in the superoxide scavenge, which resulted to be slower than superoxide dismutase (SOD), but in the range of many biologically active SOD mimetics. It seems that $C_3$ electrostatically forces $O_2^-$ to the fullerene surface, stabilising the position by hydrogen bonds with the hydrogens of the carboxylic groups. Then a second $O_2^-$ arrives and combines with the previous one and the dismutation reaction takes place. The administration of $C_3$ to $Sod2^{-/-}$ mice increased their life span by more than 300%, as typically happens using SOD mimetics.[46] The $C_3$ was used to study the inhibition of middle cerebral artery occlusion (MCAO)-induced focal cerebral ischemia. As a consequence, the MCAO presented an increasing of permeability in the blood-brain barrier (BBB). This allowed the $C_3$ to penetrate into the central nervous system following the blood stream and to deposit on neurons, directly interacting with the cells affected by MCAO. The neuroprotective effect noticed for the fullerene derivative was not only related to its ability of scavenging free radicals but also to the nitric oxide synthase inhibition, the possible suppression of toxic cytokines and interference with the increased of BBB permeability.[47]

Fullerenols $C_{60}(OH)_{2-26}$ were also mainly studied as radical scavengers. Djordjevic *et al.*[48] analysed their mechanism by ESR in presence of 2,2-diphenyl-1-picryhydrazyl (DPPH) free radical and OH radicals generated by Fenton reaction. The addition of fullerenol to the DPPH solution decreased the radical concentration as demonstrated by reduction of EPR signal of DPPH.

The same behaviour, with better results, was found in the case of OH radicals produced by Fenton reaction. Electron or hydrogen atom donation from fullerenol to free radicals was the hypothesised mechanism.

The ischemia-reperfusion (IR) oxidative stress is related to neutrophil migration, tissue inflammation, alteration of NO synthase, ROS production and lipid preoxidation. Experiments on $C_{60}(OH)_{7\pm2}$ demonstrated its ability to prevent, at low concentrations, IR damage on lungs and cell damage due to NO release or $H_2O_2$.[49] High fullerenol concentration, on the contrary, as already reported by Bogdanovic et al.,[50] had instead a negative effect. In this case, the authors reported an inhibition of mitochondrial membrane potential.

Fullerenol $C_{60}(OH)_{22}$ exhibited a dose- and time-dependent antiproliferative action vs human breast cancer cell lines T47D, MFC-7 and MDA-MB-231. This activity took place at nanomolar concentration and, above all, without photoirradiation. So far, the proposed mechanism does not involve the production of reactive oxygen species. On the contrary, it seems that the antiproliferative effect has to be attributed to the antioxidant and free radical scavenger actions of fullerenols, which can alter the redox state of the cell. The administration of $C_{60}(OH)_{22}$ together with doxorubicin, a well-known drug used in cancer treatment, displayed a cell protective action. Two mechanism of action were proposed for doxorubicin: the compound could be activated by NADPH cytochrome-P450-reductase and inactivate replication and transcription of DNA or/and could induce oxygen reactive species (ROS) production causing oxidative stress in cells and nucleic acid cleavage. Considering that mitochondria are the first target of ROS, with the uncoupling of the respiratory chain, the presence of fullerenol as radical scavenger could explain the decrease of doxorubicin toxicity.[50]

Another fullerenol, $C_{60}(OH)_{24}$, has been tested as scavenger of ROS and nitric oxide.[51] The photoinduced release of NO using sodium nitroprusside (SNP) reduces catalase, glutathione transferase and glutathione peroxidase activities. The presence of fullerenol by itself did not alter these enzymes except for glutathione transferase. The pretreatment with $C_{60}(OH)_{24}$ completely nullify the reduction of the enzymatic activity due to SNP, showing that fullerenols possess NO scavenging activity in vitro. Moreover, the same authors analysed the capability of $C_{60}(OH)_{24}$ to scavenge lipid peroxides and superoxide radical, finding a very good action comparable to that of the very well-known antioxidant butylated hydroxytoluene.

The radical scavenger and antioxidant properties of five different fullerene preparations (fullerenols, isostearic acid and PEG amine fullerene derivatives, polyvinylpyrrolidone and cyclodextrin–fullerene complexes) have been studied by ESR, obtaining similar results. The little variability of the results was attributed to the different water-soluble portion associated to fullerene.[52] The same compounds have been also used to prevent the intracellular oxidative stress due to UVB. After $C_{60}/PVP$ pretreatment, the effects of UVB irradiation or tert-butyl hydroperoxide was decreased with respect to control.

The ability of β-alanine fullerene derivatives **17–19** in scavenging OH radicals has been studied by chemiluminescence.[53] Three derivatives have been prepared

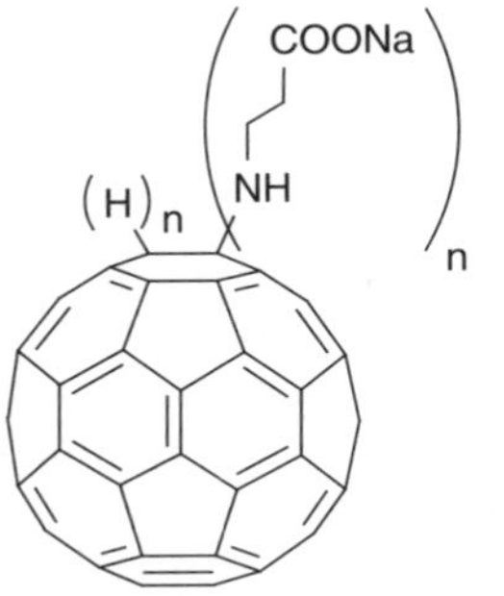

**17:** n = 1
**18:** n = 5
**19:** n = 9

**Figure 13**  *Molecular structures of compounds 17–19*

presenting one, five or nine β-alanine chains (Figure 13). The radical sponge action is dose-dependent and the three adducts presented a variable scavenger activity. Compound **18** was the best, followed by **19**, while **17** was the less effective in the series. The authors hypothesised that in the case of **18** the presence of amine groups could increase the electron cloud density of π-bonds in the fullerene cage, while the presence of only one NH was not sufficient to appreciate the difference. On the other side, the presence of nine chains could be detrimental for the radical scavenger activity because of a large number of double bond already missed and of a possible steric hindrance, which obstacles the OH radical addition.

The interesting radical scavenging and antioxidant properties of fullerene derivatives could find application in neuroprotection, but their capability to impair the mitochondrial respiratory chain has to be carefully considered.

## 10.5 DNA Photocleavage and Photodynamic Approach Using Fullerenes

A potential application of fullerene and its derivatives is related to the easy photoexcitation of $C_{60}$ from the ground state to $^1C_{60}$. This short-lived species is readily converted to the long-lived $^3C_{60}$ *via* intersystem crossing. In the presence of molecular oxygen, the fullerene can decay from its triplet to the ground state, transferring its energy to $O_2$, generating $^1O_2$, known to be a highly cytotoxic species. In addition, $^1C_{60}$ and $^3C_{60}$ are excellent acceptors and, in the presence of a donor, can undergo a different process, being easily reduced to $C_{60}^-$ by electron transfer. There are two different pathways of DNA photocleavage acting mainly at guanine sites. The generation of singlet oxygen (type II photosensitization) as well as the energy transfer from the triplet state of fullerene to bases (type I photosensitization) can be responsible of the oxidation of guanosines and these modifications increase the instability of the phosphodiesteric bond that becomes easily susceptible to alkaline hydrolysis. Molecular dynamics simulations demonstrated that fullerene could bind to the nucleotides, forming an energetically favoured structure by interactions with the free ends of the considered double strand or with the minor groove. Depending on the DNA form, the effect can be quite different. In the case of B-form double helix, the fullerene interaction left the structure unchanged, while in the case of A-form there was a disruption of hydrogen bonds between the end base pairs.[54]

Despite these docking results, the triple-helix formation between double helix and a fullerene–oligonucletide conjugate resulted disfavoured by the presence of the fullerene.[55] In fact, strong sterical and electrostatic repulsions between the negatively charged phosphate groups and the fulleropyrrolidine moiety played a crucial role in determining incorrect positioning of fullerene in the ternary oligonucleotide–fullerene-minor groove binder, with detrimental effect on selective photocleavage.

Fullerene derivative presenting a water-soluble β-cyclodextrin as appendage was able to act as DNA-cleavage agent after photoirradiation (Figure 14).[56] The presence of the cyclodextrin could partially shield the fullerene and its water solubility leads to biologically compatible fullerene products. Moreover, it has been demonstrated that the cyclodextrin cavity was able to recognise various model substrate or could bind specific guest fragments conferring selective recognition. In the present case, β-cyclodextrin was not complexing the fullerene, being too small. The photo-irradiation of this derivative transformed supercoiled DNA into nicked and linear forms. The author studied the mechanism of action and they demonstrated the necessity of oxygen in the system. Moreover, EPR studies evidenced the presence of $^1O_2$ as active species in the cleavage process.

A thymocyte suspension has been irradiated by UV light ($\lambda = 365$ nm, light power $10^{16}$ photon cm$^{-2}$ s) in the presence of the so-called fullerene aqueous solution ($C_{60}$FAS) and a strong variation of absorption density has been reported and attributed to a change of the oxidation level and structural conformation of the biological target molecules as lipids.[57] The irradiation with X-ray (absorbed dose 4.5 Gy) for 1 h at 310 K without fullerene did not influence the DNA fragmentation, while DNA damage was strongly increased

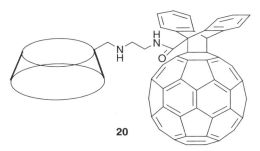

**Figure 14** CD-fullerene derivative 20

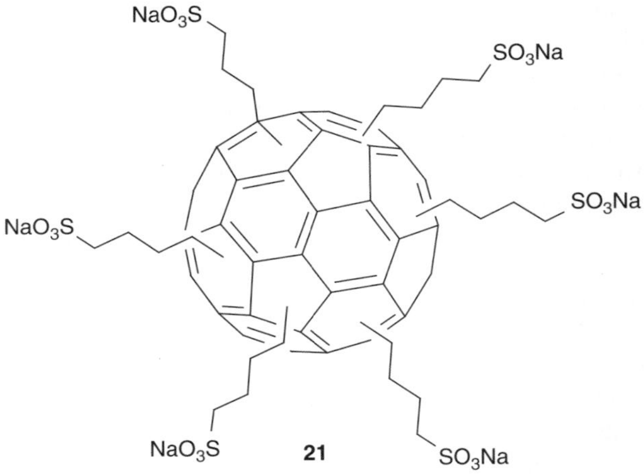

**Figure 15** *Molecular structure of fullerene hexa-adduct* **21**

if $10^{-5}$ M of $C_{60}$FAS was used. Cell viability dramatically decreased: 45% after 10 min of irradiation, 20% after 20 min and 7% after 30 min. This result is interesting in the context of cancer photodynamic therapy (PDT).

Chiang and co-workers[58] prepared hexa(sulfo-*n*-butyl)[60]fullerene **21** (Figure 15) and they tested its ability to produce singlet oxygen, finding levels comparable to photofrin $^1O_2$ generation despite the lower absorption at 600 nm. This compound presents six negative charges but they seemed not to be sufficient to avoid aggregation. Indeed, the authors found that, independently from the concentration, spherical nanoparticles of 50–60 Å diameter were present. These nanoparticles have been incorporated into coated polymer films and the new materials, together with hexa(sulfo-*n*-butyl)[60]fullerene in solution, have been tested as potential bactericidal. In both cases, Gram-positive and Gram-negative microorganisms died, with an efficiency depending on the irradiation dose. The effect on tissue was the physical breakdown of microstructures inducing apoptosis.

Ikeda and co-workers[59] prepared liposomes incorporating fullerene. To obtain a good preparation, they utilised a complex $C_{60}/\gamma$-cyclodextrin (1:2 ratio) as fullerene source. The process led to the incorporation of uncomplexed fullerene and its presence did not considerably modify the liposome structure, except for the diameters, which were increased in all cases (cationic, anionic or neutral lipids). ColE1 supercoiled plasmid has been photoirradiated in the presence or absence of the liposomes. The fullerene efficiently acted as DNA photocleaver when incorporated into positively charged liposomes, which could be attracted by the negatively charged phosphates, while the neutral system was less effective. No nicked DNA form was observed in the absence of light or fullerene, or in the presence of negatively charged liposomes. The experiment performed using cyclodextrin–fullerene complex as reference led to a DNA cleavage less effective than cationic liposome incorporating $C_{60}$ (6% *vs* 44%).

Fullerene was also incorporated into liposomes containing bis(di-isobutyl octadecylsiloxy)silicon 2,3-naphthalocyanine (isoBOSiNc), a naphthalocyanine derivative with spectral and photophysical properties that makes it attractive candidate for PDT. The irradiation with nanosecond near-infrared laser pulse (1024 nm) led to explosive release of the encapsulated material as the energy was adsorbed by $C_{60}$. The subsequent irradiation of isoBOCSiNc at 776 nm activated the usual photodynamic process. The two combined irradiations had synergic effect, increasing the survival of treated mice.[60]

Vileno et al.[61] studied hexafullerene adduct **22** (Figure 16) and a commercially available fullerenol by EPR, demonstrating that the Type-II photo process prevailed for the fullerenol while the Type-II process turned to Type I in the case of compound **22** after illumination with white light for more than 45 min. The authors used also the AFM to determine the local elastic properties of cells and to observe the changes due to the toxic action of $^1O_2$. It seemed that the toxic action of $^1O_2$ was more effective on living cells compared to cells fixed with glutaraldehyde. Commercial fullerenol under polychromatic steady-state irradiation produced a mixture of reactive oxygen species under ultraviolet and visible irradiation even at low concentrations.[62]

The introduction of sugar moieties potentially improves the biological properties of fullerene because they increase its solubility and they play an important role in cell–cell interaction. The synthesis of fullerene derivatives

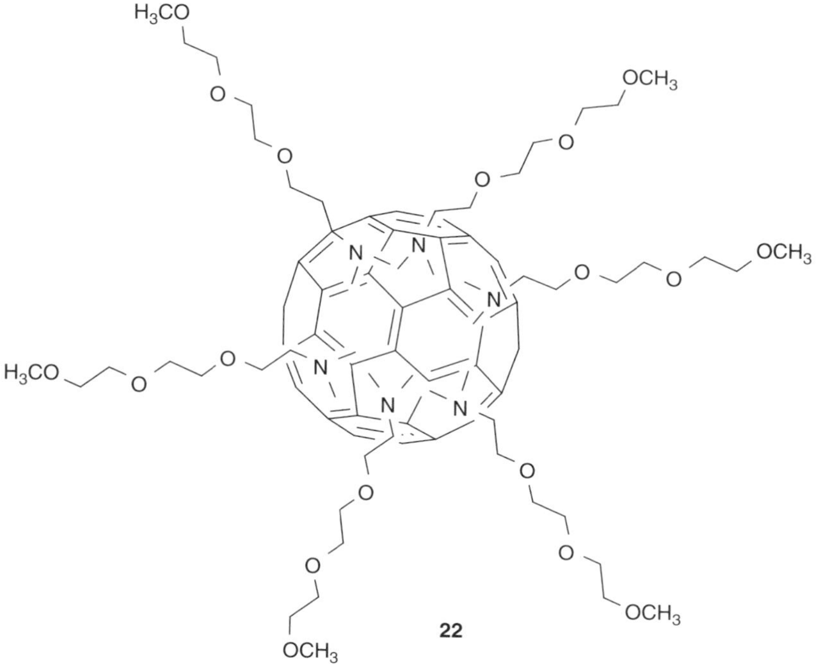

**Figure 16** *Compound 22*

containing mono- and disaccharides have been reported and their production of singlet oxygen was analysed by measuring near IR emission at 1270 nm.[15] As postulated, the bisadducts were less effective in generating singlet oxygen. The treatment of HeLa cells with these derivatives was almost not effective in dark condition, while phototoxicity was more efficient upon incubation with monoadducts. In some cases, the bis-adducts increased cell survival. It is known that the singlet oxygen yield decreases with the increase of substitution on the carbon cage, but in the reported experiment, the protection was attributed to a lack of uptake of bis-adducts into the cells.

Mono- and hexa-adduct derivatives of fullerene bearing two pyropheophorbide *a* moieties have been studied by Rancan *et al.*[63] The authors analysed the uptake and localisation in Jurkat cells. Derivative **23** (Figure 17) and the corresponding hexa-adduct showed difficulties in concentrating into cells because they were uptaken by endocytosis or pinocytosis, with kinetics slower than the passive diffusion through membrane, which was invoked as the mechanism of penetration of the control analogues devoid of fullerene moiety. The hexa-adduct acted as better photosensitiser because the presence of five extra addends on the carbon cage decreased the quantum yield of single-oxygen production, avoided the fullerene aggregation and slowed down the electron transfer from pyropheophorbide *a* to fullerene. The phototoxicity provoked cell death by inducing apoptosis, as determined by analysing the degree of activity of caspase 3 and 7.

Comparison among porphyrin, porphyrin–fullerene dyad (P-$C_{60}$) and metallated porphyrin–fullerene dyad (ZnP-$C_{60}$) (Figure 18) showed that the phototoxic activity decreased from P-$C_{60}$ **24** to ZnP-$C_{60}$ **25** and to porphyrin alone. This behaviour was found also in anaerobic conditions, demonstrating that in this case both Type I and Type II mechanisms were involved.[64]

Although it is really difficult to think about immediate application in photodynamic therapy, the peculiar photophysical properties of fullerene lead to good opportunity in DNA photocleavage and in local anticancer treatments, taking into account also the demonstrated accumulation of fullerene derivatives for tumoral tissues.[65]

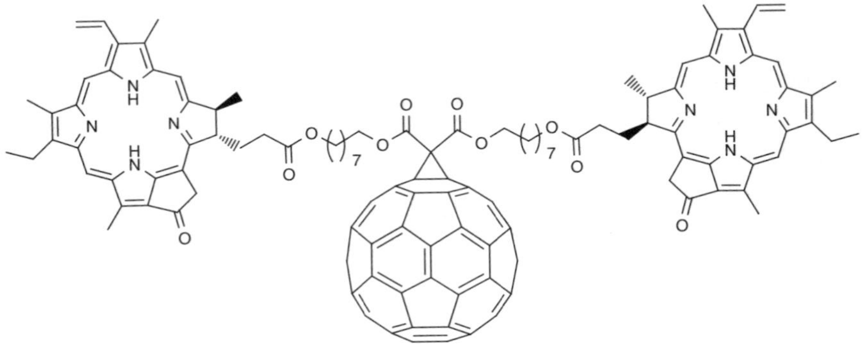

**Figure 17** *Derivative 23*

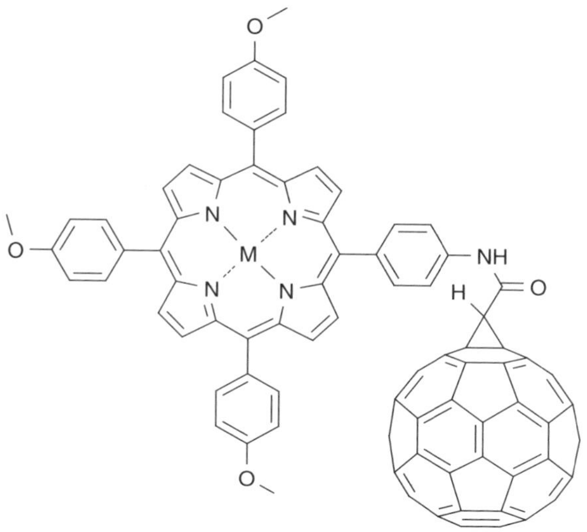

**24** : P-C$_{60}$    M = H$_2$
**25** : ZnP-C$_{60}$   M = Zn

**Figure 18** *Structure of compounds 24 and 25*

## 10.6 Antibacterial, Antiviral Activity and Enzymatic Inhibition

The antibacterial activity of fullerene derivatives was first reported in 1996.[66] The hypothesised mechanism of the water-soluble fulleropyrrolidinium salts was attributed to the cell-membrane disruption by the bulky carbon cage, which seemed not really adaptable to planar cellular surface.

More recently Mashino *et al.*[67] studied the action of some derivatives on *E. coli*. Both *trans-2* (**26**) and *trans-4* (**27**) isomers (Figure 19) inhibited bacterial growth at 1 µM concentration being the *trans-2* isomer more effective. In this case, the mechanism of action seemed to be the inhibition of the respiratory chain. A biphasic effect of fullerene on the oxygen uptake was pointed out, with a first decreasing of the oxygen uptake at low fullerene concentration, followed by an increasing of the uptake rate with the increasing of fullerene concentration. In the latter condition, the oxygen was converted to H$_2$O$_2$ and the respiratory chain was inhibited. Apparently an electron transfer from respiratory chain to fullerene took place independently from the presence of light, impairing the cellular electron path.

Compounds **26**, **27** and the *trans-3* isomer (**28**) were also studied and interesting antibacterial and antiproliferative activities were found.[68] The average antibacterial activity of these compounds was comparable with vancomycin or even better against some strands of *E. foecalis*, while the antiproliferation was analogous to cisplatin. In this case, the mechanism was not identified and it is accepted that the inhibition of the respiratory chain was not involved.

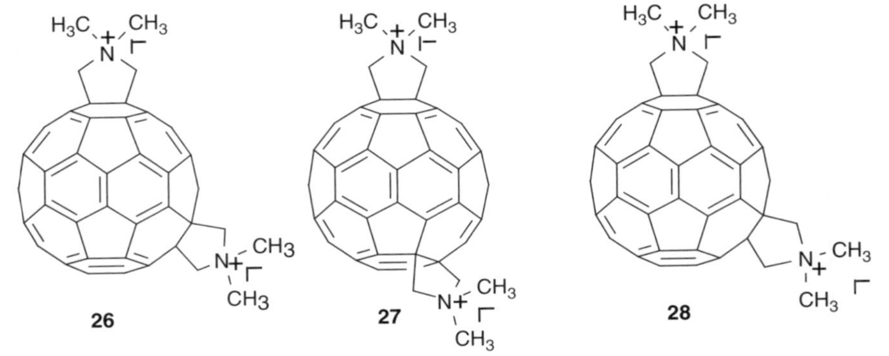

**Figure 19** *Bis-pyrrolidinium salts **26**, **27** and **28** (trans-4, trans-2 and trans-3)*

In the latest years fullerene aggregates, *i.e.* nano-$C_{60}$ particles, have raised the attention of many researchers, devoted to study their antimicrobial activity. It has been found that an association between the cells and the particles occurred but it was not clear which kind of interaction was taking place. The presence of the bacterium growth media rich in salts (*i.e.* Luria broth, PBS) increased the nanoparticle size and these aggregates did not inhibit the growth of *E. coli* or *B. subtilis*. Changing the growing medium, a completely different behaviour was observed with bacteriostatic action *vs B. subtilis* with a nano-$C_{60}$ concentration of 4.8 mg $L^{-1}$, as demonstrated by studies on $CO_2$ production. The addition of fullerenols did not affect the respiration, while a 7.5 mg $L^{-1}$ concentration of unmodified fullerene aggregates was bactericidal.[69]

A $C_{60}$/PVP complex was demonstrated to inhibit the influenza virus type A (a RNA virus) with an efficacy comparable with rimantadine. The same action has been reported *vs* the influenza virus type B and herpes simplex virus (DNA-containing virus). The activity was dependent on the percentage of fullerene contained into PVP (0.6% gave good results while lower percentages were ineffective) and on the molecular mass (m.m.) of the carrier polymer, with the better results obtained using PVP of m.m. 25,000. The complex displayed low cytotoxicity, without affecting the metabolic and morphological properties of the cultured cells. Moreover, the activity was not related to any specific decreasing in viral protein production and the action was probably exerted during the latest stage of the viral cycle.[70]

In 1993, Wuld and co-workers[71,72] reported different antiviral activity. For the first time the authors suggested the possible inhibition of the proteases of HIV (HIV-P) by interaction of fullerene with the hydrophobic active site of the enzyme. Since then, many studies on different fullerene derivatives have been performed.[73–77] A computational study on the molecular and electronic properties of 10 fullerene derivatives able to inhibit HIV-P has recently been reported.[78] Recently Marchesan *et al.* obtained good results against HIV on CEM cells infected by HIV-1(III$_B$) or HIV-2 (ROD) using a series of N,N-dimethyl bis-fulleropyrrolidinium salts. However, the mechanism involved in the antiviral activity was not elucidated. Analyses in this sense have been performed

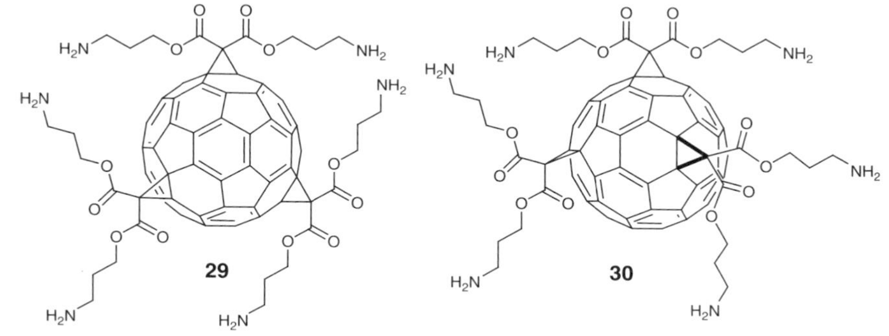

**Figure 20** *Compounds **29** ($D_3$) and **30** ($C_3$)*

by Mashino and co-workers. They identified efficient inhibitors of the HIV reverse transcriptase, more active than nevirapine. The same compounds were found to be active also against hepatitis C virus RNA polymerases.[79]

Another enzyme that can be affected by fullerene is the NO synthase.[80–82] Its role in the catalysis of the nitric oxide production is very important. The alteration of NO concentration is implicated in pathological conditions as neurodegenerative diseases. Starting from monomalonate adducts, different derivatives have been studied, with special attention to the *equatorial, equatorial, equatorial* ($C_3$) and *trans-3, trans-3, trans-3* ($D_3$) hexaamino fullerenes. Derivatives **29** and **30** (Figure 20) inhibited the three NO synthase isoforms, with a time-dependent decreasing of NO formation, which can be reverted by addition of calmodulin. Actually, the fullerene derivatives bound calmodulin as demonstrated by experiments in which displacement of trifluoperazine was evaluated. Though, the precise binding site was not identified. These fullerene derivatives were also able to bind some proteins similar to calmodulin, presenting $Ca^{2+}$-binding domain, as troponin C, but not parvalbumin.

An adamantyl fullerene derivative was prepared by Zaitsu and co-workers.[83] In principle this molecule should be hydrolysed to carboxyfullerene and amantadine, which is an antidyskinetic dopamine-releasing drug, used to treat Parkinson's disease. The hydrolysis should be catalysed by anadamine aminohydrolase, but the fullerene derivative seemed instead to inhibit the enzyme (Figure 21).

Despite the mechanisms of bacterial and viral inhibition can be various, the use of fullerene, both in presence or in absence of light, represents one of the most appealing and encouraging field of application.

## 10.7 Immunological Properties of Fullerenes

The use of fullerene derivatives and fullero-peptides in immunology is a domain that only recently started to be explored. Few years ago, Erlanger and co-workers[84] succeeded on the induction of specific antibodies against $C_{60}$ conjugated to a carrier protein. The antibodies were subsequently isolated and

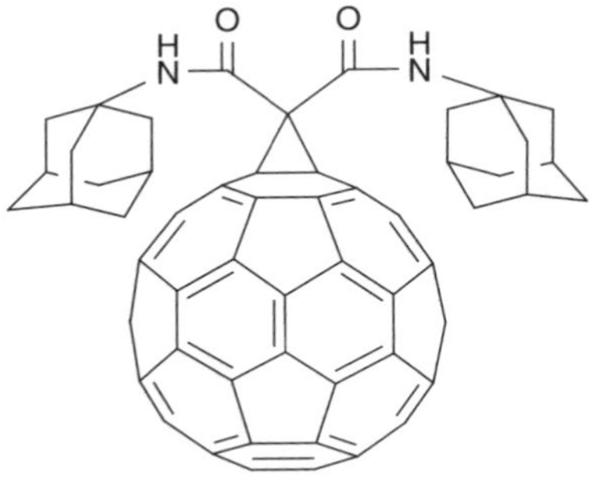

**Figure 21** Compound 31

crystallised, although in the absence of the fullerene.[85] The hydrophobic pocket in which the $C_{60}$ cage can nicely accommodate has been identified at the interface of the light and heavy chains of the crystal structure. Later on, molecular modelling studies allowed elucidating the interactions and the recognition modes between the antibody and the fullerene.[86] Therefore, it was evident that fullerene derivatives were able to stimulate the immune system and to elicit antibody responses of the right specificity. Along this line, it has been explored the possibility to modulate the immune response by using fullerene-containing peptides. For this purpose, a series of antigenic peptides have been selected and modified by insertion of suitable fullerene derivatives or fullero-amino acids. A proline-rich peptide was covalently conjugated to a $C_{60}$ moiety *via* a solubilising triethylene glycol chain (Figure 22, **32**). This peptide is one of the main target of the anti-Sm and anti-U1RNP auto-antibodies in sera of patients with autoimmune diseases such systemic lupus erythematosus (SLE) or mixed-connective tissue disease (MCTD).[87] Once modified at the *C*-terminal part with the fullerene moiety, the peptide was still recognised by anti-Sm/U1RNP sera from SLE and MCTD. This fullero-peptide can be considered a promising precursor for the design of new analogues with more potent activity and increased selectivity.

In an alternative approach, fullero-pyrrolidino-L-glutamic acid (Fgu) has been conceived and synthesised to replace certain amino acid residues into antigenic peptides. The final goal of this type of fullerene-peptide analogues is their use to modulate the immunological activity of the natural epitopes.[12,88] Peptide sequence 64–78 derived from histone H3 has been identified as a dominant T-cell epitope that plays an important role in the development of the immune response in SLE. The modification of this peptide by the introduction of non-coded amino acids would permit to obtain new ligands able to restore self-tolerance in the autoimmune response. Glutamine in position 68 of the natural peptide was replaced by Fgu (Figure 22, **33**).

The peptide analogue was prepared by solid-phase synthesis using the Fmoc/*t*Bu strategy.[88] Initially, it has been verified by molecular modelling that the

Biological Applications of Fullerenes 321

**32** H-Pro-Pro Gly-Met(O)-Arg-Pro-Pro-O⁀O⁀N-[C60]

**33** H-⁶⁴Lys-Leu-Pro-Phe-NH-CH(CH₂CH₂-C(O)-NH-CH₂-CH₂-N[C60])-C(O)-NH-Arg-Leu-Val-Arg-Glu-Ile-Ala-Gln-Asp-Phe⁷⁸-OH

**34** H-⁶⁴Asp-Phe-NH-CH(CH₂CH₂-C(O)-NH-CH₂-CH₂-N[C60])-C(O)-NH-Arg-Leu-Ile-Lys-Pro-Glu-Phe-Val-Gln-Leu-Arg-Ala⁷⁸-OH

**35** H-⁶⁴Lys-Leu-Pro-NH-CH(CH₂CH₂-C(O)-NH-CH₂-CH₂-N[C60])-C(O)-NH-Gln-Arg-Leu-Val-Arg-Glu-Ile-Ala-Gln-Asp-Phe⁷⁸-OH

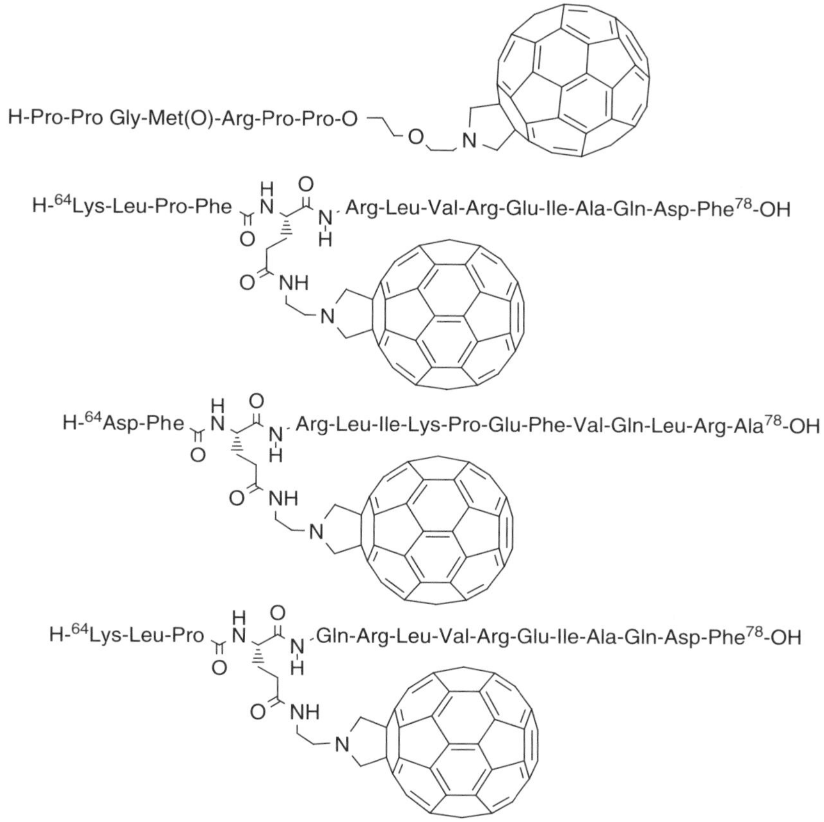

**Figure 22** *Molecular structures of fullero-peptides*

fullero-peptide was able to conserve the binding capacity to the MHC (major histocompatibility complex) molecule similar to that of the epitope. Indeed, the amino acid replacement most probably did not involved one of the MHC anchor residues. This result was very important and encouraging to demonstrate the interaction of the fullero-peptide with MHC molecules and its modulation of the T-cell response. Following the first examples of synthesis of $C_{60}$-containing peptides on a resin,[14,88] a new strategy for the solid-phase synthesis of fullero-peptides has been developed.[12] Fgu amino acid residue has been protected to the amino function with a *tert*-butyloxycarbonyl group, which permitted to prepare fullero-peptide analogues on a solid support using Merrifield approach. Two new peptides derived from 64–78 epitope sequence have been designed. Fgu was inserted in different positions in comparison to the fullero-peptide **33** (Figure 22, **34** and **35**). Fullero-peptide **34** is constituted of a scrambled sequence of the parent peptide where the residue in position 66 (corresponding to a glutamine) was substituted by Fgu, while the analogue **35** has the same sequence of the natural epitope and contains Fgu in position 67 instead of phenylalanine. Fullero-peptide **34** will permit to exclude a non-specific interaction to MHC molecule due to the hydrophobic $C_{60}$ moiety. Fullero-peptide **35** should display a reduced capacity of MHC binding since Phe67 is a key anchor residue. These analogues should allow to better understand the influence of the fullerene moiety on MHC binding and T-cell response.

Overall these results highlight the potential of fullerene derivatives and fullerene-based peptides for immunological applications and may stimulate research towards the design and preparation of other biologically relevant conjugates.

## 10.8 Biological Applications of Radio Labelled Fullerenes

Carbon nanomaterials are becoming interesting nano-vehicles for imaging, diagnosis, therapy and microsurgery.[89] Within this family, endohedral metallofullerenes have been explored for different applications depending on the type of metal ion inserted into the fullerene cage and including: (i) nuclear medicines;[90] (ii) fluorescent tracers;[91] and (iii) MRI contrast agents.[92] The great advantage would be the prevention of the release *in vivo* of the extremely toxic metal ions because entrapped into the closed-fullerene sphere. The MRI radiotracer applications using metal ion encapsulation into highly soluble modified fullerenes seem to be the most promising.

As radiotracers, endohedral fullerenes have been initially used to monitor the biodistribution of this material in animals. $C_{82}$ has been, for example, doped with holmium and transformed into a fullerenol by functionalising its surface with hydroxyl groups, which induced a remarkable water-solubility to the final compound **36** (Figure 23).[93] After neutron activation, the fullerene derivatives were localised into the liver and the bones. It was found that the metallofullerenols were slowly released from the organs but they were not acutely toxic *in vivo*. A possible explanation for their long accumulation into certain

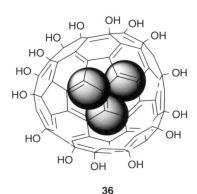

**36**

**Figure 23** *Molecular structure of the polyhydroxylated metallofullerenol $^{166}Ho_x@$-$C_{82}(OH)_y$ **36***

organs could be attributed to the formation of aggregates in aqueous solution depending on the concentration of salts.[94,95] This phenomenon modifies the behaviour of the fullerene derivatives when in contact with biological fluids and may influence the fullerene-based drug delivery.

In a similar study, a water-soluble fullerenol was radiolabelled with gallium.[96] The biodistribution studies of this endohedral $C_{60}$ showed high localisation in the bone marrow, liver and spleen. Again, slow clearance was observed. It has been suggested that such derivative might be used to target lymphatic organs similarly to other microcolloidal preparations.

An alternative approach to follow the fullerenes into living systems was their functionalization with a contrast agent at their surface. A highly water-soluble fullerene derivative based on a X-ray contrast agent has been prepared (Figure 24).[97] $C_{60}$ was derivatised using the Bingel reaction with a malonic acid modified with an iodinated aromatic ring. The final product **37** contained 24% of iodine for application as a contrast agent.

Advantages by using this type of fullerene derivatives for imaging are expected and include, for example, the facility of their intravenous administration since the globular shape of the molecule would reduce viscosity in the clinical formulations.

Multihydroxylated endohedral fullerenes containing gadolinium and forming particles in the nanomolar scale were demonstrated to exert an efficient anticancer activity.[98] These derivatives have the capacity to interfere with cancer invasion in normal cells, with an almost complete absence of toxicity both *in vitro* and *in vivo*. The fullerenols based on $C_{82}$ did not kill the tumour

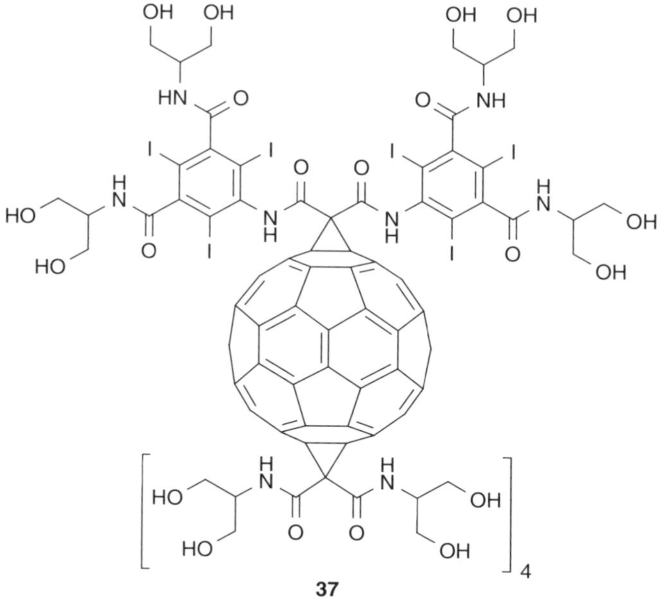

**Figure 24** *Molecular structure of the iodinated $C_{60}$ contrast agent* **37**

cells directly as they were not detected into the tumour tissues, but it was found that the gadofullerenols improved the immune response and regulated the oxidative defence system. In comparison to other chemotherapeutics such as cyclophosphamide and cisplatin, the efficiency of the gadofullerenes to achieve the same antitumoral effect proved much higher.

The results on the utilisation of radiolabelled and endohedral metallofullerenes reported until now are very promising. However, this area of research is still in his infancy and more studies will be necessary to reinforce the potential of such derivatives for chemotherapeutic applications.

## 10.9 General Conclusions

The appealing structure of fullerene, its electrochemical and photophysical properties aroused the interest of the scientists to study the biological application of $C_{60}$ and its derivatives. Up to now, no real application in medicine has been obtained and a part of the scientific community shows unbelief and scepticism. Despite this negative attitude there is wide room for successful applications of fullerene in medicine and biology.

## References

1. D. Pantarotto, N. Tagmatarchis, A. Bianco and M. Prato, *Mini Rev. Med. Chem.*, 2004, **4**, 805.
2. E. Nakamura and H. Isobe, *Acc. Chem. Res.*, 2003, **36**, 807.
3. S. Bosi, T. Da Ros, G. Spalluto and M. Prato, *Eur. J. Med. Chem.*, 2003, **38**, 913.
4. N. Tagmatarchis and H. Shinohara, *Mini Rev. Med. Chem.*, 2001, **1**, 339.
5. A. Bianco, T. Da Ros, M. Prato and C. Toniolo, *J. Pept. Sci.*, 2001, **7**, 208.
6. T. Da Ros and M. Prato, *Chem. Commun.*, 1999, 663.
7. A.W. Jensen, S.R. Wilson and D.I. Schuster, *Bioorg. Med. Chem.*, 1996, **4**, 767.
8. A. Hirsch, The chemistry of the fullerenes, *Thieme*, 1994.
9. G.N. Bogdanov, R.A. Kotel'nikova, E.S. Frog, V.N. Shtol'ko, V.S. Romanova and Y.N. Bubnov, *Dokl. Biochem. Biophys.*, 2004, **396**, 165.
10. R.F. Enes, A.C. Tome and J.A.S. Cavaleiro, *Tetrahedron*, 2005, **61**, 1423.
11. J.Z. Yang and A.R. Barron, *Chem. Commun.*, 2004, 2884.
12. A. Bianco, *Chem. Commun.*, 2005, 3174.
13. L.A. Watanabe, M.P.I. Bhuiyan, B. Jose, T. Kato and N. Nishino, *Tetrahedron Lett.*, 2004, **45**, 7137.
14. D. Pantarotto, A. Bianco, F. Pellarini, A. Tossi, A. Giangaspero, I. Zelezetsky, J.-P. Briand and M. Prato, *J. Am. Chem. Soc.*, 2002, **124**, 12543.
15. Y. Mikata, S. Takagi, M. Tanahashi, S. Ishii, M. Obata, Y. Miyamoto, K. Wakita, T. Nishisaka, T. Hirano, T. Ito, M. Hoshino, C. Ohtsuki, M. Tanihara and S. Yano, *Bioorg. Med. Chem. Lett.*, 2003, **13**, 3289.

16. Y. Nishida, A. Mizuno, H. Kato, A. Yashiro, T. Ohtake and K. Kobayashi, *Chem. Biodiver.*, 2004, **1**, 1452.
17. H. Kato, N. Kaneta, S. Nii, K. Kobayashi, N. Fukui, H. Shinohara and Y. Nishida, *Chem. Biodiver.*, 2005, **2**, 1232.
18. S. Abe, H. Moriyama, K. Niikura, F. Fei, K. Monde and S. Nishimura, *Tetrahedron-Asymmetr.*, 2005, **16**, 15.
19. Y. Liu, H. Wang, P. Liang and H.Y. Zhang, *Angew. Chem. Int. Ed.*, 2004, **43**, 2690.
20. J. Yang, Y. Wang, A. Rassat, Y. Zhanga and P. Sinay, *Tetrahedron*, 2004, **60**, 12163.
21. E. Oberdörster, *Environ. Health Persp.*, 2004, **112**, 1058.
22. V. Colvin, *Nature Biotechnol.*, 2003, **21**, 1166.
23. R. Service, *Science*, 2003, **300**, 243.
24. J.D. Fortner, D.Y. Lyon, C.M. Sayes, A.M. Boyd, J.C. Falkner, E.M. Hotze, L.B. Alemany, Y.J. Tao, W. Guo, K.D. Ausman, V.L. Colvin and J.B. Hughes, *Environ. Sci. Technol.*, 2005, **39**, 4307.
25. C.M. Sayes, A. Gobin, K. Ausman, J. Mendez, J. West and V.L. Colvin, *Biomaterials*, 2005, **26**, 7587.
26. R. Anderson and A.R. Barron, *J. Am. Chem. Soc.*, 2005, **127**, 10458.
27. C.M. Sayes, J.D. Fortner, W. Guo, D. Lyon, A.M. Boyd, K.D. Ausman, Y.J. Tao, B. Sitharaman, L.J. Wilson, J.B. Hughes, J.L. West and V.L. Colvin, *Nano Lett.*, 2004, **4**, 1881.
28. F. Rancan, S. Rosan, F. Boehm, A. Cantrell, M. Brellreich, H. Schoenberger, A. Hirsch and F. Moussa, *J. Photochem. Photobiol. B*, 2002, **67**, 157.
29. S. Bosi, L. Feruglio, T. Da Ros, G. Spalluto, B. Gregoretti, M. Terdoslavich, G. Decorti, S. Passamonti, S. Moro and M. Prato, *J. Med. Chem.*, 2004, **47**, 6711.
30. G. Andrievsky, V. Klochkov and L. Derevyanchenko, *Fuller. Nanotub. Car. N.*, 2005, **13**, 363.
31. G. Jia, H.F. Wang, L. Yan, X. Wang, R.J. Pei, T. Yan, Y.L. Zhao and X.B. Guo, *Env. Sci. Technol.*, 2005, **39**, 1378.
32. N. Gharbi, M. Pressac, M. Hadchouel, H. Szwarc, S. Wilson and F. Moussa, *Nano Lett.*, 2005, **5**, 2578.
33. V. Georgakilas, F. Pellarini, M. Prato, D. Guldi, M. Melle-Franco and F. Zerbetto, *P. Natl. Acad. Sci. USA*, 2002, **99**, 5075.
34. A. Bianco, *Exp. Opin. Drug Deliv.*, 2004, **1**, 57.
35. S. Foley, C. Crowley, M. Smaihi, C. Bonfils, B. Erlanger, P. Seta and C. Larroque, *Biochem. Biophys. Res. Commun.*, 2002, **294**, 116.
36. T. Zakharian, A. Seryshev, B. Sitharaman, B. Gilbert, V. Knight and L. Wilson, *J. Am. Chem. Soc.*, 2005, **127**, 12508.
37. N. Venkatesan, J. Yoshimitsu, Y. Ito, N. Shibata and K. Takada, *Biomaterials*, 2005, **26**, 7154.
38. F. Perret, M. Nishihara, T. Takeuchi, S. Futaki, A.N. Lazar, A.W. Coleman, N. Sakai and S. Matile, *J. Am. Chem. Soc.*, 2005, **127**, 1114.
39. M. Nishihara, F. Perret, T. Takeuchi, S. Futaki, A.N. Lazar, A.W. Coleman, N. Sakai and S. Matile, *Org. Biomol. Chem.*, 2005, **3**, 1659.

40. E. Nakamura, H. Isobe, N. Tomita, M. Sawamura, S. Jinno and H. Okayama, *Angew. Chem. Int. Edit.*, 2000, **39**, 4254.
41. P.J. Krusic, E. Wasserman, P.N. Keizer, J.R. Morton and K.F. Preston, *Science*, 1991, **254**, 1184.
42. L.L. Dugan, E. Lovett, S. Cuddihy, B.-W. Ma, T.-S. Lin and D.W. Choi, in *Carboxyfullerenes as Neuroprotective Antioxidants*, K.M. Kadish and R.S. Ruoff (eds), John Wiley, NY, 2001.
43. L.L. Dugan, D.M. Turetsky, C. Du, D. Lobner, M. Wheeler, C.R. Almli, C.K.-F. Shen, T.-Y. Luh, D.W. Choi and T.-S. Lin, *P. Natl. Acad. Sci. USA*, 1997, **94**, 9434.
44. L. Dugan, J. Gabrielsen, S. Yu, T. Lin and D. Choi, *Neurobiol. Dis.*, 1996, **3**, 129.
45. A.M.-Y. Lin, C.H. Yang, Y.-F. Ueng, T.-Y. Luh, T.Y. Liu, Y.P. Lay and L.-T. Ho, *Neurochem. Int.*, 2004, **44**, 99.
46. S.S. Ali, J.I. Hardt, K.L. Quick, J.S. Kim-Han, B.F. Erlanger, T.T. Huang, C.J. Epstein and L.L. Dugan, *Free Radical Bio. Med.*, 2004, **37**, 1191.
47. Y.H. Wang, E.J. Lee, C.M. Wu, T.Y. Luh, C.K. Chou and H.Y. Lei, *J. Drug Del. Sci. Tech.*, 2004, **14**, 45.
48. A.N. Djordjevic, J.M. Canadanovic-Brunet, M. Voijnovic-Miloradov and G.M. Bogdanovic, *Oxid. Commun.*, 2004, **27**, 806.
49. Y.W. Chen, K.C. Hwang, C.C. Yen and Y.L. Lai, *Am. J. Physiol.-Reg. I.*, 2004, **287**, R21.
50. G.M. Bogdanovic, V. Kojic, A. Dordevic, J.M. Canadanovic-Brunet, M.B. Vojinovic-Miloradov and V.V. Baltic, *Toxicol. In vitro*, 2004, **18**, 629.
51. S.M. Mirkov, A.N. Djordjevic, N.L. Andric, S.A. Andric, T.S. Kostic, G.M. Bogdanovic, M.B. Vojinovic-Miloradov and R.Z. Kovacevic, *Nitric Oxide-Biol. Ch.*, 2004, **11**, 201.
52. L. Xiao, H. Takada, K. Maeda, M. Haramoto and N. Miwa, *Biomed. Pharmacother.*, 2005, **59**, 351.
53. T. Sun, Z. Jia and Z. Xu, *Bioorg. Med. Chem. Lett.*, 2004, **14**, 1779.
54. X. Zhao, A. Striolo and P.T. Cummings, *Biophys. J.*, 2005, **89**, 3856.
55. T. Da Ros, M. Bergamin, E. Vázquez, G. Spalluto, B. Baiti, S. Moro, A. Boutorine and M. Prato, *Eur. J. Org. Chem.*, 2002, 405.
56. Y. Liu, Y.L. Zhao, Y. Chen, P. Liang and L. Li, *Tetrahedron Lett.*, 2005, **46**, 2507.
57. P. Scharff, L. Carta-Abelmann, C. Siegmund, O.P. Matyshevska, S.V. Prylutska, T.V. Koval, A.A. Golub, V.M. Yashchuk, K.M. Kushnir and Y.I. Prylutskyy, *Carbon*, 2004, **42**, 1199.
58. C. Yu, T. Canteenwala, L.Y. Chiang, B. Wilson and K. Pritzker, *Synth. Metals*, 2005, **153**, 37.
59. A. Ikeda, T. Sato, K. Kitamura, K. Nishiguchi, Y. Sasaki, J. Kikuchi, T. Ogawa, K. Yogo and T. Takeya, *Org. Biomol. Chem.*, 2005, **3**, 2907.
60. M. Babincova, P. Sourivong, D. Leszczynska and P. Babinec, *Laser Phys. Lett.*, 2004, **1**, 476.
61. B. Vileno, A. Sienkiewicz, M. Lekka, A.J. Kulik and L. Forro, *Carbon*, 2004, **42**, 1195.

62. K.D. Pickering and M.R. Wiesner, *Environ. Sci. Technol.*, 2005, **39**, 1359.
63. F. Rancan, M. Helmreich, A. Molich, N. Jux, A. Hirsch, B. Roder, C. Witt and F. Bohm, *J. Photoch. Photobio. B*, 2005, **80**, 1.
64. M. Milanesio, M. Alvarez, V. Rivarola, J. Silber and E. Durantini, *Photochem. Photobiol.*, 2005, **81**, 891.
65. Y. Tabata, Y. Murakami and Y. Ikada, *Fuller. Sci. Technol.*, 1997, **5**, 989.
66. T. Da Ros, M. Prato, F. Novello, M. Maggini and E. Banfi, *J. Org. Chem.*, 1996, **61**, 9070.
67. T. Mashino, N. Usui, K. Okuda, T. Hirota and M. Mochizuki, *Bioorg. Med. Chem.*, 2003, **11**, 1433.
68. T. Mashino, D. Nishikawa, K. Takahashi, N. Usui, T. Yamori, M. Seki, T. Endo and M. Mochizuchi, *Bioorg. Med. Chem. Lett.*, 2003, **13**, 4395.
69. D.Y. Lyon, J.D. Fortner, C.M. Sayes, V.L. Colvin and J.B. Hughes, *Environ. Toxicol. Chem.*, 2005, **24**, 2757.
70. L.B. Piotrovsky and O.I. Kiselev, *Fuller. Nanotub. Car. N.*, 2004, **12**, 397.
71. R.F. Schinazi, R.P. Sijbesma, G. Srdanov, C.L. Hill and F. Wudl, *Antimicrob. Agents Ch.*, 1993, **37**, 1707.
72. R. Sijbesma, G. Srdanov, F. Wudl, J.A. Castoro, C. Wilkins, S.H. Friedman, D.L. DeCamp and G.L. Kenyon, *J. Am. Chem. Soc.*, 1993, **115**, 6510.
73. E. Nakamura, H. Tokuyama, S. Yamago, T. Shiraki and Y. Sugiura, *Bull. Chem. Soc. Jpn.*, 1996, **69**, 2143.
74. S.H. Friedman, P.S. Ganapathi, Y. Rubin and G.L. Kenyon, *J. Med. Chem.*, 1998, **41**, 2424.
75. G. Marcorin, T. Da Ros, S. Castellano, G. Stefancich, I. Borin, S. Miertus and M. Prato, *Org. Lett.*, 2000, **2**, 3955.
76. D.I. Schuster, L.J. Wilson, A.N. Kirschner, R.F. Schinazi, S. Schlueter-Wirtz, P. Tharnish, T. Barnett, J. Ermolieff, J. Tang, M. Brettreich and A. Hirsch, in: N. Martin, M. Maggini, D.M. Guldi (eds), *Fullerene – Functionalised Fullerenes*, vol. 9, The Electrochemical Society Inc., Pennington, NJ, 2000, pp. 267.
77. S. Bosi, T. Da Ros, G. Spalluto, J. Balzarini and M. Prato, *Bioorg. Med. Chem. Lett.*, 2003, **13**, 4437.
78. S. Promsri, P. Chuichay, V. Sanghiran, V. Parasuk and S. Hannongbua, *J. Mol. Struc.-Theochem.*, 2005, **715**, 47.
79. T. Mashino, K. Shimotohno, N. Ikegami, D. Nishikawa, K. Okuda, K. Takahashi, S. Nakamura and M. Mochizuki, *Bioorg. Med. Chem. Lett.*, 2005, **15**, 1107.
80. D. Wolff, C. Barbieri, C. Richardson, D. Schuster and S. Wilson, *Arch. Biochem. Biophys.*, 2002, **399**, 130.
81. D. Wolff, K. Mialkowski, C.F. Richardson and S.R. Wilson, *Biochemistry*, 2001, **40**, 37.
82. D. Wolff, A. Papoiu, K. Mialkowski, C. Richardson, D. Schuster and S. Wilson, *Arch. Biochem. Biophys.*, 2000, **378**, 216.
83. M. Nakazono, S. Hasegawa, T. Yamamoto and K. Zaitsu, *Bioorg. Med. Chem. Lett.*, 2004, **14**, 5619.

84. B.-X. Chen, S.R. Wilson, M. Das, D.J. Coughlin and B.F. Erlanger, *P. Natl. Acad. Sci. USA*, 1998, **95**, 10809.
85. B. Braden, F. Goldbaum, B. Chen, A. Kirschner, S. Wilson and B. Erlanger, *P. Natl. Acad. Sci. USA*, 2000, **97**, 12193.
86. W. Noon, Y. Kong and J. Ma, *P. Natl. Acad. Sci. USA*, 2002, **99**, 6466.
87. P. Sofou, Y. Elemes, E. Panou-Pomonis, A. Stavrakoudis, V. Tsikaris, C. Sakarellos, M. Sakarellos-Daitsiotis, M. Maggini, F. Formaggio and C. Toniolo, *Tetrahedron*, 2004, **60**, 2823.
88. A. Bianco, D. Pantarotto, J. Hoebeke, J.P. Briand and M. Prato, *Org. Biomol. Chem.*, 2003, **1**, 4141.
89. H.B.S. Chan, B.L. Ellis, H.L. Sharma, W. Frost, V. Caps, R.A. Shields and S.C. Tsang, *Adv. Mater.*, 2004, **16**, 144.
90. D.W. Cagle, S.J. Kennel, S. Mirzadeh, J.M. Alford and L.J. Wilson, *P. Natl. Acad. Sci. USA*, 1999, **96**, 5182.
91. R. Macfarlane, G. Wittmann, P. vanLoosdrecht, M. deVries, D. Bethune, S. Stevenson and H. Dorn, *Phys. Rev. Lett.*, 1997, **79**, 1397.
92. E. Toth, R.D. Bolskar, A. Borel, G. Gonzalez, L. Helm, A.E. Merbach, B. Sitharaman and L.J. Wilson, *J. Am. Chem. Soc.*, 2005, **127**, 799.
93. D.W. Cagle, T.P. Thrash, M. Alford, L.P.F. Chibante, G.J. Ehrhardt and L.J. Wilson, *J. Am. Chem. Soc.*, 1996, **118**, 8043.
94. B. Sitharaman, R. Bolskar, I. Rusakova and L.J. Wilson, *Nano Lett.*, 2004, **4**, 2373.
95. S. Laus, B. Sitharaman, V. Toth, R.D. Bolskar, L. Helm, S. Asokan, M.S. Wong, L.J. Wilson and A.E. Merbach, *J. Am. Chem. Soc.*, 2005, **127**, 9368.
96. L. Yu-guo, H. Xuan, L. Riu-li, L. Qing-nuan, Z. Xiao-dong and L. Wen-xin, *J. Radioanal. Nucl. Chem.*, 2005, **265**, 127.
97. T. Wharton and L. Wilson, *Bioorg. Med. Chem.*, 2002, **10**, 3545.
98. C. Chen, G. Xing, J. Wang, Y. Zhao, B. Li, J. Tang, G. Jia, T. Wang, J. Sun, L. Xing, H. Yuan, Y. Gao, H. Meng, Z. Chen, F. Zhao, Z. Chai and X. Fang, *Nano Lett.*, 2005, **5**, 2050.

CHAPTER 11

# Covalent and Non-Covalent Approaches Toward Multifunctional Carbon Nanotube Materials

VITO SGOBBA, G.M. AMINUR RAHMAN, CHRISTIAN EHLI AND DIRK M. GULDI

Friedrich-Alexander-Universität Erlangen-Nürnberg Institute for Physical Chemistry Egerlandstr. 3, D-91058, Erlangen, Germany

## 11.1 Introduction

A very appealing cast of *nanometer scale structures* that has recently drawn a lot of attention is carbon nanostructures. Of the wide range of carbon nanostructures available, carbon nanotubes (CNTs), in general, and single-wall carbon nanotubes (SWCNTs), in particular, stand out as truly unique materials.[1]

SWCNT are one-dimensional nanostructures, which are composed of interlinked hexagon networks of carbon atoms. A SWCNT is formed when one single layer of graphite is wrapped onto itself and the resulting edges are joined. Hereby, the hexagons create seamless cylinders that are oriented along a chiral vector. While their diameters are typically in the range of nanometers, individually SWCNT reach lengths of up to 4 cm leading to high aspect ratios. Their shell of $sp^2$-hybridized carbon atoms forms a highly aromatic hexagonal network, which is the beginning for chemical modification.[2] One of the most important aspects in contemporary CNT chemistry is their purification.[3] Another hot topic of interest, which, however, relates to the intrinsic characteristics of CNT, is the fact that their overall conductivity varies between semiconducting and metallic behavior. The electrical conductivity is up to 1000 times greater than, for instance, in copper. Moreover, the conductivity depends, in large, on the chiral angle of the tubes and their diameter – measurements have become possible for individual multi-wall carbon nanotube (MWCNT) and individual SWCNT. Therefore, not surprising the chemistry of CNT emerged as a powerful tool for separating metallic from semiconducting species.[4]

CNT have found noteworthy impact in the fields of electronics,[5] optoelectronics,[6] sensing applications,[7] and for optimizing interfaces with polymer composites that might ultimately pave the way to reinforced fibers.[8] In addition, they readily accept electrons, which can then be transported under nearly ideal conditions along the tubular axis. The fact that CNTs appear in structurally defined semiconductive or conductive forms turns the CNT into ideal components for different electronic applications.[9]

On the other hand, multiple concentric graphene cylinders – MWCNTs – exhibit metallic or semiconducting properties, which depend solely on their outermost shell. On account of the large number of concentric cylindrical graphitic tubes present in MWCNT, they are considered even more suitable in electron-donor–acceptor ensembles than SWCNT.[10]

In the context of chemical reactivity, CNT are regarded as either sterically bulky π-conjugated ligands or as electron deficient alkenes. Joselevich, for example, discussed in a qualitative description the electronic structure of SWCNT from a chemical perspective. In fact, this work was based on using real-space orbital representations and traditional concepts of aromaticity, orbital symmetry and frontier orbitals.[11] Important for the chemical reactivity is the fact that metallic CNT are slightly less aromatic than semiconducting ones[12] and, in turn, more reactive than the latter.[11,4a,4c]

Relative to planar graphene the curvature that the carbon frameworks in CNT possess leads to (i) an appreciable change in pyramidalization ($\theta_p$) and (ii) a misalignment ($\phi$) of the π-orbitals (see Figure 1). Consequently, CNT exhibit an increased tendency to undergo addition reactions. It is mainly this reason why the sidewalls of CNT are much more reactive than planar sheets of graphene–graphite. However, at the same time, they are found to be less reactive than fullerenes – here the spherical geometry and the smaller diameters engender significant strain.[13] As aforementioned, the pyramidalization strain is

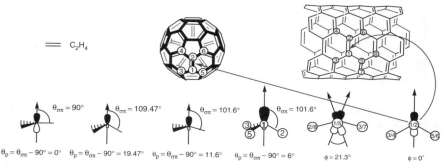

**Figure 1** *Pyramidalization angles ($\theta_p$) for ethylene and methane; some examples of pyramidalization ($\theta_p$) and misalignment ($\phi$) angles for $C_{60}$ fullerene and for (5,5) SWNT*

in SWCNT by no means comparable to that found in fullerenes. In other words, the π-orbital misalignment must exert a greater influence on the reactivity.[14]

Since both π-orbitals misalignment and pyramidalization scale inversely with the tube diameter, SWCNT produced by the HiPco process – with average diameters of around 1 nm[15] – are more reactive than those produced by laser ablation of a graphite target[16] – with average diameters of around 1.38 nm. The reactivity of the inner surface, due to the concave curvature, is markedly lower that what is typically found for the outer surface.[17] Still, the inside of SWCNT is reactive to hydrogen and fluorine radicals. Note that fullerenes, on the other hand, lack any considerable reactivity of their internal surface.[18]

The different production methodologies – laser ablation, electric arc discharge and catalytic decomposition of gaseous hydrocarbon – supply large quantities of CNT. Here, however, hundreds of individual CNT bundle up to form largely interwoven aggregates/bundles. These bundles are responsible for the low solubility of SWCNT in common organic solvents. Moreover, the bundling effects impact the feasibility of functionalization, their rates and yields. The success of the functionalization implies a reasonable access of the chemical reagents. Obviously, it requires a debundling of the bundles – ultrasonication is hereby the most appropriate and most widely applied method.[19] Notably, controlled conditions are a key requisite to limit damages occurring during the ultrasonication to the sidewalls.[20] UV/Vis/NIR, Raman, X-ray photoelectron spectroscopy (XPS), nuclear magnetic resonance (NMR) spectroscopies, near-edge X-ray adsorption fine structure (EXAFS), electron microscopies (SEM, TEM), atomic force microscopy (AFM), scanning tunneling microscopy (STM) and X-ray diffractometry (XRD) are the most often used techniques for the characterization of CNT.[21]

This chapter overviews recent advances in the rapidly developing areas of covalent and non-covalent approaches of functionalizing CNT toward multifunctional CNT materials. The exploitation of the many promising properties of CNT in technical applications requires an efficient manipulation of the tubes. As far as covalent methodologies are concerned (*i.e.*, thermally activated, electrochemical modification, photochemical or mechanochemical functionalization) we will first focus on the end tip/defect functionalization and then later on the sidewall functionalization. In the context of non-covalent methodologies we will mainly elucidate π-π interactions as a powerful means to add additional functionalities to CNT. The chapter will conclude with a survey on synthetic strategies to separate metallic and semiconducting CNT.

## 11.2 End Tips and Defect Functionalization

A simple structural consideration suggests that CNT might be far more reactive at their ends than in areas along the sidewalls. It is mainly the increased curvature that is accountable for this conclusion.[2c,22] Notwithstanding, the sidewalls show upon closer inspection some fairly reactive sites. The reactive

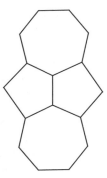

**Figure 2** *Stone–Wales defect on the sidewall of CNT*

sites are mainly localized at/or close to structural defects, which are formed during the growth of CNT or introduced during post processing applications (*i.e.*, purification, separation, *etc.*) – see below. Depending on the actual conditions of such post treatment the defect sites might amount to around 1–3% of all carbon atoms present.[23]

Very often such defects are due to pentagon-heptagon pairs, which are typically referred to Stone–Wales or 7/5/5/7 defects (see Figure 2). Important is hereby that they augment the curvature along the sidewalls.[24] Alternatively, the defects might be caused by (i) substitutional dopant impurities, (ii) structural deformation, especially bending or twisting of CNT, and (iii) vacancies in the CNT lattice.[25]

These defects influence not only the chemical but also the physical properties of the nanotubes.[26] A remarkable demonstration is the rectifying behavior that originates from naturally occurring defects.[27]

### 11.2.1 Oxidation

Oxidation has been among the first chemical reactions that were tested with CNT.[28] Not surprisingly, the oxidation has been widely employed as a key step in their purification – oxidative removal of Fe, Co or Ni catalyst and of amorphous/graphitic carbon impurities.[29] Less success was, however, noted in the oxidation-induced separation of metallic from semiconducting specimen.[30] Notably, physical modification of CNT – opening of the hemispherical caps, shortening and chemical functionalization – accompanies in many instances the oxidative treatment.

Importantly, SWCNT and MWCNT behave differently toward oxidants. As SWCNT are known to exist in bundles, which are composed of up to hundreds of individual tubes, their oxidation takes place mainly at the end tips and to a lesser amount at the defect sites. The reactivity is, nonetheless, limited to the outer layer of the bundles.[31] Although the oxidation of MWCNT acts, in principle, on the same sites, it is routinely employed in the thinning of MWCNT through the systematic layer by layer removal of the outer layers.[32]

The oxidation of CNT was first observed in air at 700°C by Iijima et al.[33] in 1993. Nowadays the oxidation in air is still operative as the first step in the purification of CNT, namely, eliminating amorphous carbon. The reaction temperature is, however, around 350°C.[28a,34]

Oxidative treatment of CNT was performed under a wide variety of experimental conditions. Hereby, sonications in oxygen-containing acids (*i.e.*, $HNO_3$,[31,35] $H_2SO_4$,[36] $H_2SO_4$ + $K_2S_2O_8$,[37] $HNO_3$ + $H_2SO_4$,[30,38] $HNO_3$ + supercritical water,[39] trifluromethanesulfonic and chlorosulfonic acid,[40] peroxytrifluoroacetic acid,[41] $H_2SO_4$ + $KMnO_4$,[42] $H_2SO_4$ + $H_2O_2$,[43] $HClO_4$,[42] $H_2SO_4$ + $K_2Cr_2O_7$[44]) play the most dominant roles. As a consequence of such treatments carboxylate groups are generated that are distributed over the CNT surface. In the presence of $H_2SO_4$ also sulfate[38a] ketone, phenol, alcohol and ether groups were registered.[45] Other often utilized oxidants are dilute ceric sulfate,[46] $CO_2$,[47] $H_2O_2$,[42] $O_2$ plasma[48] and $RuO_4$.[49] Considering the high reactivity that is associated with the opened tips oxidized nanotubes often display much larger internal diameters in the area near the open ends than what is typically seen in areas far from these ends.[50]

The concentrations of carboxylic acid that are formed during the oxidative treatment were determined by evaluating the $CO_{(g)}$ and $CO_{2(g)}$ concentrations at high temperatures[38c] or measuring the atomic oxygen percentage with calibrated energy-dispersive X-ray spectroscopy (EDX).[51] Chemical titration assays emerged as alternative tests to macroscopically estimate the defect density in CNT. Titration of the purified SWCNT with NaOH and $NaHCO_3$ solutions was used to determine the total percentage of acidic sites and carboxylic acid groups, respectively.[52] A recent study also highlighted the opportunity to label the carboxylic groups with $TiO_2$ nanoparticles as markers.[53]

## 11.2.2 Derivatization of the Carboxylic Functionalities

Carboxylic functionalities, which have been introduced as a matter of oxidation, are the starting point for further derivatization with amide[54] and ester functionalities.[54c,55] The coupling occurs *via* acyl chloride intermediates or, alternatively, in a single step using carbodiimide mediated linking of the carboxylate (1-ethyl-3-(3-dimethylaminopropyl) carbodiimide). In water, carbodiimide hydrochloride (EDC) is used, while dry organic solvents necessitate 1,3-dicyclohexylcarbodiimide (DCC). The chart, shown in Figure 3, summarizes some representative cases.

Attachment of long hydrocarbon chains onto CNT is particularly important, since it helps to improve their solubility[54a,b] without, however, compromising the electrical and mechanical properties.[2a,56] As a result of the increased solubility, the modified SWCNT were purified: purification methods rely on filtration, centrifugation, and chromatographic techniques to separate them from the residual impurities, which usually consist of trace amounts of catalyst, nanoparticles, and amorphous carbon after the as-prepared SWCNT have been

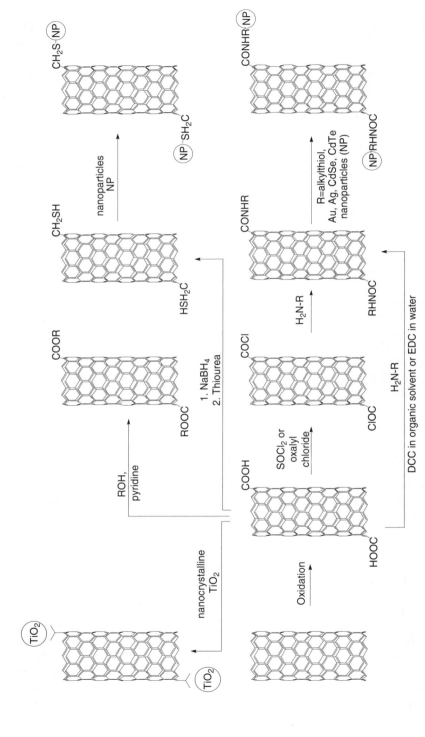

**Figure 3** *Further derivatization of carboxylic functionalities in CNT*

treated with acid or oxidizing reagents.[37,57] Functionalized CNT can also be separated by electrophoresis.[58]

As mentioned above CNT are cut into smaller segments by sonication – agitation using ultrasound – in concentrated acid mixtures, and the resulting fragments can be separated into tubes of narrow length distributions using chromatography. What is been missing is an extension of these techniques to sort SWCNT by diameter, and chirality.

This widely applicable approach has also been used to achieve water solubility of SWCNT, a key material feature, through functionalization with glucosamine[59] or 2-aminoethanesulfonic acid.[60] A viable alternative toward water solubility implies the use of oligomeric or polymeric building blocks, which once attached provides decent water solubility to the correspondingly functionalized SWCNT.[59,61] But not only these water-soluble, hydrophilic functionalities were linked to CNT, in addition, metal complexes,[62] imidazolium-based ionic liquids,[63] lipophilic,[64] and hydrophilic dendrons were successfully added.[65] Remarkably, the hydrophilic dendrons were afterwards removed, that is, under either basic or acidic hydrolysis, which led to the recovery of the pristine CNT structures.[66]

In the context of bioinspired materials, peptide nucleic acids (PNA),[67] deoxyribonucleic acid (DNA),[68] enzymes,[69] sugars,[70] proteins,[71] bacteria,[72] viruses,[73] biotin[74] and fluorescent dyes[75] were useful targets for linking to CNT.[76] The electrical contacting of redoxactive enzymes by their reconstitution on aligned CNT linked perpendicularly to an electrode support has demonstrated an unprecedented long-range electron transport without the need of redox mediators. These studies bear particular promises, when considering that functionalized CNT are able to cross cell membranes and accumulate in the cytoplasm, and even reach the nucleus, without being cytotoxic – in concentrations up to 10 mM.

Totally different is the outcome when treating the carboxylic functionalities at the end tips in the presence of DCC in dry dimethylformamide. In particular, ring-shaped nanostructured materials were isolated. Implicit in the formation process is an end-to-end self coupling, involving complementary anhydride and ester bonds.[77] Similarly, intermolecular and intramolecular end-to-end and end-to-side nanostructures were obtained when aliphatic diamine was linked to SWCNT.[78]

Figures 4 and 5 illustrate a number of important materials that are based on SWCNT. On one hand, catalytically active vanadyl complexes were bound through this synthetic route.[79] On the other hand, photo- and redox-active porphyrins[80] and phthalocyanines[81] have been linked through esterification of the carboxylate functionalities that were introduced to the nanotubes.

Covalent modification of SWCNT was utilized to create high-resolution, chemically sensitive probe microscopy tips. Hereby, carboxylic acid groups at the open ends of SWCNT were coupled to amines to create additional probes with basic or hydrophobic functionalities. Force titrations recorded between the ends of the SWCNT tips and hydroxy-terminated self-assembled monolayers confirmed the chemical sensitivity and robustness of these SWCNT

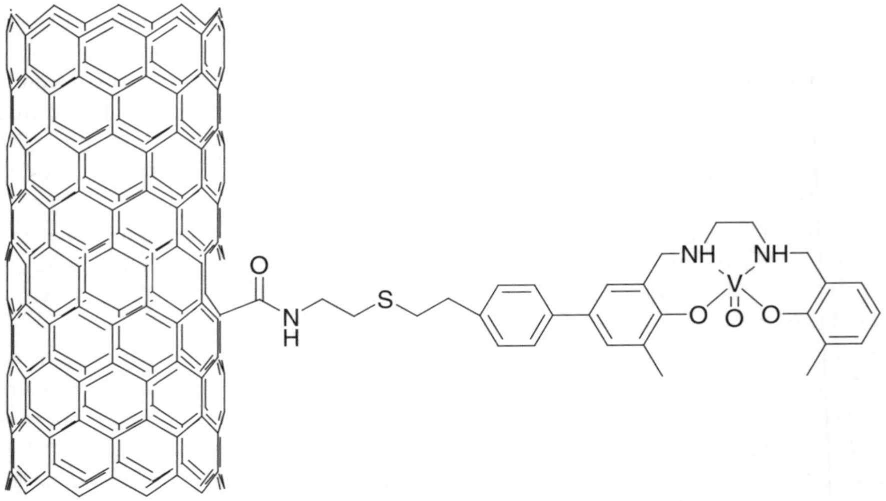

**Figure 4** SWNT linked with a vanadyl complex as novel catalysis material

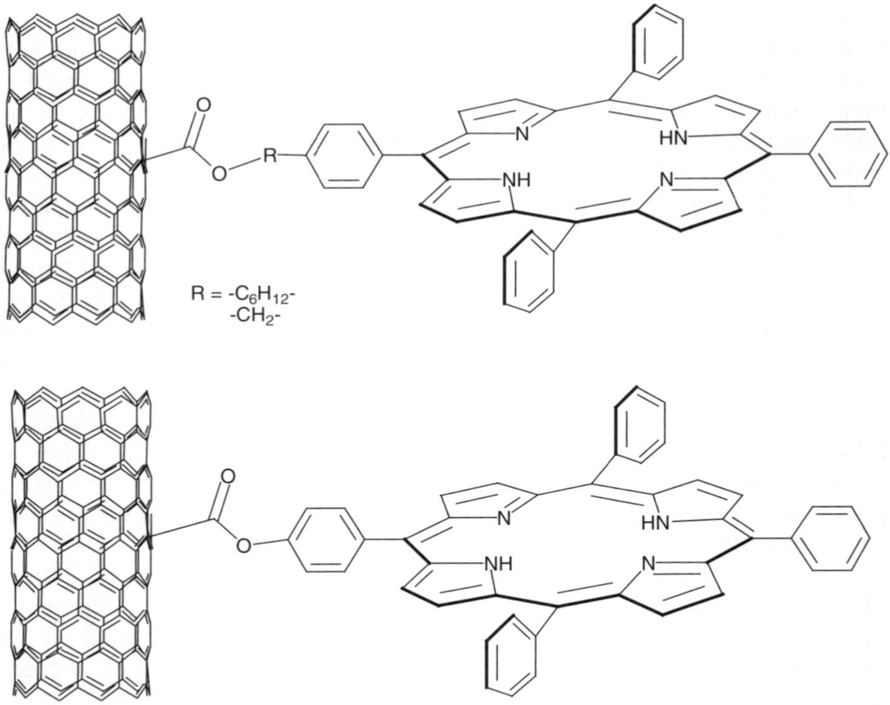

$R = -C_6H_{12}-$
$-CH_2-$

**Figure 5** SWNT linked to porphyrins as novel photoactive and redoxactive material

tips. In addition, images recorded on patterned self assembled monolayers (SAM) and partial bilayer surfaces have demonstrated chemically sensitive imaging with nanometer-scale resolution. These studies show that well-defined covalent chemistry can be exploited to create functionalized SWCNT probes that have the potential for true molecular-resolution, chemically sensitive imaging.[82]

In recent years, organic composites or organic/inorganic mixed-nanocomposites emerged as extremely valuable classes of materials. A particular challenge aims at the immobilization of tailorable metallic and/or semiconducting nanoparticles onto the surface of CNT. One successful strategy involves the tethering of gold nanoparticles, for example, to SWCNT and MWCNT. This was enabled through the treatment of oxidized SWCNT, namely, thiol-derivatized by reacting with $NH_2$-$(CH_2)_n$-SH.[38b,83] The spontaneous chemical adsorption to gold occurs via Au-S chemical bonding. Although the more nucleophilic thiol presumably results in the predominate formation of thioamides, free thiols were shown to exist de facto by AFM imaging of attached 10-nm gold nanoparticles.[84]

A multistep strategy involves the reduction of carboxylates at the open ends of CNT, their chlorination and subsequent thiolation. The corresponding thiol groups were afterwards conveniently assembled to silver[85] and gold nanoparticles,[86] or even to CdS nanoparticles[87] Moreover, SWCNT that are oxidized by either acid or ozone treatment (for details see Section 11.3) – have been assembled on amine modified gold surfaces[88] and were employed as electrodes to investigate the charge transfer process between SWCNT and the underlying substrates.[89] When the oxidized ends of CNT are functionalized asymmetrically[90] onto gold and mercury surfaces the modulation of current rectification has been recently demonstrated.[91]

Oxidized SWCNT and MWCNT have been linked to carboxylic acid terminated cadmium selenide (CdSe) nanocrystals that are capped with mercaptothiol derivatives.[92] Similarly, titanium dioxide ($TiO_2$) nanocrystals, functionalized with 11-aminoundecanoic acid, were probed. Hereby, the aid of intermediary linking agents, which were either ethylenediamine or semicarbazide, were extremely beneficial to form nanoscale heterostructured materials.[93] Oxidized MWCNT have also been covalently linked to thiol-stabilized ZnS-capped CdSe quantum dots containing amine terminal groups (QD-$NH_2$).[94] For SWCNT with lengths of <200 nm, binding of quantum dots occurred at the nanotube ends, whereas longer nanotubes showed evidence of sidewall attachment.[95]

SWCNT-$NH_2$, with amino groups directly attached to tube open ends from purified SWCNT, from edge-terminated with carboxylic groups were produced. The first steps of the transformation were based on the selective functionalization of carboxylic groups. The final transformation steps consisted of direct introduction of amino groups instead of carboxylic moieties, and included the first application of the Hofmann rearrangement of carboxylic acid amides and Curtius rearrangement of carboxylic acid azides in CNT chemistry.[96]

## 11.2.3 Solvent-Free Amination and Thiolation

Direct solvent-free amination of CNT with aromatics (*i.e.*, pyridine)[97] and aliphatic primary amines (*i.e.*, octadecylamine) and thiolation has been reported.[98] Mainly, the closed caps of MWCNT were functionalized. This reaction resembles essentially the amination or the thiolation of spherical fullerenes. Oxidized SWCNT were also found to react with vaporous aliphatic amines.[99]

## 11.3 Sidewall Chemistry

In general, addition reactions, as we will describe them in the following section are thought to occur predominantly at the intact sidewalls. However, additions to sites of defects or close by to these are not ruled out. To avoid possible side reactions, purified CNT that bear low amounts of oxygenated species are examined.

### 11.3.1 Fluorination

Considering the inertness that characterize the sidewalls of SWCNT it is obvious that very aggressive reaction conditions are needed.[16d,100] In initial work, fluorination conditions already successful in the chemical transformation of graphite were applied.[101] When testing, for example, HF as catalyst in the wide temperature range between 250 and 400°C, fluorine reacts with SWCNT to afford fluorinated SWCNT (F-SWCNT).[16c,d,102] In these novel, fluorinated SWCNT materials the presence of covalently linked fluorine was confirmed by FT-IR measurements with a characteristic peak at 1220–1250 cm$^{-1}$. Working in higher temperature regimes helped to exceed the fluorination level past 50%, before, however, the collapse of the tubular structure was noted. With the same method, fluorine atoms were selectively attached to the sidewall of the outer shell of DWNTs without disrupting the double-layered morphology.[103] An alternative approach for performing fluorination involves the use of $BrF_3$ and $Br_2$ at room temperature.[104] SWCNT have also been successfully fluorinated with great success upon exposure to a $CF_4$ plasma exposure.[105]

An important feature of fluorinated SWCNT is their superior solubility, when compared to pristine SWCNT. They were solubilized, for instance, through ultrasonication in various alcohols and other solvents.[106] Notably, the partial destruction of the graphitic structure lowers the conductivity of SWCNT.

Fluorinated CNT were used as cathodes in lithium electrochemical cells. Valuable is that higher cell potentials, compared with commercially used fluorographite cells, point to promising future applications.[107] A superior electrocatalytic performance is in general expected from CNT materials compared with other carbon forms because of the much higher surface-to-volume ratios.[108] These materials showed also good performances as solid lubricants.[109]

In summary, fluorination provides a convenient access to a fascinating class of functionalized CNT materials. With an overall stoichiometry close to $C_2F$, the highest degree of functionalization observed to date is available.

## 11.3.2 Derivatization of Fluorinated Carbon Nanotubes

From a more general point of view, this new form of fluorocarbon material can serve as a benchmark for in in-depth studies of the implications of functionalization on the physical, electrical, mechanical, and chemical properties of the carbon-nanotube framework.

In general the C-F bonds in fluorinated CNT are particularly weak – an eclipsing strain effect is mainly responsible for this effect.[100b] Based on this increased reactivity, fluorinated SWCNT react with many nucleophilic reagents, for which an overview is given in Figure 6.[110] Leading examples are organolithium and Grignard reagents[111] that give rise to alkyl functionalized nanotubes.[112] In this particular case the alkyl functionalities were removed at a temperature of around 250 °C. Analogously, thiol- and thiophene-functionalized SWCNT were prepared by reacting fluorinated SWCNT with substituted amines.[113] A reaction with hydrazine, on the other hand, affords at room temperature pristine CNT, $N_2$ and HF.[16c,d] When alkoxy groups[114] or aliphatic terminal diamines[115] are probed, alkoxylated nanotubes or nanotubes bearing

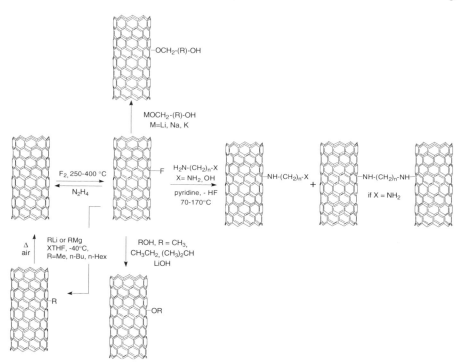

**Figure 6** *Fluorination of CNT and subsequent fluorine replacements*

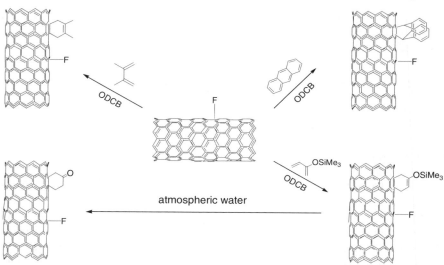

**Figure 7** Diels–Alder cycloaddition to fluorinated SWNT

N-alkylidene amino groups – covalently attached and also cross-linked to their side walls – are isolated, respectively. These amino-terminated groups are then available for subsequent binding of (i) amino acids, (ii) DNA and (iii) polymers. Implicit is here that these bindings occur almost exclusively to the sidewalls of SWCNT. As an example, the reaction with adipoyl chloride should be considered, which is shown to afford new nylon-SWCNT polymer materials.[115] Fluorinated SWCNT also facilitate Diels–Alder cycloaddition with dienes – see Figure 7.

### 11.3.3  1,3 Dipolar Cycloadditions – Ozonolysis, Cycloaddition of Azomethine Ylide, and Cycloaddition of Nitrile Ymine

In 1,3 dipolar cycloadditions,[116] 1,3-dipoles – ozone, azomethine ylides and nitrile imines – were shown to add to double bonds (*i.e.*, dipolarophile) that are located along the sidewalls of CNT. As Figure 8 illustrates, five member rings are formed.

Ozone undergoes 1,3 dipolar cycloaddition with olefins. This initial step is followed by sequential decomposition and isomerization steps that occur with the primary formed ozonide. The Criegee mechanism, as shown in Figure 9, summarizes the reaction sequence.[117] Previous experiments with $C_{60}$ and $C_{70}$ confirmed its applicability for the chemistry of carbon nanostructures.[118]

First work on the cycloaddition of ozone to SWCNT and MWCNT reports methanolic dispersions at around − 78°C as working conditions.[2d,119] SWCNT-ozonides emerge as the primary product of this procedure.[120] In follow-up reactions, cleavage with hydrogen peroxide, dimethyl sulfide or sodium borohydride leads to the carboxylic, the ketone/aldehyde or the alcohol

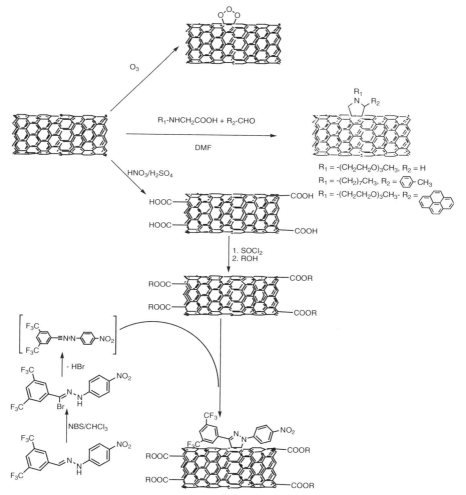

**Figure 8** *(1,3) dipolar cycloaddition to SWNT*

appended CNT, respectively. This methodology – involving the oxidatively-induced cleavage – leads to some opening of the sidewall and, hence, bears great similarity with the derivatization described in the first part of this contribution (*i.e.*, end tips and defect functionalization).[119a,121] Critical to these experiments is that the electrical resistance of the ozonized SWCNT depends largely on the degree of oxidation. In fact, it is about 20–2000 times higher than that found for pristine SWCNT. A likely rationale for the resistance enhancement is the structural deformation of the π-conjugation, especially along the tubular axis.[88c]

The cycloaddition of azomethine ylides – generated *in situ* by the condensation of α-amino-acids and aldehydes – is widely known as the Prato reaction. It

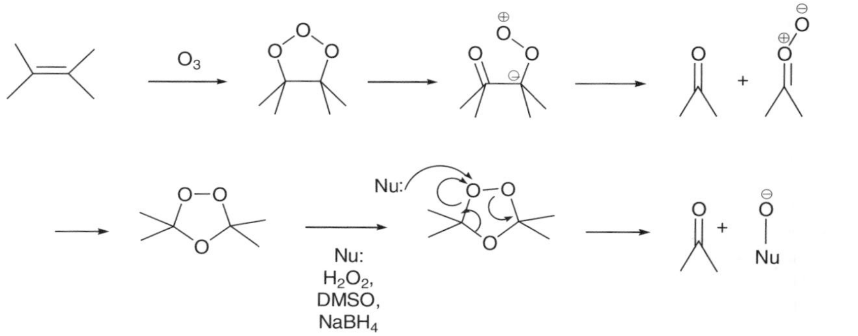

**Figure 9** *Criegee Ozonization mechanism*

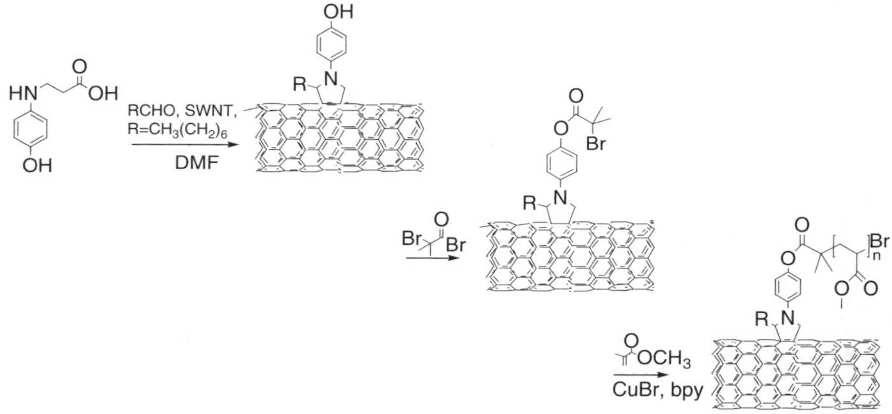

**Figure 10** *Polymerization of the side group of SWNT*

has been shown to play a dominant role in the chemistry of fullerenes.[122] Furthermore, this methodology is also applicable in modifying the sidewalls of SWCNT and MWCNT toward multifunctional materials.[3a,123] CNT that were subjected to cycloaddition with azomethine ylides are remarkably soluble in a variety of organic solvents.[123a,124] The largely improved solubility also plays a major role in their purification. Noteworthy, the organic groups that were covalently attached to the sidewalls of SWCNT were removed by simple heat treatment under ambient conditions and 350°C. This led to a nearly quantitative recovery of SWCNT that exhibit pristine-like electronic properties.[3a]

In the course of the years, a large variety of functionalities have been linked to the sidewalls of CNT toward novel multifunctional materials.[2,125] Figures 10–12 reveal the most significant cases. In these examples CNT bear polymers,[126] ferrocenes, which were probed in optoelectronic applications[127] or as exoreceptors for the redox recognition of $H_2PO_4^-$ [128] or for the detection of the glucose,[129] bio-inspired N-protected aminoacids,[130] and bioactive peptides for

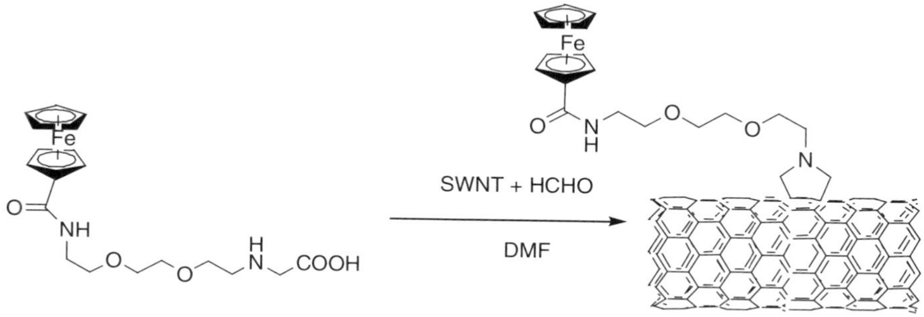

**Figure 11** *Amidoferrocenyl functionalization of SWNT*

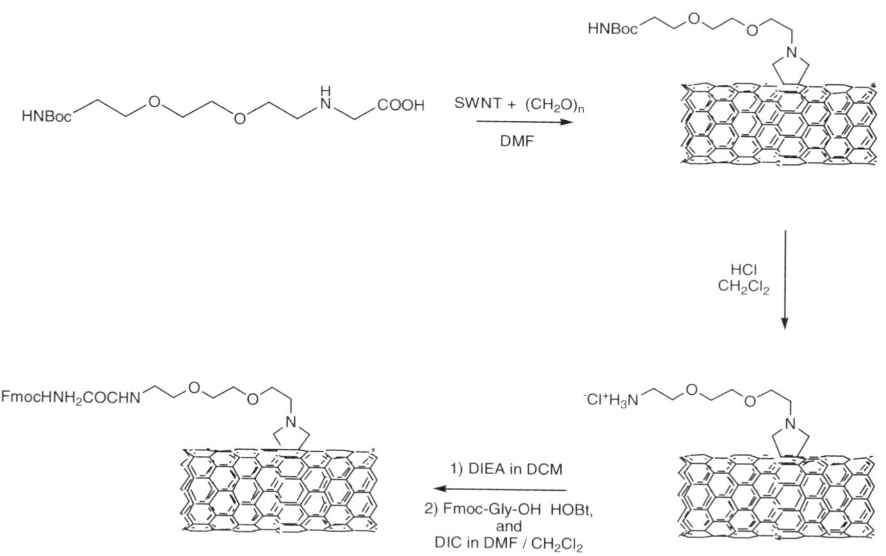

**Figure 12** *Examples of aminoacid functionalization of SWNT. DIEA: diisopropylethylamine, DIC, diisopropylcarbodiimide, HOBt, N-hydroxybenzotriazole, Boc, N-tertbutoxycarbonyl, Fmoc, Fluorenylmethoxycarbonyl*

biomedical application, which were tested for with great success for disease diagnosis and vaccine delivery.[125c,131]

Similarly, a 1,3 dipole that stems from nitrile imines – generated *in situ* from its molecular precursors – to ester modified SWCNT has been used to synthesize two soluble, photoactive single-walled carbon nanotubes containing *n*-pentyl esters at the tips and 2,5-diarylpyrazoline units at the walls of the tubes. The presence of an electron-poor 3,5-*bis*(trifluoromethyl)phenyl or electron-rich 4-(*N,N*-dimethylamino)phenyl on the C atom of the pyrazoline ring give rise to electron transfer from the electron rich substituents of the pyrazoline unit to electron acceptor termini including electron poor aryl rings and nanotube walls.[132]

## 11.3.4 Diels–Alder Cycloaddition

The first Diels–Alder cycloadducts were reported employing o-quinodimethane as described in Figure 13. The latter reagent, which was generated *in situ* from 4,5-benzo-1,2-oxathiin-2-oxide, reacted with ester functionalized SWCNT. Important in this study was the use of microwave irradiation.[133]

## 11.3.5 Osmylation

Sidewall osmylation of individual, but metallic SWCNT has been achieved by exposing them to osmium tetroxide vapor under UV photoirradiation. The covalent attachment of the osmium oxide leads to novel materials that exhibit an increase in electrical resistance. The increase is at times several orders of magnitude higher than what has been found in the starting material. UV light helps to cleave the cycloaddition product. Typical reaction conditions for the photocleavage are vacuum or oxygen atmosphere – following the cleavage, the original resistance is restored.[134] Comparable results were seen for the same reaction when performed in solution.[4c] In analogy to the photoactivated osmylation of benzenoids hydrocarbons, the formation of an intermediate charge-transfer complex is likely the basis for the observed selectivity and reactivity of metallic tubes.[135]

## 11.3.6 [2+1] Cycloadditions – Carbenes or Nitrenes

1,1-Dichlorocarbene are so reactive that they attack carbon-carbon double bonds in CNT that connect two adjacent six-membered rings to yield a corresponding 1,1-dichlorocyclopropane structure on both the sidewalls of SWCNT and MWCNT – (Figure 14).[54a,136]

1,1-Dichlorocarbene has been generated from different sources, including NaOH/CHCl$_3$, mercury complexes and dipyridyl imidazolidenes.[137] The presence of chloride on the side chains was manifested in energy dispersive X-ray diffraction measurements. An estimate suggested an overall chloride yield of about 2%.[138] Moreover, in the absorption spectrum the far infrared bands are almost completely removed. This finding is indicative for a transformation from metallic to semiconducting CNT.[136b] A more recently tested carbene addition is the Bingel reaction (Figure 15). It is based on diethyl bromomalonate as carbene precursor materials and takes place under more moderate experimental conditions.[139]

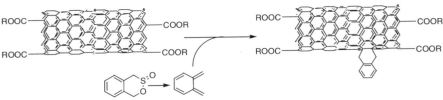

**Figure 13** *Diels–Alder cycloaddition on the SWNT sidewall*

**Figure 14** Cycloaddition of dichlorocarbene

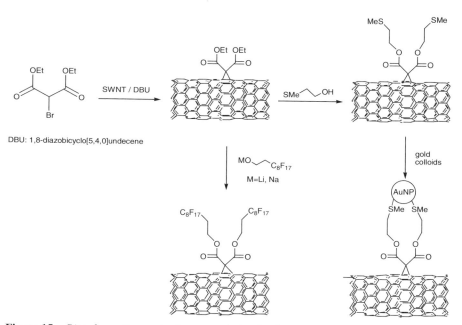

**Figure 15** Bingel reaction and subsequent derivatization

Subsequent reaction of the Bingel product of SWCNT with an excess of 2-(methylthio)ethanol and in the presence of 5 nm gold nanocrystals gave a new product that was microscopically discernable in AFM measurements. Alternatively, the Bingel-SWCNT were transesterificated with lithium or sodium salts of $1H,1H,2H,2H$-perfluorodecan-1-ol. These materials were characterized by $^{19}F$ NMR and XPS.

Nitrenes, as shown in Figure 16, in form of different $(R)$-oxycarbonylnitrene compounds also gives rise to [2+1] cycloaddition with CNT. The different $(R)$-oxycarbonylnitrenes were synthesized *in situ* starting with the thermal decomposition of the corresponding $(R)$-azido-formate precursors. Since the reaction proceeds mainly with the sidewalls it affords $(R)$-oxcarbonylaziridino-SWCNT with increased solubility in organic solvents. This opens the way for purifying SWCNT from insoluble contaminants.[137a,140]

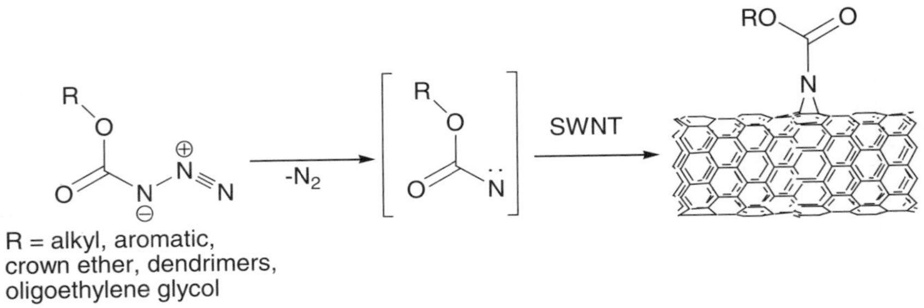

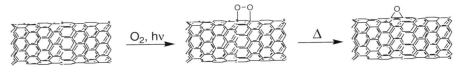

R = alkyl, aromatic, crown ether, dendrimers, oligoethylene glycol

**Figure 16**  *[2 + 1] cycloaddition of nitrene*

**Figure 17**  *Reversible [2 + 2] O$_2$ cycloaddition*

### 11.3.7  [2+2] Cycloaddition of Singlet O$_2$ and Sidewall Oxidation

When considering photoinduced reactions in the presence of CNT the following reactivity is of great concern. O$_2$, in its triplet ground state, physisorbs well to the sidewall of CNT. In the process semiconducting CNT are converted to conducting CNT. Upon photoexcitation physisorbed O$_2$ desorbs from the CNT sidewalls.[141] On the other hand, the much more reactive O$_2$ singlet – generated, for example, by UV photoexcitation – affords dioxetane through [2+2] cycloaddition (*i.e.*, Figure 17). The course of the reaction is accompanied by an appreciable luminescence quenching. Upon thermal activation, the [2+2] adducts readily undergo O–O bond cleavage. Surface bound epoxide species are identified as the product of the O–O bond cleavage.[142]

### 11.3.8  Reductive Hydrogenation/Alkylation/Arylation of CNT Sidewall

Birch reductions[143] of SWCNT and MWCNT using metal Li and methanol dissolved in liquid ammonia were conducted for the sidewall hydrogenation of CNT (see Figure 18).[144] The hydrogenated derivatives are thermally stable up to 400°C. Above 400°C, however, a characteristic decomposition takes place accompanied with the simultaneous formations of hydrogen and a small amount of methane. Transmission electron micrographs show corrugation and disorder of the nanotube walls and the graphite layers due to hydrogenation. As a practicable alternative to the Birch reduction the reaction with atomic hydrogen (*i.e.*, generated in a cold plasma) can be considered. This also causes the hydrogenation of sidewall CNT.[145]

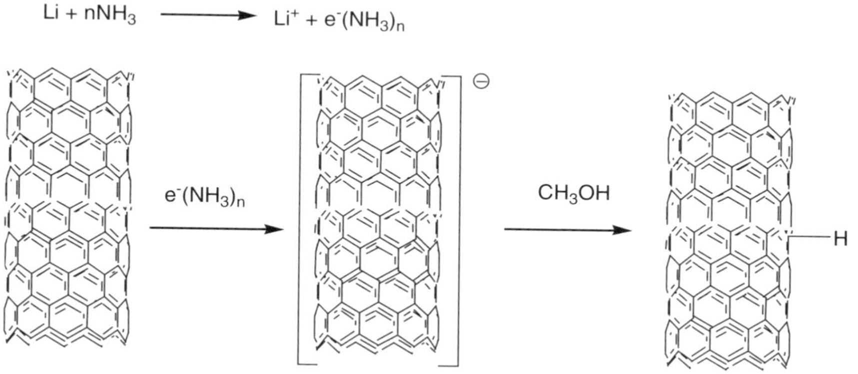

Li + nNH₃ ⟶ Li⁺ + e⁻(NH₃)ₙ

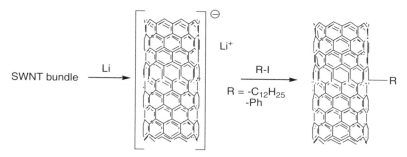

**Figure 18** *Birch reduction of SWNT*

**Figure 19** *Nucleophilic alkylation of SWNT*

Recently SWCNT were reduced by lithium, sodium or potassium in liquid ammonia[146] (Figure 19) and the resulting nanotube salts were treated with alkyl iodides to afford, for instance, dodecylated SWCNT, which are quite soluble in organic solvents. Microscopic studies of dodecylated SWCNT prepared from HiPco nanotubes and 1-iodododecane show that extensive debundling results from intercalation of the alkali metal into the SWCNT ropes. TGA-FTIR analyses of samples prepared from the different metals revealed radically different thermal behavior during detachment of the dodecyl groups. The SWCNT prepared using lithium can be converted into the pristine SWCNT at 180–330°C, whereas the dodecylated SWCNT prepared using sodium require much higher temperatures (*i.e.*, 380–530°C) for dealkylation. In stark contrast, SWCNT prepared using potassium behave differently, leading to detachment of the alkyl groups over a temperature range of 180–500°C.[147]

In analogy to the chemistry of fullerenes[148] CNT sidewalls react with organolithium compounds giving SWCNT anions as key intermediates for the subsequent alkylations and/or arylations. For example, the *in situ* composite synthesis by attachment of polystyrene (PS) chains to full-length SWCNT without disrupting the original structure, is based on an established anionic polymerization scheme.[149] Carbanions of SWCNT react as well with carbon

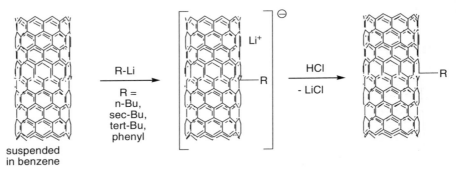

**Figure 20**  *Functionalization of CNT with organolithium compounds*

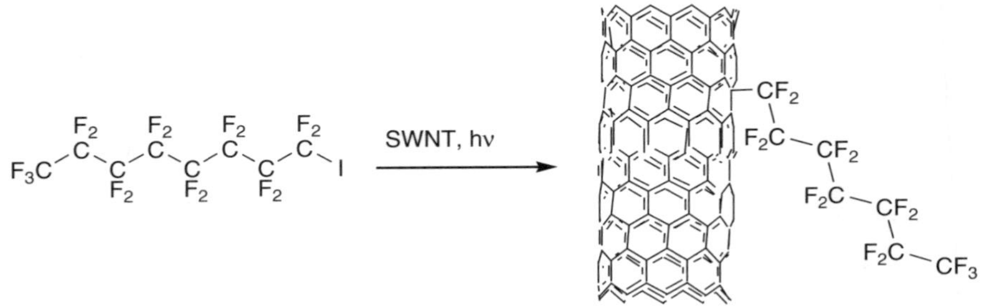

**Figure 21**  *Photoactivated alkylation of the sidewall of SWNT*

dioxide (*i.e.*, electrophile) leading to SWCNT functionalized with both alkyl and carboxyl groups (Figure 20).[150]

## 11.3.9 Addition of Radicals

Photolysis of heptadecafluorooctyl iodide generates reactive alkyl radicals. These alkyl radicals are the start of a straightforward alkylation process with, for example, the sidewalls of CNT. Figure 21 illustrates a representative case for such a reaction.[137a,151]

An alternative concept toward radical addition to the CNT sidewalls involves γ-radiolysis in ethanol.[152] Conceptually similar is the thermolysis of benzoyl peroxides, which leads to the generation of free phenyl radicals. These phenyl radicals are sufficiently potent to react with alkyl iodides and to produce free alkyl radicals. Finally, the precursors alkylate SWCNT.[153] Following a comparable reaction mechanism, grafting of 1-vinylimidazole onto SWCNT occurred in an argon plasma-assisted ultraviolet radiation.[154]

In alternative approaches, the necessary radicals are generated either electrochemically[155] or thermally.[156] Figures 22 and 23 schematize the radical-induced arylation of SWCNT. Hereby it is important to note that the arylation proceeds preferably with CNT that have smaller diameter[15] and are metallic (Figure24).[4a,156]

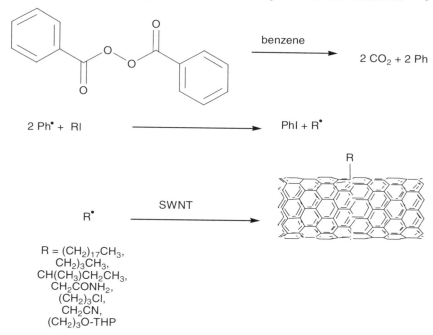

**Figure 22** *Thermoactivated alkylation of the sidewall of SWNT*

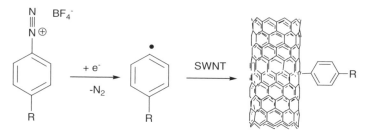

**Figure 23** *Electrochemical arylation of SWNT*

Thermal dearylation of these derivatized SWCNT materials also occurs at around ~500°C in an argon atmosphere.

Aryl diazonium salts, which were employed either as ex-situ prepared[155a] or as *in situ* generated intermediates,[156,157] can react with CNT also in ionic liquids[158] or even without solvent.[157g]

## 11.3.10 Replacements of Carbon Atoms – Chemical Doping

Alternative ways to functionalize CNT are substitution reactions, involving replacement of carbon atoms by boron or nitrogen. Such reactions are typically conducted by thermal treatment with a mixture of boron trioxide and in a nitrogen

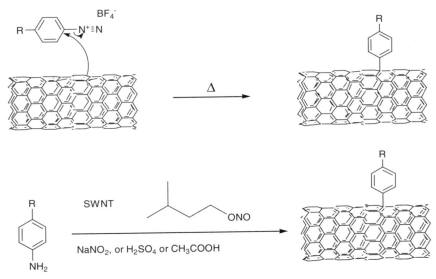

**Figure 24** *Thermal arylation of SWNT*

flow at 1523–1623 K.[159] Alternative reaction conditions involve exposing CNT to a microwave-generated $N_2$ plasma.[160] The Fermi level shift, which is induced by the dopants, governs the conductivity features, that is, p- or n-conductivity.[161] Interestingly, doping of a semiconducting SWCNT is n-type for half of its length and p-type for the other half. In its entirety an intramolecular wire is formed that exhibit rectifying characteristics.[162] SWCNT bulk materials are also reversibly doped through a range of different redox reactions,[163] initiated, for instance, in solutions of organic radical anions and $Li^+$ as counterions.[164]

### 11.3.11 Mechanochemical Functionalization

The use of ball-milling to CNT (*i.e.*, specific atmospheres include $H_2S$, $NH_3$, $Cl_2$, CO, $CH_3SH$, $COCl_2$, *etc.*) resulted in shorter tubes. Such ball-milling proved furthermore to be beneficial in the introduction of functional groups, such as thiol, amine, amide, carbonyl and chlorine onto the CNT sidewall.[165] Upon milling SWCNT with potassium hydroxide the presence of hydroxyl groups on the CNT surface were attested by spectroscopic means.[166] Fullerenes were also attached with the same technique.[167]

## 11.4 Non-Covalent Functionalization of Carbon Nanotubes

In contrast to the preceding section, which involves linking functional groups directly to the surface or to the defect sites of CNT, the following section covers non-covalent modifications, which primarily involve utilization of surfactants, oligomers, biomolecules, and polymers together with CNT. The advantage of most non-covalent approaches is that this method does not destroy or alter the intrinsic electronic structures of CNT – instead it guarantees structurally intact CNT, to which additional functionalities have been added.[2c,124a,168]

### 11.4.1 π-π Interactions with π-Electron Rich Polymeric Blocks – Polymer Wrapping

The optical and electronic properties of CNT polymer mixed composites have attracted much attention due to potential communications between the highly delocalized π-electrons of CNT and the π-electrons correlated with the lattice of the polymer skeleton. CNT that, for example, are doped with conducting polymers might – among many novel features – exhibit significantly improved solubilities. Note in this respect the impressing progress in the field of fullerene-doped conducting polymer systems.[169]

Poly(*m*-phenylenevinylene-*co*-2,5-dioctoxy-*p*-phenylene vinylene) (PmPV-*co*-DOctOPV), whose structure is a simple variation of the more common PPV, were first shown to form stable composites with MWCNT through simple π–π interactions.[170] MWCNT doping of such polymers exerts a notable impact on a broad variety of properties: for instance, the electric conductivity increases somewhere between eight and ten orders of magnitudes. A likely rationale involves the introduction of conducting paths within the polymer. Although a noteworthy heat generation occurs, MWCNT act as nanometric heat sinks. When applying the resultant MWCNT/PmPV-*co*-DOctOPV as emissive layers in a device remarkable electroluminescence was achieved. Comparing the device (*i.e.*, organic light-emitting diode) lifetime of the composite with that of the pure polymer, an improvement (*i.e.*, factor of 5) was noted. Raman spectroscopy was employed to test the purity of MWCNT/PmPV-*co*-DOctOPV, indicating that MWCNT/PmPV-*co*-DOctOPV acts as a purification filter. In particular, composite formation assists in separating the amorphous materials from the CNT surfaces.

Homogeneous nanocomposites of poly(phenyleneethynylene) (PPE)-SWCNT/polystyrene and PPE-SWCNT/polycarbonate that were non-covalently functionalized revealed dramatic improvements in the electric conductivity with very low percolation thresholds (*i.e.*, 0.05–0.1 wt% SWCNT loading).[171] Similarly, fairly strong contacts between SWCNT and PmPV-*co*-DOctOPV were attested through a variety of microscopic and spectroscopic techniques. The polymer undergoes structural wrapping around the SWCNT lattice during the composite formation. The polymer coating is thirty times the diameter of SWCNT, suggesting promising applications as novel composite materials in mechanical reinforcement and thermal stabilization.[172] Given the promises that rest on solubilzing SWCNT with PmPV-*co*-DOctOPV polymer molecular dynamics simulation were carried out.[173] The results demonstrated strong binding interactions between SWCNT and PmPV-*co*-DOctOPV. This work also indicated that sidewall functionalized composites are stable and could be used to separate individual SWCNT from bigger bundles. Among the different conformations of PmPV-co-DOctOPV, the flat conformer was found to be the most stable one for sidewall functionalization.

Star et al.[174] demonstrated that poly{(m-phenylenevinylene)-co-[2,5-dioctoxy-p-phenylene)-vinylene]} (PmPV) help to solubilize SWCNT in organic solvents. Both absorption and $^1$H NMR spectra of SWCNT/PmPV are significantly broadened, due to π–π forces between SWCNT and PmPV. The average diameters of aggregated SWCNT are ca. 7.1 nm – as discernable from AFM measurements. This indicates that non-covalent modification is very effective in disrupting the inherently strong van der Waals interactions that cause them to aggregate into bundles. An interesting observation is that when higher polymer concentrations are used individual SWCNT are exfoliated from the bundles. The optoelectronic properties, namely, a photoamplification of nearly thousand, indicate that the polymer must be in intimate electrical contact with the SWCNT.

Another fascinating class of polymers is poly{(2,6- phenylenevinylene)-co-[2,5-dioctoxy-p-phenylene)-vinylene]} (PPyPV).[175] PpyPV, similar to PmPV, promotes the SWCNT solubilization in chloroform. In contrast to PmPV, PpyPV is a base, which readily undergoes protonation by HCl. In the case of SWCNT/PpyPV the electronic interaction is at the single structure level, involving the protonated form of the polymer, while PmPV interacts with SWCNT under neutral charge condition.

Similarly, the conducting poly{(5-alkoxy-m-phenylenevinylene)-co-[2,5-dioctoxy-p-phenylene)-vinylene]} (PAmPV) functionalize SWCNT leading to the intriguing prospects of developing future arrays of molecular actuators and switches (Figure 25).[176]

Among the many conducting polymers, polypyrrole (PPY) stands out as material with high promises in terms of commercial applications. Notable properties include high conductivity, air stability and ease of preparation. Nanotubular CNT/PPY material were prepared with MWCNT using in-situ polymerization.[177] The diameters of CNT/PPY start at 80 nm and reach up to 100 nm, depending primarily on the synthetic conditions. The resulting CNT/PPY were characterized by elemental analysis, XPS, Raman spectra, and X-ray diffraction. While no meaningful chemical interactions were noted to arise between CNT and PPY, the conducting PPY modified the physical properties of CNT (i.e., electrical, magnetic and thermal properties).

In the presence of MWCNT, catalytically polymerized poly(phenylacetylene) PPA, afford MWCNT/PPA with helically wrapped PPA chains.[178] These composite materials are soluble in organic solvents like: THF, toluene, chloroform, 1,4-dioxane etc. After applying mechanical forces to CNT/PPA, the composites readily align to the direction of the applied force. Efficient optical limiting property was observed in these nanocomposite materials.

Enhanced electrical properties were obtained when interacting MWCNT with polyaniline (PANI).[179] Raman and transport measurements suggest selective interactions between the quinoid rings of PANI and MWCNT

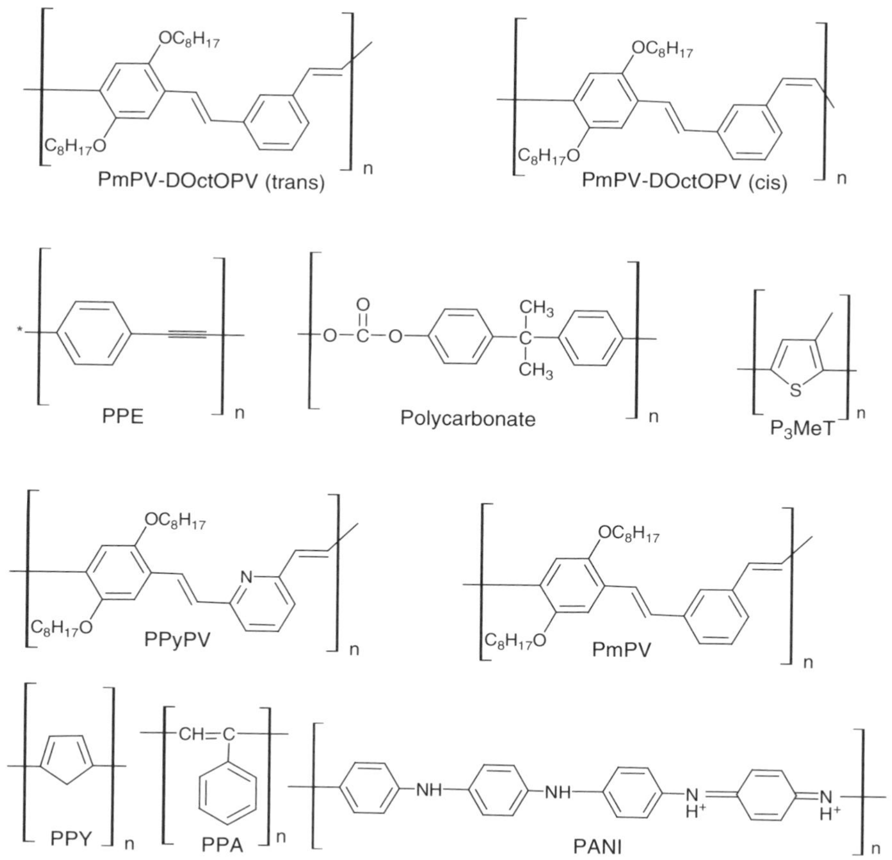

**Figure 25** *Polymers that non-covalently functionalize CNT*

materializing in charge-transfer features between the two components. These facts rationalize that the electrical conductivity of MWCNT/PANI increased by an order of magnitude relative to neat PANI. Doping effects of MWCNT are also considered.

When applying electrochemical polymerization conditions, MWCNT were simultaneously co-deposited with PPY and poly(3-methylthiophene) (P3MeT) to yield nanoporous composite films.[180] In the absence of any supporting electrolytes, increasing the MWCNT concentration at the polymerization electrode decreases the thickness of polymer coating on each MWCNT. Obviously, it is feasible to minimize the ionic diffusion distances and, thereby, to improve the electrolyte access within the nanoporous MWCNT/PPY films. This eventually amplifies the capacitance relative to similarly prepared PPY films (Figure 25).

In addition, a good deal of research has been performed with non-conducting polymers. Effective dispersibility, strong interfacial binding, and good alignment of CNT are common features of these CNT/polymer. Impressive solubilizations of SWCNT have been achieved in a new non-wrapping approach to non-covalent engineering of SWCNT surfaces by short, rigid functional conjugated polymers, poly(aryleneethynylene). The technique not only enables the dissolution of various types of carbon nanotubes in organic solvents, which represents the first example of solubilization of SWCNT *via* π-stacking without polymer wrapping, but could also introduce numerous neutral and ionic functional groups onto SWCNT surfaces.[181] Amphiphilic *m*-linked rigid polymers based on oligo(ethylene oxide) chains self-assemble into a lamellar structure based on an unfolded conformation. In contrast, polymers based on bulky dendritic chains adopt helical conformations that self-organize into a tetragonal columnar structure. Interestingly, the polymer based on a dendritic side group can encapsulate SWCNT by wrapping with helical conformation. The results suggest that this class of self-assembling system may allow the design of well-defined helical tubules with desired functions.[182]

It is essential to solubilize CNT in aqueous media to improve their biocompatibility and enabling their environmental friendly characterization, separation and self-assembly. To exploit the unique properties of CNT into biological relevant system, SWCNT were suspended in water soluble amylose in the presence of iodine.[183] Very likely, iodine plays a key role for the initial preorganization of amylose in support of a helical conformation. The resulting SWCNT/starch were investigated by AFM, which discloses small bundles that are covered with amorphous polysaccharide. Since this process is reversible at high temperatures, it opens the exciting opportunity to employ this strategy to separate SWCNT from amorphous carbon.

Linear polymers that bear polar octyloxy alkyl side-chains – such as polyvinyl pyrrolidine (PVP) and polystyrene sulfonate (PSS) (Figure 26) – also form stable composite materials with CNT.[184] For this complex formation to succeed, the thermodynamic force have to disrupt both the hydrophobic interfaces with water and CNT interactions in the aggregates. Nevertheless, the aqueous environment is very effective for the precise manipulation, purification and fabrication of CNT.

The composite material of MWCNT/PPV was prepared and the electronic exchange between PPV and MWCNT were studied based on absorption,

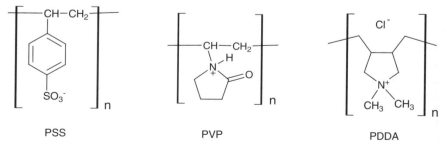

**Figure 26** *Polyelectrolytes that non-covalently functionalize CNT*

photoluminescence, and photoinduced absorption spectroscopy.[185] PPV strongly interacts with MWCNT in the photoexcited state, while no significant interactions were observed in the ground state. Drastically quenched signals in photoluminescence and photoinduced absorption spectroscopy for the MWCNT/PPV composite are regarded as evidence for some electronic communication between PPV and MWCNT. Actually, the noted phenomenon implies an energy transfer scenario coupled to a potential hole transfer from PPV to MWCNT. Overall, the findings from these studies suggest a possible application of CNT as new electrode material in macroscopic devices.

Various approaches have been developed to improve the solubility of CNT in different polymer matrices. The addition of 1 wt% non-ionic surfactant improves, for instant, the glass transition temperature. Moreover, the elastic modulus increased by more than 30% relative to the scenario in the absence of a surfactant.[186] In the presence of 0.5 wt% poly(vinylene fluoride) (PVDF) the storage modulus of MWCNT/poly(methyl methacrylate) composite was significantly improved at low temperatures.[187]

MWCNT have been non-covalently functionalized by pyrene-linked polymers (1-pyrene)methyl 2-methyl-2-propenoate (PyMMP) and poly(methyl methacrylate-co-PyMMP).[188] The MWCNT/ PyMMP were characterized by TGA, TEM and AFM and reveal that the grafted polymer layer is homogeniously deposited onto the MWCNT surface. Depending on the PyMMP the composites were dispersible either in common organic solvents or in aqueous media. Taking advantage of the large aromatic pyrene moieties in poly(pyrenebutyric acid) (PPBA), interactions with SWCNT sidewalls helps to disperse them in common organic solvents like chloroform, THF, $N,N$-dimethylformamide, dimethyl sulfoxide and alkaline aqueous media.[189] Much stronger fluorescence intensity was observed for SWCNT/PPBA, when compared with the α-pyrenebutyric acid monomer. Such a flexible methodology, involving a non-covalent modification of SWCNT via adsorbed pyrene-linked polymers, have been tested as initiators for ring-opening metathesis polymerization with CNT.[190]

Recently, we succeeded in the facile supramolecular association of pristine SWCNT with linearly polymerized porphyrin polymers.[191] The target SWCNT nanohybrids, which are dispersable in organic media, were realized through the usage of soluble and redox-inert poly(methylmethacrylate) (PMMA) bearing surface immobilized porphyrins (i.e., $H_2P$-polymer). Conclusive evidence for SWCNT/$H_2P$-polymer interactions came from absorption spectroscopy: The fingerprints of SWCNT and $H_2P$-polymer are discernable throughout the UV, VIS, and NIR part of the spectrum. A similar conclusion was also derived from TEM and AFM. AFM images illustrate, for example, the debundling of individual SWCNT. An additional feature of SWCNT/$H_2P$-polymer is an intrahybrid charge separation, which has been shown to last for $2.1 \pm 0.1$ μs.

Stable SWCNT/$H_2P$ composites were also realized by condensing tetraformylporphyrins and diaminopyrenes on SWCNT.[192] The degree of interaction between SWCNT and $H_2P$ was evaluated by UV-Vis and fluorescence spectra, and chemical removal of $H_2P$ from SWCNT. In SWCNT/$H_2P$, the

Soret and Q-bands of $H_2P$ moieties were significantly broadened and their fluorescence was almost completely quenched (Figure 27).

Donor–acceptor interactions dominate between amines and CNT. Such charge-transfer interactions are the reason why CNT interact with PEI, poly(allylamine) and poly(4-vinylpyridine).[193]

Water soluble quaternary ammonium polyelectrolytes, poly(diallyldimethyl ammonium) chlorides (PDDA) (Figure 26) are useful for water treatment, paper industry, mining industry as well as biological applications. Chemically oxidized shortened MWCNT have been self-aligned on polyelectrolyte layers, where positively charged PDDA bound to the oppositely charged carboxylate anions of MWCNT.[194] In contrast, electrostatic repulsions govern possible interactions between poly(sodium 4-styrenesulfonate) (PSS) and oxidized MWCNT and no layer deposition was observed. The shortened MWCNT are stable in aqueous media, because of the electrostatic charges that were introduced onto the MWCNT surface.

Very recently water-soluble MWCNT/PDDA were also obtained from unoxidized MWCNT, where mild sonication improved the dispersibility and enhance the adhesion property significantly.[195] The characterization of MWCNT/PDDA was carried out using XPS and photoacoustic Fourier transform infrared (PA-FTIR) spectroscopies. In this context, it has been suggested that the interaction between PDDA and MWCNT is not electrostatic in nature, but the presence of unsaturated impurities in the PDDA chain drives π–π interactions. Thus, CNT coated with PDDA function as positively charged

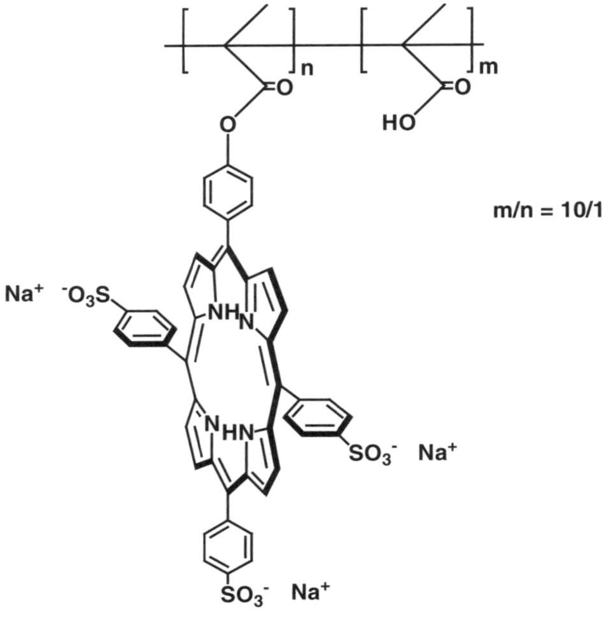

**Figure 27** *Partial structure of $H_2P$-polymer*

polyelectrolytes, which exhibit electrostatic repulsion between CNT/PDDA and likewise guarantees CNT hydrophilicity.

Such CNT/PDDA and CNT/PSS are important building blocks for the LBL approach, that is, the electrostatic integration of SWCNT into novel CNT films. Hereby, a film thickness that ranges between a few nanometers and tens or hundreds of nanometer has been achieved. Implicit is that these films contain several tens or hundreds of nanotube layers. A variety of substrates were assembled onto SWCNT by the LBL technique and characterized with vis-near-IR, AFM, and imaging ellipsometry techniques.[196] Alternatively to CNT/PDDA and CNT/PSS amphiphilic pyrene derivatives (*i.e.*, bearing either positively or negatively charged headgroups) create a charged template and facilitate anchoring oppositely charged polymer deposition. The surface of the polymer layer provides a robust and synthetically flexible platform for numerous possibilities in material science, biophysics, and nanotechnology.

### 11.4.2 π-π Interactions with π-Electron Rich Molecular Building Blocks

Due to the highly aromatic nature of polynuclear aromatic hydrocarbons – pyrene molecules – they strongly interact with the basal plane of graphite as well as with the sidewalls of SWCNT in a similar manner *via* π-π stacking.[197] Dai and co-workers have demonstrated that pyrene derivatives served indeed as anchors for selectively immobilizing various biomolecules onto SWCNT surfaces.[198] The immobilization of proteins involves the highly reactive nucleophilic substitution of N-hydroxysuccinimide by amides group of the protein structure, while the pyrene molecules bind to the inherently hydrophobic CNT surface in organic solvents (*i.e.*, dimethylformamid or methanol). The process was achieved by the deposition of pyrene derivatives onto the surface of SWCNT, followed by the incubation of proteins. The TEM and AFM images confirmed the dense immobilization of proteins on the surface of individual and bundles of SWCNT. The role of pyrene derivatives for anchoring process was observed by control experiments, where without pretreatment with pyrene derivatives, the protein molecules were not adsorbed onto the surface of SWCNT (Figure 28).

A key requirement to test biological applications and/or biophysical processing is, however, the unlimited dispersability of CNT in aqueous media. With that objective in mind, water-soluble SWCNT and MWCNT were obtained in aqueous solutions of 1-(trimethylammonium acetyl) pyrene (pyrene$^+$) or 1-pyreneacetic acid, 1-pyrenecarboxylic acid, 1-pyrenebutyric acid, 8-hydroxy-1,3,6-pyrenetrisulfonic acid (pyrene$^-$).[10,199] Important is that excess free pyrene was removed from solution using vigorous centrifugation and the final CNT/pyrene solids were resuspended in water. The solubility of CNT in the resulting black suspensions is as high as 0.2 mg/ml and the stability reaches several months under ambient conditions without showing any apparent precipitations. TEM and AFM images revealed the coexistence of individual and bundles of CNT. Now that the surface of CNT is covered with positively or

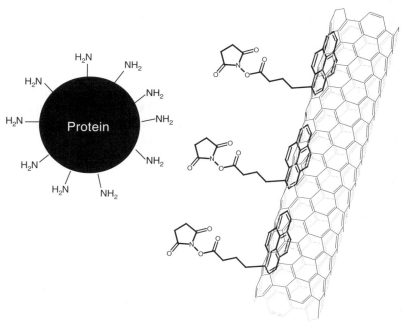

**Figure 28** *Immobilization of proteins on the surface of CNT*

negatively charged ionic head groups, van der Waals and electrostatic interactions are utilized to complex oppositely charged electron-donors. Water-soluble porphyrins (*i.e.*, octapyridinium $ZnP/H_2P$ salts or octacarboxylate $ZnP/H_2P$ salts) emerged as ideal candidates to realize novel *electron-donor–acceptor* nanohybrids, which were characterized by spectroscopic (absorption and fluorescence) and microscopic (TEM and AFM) means. In absorption experiments, the successful complex formation – for instance, of CNT and $ZnP^{8-}$ – were confirmed by red-shifted Soret- and Q-bands and the development of a series of isosbestic points. Photoexcitation of all the resulting *electron-donor–acceptor* nanohybrids with visible light causes reduction of the electron accepting CNT and oxidation of the electron donating ZnP or $H_2P$. In fact, long-lived radical ion pairs – with lifetimes that are in the range of microseconds – were confirmed by transient absorption measurements. Interesting is the fact that a better delocalization of electrons in MWCNT helps to significantly enhance the stability of the radical ion pair (5.8 ± 0.2 microseconds) relative to the analogous SWCNT (0.4 ± 0.05 microseconds). Percolation of the charge inside the concentric wires in MWCNT decelerates the decay dynamics associated with charge recombination (Figure 29).

The pyrene approach toward functional nanohybrids does not form any covalent bonds but only π-π interactions, and perturbs the CNT conjugated system weakly. Pyrene derivatives with a large variety of functional groups can be easily prepared, so that the approach is very general and easy to exploit (Figure 30).

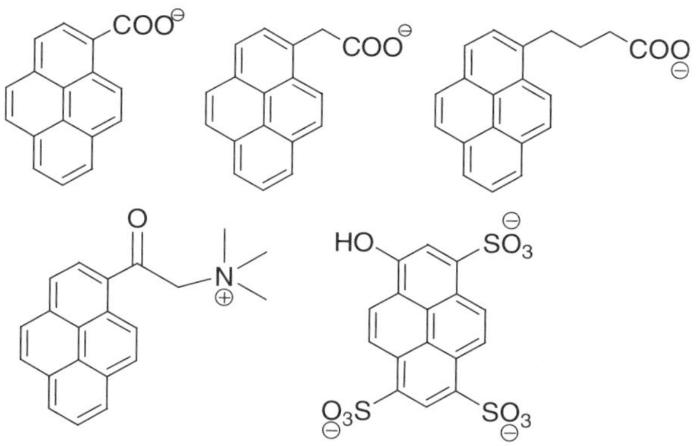

**Figure 29** *Pyrene derivatives that non covalently functionalize CNT*

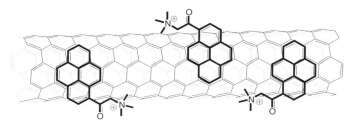

**Figure 30** *π–π interaction between SWNT and pyrene*

Following the same non-covalent protocol the integration of gold nanoparticles was accomplished onto the CNT surface.[200] Significant quenching of the luminescence upon binding gold nanoparticles was observed. The results imply that either energy transfer or charge transfer evolve from the photoexcited pyrene to the gold nanoparticles through the interlinkers.

Methylene blue (MB), a polynuclear aromatic electroactive dye, form electrochemically functionalized SWCNT through π-π stacking interaction between the aromatic compond and the graphitic surface of CNT.[201] MB and SWCNT showed not only hydrophobic interactions. UV-vis, IR spectroscopy, and electrochemical studies suggested also the existence of charge transfer interactions – SWCNT act as electron-donor while MB act as acceptor.

An impressive SWCNT solubility has been achieved upon treatment with zinc [5,10,15,20-tetraphenyl-21$H$,23$H$-porphine] in DMF.[202] TEM and AFM images revealed the existence of high density SWCNT and exfoliation of individual SWCNT as a result of the tedious work-up procedure. The CNT diameter ranged typically from 0.9 to 1.5 nm, underlining the successful debundling properties of ZnP. Evidence for interactions between porphyrins and SWCNT was obtained from fluorescence spectra, where the porphyrin fluorescence was significantly quenched compared to the bare porphyrin. The fluorescence quenching has been ascribed to energy transfer between the photoexcited porphyrin and SWCNT.

In a similar fashion SWCNT were non-covalently functionalized with porphyrins (*i.e.*, H$_2$P and ZnP) and electronic interactions between CNT and porphyrin have been demonstrated.[203] A series of microscopic and spectroscopic investigations have been carried out to characterize SWCNT/ZnP and SWCNT/H$_2$P. The absorption and fluorescence spectra of SWCNT/H$_2$P showed efficient electronic communication between the π-systems of SWCNT and H$_2$P. While weaker interactions were observed for SWCNT/ZnP, which might be caused from the interference between the zinc metal and the surface of SWCNT. Clear evidence for the π-π interactions between SWCNT and H$_2$P were obtained from AFM images. The AFM images of SWCNT/H$_2$P and of SWCNT/ZnP showed smaller bundles and some individual nanotubes (diameter 1.5 ± 0.2 nm) are also possible to see, while SWCNT/ZnP showed more aggregates. And the suspensions of SWCNT/H$_2$P are more stable than that of SWCNT/ZnP (Figure 31).

Water-soluble H$_2$P [meso-(tetrakis-4-sulfonatophenyl) porphines dihydrochloride] were employed to suspend SWCNT in aqueous solutions, which were shown to be stable for several weeks.[204] The resulting materials have been characterized by absorption and fluorescence spectroscopy, further

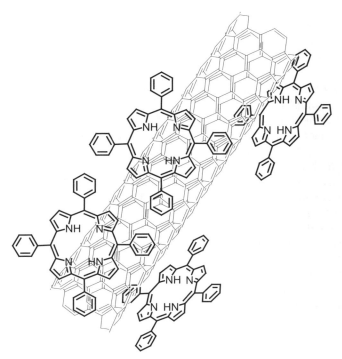

**Figure 31** *π–π interaction between SWNT and porphyrin*

complemented by AFM. H$_2$P and SWCNT interact selectively and the interaction stabilizes the free base against protonation to the diacid. Under mildly acidic conditions SWCNT mediated J-aggregates are formed, which are unstable in solution and result in precipitation of SWCNT over the course of a few days. H$_2$P coated SWCNT can be precisely aligned on hydrophilic poly(dimethylsiloxane) surfaces by combing SWCNT solution along a desired direction and then transferred to silicon substrates by stamping.

Pyrene derivatives strongly interact with the sidewalls of SWCNT, which facilitates their dispersion in aqueous media.[205] The presence of pyrene derivatives on the SWCNT surface might affect the S$_{11}$ band, similar to recent work, which shows that the surface modification *via* covalent functionalization completely eliminates the semiconducting transitions in the absorption spectrum (*i.e.*, S$_{11}$ and S$_{22}$).[2e] But markedly, the non-covalent functionalization is thought to preserve the CNT electronic properties – as evidence for that the van Hove singularities of CNT should be viewed. A notable exception was observed upon treating SWCNT with 1-docosyloxymethyl-pyrene (DomP). This work-up resulted in a diminishing of the semiconducting bands in the near-IR region.[206] The successful removal of DomP from carbon nanotubes could be achieved when polyvinylidene fluoride (PVDF) was added. Alternatively, refluxing and repeated washing cycles with cyclohexane lead to the same removal. The recovered SWCNT exhibited once again the characteristic S$_{11}$ and S$_{22}$ bands. The mechanism is not clear yet, but it was suggested that after complexation the surface of SWCNT is non-covalently covered by a layer of pyrene moieties and might be viewed as a highly defected double-walled carbon nanotubes (DWCNT).

## 11.5 Separation of Metallic and Semiconducting CNT

Although CNT have become the subject of intense research activities around the globe, still many intriguing challenges lie ahead of us. Particularly critical for basic and applied research on SWCNT is the diversity of tube diameter and chiral angles. On the basis of their electronic structures CNT are classified into two categories: metallic and semiconducting tubes. Semiconducting CNT are further classified by their diameter. Notable, for the application in nanoelectronics, the band gap of semiconducting tube is a critical parameter, which needs to be controlled. All the synthetic methods for generating CNT (*i.e.*, carbon arc, laser ablation, and chemical vapor deposition) yield them in a range of diameter and chiral angles. Consequently, wide mixtures of metallic and semiconducting CNT are present. This, however, hinders the applications of CNT as versatile building blocks for molecular electronics because semiconducting and metallic tubes have different function in electronic devices. Semiconducting SWCNT, for example, exhibit a significant electronic response to field gating effects and chemical doping effects. These characteristics are essential for FET and chemical sensors. Metallic SWCNT, on the other hand, are needed as nanometer-sized conductors. Targeting their versatile applications in nanotechnology many researchers have focused on the selective purification of metallic and semiconductive SWCNT.

The separation of metallic and semiconducting SWCNT has been achieved through wrapping them with single-stranded DNA that involves anion exchange chromatography.[207] Here the specific DNA sequence, which has been shown to be diameter dependent, enables the separation of CNT – the larger diameters elute later than the smaller diameters. This conclusion was derived from a systematic shifting of the Raman tangential mode with increasing elution time as well as changes in the relative intensities of the radial breathing modes. Anion exchange chromatography, as a CNT separation technique, bears, however, the disadvantage that it is expensive and requires removal of DNA from SWCNT. Spectrofluorimetry was used to examine the detailed composition of the bulk SWCNT samples, providing distributions in both tube diameter and chiral angle. But also the direct characterization of semiconducting components in a mixture can be achieved by measuring the optical band gap.[208]

Krupke et al., presented a method to separate metallic SWCNTs from their semiconducting counterpart using alternating current dielectrophoresis.[209] Applying different relative dielectric constants leads to opposite movement of metallic and semiconducting tubes. More precisely, the metallic tubes are attracted toward a microelectrode array. The effectiveness of this technique was proven by using a comparative Raman spectroscopic assay on the dielectrophoretically deposited tubes with a reference sample. The enrichment of metallic tubes amounted to up to 80%. Fundamental disadvantages of this method are that it completely depends on the amount of individual SWCNT and that the efficiency decreases with increasing number of bundles.

SWCNT suspensions in the presence of $H_2P$ – [5,10,15,20-tetrakis(hexadecyloxyphenyl)-$21H,23H$-porphine] – in organic solvents yielded novel soluble SWCNT/$H_2P$.[210] The insoluble and recovered SWCNT were separated from $H_2P$ by treatment with acetic acid and vigorous centrifugation. After heating the recovered SWCNT and the free SWCNT to 800°C in a nitrogen atmosphere, the spectroscopic analysis showed that the semiconducting SWCNT and the free SWCNT are enriched in recovered and metallic SWCNT, respectively. Under ambient conditions, the bulk conductivity of the semiconducting SWCNT (i.e., recovered) is 0.007 S cm$^{-1}$, while that of the metallic SWCNT (i.e., free) is 1.1 S cm$^{-1}$. Thus, selective interactions between $H_2P$ and semiconducting SWCNT are the inception to a successful separation of metallic and semiconducting nanotubes.

In recent work a remarkable affinity of the SWCNT sidewalls toward the physisorption of amines has been demonstrated.[211] The electronic properties, like electrical conductance of semiconducting SWCNT are significantly altered upon adsorption of linear alkylamines due to charge transfer. Interestingly, the conducting properties of the metallic SWCNT remained insensitive after adsorption of a variety of adsorbates (i.e., amines, $NH_3$, $NO_2$, etc.).

Substantial separation of SWCNT according to type, that is, metallic versus semiconducting, has been achieved for HiPco and laser-ablated SWCNT.[4b]

Stable dispersions of SWCNT with octadecylamine in THF originate from the physisorption and organization of octadecylamine along the SWCNT sidewalls in addition to the originally proposed zwitterion model. Furthermore, the reported affinity of amine groups for semiconducting SWCNT, as opposed to their metallic counterparts, contributes additional stability to the physisorbed octadecylamine. This provides a venue for the selective precipitation of metallic SWCNT upon increasing dispersion concentration, as indicated by Raman investigations.

In extension of earlier work, the bulk-scale separation of metallic and semiconducting SWCNTs has been reported – starting from as-prepared SWCNT rather than from carboxy-functionalized SWCNT.[212] More precisely, SWCNT were reacted with 1-octylamine, propylamine, and isopropylamine in THF and THF/hexane mixture. A full-fledged experimental characterization involved absorption spectroscopy, Raman spectroscopy, SEM, AFM, I-V characterization. Futhermore, local density functional theory and ultrasoft pseudopotential plane-wave method were employed to correlate the experimental observations. The theoretical calculation suggested that the amine groups have a greater tendency to adsorb to metallic SWCNT than to semiconducting SWCNT. The mode of interactions is, however, based on hydrogen atom interactions rather than reactions with the nitrogen lone pair. After functionalization, SWCNT were separated into two fractions: supernatant and deposit. The enrichment of amine-functionalized metallic SWCNT (from 41% to 87%) was obtained using repeated dispersion and centrifugation cycles carried out with the supernatant solutions. Semiconducting SWCNT, on the other hand, were enriched (from 59% to 64%) in the deposit fraction also by applying repeated dispersion and centrifugation cycles. The great advantage of this technique is that the weakly adsorbed amines are easily removed from the reaction mixture and from the SWCNT surface, once the enrichment process comes to an end.

Most of the aforementioned approaches require the suspension of CNT in a solvent. Network field effect transistors have been fabricated by selectively reacting and separating all the metallic SWCNT in the devices with diazonium reagents in a controlled manner.[213] It has been shown that the concentration of diazonium reagents used is crucial for selectively eliminating metallic SWCNT and keeping semiconducting ones intact. Excessive amounts of diazonium reagents can indiscriminately react with both metallic and semiconducting SWCNT and thus degrade the performance of the devices. This new technique will facilitate the process of fabricating of high-performance SWCNT-based electronic devices.

# References

1. (*a*) S. Reich, C. Thomsen and J. Maultzsch, *Carbon Nanotubes: Basic Concepts and Physical Properties*, Wiley-VCH, Weinheim, Germany, 2004; (*b*) M.S. Dresselhaus, G. Dresselhaus and P. Avouris, *Carbon*

Nanotubes: Synthesis, Structure, Properties and Applications, Springer, Berlin, Germany, 2001; (c) P.J.F. Harris, Carbon Nanotubes and Related Structures: New Materials for the Twenty-First Century, Cambridge University Press, Cambridge, UK, 2001; (d) S. Roth and D. Carroll, One-Dimensional Metals: Conjugated Polymers, Organic Crystals, Carbon Nanotubes, Wiley-VCH, Weinheim, Germany, 2004; (e) M. Meyyappan, Carbon Nanotubes, Science and Application, Wiley-VCH, Weinheim, Germany, 2006; (f) Special issue on Carbon Nanotubes. Acc. Chem. Res., 2002, 35, 997; (g) J.A. Schwarz, C.I. Contescu and K. Putyera, Dekker Encyclopedia of Nanoscience and Nanotechnology, Marcel Dekker, New York, USA, 2004; (h) C.N.R. Rao, A. Müller and A.K. Cheetham, The Chemistry of Nanomaterials, Wiley-VCH, Weinheim, Germany, 2005.
2. for reviews see:(a) K. Balasubramanian and M. Burghard, Small, 2005, 1, 180; (b) S. Baneriee, T. Hemraj-Benny and S.S. Wong, Adv. Mater., 2005, 17, 17; (c) A. Hirsch, Angew. Chem. Int. Ed., 2002, 41, 1853; (d) S. Banerjee, M.G.C. Kahn and S.S. Wong, Chem. Eur. J., 2003, 9, 1898; (e) X. Lu and Z. Chen, Chem. Rev., 2005, 105, 3643; (f) J.L. Bahr and J.M. Tour, J. Mater. Chem., 2002, 12, 1952; (g) H.S. Nalwa, Encyclopedia of Nanoscience and Nanotechnology, ACS, Stevenson Ranch, CA, 2004, 1, 761; (h) A. Hirsch and O. Vostrowsky, Top. Curr. Chem., 2005, 245, 193; (i) H. Kuzmany, A. Kukovecz, F. Simona, M. Holzweber, C. Kramberger and T. Pichler, Synth. Met., 2004, 141, 113; (j) D. Tasis, N. Tagmatarchis, A. Bianco and M. Prato, Chem. Rev., 2006, 106, 1105; (k) S. Ciraci, S. Dag, T. Yildirim, O. Gülseren and R.T. Senger, J. Phys.: Condens. Matter, 2004, 16, R901.
3. (a) V. Georgakilas, D. Voulgaris, E. Vázquez, M. Prato, D.M. Guldi, A. Kukovecz and H. Kuzmany, J. Am. Chem. Soc., 2002, 124, 14318; (b) M.E. Itkis, D.E. Perea, R. Jung, S. Niyogi and R.C. Haddon, J. Am. Chem. Soc., 2005, 127, 3439; (c) M. Holzinger, J. Abraham, P. Whelan, R. Graupner, L. Ley, F. Hennrich, M. Kappes and A. Hirsch, J. Am. Chem. Soc., 2003, 125, 8566; (d) Y. Lian, Y. Maeda, T. Wakahara, T. Akasaka, S. Kazaoui, N. Minami, N. Choi and H. Tokumoto, J. Phys. Chem. B., 2003, 107, 12082; (e) H. Jia, Y. Lian, M.O. Ishitsuka, T. Nakahodo, Y. Maeda, T. Tsuchiya, T. Wakahara and T. Akasaka, Sci. and Techn. of Adv. Mat., 2005, 6, 571; (f) E. Vázquez, V. Georgakilas and M. Prato, Chem. Commun., 2002, 2308.
4. (a) M.S. Strano, C.A. Dyke, M.L. Usrey, P.W. Barone, M.J. Allen, H. Shan, C. Kittrell, R.H. Hauge, J.M. Tour and R.E. Smalley, Science, 2003, 301, 1519; (b) D. Chattopadhyay, I. Galeska and F. Papadimitrakopoulos, J. Am. Chem. Soc., 2003, 125, 3370; (c) S. Baneriee and S.S. Wong, J. Am. Chem. Soc., 2004, 126, 2073; (d) C. Wang, Q. Cao, T. Ozel, A. Gaur, J.A. Rogers and M. Shim, J. Am. Chem. Soc., 2005, 127, 11460; (e) M.L. Usrey, E.S. Lippmann and M.S. Strano, J. Am. Chem. Soc., 2005, 127, 16129; (f) K.H. An, J.S. Park, C.-M. Yang, S.Y. Jeong, S.C. Lim, C. Kang, J.-H. Son, M.S. Jeong and Y.L. Hee, J. Am. Chem. Soc., 2005, 127, 5196; (g) M. Yudasaka, M. Zhang and S. Iijima, Chem.

*Phys. Lett.*, 2003, **374**, 132; (*h*) R. Krupke and F. Hennrich, *J. Phys. Chem. B.*, 2005, **109**, 17014.

5. (*a*) J. Hu, M. Ouyang, P. Yang and C.M. Lieber, *Nature*, 1999, **399**, 48; (*b*) J. Luo, L. Zhang, Y. Zhang and J. Zhu, *Adv. Mater.*, 2002, **14**, 1413; (*c*) Y. Zhang, T. Ichihashi, E. Landree, F. Nihey and S. Iijima, *Science*, 1999, **285**, 1719.

6. M.S. Gudiksen, L.J. Lauhon, J. Wang, D.C. Smith and C.M. Lieber, *Nature*, 2002, **415**, 617.

7. (*a*) H. Dai, *Acc. Chem. Res.*, 2002, **35**, 1035; (*b*) J.J. Davis, K.S. Coleman, B.R. Azamian, C.B. Bagshaw and M.L.H. Green, *Chem. Eur. J.*, 2003, **9**, 3733; (*c*) G.G. Wildgoose, C.E. Banks, H.C. Leventis and R.G. Compton, *Microchimica Acta*, 2006, **152**, 187; (*d*) C. Hu, X. Chen and S. Hu, *J. Electroanal. Chem.*, 2006, **586**, 77; (*e*) J.J. Gooding, *Electroch. Acta*, 2005, **50**, 3049; (*f*) E. Bekyarova, M. Davis, T. Burch, M.E. Itkis, B. Zhao, S. Sunshine and R.C. Haddon, *J. Phys. Chem. B.*, 2004, **108**, 19717; (*g*) G.L. Luque, N.F. Ferreyra and G.A. Rivas, *Microchim Acta*, 2006, **152**, 277; (*h*) P.W. Barone, S. Baik, D.A. Heller and M.S. Strano, *Nat. Mat.*, 2005, **4**, 86.

8. (*a*) P. Calvert, *Nature*, 1999, **399**, 210; (*b*) R.H. Baughman, A.A. Zakhidov and W.A. de Heer, *Science*, 2002, **297**, 787; (*c*) M.M.J. Treacy, T.W. Ebbesen and J.M. Gibson, *Nature*, 1996, **381**, 678; (*d*) B. Zhao, H. Hu, S.K. Mandal and R.C. Haddon, *Chem. Mater.*, 2005, **17**, 3235.

9. (*a*) D.M. Guldi, G.M.A. Rahman, F. Zerbetto and M. Prato, *Acc. Chem. Res.*, 2005, **38**, 871; (*b*) N. Tagmatarchis, M. Prato and D.M. Guldi, *Physica E*, 2005, **29**, 546; (*c*) D.M. Guldi, G.M.A. Rahman, C. Ehli and V. Sgobba, *Chem. Soc. Rev.*, 2006, **35**, 471.

10. D.M. Guldi, G.M.A. Rahman, N. Jux, D. Balbinot, N. Tagmatarchis and M. Prato, *Chem. Commun.*, 2005, 2038.

11. E. Joselevich, *Chem. Phys. Chem.*, 2004, **5**, 619.

12. J. Aihara, *J. Phys. Chem. A.*, 1999, **103**, 7487.

13. R.C. Haddon, *J. Am. Chem. Soc.*, 1990, **112**, 3385.

14. (*a*) S. Niyogi, M.A. Hamon, H. Hu, B. Zhao, P. Bhowmik, R. Sen, M.E. Itkis and R.C. Haddon, *Acc. Chem. Res.*, 2002, **35**, 1105; (*b*) M.H. Hamon, M.E. Itkis, S. Niyogi, T. Alvarez, C. Kuper, M. Menon and R.C. Haddon, *J. Am. Chem. Soc.*, 2001, **123**, 11292.

15. P. Nikolaev, M.J. Bronikowski, R.K. Bradley, F. Rohmund, D.T. Colbert, K. A. Smith and R.E. Smalley, *Chem. Phys. Lett.*, 1999, **313**, 91.

16. (*a*) K. Mylvaganam and L.C. Zhang, *J. Phys. Chem. B.*, 2004, **108**, 5217; (*b*) X. Lu, F. Tian, X. Xu, N. Wang and Q. Zhang, *J. Am. Chem. Soc.*, 2003, **125**, 10459; (*c*) E.T. Mickelson, C.B. Huffmann, A.G. Rinzler, R.E. Smalley, R.H. Hauge and J.L. Margrave, *Chem. Phys. Lett.*, 1998, **296**, 188; (*d*) H.F. Bettinger, *Chem. Phys. Chem.*, 2003, **4**, 1283; (*e*) J.H. Hafner, M.J. Bronikowski, B.R. Azamian, P. Nikolaev, A.G. Rinzler, D.T. Colbert, K.A. Smith and R.E. Smalley, *Chem. Phys. Lett.*, 1998, **296**, 195; (*f*) W. Zhou, Y. H. Ooi, R. Russo, P. Papanek, D.E. Luzzi,

J.E. Fischer, M.J. Bronikowski, P.A. Willis and R.E. Smalley, *Chem. Phys. Lett.*, 2001, **350**, 6.
17. Z. Chen, W. Thiel and A. Hirsch, *Chem. Phys. Chem.*, 2003, **1**, 93.
18. (*a*) H. Mauser, N.J.R. van Eikema Hommes, T. Clark, A. Hirsch, B. Pietzak, A. Weidinger and L. Dunsch, *Angew. Chem. Int. Ed. Engl.*, 1997, **36**, 2835; (*b*) H. Mauser, Hirsch, N.J.R. van Eikema Hommes and T. Clark, *J. Mol. Model.*, 1997, **3**, 415.
19. M.J. O'Connell, S.M. Bachilo, C.B. Huffman, V.C. Moore, M.S. Strano, E.H. Haroz, K.L. Rialon, P.J. Boul, W.H. Noon, C. Kittrell, J.P. Ma, R.H. Hauge, R.B. Weisman and R.E. Smalley, *Science*, 2002, **297**, 593.
20. M. Monthioux, B.W. Smith, B. Burteaux, A. Claye, J.E. Fischer and D.E. Luzzi, *Carbon*, 2001, **39**, 1251.
21. for reviews see:(*a*) T. Belin and F. Epron, *Mat. Sci. and Eng. B.*, 2005, **119**, 105; (*b*) S. Arepalli, P. Nikolaev, O. Gorelik, V.G. Hadjiev, W. Holmes, B. Files and L. Yowell, *Carbon*, 2004, **42**, 1783.
22. J. Zhao and P.B. Balbuena, *J. Phys. Chem. A.*, 2006, **110**, 2771.
23. H. Hu, P. Bhowmik, B. Zhao, M.H. Hamon, M.E. Itkis and R.C. Haddon, *Chem. Phys. Lett.*, 2001, **291**, 283.
24. (*a*) A.J. Stone and D.J. Wales, *Chem. Phys. Lett.*, 1986, **128**, 501; (*b*) H.F. Bettinger, *J. Phys. Chem. B.*, 2005, **109**, 6922; (*c*) X. Lu, Z. Chen and P.v.R. Schleyer, *J. Am. Chem. Soc.*, 2005, **127**, 4313.
25. (*a*) A. Hashimoto, K. Suenaga, A. Gloter, K. Urita and S. Iijima, *Nature*, 2004, **430**, 870; (*b*) P.C.P. Watts, W.-K. Hsu, H.W. Kroto and D.R.M. Walton, *Nano. Lett.*, 2003, **3**, 549; (*c*) J.-C. Charlier, *Acc. Chem. Res.*, 2002, **35**, 1063; (*d*) M. Ouyang, J.L. Huang, C.L. Cheung and C.M. Lieber, *Science*, 2001, **291**, 97; (*e*) S.G. Louie, *Top. Appl. Phys.*, 2001, **80**, 113; (*f*) D. Orlikowski, M.B. Nardelli, J. Bernholc and C. Roland, *Phys. Rev. Lett.*, 1999, **83**, 4132; (*g*) T.W. Ebbesen and T. Takada, *Carbon*, 1995, **33**, 973; (*h*) J.C. Charlier, T.W. Ebbesen and P. Lambin, *Phys. Rev. B.*, 1996, **53**, 11108; (*i*) V.H. Crespi and M.L. Cohen, *Phys. Rev. Lett.*, 1997, **79**, 2093; (*j*) L. Chico, M.P.L. Sancho and M.C. Muñoz, *Phys. Rev. Lett.*, 1998, **98**, 1278.
26. (*a*) H. Dai, *Phys. World*, 2000, **13**, 43; (*b*) M. Ouyang, J.-L. Huang and C.M. Lieber, *Acc. Chem. Res.*, 2002, **35**, 1018; (*c*) M. Grujicic, C. Cao and R. Singh, *Appl. Surf. Sci.*, 2003, **211**, 166; (*d*) M. Grujicic, G. Cao, A.M. Rao, T.M. Tritt and S. Nayak, *Appl. Surf. Sci.*, 2003, **214**, 289.
27. Z. Yao, H.W.C. Postma, L. Balents and C. Dekker, *Nature*, 1999, **402**, 273.
28. (*a*) C.A. Furtado, U.J. Kim, H.R. Gutierrez, L. Pan, E.C. Dickey and P.C. Eklund, *J. Am. Chem. Soc.*, 2004, **126**, 6095; (*b*) R. Sen, S.M. Rickard, M.E. Itkis and R.C. Haddon, *Chem. Mater.*, 2003, **15**, 4273; (*c*) S. Huang and L. Dai, *J. Phys. Chem. B.*, 2002, **106**, 3543.
29. Y. Feng, G. Zhou, G. Wang, M. Qu and Z. Yu, *Chem. Phys. Lett.*, 2003, **375**, 645.
30. C.-Min Yang, J.S. Park, K.H. An, S.C. Lim, K. Seo, B. Kim, K. Ah Park, S. Han, C.Y. Park and Y. Hee Lee, *J. Phys. Chem. B.*, 2005, **109**, 19242.

31. E. Dujardin, T.W. Ebbesen, A. Treacy and M.M.J. Krishnan, *Adv. Mater.*, 1998, **10**, 611.
32. P.M. Ajayan and S. Iijima, *Nature*, 1993, **361**, 333.
33. (*a*) P.M. Ajayan, T.W. Ebbesen, T. Ichihashi, S. Iijima, K. Tanigaki and H. Hiura, *Nature*, 1993, **362**, 522; (*b*) L. Qingwen, Y. Hao, Y. Yinchun, Z. Jin and L. Zhongfan, *J. Phys. Chem. B.*, 2002, **106**, 11085.
34. (*a*) J. Steinmetz, M. Glerup, M. Paillet, P. Bernier and M. Holzinger, *Carbon*, 2005, **43**, 2397; (*b*) J. Fan, M. Yudasaka, J. Miyawaki, K. Ajima, K. Murata and S. Iijima, *J. Phys. Chem. B.*, 2006, **110**, 1587.
35. (*a*) S.C. Tsang, Y.K. Chen, P.J.F. Harris and M.L.H. Green, *Nature*, 1994, **372**, 159; (*b*) S. Nagasawa, M. Yudasaka, K. Hirahara, T. Ichihashi and S. Iijima, *Chem. Phys. Lett.*, 2000, **328**, 374; (*c*) G.-W. Lee and S. Kumar, *J. Phys. Chem. B.*, 2005, **109**, 17128; (*d*) I.D. Rosca, F. Watari, M. Uo and T. Akasaka, *Carbon*, 2005, **43**, 3124; (*e*) T. Kyotani, S. Nakazaki, W.-H. Xu and A. Tomita, *Carbon*, 2001, **39**, 771; (*f*) M. Zhang, M. Yudasaka and S. Iijima, *J. Phys. Chem. B.*, 2004, **108**, 149; (*g*) F. Hennrich, R. Wellmann, S. Malik, S. Lebedkin and M.M. Kappes, *Phys. Chem. Chem. Phys.*, 2003, **5**, 178.
36. W. Zhou, J.E. Fischer, P.A. Heiney, H. Fan, V.A. Davis, M. Pasquali and R.E. Smalley, *Phys. Rev. B.*, 2005, **72**, 045440.
37. Y. Lian, Y. Maeda, T. Wakahara, T. Akasaka, S. Kazaoui, N. Minami, T. Shimizu, N. Choi and H. Tokumoto, *J. Phys. Chem. B*, 2004, **108**, 8848.
38. (*a*) R. Yu, L. Chen, Q. Liu, J. Lin, K. L. Tan, S.C. Ng, H.S.O. Chan, G.Q. Xu and T.S.A. Hor, *Chem. Mater.*, 1998, **10**, 718; (*b*) J. Liu, A.G. Rinzler, H. Dai, J.H. Hafner, R.K. Bradley, P.J. Boul, A. Lu, T. Iverson, K. Shelimov, C.B. Huffman, F. Rodrigues-Macias, Y.-S. Shon, T.R. Lee, D.T. Colbert and R.E. Smalley, *Science*, 1998, **280**, 1253; (*c*) D.B. Mawhinney, V. Naumenko, A. Kuznetsova, J.T. Yates Jr., J. Liu and R.E. Smalley, *Chem. Phys. Lett.*, 2000, **324**, 213; (*d*) Y. Wang, Z. Iqbal and S. Mitra, *J. Am. Chem. Soc.*, 2006, **128**, 95.
39. K.C. Park, T. Hayashi, H. Tomiyasu, M. Endo and M.S. Dresselhaus, *J. Mater. Chem.*, 2005, **15**, 407.
40. S. Ramesh, L.M. Ericson, V.A. Davis, R.K. Saini, C. Kittrell, M. Pasquali, W.E. Billups, W.W. Adams, R.H. Hauge and R.E. Smalley, *J. Phys. Chem. B.*, 2004, **108**, 8794.
41. M. Liu, Y. Yang, T. Zhu and Z. Liu, *Carbon*, 2005, **43**, 1470.
42. (*a*) H. Hiura, T.W. Ebbesen and K. Tanigaki, *Adv. Mater.*, 1995, **7**, 275; (*b*) K. Hernadi, A. Siska, L. Thien-Nga, L. Forro and I. Kiricsi, *Solid State Ionics*, 2001, **141–142**, 203.
43. A. Kuznetsova, D.B. Mawhinney, V. Naumenko, J.T. Yates Jr., J. Liu and R.E. Smalley, *Chem. Phys. Lett.*, 2000, **321**, 292.
44. Z. Liu, X. Lin, J.Y. Lee, W. Zhang, M. Han and L.M. Gan, *Langmuir*, 2002, **18**, 4054.
45. A. Kuznetsova, I. Popova, J.T. Yates Jr., M.J. Bronikowski, C.B. Huffman, J. Liu, R.E. Smalley, H.H. Hwu and J.G. Chen, *J. Am. Chem. Soc.*, 2001, **123**, 10699.

46. J.H.T. Luong, S. Hrapovic, Y. Liu, D.-Q. Yang, E. Sacher, D. Wang, C.T. Kingston and G.D. Enright, *J. Phys. Chem. B.*, 2005, **109**, 1400.
47. M.R. Smith, S.W. Hedges, R. LaCount, D. Kern, N. Shah, G.P. Huffman and B. Bockrath, *Carbon*, 2003, **41**, 1221.
48. Y. Fan, M. Burghard and K. Kern, *Adv. Mater.*, 2002, **14**, 130.
49. K.C. Hwang, *J. Chem. Soc., Chem. Commun.*, 1995, **2**, 173.
50. Z.-J. Liu, Z.-Y. Yuan, W. Zhou, L.-M. Peng and Z. Xu, *Phys. Chem. Chem. Phys.*, 2001, **3**, 2518.
51. J. Chen, A.M. Rao, S. Lyuksyutov, M.E. Itkis, M.A. Hamon, H. Hu, R.W. Cohn, P.C. Eklund, D.T. Colbert, R.E. Smalley and R.C. Haddon, *J. Phys. Chem. B.*, 2001, **105**, 2525.
52. H. Hu, P. Bhowmik, B. Zhao, M.A. Hamon, M.E. Itkis and R.C. Haddon, *Chem. Phys. Lett.*, 2001, **345**, 25.
53. X. Li, J. Niu, J. Zhang, H. Li and Z. Liu, *J. Phys. Chem. B.*, 2003, **107**, 2453.
54. (*a*) J. Chen, M.A. Hamon, H. Hu, Y. Chen, A.M. Rao, P.C. Eklund and R.C. Haddon, *Science*, 1998, **282**, 95; (*b*) M.A. Hamon, J. Chen, H. Hu, Y. Chen, M.E. Itkis, A.M. Rao, P.C. Eklund and R.C. Haddon, *Adv. Mater.*, 1999, **11**, 834; (*c*) B. Zhou, Y. Lin, H. Li, W. Huang, J.W. Connell, L.F. Allard and Y.-P. Sun, *J. Phys. Chem. B.*, 2003, **107**, 13588; (*d*) M. Alvaro, P. Atienzar, J.L. Bourdelande and H. García, *Chem. Phys. Lett.*, 2004, **384**, 119.
55. (*a*) M.H. Hamon, H. Hui, P. Bhowmik, H.M.E. Itkis and R.C. Haddon, *Appl. Phys. A.: Mater. Sci. Process.*, 2002, **74**, 333; (*b*) L. Qu, R.B. Martin, W. Huang, K. Fu, D. Zweifel, Y. Lin, Y.-P. Sun, C.E. Bunker, B.A. Harruff, J.R. Gord and L.F. Allard, *J. Chem. Phys.*, 2002, **117**, 8089; (*c*) J. Zhang, G. Wang, Y.-S. Shon, O. Zhou, R. Superfine and R.W. Murray, *J. Phys. Chem. B.*, 2003, **107**, 3726; (*d*) Y. Lin, S. Taylor, W. Huang and Y.-P. Sun, *J. Phys. Chem. B.*, 2003, **107**, 914.
56. Y. Yang, S. Chen, Q. Xue, A. Biris and W. Zhao, *Electrochimica Acta*, 2005, **50**, 3061.
57. (*a*) S. Niyogi, H. Hu, M.A. Hamon, P. Bhowmik, B. Zhao, S.M. Rozenhak, J. Chen, M.E. Itkis, M.S. Meier and R.C. Haddon, *J. Am. Chem. Soc.*, 2001, **123**, 733; (*b*) B. Zhao, H. Hu, S. Niyogi, M.E. Itkis, M.A. Hamon, P. Bhowmik and R.C. Haddon, *J. Am. Chem. Soc.*, 2001, **123**, 11673; (*c*) W. Huang, S. Fernando, Y. Lin, B. Zhou, L.F. Allard and Y.-P. Sun, *Langmuir*, 2003, **19**, 7084.
58. (*a*) S. Baik, M. Usrey, L. Rotkina and M. Strano, *J. Phys. Chem. B.*, 2004, **108**, 15560; (*b*) R. Krupke and F. Hennrich, *J. Phys. Chem. B.*, 2005, **109**, 17014; (*c*) N. Nair and M.S. Strano, *J. Phys. Chem. B.*, 2005, **109**, 17016.
59. F. Pompeo and D.E. Resasco, *Nano Lett.*, 2002, **2**, 369.
60. B. Li, Z. Shi, Y. Lian and Z. Gu, *Chem. Lett.*, 2001, **7**, 598.
61. (*a*) W. Huang, S. Fernando, L.F. Allard and Y.-P. Sun, *Nano Lett.*, 2003, **3**, 565; (*b*) J.E. Riggs, Z. Guo, D.L. Carroll and Y.-P. Sun, *J. Am. Chem. Soc.*, 2000, **122**, 5879; (*c*) Y. Lin, A.M. Rao, B. Sadanadan, E.A. Kenik and Y.-P. Sun, *J. Phys. Chem. B.*, 2002, **106**, 1294; (*d*) Y. Lin, D.E. Hill,

J. Bentley, L. F. Allard and Y.-P. Sun, *J. Phys. Chem. B.*, 2003, **107**, 10453; (*e*) Y. Lin, B. Zhou, S. Fernando, P. Liu, L.F. Allard and Y.-P. Sun, *Macromolecules*, 2003, **36**, 7199; (*f*) W. Huang, S. Taylor, K. Fu, Y. Lin, D. Zhang, T.W. Hanks, A.M. Rao and Y.-P. Sun, *Nano Lett.*, 2002, **2**, 311; (*g*) S.E. Baker, W. Cai, T.L. Lasseter, K.P. Weidkamp and R.J. Hamers, *Nano Lett.*, 2002, **2**, 1413; (*h*) K.A.S. Fernando, Y. Lin and Y.P. Sun, *Langmuir*, 2004, **20**, 4777; (*i*) J. Gao, M.E. Itkis, A. Yu, E. Bekyarova, B. Zhao and R.C. Haddon, *J. Am. Chem. Soc.*, 2005, **127**, 3847; (*j*) G.-X. Chen, H.-S. Kim, B.H. Park and J.-S. Yoon, *J. Phys. Chem. B.*, 2005, **109**, 22237; (*k*) M. Baibarac, I. Baltog, C. Godon, S. Lefrant and O. Chauve, *Carbon*, 2004, **42**, 3143; (*l*) Z. Yang, H. Pu and J. Yin, *Mat. Lett.*, 2005, **59**, 2838; (*m*) S. Lefrant, M. Baibarac, I. Baltog, J.Y. Mevellec, C. Godon and O. Chauvet, *Diam. & Rel. Mat.*, 2005, **14**, 867; (*n*) Y. Wang, Z. Iqbal and S.V. Malhotra, *Macromolecules*, 2005, **38**, 7670; (*o*) K. Ro Yoon, W.-J. Kim and I.S. Choi, *Macromol. Chem. Phys.*, 2004, **205**, 1218; (*p*) D.E. Hill, Y. Lin, A.M. Rao, L.F. Allard and Y.-P. Sun, *Macromolecules*, 2002, **35**, 9466; (*q*) H. Kong, C. Gao and D. Yan, *Macromolecules*, 2004, **37**, 4022; (*r*) X. Lou, C. Detrembleur, V. Sciannamea, C. Pagnoulle and R. Jérôme, *Polymer*, 2004, **45**, 6097; (*s*) C. Gao, C.D. Vo, Y.Z. Jin, W. Li and S.P. Armes, *Macromolecules*, 2005, **38**, 8634; (*t*) C.-Y. Hong, Y.-Z. You, D. Wu, Y. Liu and C.-Y. Pan, *Macromolecules*, 2005, **38**, 2606; (*u*) J.J. Ge, D. Zhang, Q. Li, H. Hou, M.J. Graham, L. Dai, F.W. Harris and S.Z.D. Cheng, *J. Am. Chem. Soc.*, 2005, **127**, 9984; (*v*) H. Hu, Y. Ni, S.K. Mandal, V. Montana, B. Zhao, R.C. Haddon and V. Parpura, *J. Phys. Chem. B.*, 2005, **109**, 4285; (*w*) Y. Liu, D.-C. Wu, W.-D. Zhang, X. Jiang, C.-B. He, T.S. Chung, S.H. Goh and K.W. Leong, *Angew. Chem. Int. Ed.*, 2005, **44**, 4782; (*x*) C. Gao, Y.Z. Jin, H. Kong, R.L.D. Whitby, S.F.A. Acquah, G.Y. Chen, H. Qian, A. Hartschuh, S.R.P. Silva, S. Henley, P. Fearon, H.W. Kroto and D.R.M. Walton, *J. Phys. Chem. B.*, 2005, **109**, 11925.

62. (*a*) R.F. Khairoutdinov, L.V. Doubova, R.C. Haddon and L. Saraf, *J. Phys. Chem. B.*, 2004, **108**, 19976; (*b*) F. Frehill, J.G. Vos, S. Benrezzak, A.A. Koos, Z. Konya, M.G. Ruether, W.J. Blau, A. Fonseca, J.B. Nagy, L.P. Biro, A.I. Minett and M. in het Panhuis, *J. Am. Chem. Soc.*, 2002, **124**, 13694; (*c*) T.Y. Lee and J.-B. Yoo, *Diamond & Related Materials*, 2005, **14**, 1888; (*d*) S.-R. Jang, R. Vittal and K.-J. Kim, *Langmuir*, 2004, **20**, 9807.
63. M.J. Park, J.K. Lee, B.S. Lee, Y.-W. Lee, I.S. Choi and S. Lee, *Chem. Mater.*, 2006, **18**, 1546.
64. E.V. Basiuk, M. Monroy-Pelaez, I. Puente-Lee and V.A. Basiuk, *Nano Lett.*, 2004, **4**, 863.
65. Y.-P. Sun, W. Huang, Y. Lin, K. Fu, A. Kitaygorodskiy, L.A. Riddle, Y.J. Yu and D.L. Carroll, *Chem. Mater.*, 2001, **13**, 2864.
66. K. Fu, W. Huang, Y. Lin, L.A. Riddle, D.L. Carroll and Y.-P. Sun, *Nano Lett.*, 2001, **1**, 439.
67. K.A. Williams, P.T.M. Veenhuizen, B.G. de la Torre, R. Eritja and C. Dekker, *Nature*, 2002, **420**, 761.

68. (a) S. Li, P. He, J. Dong, Z. Guo and L. Dai, *J. Am. Chem. Soc.*, 2005, **127**, 14; (b) P. He, S. Li and L. Dai, *Synth. Met.*, 2005, **154**, 17; (c) M. Hazani, R. Naaman, F. Hennrich and M.M. Kappes, *Nano. Lett.*, 2003, **3**, 153; (d) D.-H. Jung, B.H. Kim, Y.K. Ko, M.S. Jung, S. Jung, S.Y. Lee and H.-T. Jung, *Langmuir*, 2004, **20**, 8886; (e) N. Wong S. Kam, Z. Liu and H. Dai, *J. Am. Chem. Soc.*, 2005, **127**, 12492.
69. (a) Y. Wang, Z. Iqbal and S.V. Malhotra, *Chem. Phys. Lett.*, 2005, **402**, 96; (b) Y. Zhang, Y. Shen, J. Li, L. Niu, S. Dong and A. Ivaska, *Langmuir*, 2005, **21**, 4797.
70. (a) L. Gu, Y. Lin, L. Qu and Y.-P. Sun, *Biomacromol.*, 2006, **7**, 400; (b) V.B. Kandimalla and H. Ju, *Chem. Eur. J.*, 2006, **12**, 1074.
71. Y.-P. Sun, K. Fu, Y. Lin and W. Huang, *Acc. Chem. Res.*, 2002, **35**, 1096.
72. T. Elkin, X. Jiang, S. Taylor, Y. Lin, L. Gu, H. Yang, J. Brown, S. Collins and Y.-P. Sun, *Chem. Bio. Chem.*, 2005, **6**, 640.
73. N.G. Portney, K. Singh, S. Chaudhary, G. Destito, A. Schneemann, M. Manchester and M. Ozkan, *Langmuir*, 2005, **21**, 2098.
74. M. Li, E. Dujardinb and S. Mann, *Chem. Commun.*, 2005, 4952.
75. (a) N.W.S. Kam, T.C. Jessop, P.A. Wender and H. Dai, *J. Am. Chem. Soc.*, 2004, **126**, 6850; (b) W. Zhu, N. Minami, S. Kazaoui and Y. Kim, *J. Mater. Chem.*, 2003, **13**, 2196; (c) S.A. Curran, A.V. Ellis, A. Vijayaraghavan and P.M. Ajayan, *J. Chem. Phys.*, 2004, **120**, 4886; (d) R.B. Martin, L. Qu, Y. Lin, B.A. Harruff, C.E. Bunker, J.R. Gord, L.F. Allard and Y.-P. Sun, *J. Phys. Chem. B.*, 2004, **108**, 11447; (e) M. Alvaro, C. Aprile, P. Atienzar and H. Garcia, *J. Phys. Chem. B.*, 2005, **109**, 7692.
76. for reviews see:(a) E. Katz and I. Willner, *Chem. Phys. Chem.*, 2004, **5**, 1085; (b) Y. Lin, S. Taylor, H. Li, K.A. Shiral Fernando, L. Qu, W. Wang, L. Gu, B. Zhou and Y.-P. Sun, *J. Mater. Chem.*, 2004, **14**, 527.
77. M. Sano, A. Kamino, J. Okamura and S. Shinkai, *Science*, 2001, **293**, 1299.
78. P.W. Chiu, G.S. Duesberg, U. Dettlaff-Weglikowska and S. Roth, *Appl. Phys. Lett.*, 2002, **80**, 3811.
79. C. Bailezão, B. Gigante, H. Garcia and A. Corma, *J. Catal.*, 2004, **221**, 77.
80. (a) D. Baskaran, J.W. Mays, X.P. Zhang and M.S. Bratcher, *J. Am. Chem. Soc.*, 2005, **127**, 6916; (b) H. Li, R.B. Martin, B.A. Harruff, R.A. Carino, L.F. Allard and Y.-P. Sun, *Adv. Mat.*, 2004, **16**, 896.
81. G. de la Torre, W. Blau and T. Torres, *Nanotechnology*, 2003, **14**, 765.
82. (a) Y. Yang, J. Zhang, X. Nan and Z. Liu, *J. Phys. Chem. B.*, 2002, **106**, 4139; (b) S.S. Wong, E. Joselevich, A.T. Woolley, C.L. Cheung and C.M. Lieber, *Nature*, 1998, **394**, 52; (c) S.S. Wong, A.T. Woolley, E. Joselevich, C.L. Cheung and C.M. Lieber, *J. Am. Chem. Soc.*, 1998, **120**, 8557.
83. (a) T. Sainsbury, J. Stolarczyk and D. Fitzmaurice, *J. Phys. Chem. B.*, 2005, **109**, 16310; (b) B.R. Azamian, K.S. Coleman, J.J. Davis, N. Hanson and M. L. Green, *Chem. Commun.*, 2002, **4**, 366; (c) J. Hu, J. Shi, S. Li, Y. Qin, Z.-X. Guo, Y. Song and D. Zhu, *Chem. Phys. Lett.*, 2005, **401**, 352.
84. Z. Liu, Z. Shen, T. Zhu, S. Hou and L. Ying, *Langmuir*, 2000, **16**, 3569.

85. A. Zamudio, A.L. Elías, J.A. Rodríguez-Manzo, F. López-Urías, G. Rodríguez-Gattorno, F. Lupo, M. Rühle, D.J. Smith, H. Terrones, D. Díaz and M. Terrones, *Small*, 2006, **2**, 346.
86. J.K. Lim, W.S. Yun, M.-H. Yoon, S.K. Lee, C.H. Kim, K. Kim and S.K. Kim, *Synth. Met.*, 2003, **139**, 521.
87. L. Sheeney-Haj-Ichia, B. Basnar and I. Willner, *Angew. Chem., Int. Ed.*, 2005, **44**, 78.
88. (*a*) P. Diao and Z. Liu, *J. Phys. Chem. B.*, 2005, **109**, 20906; (*b*) Z. Liu, Z. Shen, T. Zhu, S. Hou, L. Ying, Z. Shi and Z. Gu, *Langmuir*, 2000, **16**, 3569; (*c*) L. Cai, J.L. Bahr, Y. Yao and J.M. Tour, *Chem. Mater.*, 2002, **14**, 4235; (*d*) P. Diao, Z. Liu, B. Wu, X. Nan, J. Zhang and Z. Wei, *Chem. Phys. Chem.*, 2002, **12**, 898.
89. P. Diao and Z. Liu, *J. Phys. Chem. B.*, 2005, **109**, 20906.
90. (*a*) M. Burghard, *Small*, 2005, **1**, 1148; (*b*) K.M. Lee, L. Li and L. Dai, *J. Am. Chem. Soc.*, 2005, **127**, 4122; (*c*) N. Chopra, M. Majumder and B.J. Hinds, *Adv. Funct. Mater.*, 2005, **15**, 858.
91. Z. Wei, M. Kondratenko, L.H. Dao and D.M. Perepichka, *J. Am. Chem. Soc.*, 2006, **128**, 3134.
92. B. Pan, D. Cui, R. He, F. Gao and Y. Zhang, *Chem. Phys. Lett.*, 2006, **417**, 419.
93. S. Banerjee and S.S. Wong, *Nano Lett.*, 2002, **2**, 195.
94. S. Ravindran, S. Chaudhary, B. Colburn, M. Ozkan and C.S. Ozkan, *Nano Lett.*, 2003, **3**, 447.
95. J.M. Haremza, M.A. Hahn, T.D. Krauss, S. Chen and J. Calcines, *Nano Lett.*, 2002, **2**, 1253.
96. A. Gromov, S. Dittmer, J. Svensson, O.A. Nerushev, S.A. Perez-García, L. Licea-Jiménez, R. Rychwalskib and E.E.B. Campbella, *J. Mater. Chem.*, 2005, **15**, 3334.
97. Y. Sun, S.R. Wilson and D.I. Schuster, *J. Am. Chem. Soc.*, 2001, **123**, 5348.
98. (*a*) E.V. Basiuk, M. Monroy-Pelaez, I. Puente-Lee and V.A. Basiuk, *Nano Lett.*, 2004, **4**, 863; (*b*) R. Zanella, E.V. Basiuk, P. Santiago, V.A. Basiuk, E. Mireles, I. Puente-Lee and J.M. Saniger, *J. Phys. Chem. B.*, 2005, **109**, 16290.
99. (*a*) E.V. Basiuk, V.A. Basiuk, J.-G. Banuelos, J.-M. Saniger-Blesa, V.A. Pokrovskiy, T.Y. Gromovoy, A.V. Mischanchuk and B.G. Mischanchuk, *J. Phys. Chem. B.*, 2002, **106**, 1588; (*b*) V.A. Basiuk, E.V. Basiuk and J.-M. Saniger-Blesa, *Nano Lett.*, 2001, **1**, 657.
100. (*a*) T. Nakajima, S. Kasamatsu and Y. Matsuo, *Eur. J. Inorg. Chem.*, 1997, **33**, 831; (*b*) V.N. Khabashesku, W.E. Billups and J.L. Margrave, *Acc. Chem. Res.*, 2002, **35**, 1087 and references therein.
101. R.J. Lagow, R.B. Badachhape, J.L. Wood and J.L. Margrave, *J. Chem. Soc., Dalton Trans.*, 1974, **12**, 1268.
102. E.T. Mickelson, C.B. Huffman, A.G. Rinzler, R.E. Smalley, R.H. Hauge and J.L. Margrave, *Chem. Phys. Lett.*, 1998, **296**, 198.

103. H. Muramatsu, Y.A. Kim, T. Hayashi, M. Endo, A. Yonemoto, H. Arikai, F. Okino and H. Touhara, *Chem. Commun.*, 2005, 2002.
104. N.F. Yudanov, A.V. Okotrub, Y.V. Shubin, L.I. Yudanova, L.G. Bulusheva, A.L. Chuvilin and J.-M. Bonard, *Chem. Mater.*, 2002, **14**, 1472.
105. N.O.V. Plank, G.A. Forrest, R. Cheung and A.J. Alexander, *J. Phys. Chem. B.*, 2005, **109**, 22096.
106. E.T. Mickelson, I.W. Chiang, J.L. Zimmerman, P.J. Boul, J. Lozano, J. Liu, R.E. Smalley, R.H. Hauge and J.L. Margrave, *J. Phys. Chem. B.*, 1999, **103**, 4318.
107. H. Peng, Z. Gu, J. Yang, J.L. Zimmerman, P.A. Willis, M.J. Bronikowski, R.E. Smalley, R.H. Hauge and J.L. Margrave, *Nano Lett.*, 2001, **1**, 625.
108. R.L. McCreery, in *Electroanalytical Chemistry*, Vol 17, A.J. Bard, New York, 1991, 221.
109. R.L. Vander, M.K. Wal, K.W. Street, A.J. Tomasek, H. Peng, Y. Liu, J.L. Margrave and V.N. Khabashesku, *Wear*, 2005, **259**, 738.
110. V.N. Khabasheskua, J.L. Margrave and E.V. Barrera, *Diam. & Rel. Mat.*, 2005, **14**, 859.
111. R.K. Saini, I.W. Chiang, H. Peng, R.E. Smalley, W.E. Billups, R.H. Hauge and J.L. Margrave, *J. Am. Chem. Soc.*, 2003, **125**, 3617.
112. P.J. Boul, J. Liu, E.T. Mickelson, C.B. Huffman, L.M. Ericson, I.W. Chiang, K.A. Smith, D.T. Colbert, R.H. Hauge, J.L. Margrave and R.E. Smalley, *Chem. Phys. Lett.*, 1999, **310**, 367.
113. L. Zhang, J. Zhang, N. Schmandt, J. Cratty, V.N. Khabashesku, K.F. Kelly and A.R. Barron, *Chem. Commun.*, 2005, 5429.
114. L. Zhang, V.U. Kiny, H. Peng, J. Zhu, R.F.M. Lobo, J.L. Margrave and V.N. Khabashesku, *Chem. Mater.*, 2004, **16**, 2055.
115. J.L. Stevens, A.Y. Huang, H.Q. Peng, I.W. Chiang, V.N. Khabashesku and J.L. Margrave, *Nano Lett.*, 2003, **3**, 331.
116. (*a*) A. Padwa, *1,3-Dipolar Cycloaddition Chemistry*, Wiley, New York, 1984; (*b*) R.B. Woodward and R. Hoffmann, *Angew. Chem., Int. Ed. Engl.*, 1969, **8**, 781; (*c*) R.B. Woodward and R. Hoffmann, *Angew. Chem., Int. Ed. Engl.*, 1969, **8**, 817; (*d*) K. Fukui, in *Molecular Orbitals in Chemistry, Physics and Biology*, P.O. Lowdin and B. Pullman (eds), Academic Press, New York, 1964.
117. (*a*) R. Criegee, *Angew. Chem., Int. Ed. Engl.*, 1975, **14**, 745; (*b*) R. Atkinson and W.P.L. Carter, *Chem. Rev.*, 1984, **84**, 437; (*c*) J.M. Anglada, R. Crehuet and J.M. Bofill, *Chem. Eur. J.*, 1999, **5**, 1809.
118. (*a*) D. Heymann, S.M. Bachilo, R.B. Weisman, F. Cataldo, R.F. Fokkens, N.M.M. Nibbering, R.D. Vis and L.P. Felipe Chibante, *J. Am. Chem. Soc.*, 2000, **122**, 11473; (*b*) R. Malhotra, S. Kumar and A. Satyam, *J. Chem. Soc., Chem. Commun.*, 1994, 1339; (*c*) J.P. Deng, C.Y. Mou and C.C. Han, *Fullerene Sci. Technol.*, 1997, **5**, 1033; (*d*) J.P. Deng, C.Y. Mou and C.C. Han, *Fullerene Sci. Technol.*, 1997, **5**, 1325.
119. (*a*) S. Banerjee and S.S. Wong, *J. Phys. Chem. B.*, 2002, **106**, 12144; (*b*) S. Banerjee and S.S. Wong, *Nano Lett.*, 2004, **4**, 1445; (*c*) S. Banerjee,

T. Hemraj-Benny, M. Balasubramanian, D.A. Fischer, J.A. Misewich and S.S. Wong, *Chem. Phys. Chem.*, 2004, **5**, 1416.
120. D.B. Mawhinney, V. Naumenko, A. Kuznetsova, J.T. Yates Jr., L. Liu and R.E. Smalley, *J. Am. Chem. Soc.*, 2000, **122**, 2383.
121. S. Banerjee and S.S. Wong, *Chem. Commun.*, 2004, 772.
122. M. Prato and M. Maggini, *Acc. Chem. Res.*, 1998, **31**, 519.
123. (*a*) D. Georgakilas, K. Kordatos, M. Prato, D.M. Guldi, M. Holzinger and A. Hirsch, *J. Am. Chem. Soc.*, 2002, **124**, 760; (*b*) V. Georgakilas, N. Tagmatarchis, D. Pantarotto, A. Bianco, J.-P. Briand and M. Prato, *Chem. Commun.*, 2002, 3050; (*c*) A. Callegari, M. Marcaccio, D. Paolucci, F. Paolucci, N. Tagmatarchis, D. Tasis, E. Vázquez and M. Prato, *Chem. Commun.*, 2003, 2576.
124. D. Tasis, N. Tagmatarchis, V. Georgakilas and M. Prato, *Chem. Eur. J.*, 2003, **9**, 4000.
125. for reviews see:(*a*) N. Tagmatarchis and M. Prato, *J. Mater. Chem.*, 2004, **14**, 437; (*b*) D. Tasis, N. Tagmatarchis, V. Georgakilas and M. Prato, *Chem. Eur. J.*, 2003, **9**, 4001; (*c*) A. Bianco, K. Kostarelos, C.D. Partidos and M. Prato, *Chem. Commun.*, 2005, 571.
126. L.S. Cahill, Z. Yao, A. Adronov, J. Penner, K.R. Moonoosawmy, P. Kruse and G.R. Goward, *J. Phys. Chem. B.*, 2004, **108**, 11412.
127. D.M. Guldi, M. Marcaccio, D. Paolucci, F. Paolucci, N. Tagmatarchis, D. Tasis, E. Vázquez and M. Prato, *Angew. Chem., Int. Ed.*, 2003, **42**, 4206.
128. A. Callegari, M. Marcaccio, D. Paolucci, F. Paolucci, N. Tagmatarchis, D. Tasis, E. Vázquez and M. Prato, *Chem. Commun.*, 2003, 2576.
129. A. Callegari, S. Cosnier, M. Marcaccio, D. Paolucci, F. Paolucci, V. Georgakilas, N. Tagmatarchis, E. Vázquez and M. Prato, *J. Mater. Chem.*, 2004, **14**, 807.
130. V. Georgakilas, N. Tagmatarchis, D. Pantarotto, A. Bianco, J.-P. Briand and M. Prato, *Chem. Commun.*, 2002, 3050.
131. (*a*) D. Pantarotto, C.D. Partidos, R. Graff, J. Hoebeke, J.-P. Briand, M. Prato and A. Bianco, *J. Am. Chem. Soc.*, 2003, **125**, 6160; (*b*) A. Bianco and M. Prato, *Adv. Mater.*, 2003, **15**, 1765; (*c*) D. Pantarotto, C.D. Partidos, J. Hoebeke, F. Brown, E. Kramer, J.-P. Briand, S. Muller, M. Prato and A. Bianco, *Chem. & Biol.*, 2003, **10**, 961; (*d*) R. Singh, D. Pantarotto, D. McCarthy, O. Chaloin, J. Hoebeke, C.D. Partidos, J.-P. Briand, M. Prato, A. Bianco and K. Kostarelos, *J. Am. Chem. Soc.*, 2005, **127**, 4388; (*e*) D. Pantarotto, R. Singh, D. McCarthy, M. Erhardt, J.-P. Briand, M. Prato, K. Kostarelos and A. Bianco, *Angew. Chem. Int. Ed.*, 2004, **43**, 5242.
132. M. Alvaro, P. Atienzar, P. de la Cruz, J.L. Delgado, H. Garcia and F. Langa, *J. Phys. Chem. B.*, 2004, **108**, 12691.
133. J.L. Delgado, P. de la Cruz, F. Langa, A. Urbina, J. Casado and J.T.L. Navarrete, *Chem. Commun.*, 2004, 1734.
134. J.B. Cui, M. Burghard and K. Kern, *Nano Lett.*, 2003, **3**, 613.

135. (a) J.M. Wallis and J.K. Kochi, *J. Am. Chem. Soc.*, 1988, **110**, 8207; (b) W.B. Motherwell and A.S. Williams, *Angew. Chem. Int. Ed.*, 1995, **107**, 2207.
136. (a) Y. Chen, R.C. Haddon, S. Fang, A.M. Rao, W.H. Lee, E.C. Dickey, E.A. Grulke, J.C. Pendergrass, A. Chavan, B.E. Haley and R.E. Smalley, *J. Mater. Res.*, 1998, **13**, 2423; (b) K. Kamaras, M.E. Itkis, H. Hu, B. Zhao and R.C. Haddon, *Science*, 2003, **301**, 1501; (c) H. Hu, B. Zhao, M.A. Hamon, K. Kamaras, M.E. Itkis and R.C. Haddon, *J. Am. Chem. Soc.*, 2003, **125**, 14893.
137. (a) M. Holzinger, O. Vostrovsky, A. Hirsch, F. Hennrich, M. Kappes, R. Weiss and F. Jellen, *Angew. Chem., Int. Ed.*, 2001, **40**, 4002; (b) R. Weiss, S. Reichel, M. Handke and F. Hampel, *Angew. Chem., Int. Ed.*, 1998, **37**, 344; (c) R. Weiss and S. Reichel, *Eur. J. Inorg. Chem.*, 2000, 1935; (d) R. Weiss, N. Kraut and F. Hampel, *J. Organomet. Chem.*, 2001, **617**, 473.
138. W.H. Lee, S.J. Kim, W.J. Lee, J.G. Lee, R.C. Haddon and P.J. Reucroft, *Appl. Surf. Sci.*, 2001, **181**, 121.
139. (a) K.S. Coleman, S.R. Bailey, S. Fogden and M.L.H. Green, *J. Am. Chem. Soc.*, 2003, **125**, 8722; (b) K.A. Worsley, K.R. Moonoosawmy and P. Kruse, *Nano Lett.*, 2004, **4**, 1541.
140. M.J. Moghaddam, S. Taylor, M. Gao, S. Huang, L. Dai and M.J. McCall, *Nano Lett.*, 2004, **4**, 89.
141. (a) P.G. Collins, K. Bradley, M. Ishigami and A. Zettl, *Science*, 2000, **287**, 1801; (b) R.J. Chen, N.R. Franklin, J. Kong, J.J. Cao, T.W. Tombler, Y. Zhang and H. Dai, *Appl. Phys. Lett.*, 2001, **79**, 2258; (c) H. Ulbricht, G. Moos and T. Hertel, *Phys. Rev. B.*, 2002, **66**, 075404; (d) A. Goldoni, R. Larciprete, L. Petaccia and S. Lizzit, *J. Am. Chem. Soc.*, 2003, **125**, 11329.
142. (a) T. Savage, S. Bhattacharya, B. Sadanadan, J. Gaillard, T.M. Tritt, Y.P. Sun, Y. Wu, S. Nayak, R. Car, N. Marzari, P.M. Ajayan and A.M. Rao, *J. Phys. Condens. Matter*, 2003, **15**, 5915; (b) G. Dukovic, B.E. White, Z. Zhou, F. Wang, S. Jockusch, M.L. Steigerwald, T.F. Heinz, R.A. Friesner, N.J. Turro and L.E. Brus, *J. Am. Chem. Soc.*, 2004, **126**, 15269.
143. A.J. Birch, *Quart. Rev.*, 1950, **4**, 69.
144. (a) Y. Chen, R.C. Haddon, S. Fang, A.M. Rao, W.H. Lee, E.C. Dickey, E.A. Grulke, J.C. Pendergrass, A. Chavan, B.E. Haley and R.E. Smalley, *J. Mater. Res.*, 1998, **13**, 2423; (b) S. Pekker, J.-P. Salvelat, E. Jakab, J.-M. Bonard and L. Forro, *J. Phys. Chem. B.*, 2001, **105**, 7938.
145. B.N. Khare, M. Meyyappan and A.M. Cassell, *Nano Lett.*, 2002, **2**, 73.
146. F. Liang, A.K. Sadana, A. Peera, J. Gu, Z. Chattopadhyay, R.H. Hauge and W.E. Billups, *Nano Lett.*, 2004, **4**, 1257.
147. (a) J. Chattopadhyay, A.K. Sadana, F. Liang, J.M. Beach, Y. Xiao, R. H. Hauge and W.E. Billups, *Org. Lett.*, 2005, **7**, 4067; (b) F. Liang, L.B. Alemany, J.M. Beach and W.E. Billups, *J. Am. Chem. Soc.*, 2005, **127**, 13941.
148. A. Hirsch, A. Soi and H.R. Karfunkel, *Angew. Chem. Int. Ed.*, 1992, **31**, 766.

149. (a) G. Viswanathan, N. Chakrapani, H. Yang, B. Wei, H. Chung, K. Cho, C.Y. Ryu and P.M. Ajayan, *J. Am. Chem. Soc.*, 2003, **125**, 9258; (b) *Functionalization of Carbon Nanotubes*, J. Abraham, PhD thesis, Erlangen, 2005.
150. S. Chen, W. Shen, G. Wu, D. Chen and M. Jiang, *Chem. Phys. Lett.*, 2005, **402**, 312.
151. (a) Y. Ying, R.K. Saini, F. Liang, A.K. Sadana and W.E. Billups, *Org. Lett.*, 2003, **5**, 1471; (b) H. Peng, L.B. Alemany, J.L. Margrave and V.N. Khabashesku, *J. Am. Chem. Soc.*, 2003, **125**, 15174.
152. S. Chen, G. Wu, Y. Liu and D. Long, *Macromolecules*, 2006, **39**, 330.
153. A.K. Sadana, F. Liang, B. Brinson, S. Arepalli, S. Farhat, R.H. Hauge, R.E. Smalley and W.E. Billups, *J. Phys. Chem. B.*, 2005, **109**, 4416.
154. Y.H. Yan, M.B. C. Parka, Q. Zhou, C.M. Li and C.Y. Yue, *Appl. Phys. Lett.*, 2005, **87**, 213101.
155. (a) J.L. Bahr, J. Yang, D.V. Kosynkin, M.J. Bronikowski, R.E. Smalley and J.M. Tour, *J. Am. Chem. Soc.*, 2001, **123**, 6536; (b) K. Balasubramanian, M. Friedrich, C. Jiang, Y. Fan, A. Mews and M. Burghard, *Adv. Mater.*, 2003, **15**, 1515; (c) M. Knez, M. Sumser, A.M. Bittner, C. Wege, H. Jeske, S. Kooi, M. Burghard and K. Kern, *J. Electroanal. Chem.*, 2002, **522**, 70; (d) H.C. Leventis, G.G. Wildgoose, I.G. Davies, L. Jiang, T.G.J. Jones and R.G. Compton, *Chem. Phys. Chem.*, 2005, **6**, 590.
156. (a) M.S. Strano, *J. Am. Chem. Soc.*, 2003, **125**, 16148; (b) C.A. Dyke and J.M. Tour, *Nano Lett.*, 2003, **3**, 1215.
157. (a) J.L. Bahr and J.M. Tour, *Chem. Mater.*, 2001, **13**, 3823; (b) C.A. Dyke and J.M. Tour, *J. Am. Chem. Soc.*, 2003, **125**, 1156; (c) C.A. Dyke, M. P. Stewart, F. Maya and J.M. Tour, *Synlett*, 2004, 155; (d) J.L. Hudson, M.J. Casavant and J.M. Tour, *J. Am. Chem. Soc.*, 2004, **126**, 11158; (e) H. Li, F. Cheng, A.M. Duft and A. Adronov, *J. Am. Chem. Soc.*, 2005, **127**, 14518; (f) J.J. Stephenson, J.L. Hudson, S. Azad and J.M. Tour, *Chem. Mat.*, 2006, **18**, 374; (g) B. Chen, A.K. Flatt, H. Jian, J.L. Hudson and J.M. Tour, *Chem. Mater.*, 2005, **17**, 4832; (h) C.A. Dyke and J.M. Tour, *Chem. Eur. J.*, 2004, **10**, 812.
158. B.K. Price, J.L. Hudson and J.M. Tour, *J. Am. Chem. Soc.*, 2005, **127**, 14867.
159. D. Golberg, Y. Bando, W. Han, K. Kurashima and T. Sato, *Chem. Phys. Lett.*, 1999, **308**, 337.
160. B. Khare, P. Wilhite, B. Tran, E. Teixeira, K. Fresquez, D.N. Mvondo, C. Bauschlicher Jr. and M. Meyyappan, *J. Phys. Chem. B.*, 2005, **109**, 23466.
161. G.G. Fuentes, E. Borowiak-Palen, M. Knupfer, T. Pichler, J. Fink, L. Wirtz and A. Rubio, *Phys. Rev. B.*, 2004, **69**, 245403.
162. C. Zhou, J. Kong, E. Yenilmez and H. Dai, *Science*, 2000, **290**, 1552.
163. (a) V. Skákalová, A.B. Kaiser, U. Dettlaff-Weglikowska, K. Hrnčariková and S. Roth, *J. Phys. Chem. B.*, 2005, **109**, 7174; (b) U. Dettlaff-Weglikowska, V. Skakalova, R. Graupner, S.H. Jhang, B.H. Kim, H.J. Lee, L. Ley, Y.W. Park, S. Berber, D. Tomanek and S. Roth, *J. Am. Chem. Soc.*, 2005, **127**, 5125.

164. E. Jouguelet, C. Mathis and P. Petit, *Chem. Phys. Lett.*, 2000, **318**, 561.
165. (*a*) R. Barthos, D. Méhn, A. Demortier, N. Pierard, Y. Morciaux, G. Demortier, A. Fonseca and J.B. Nagy, *Carbon*, 2005, **43**, 321; (*b*) Z. Kónya, I. Vesselenyi, K. Niesz, A. Kukovecz, A. Demortier, A. Fonseca, J. Delhalle, Z. Mekhalif, J.B. Nagy, A.A. Koós, Z. Osváth, A. Kocsonya, L.P. Biró and I. Kiricsi, *Chem. Phys. Lett.*, 2002, **360**, 429.
166. H.L. Pan, L.Q. Liu, Z.X. Guo, L.M. Dai, F.S. Zhang, D.B. Zhu, R. Czerw and D.L. Carroll, *Nano Lett.*, 2003, **3**, 29.
167. X. Li, L. Liu, Y. Qin, W. Wu, Z.-X. Guo, L. Dai and D. Zhu, *Chem. Phys. Lett.*, 2003, **377**, 32.
168. (*a*) J. Zhao, H. Park, J. Han and J.P. Lu, *J. Phys. Chem. B.*, 2004, **108**, 4227; (*b*) E. Bekyarova, M.E. Itkis, N. Cabrera, B. Zhao, A. Yu, J. Gao and R.C. Haddon, *J. Am. Chem. Soc.*, 2005, **127**, 5990.
169. C. Wang, Z.X. Guo, S. Fu, W. Wu and D. Shu, *Prog. Polym. Sci.*, 2004, **29**, 1079.
170. (*a*) S.A. Curran, P.M. Ajayan, W.J. Blau, D.L. Carroll, J.N. Coleman, A.B. Dalton, A.P. Davey, A. Drury, B. McCarthy, S. Maier and A. Strevens, *Adv. Mater.*, 1998, **10**, 1091; (*b*) J.N. Coleman, S. Curran, A.B. Dalton, A.P. Davey, B. McCarthy, W.J. Blau and R.C. Barllie, *Phys. Rev. B.*, 1998, **58**, R7492; (*c*) R. Murphy, J.N. Coleman, M. Cadek, B. McCarthy, M. Bent, A. Drury, R.C. Barklie and W.J. Blau, *J. Phys. Chem. B.*, 2002, **106**, 3087; (*d*) J.N. Coleman, A.B. Dalton, S. Curran, A. Rubio, A.P. Davey, A. Drury, B. McCarthy, B. Lahr, P.M. Ajayan, S. Roth, R.C. Barllie and W.J. Blau, *Adv. Mater.*, 2000, **12**, 213.
171. (*a*) R. Ramasubramanian and J. Chen, *Appl. Phys. Lett.*, 2003, **83**, 2928; (*b*) S. Curran, A.P. Davey, J.N. Coleman, R. Czerw, A.B. Dalton, B. McCarthy, S. Maier, A. Drury, D. Gray, M. Brennan, K. Ryder, M. Lamy de la Chapelle, C. Journet, P. Bernier, H. J. Byrne, D. Carroll, P.M. Ajayan, S. Lefrant and W.J. Blau, *Synth. Met.*, 1999, **103**, 2559.
172. (*a*) B. McCarthy, J.N. Coleman, R. Czerw, A.B. Dalton, D.L. Carroll and W.J. Blau, *Synth. Met.*, 2001, **121**, 1225; (*b*) B. McCarthy, A.B. Dalton, J.N. Coleman, H.J. Byrne, P. Bernier and W.J. Blau, *Chem. Phys. Lett.*, 2001, **350**, 27.
173. M. Grujicic, G. Cao and W.N. Roy, *Appl. Surf. Sci.*, 2004, **227**, 349.
174. A. Star, J.F. Stoddart, D. Steuerman, M. Diehl, A. Boukai, E.W. Wong, X. Yang, S.-W. Chung, H. Choi and J.R. Heath, *Angew. Chem., Int. Ed.*, 2001, **40**, 1721.
175. D.V. Steuerman, A. Star, R. Narizzano, H. Choi, R.S. Ries, C. Nicolini, J.F. Stoddart and J.R. Heath, *J. Phys. Chem. B.*, 2002, **106**, 3124.
176. A. Star, Y. Liu, K. Grant, L. Ridvan, J.F. Stoddart, D.W. Steuerman, M.R. Diehl, A. Boukai and J.R. Heath, *Macromolecules*, 2003, **36**, 553.
177. (*a*) J. Fan, M. Wan, D. Zhu, B. Chang, Z. Pan and S. Xie, *Synth. Met.*, 1999, **32**, 2569; (*b*) J. Fan, M. Wan, D. Zhu, B. Chang, Z. Pan and S. Xie, *J. Appl. Polym. Sci.*, 1999, **74**, 2605.
178. B.Z. Tang and H. Xu, *Macromolecules*, 1999, **32**, 2569.

179. (a) M. Cochet, W.K. Maser, A.M. Benito, M.A. Callejas, M.T. Martinez, J.M. Benoit, J. Schreiber and O. Chauvet, *Chem. Commun.*, 2001, 1450; (b) H. Zengin, W. Zhou, J. Jin, R. Czerw, D.W. Smith, L. Echegoyen, D.L. Carroll, S.H. Foulger and J. Ballato, *Adv. Mater.*, 2002, **14**, 1480.
180. M. Hughes, G.Z. Chen, M.S.P. Shaffer, D.J. Fray and A.H. Windle, *Comp. Sci Tech.*, 2004, **64**, 2325.
181. J. Chen, H. Liu, W.A. Weimer, M.D. Halls, D.D. Waldeck and G.C. Walker, *J. Am. Chem. Soc.*, 2002, **124**, 9034.
182. J.H. Ryu, J. Bae and M. Lee, *Macromolecules*, 2005, **38**, 2050.
183. A. Star, D.W. Steuerman, J.R. Heath and J.F. Stoddart, *Angew. Chem. Int. Ed.*, 2002, **41**, 2508.
184. R.J. O'Connel, P.J. Boul, L.M. Ericson, C. Huffman, Y. Wang, E. Haroz, C. Kuper, J. Tour, K.D. Ausman and R.E. Smalley, *Chem. Phys. Lett.*, 2001, **342**, 265.
185. (a) H. Ago, K. Petritsch, M.S.P. Shaffer, A.H. Windle and R.H. Friend, *Adv. Mater.*, 1999, **11**, 1281; (b) H. Ago, M.S.P. Shaffer, D.S. Ginger, A.H. Windle and R.H. Friend, *Phys. Rev. B.*, 2000, **61**, 2286.
186. X. Gong, J. Liu, S. Baskaran, R.D. Voise and J.S. Young, *Chem. Mater.*, 2000, **12**, 1049.
187. Z. Jin, K.P. Pramoda, S.H. Goh and G. Xu, *Mater. Res. Bull.*, 2002, **37**, 271.
188. (a) P. Petrov, F. Stassin, C. Pagnoulle and R. Jernome, *Chem. Commun.*, 2003, 2904; (b) X. Lou, R. Daussin, S. Cuenot, A.-S. Duwez, C. Pagnoulle, C. Detrembleur, C. Bailly and R. Jernome, *Chem. Mater.*, 2004, **16**, 4005.
189. X. Wu and G. Shi, *J. Mater. Chem.*, 2005, **15**, 1833.
190. F.J. Gomez, R.J. Chen, D. Wang, R.M. Waymouth and H. Dai, *Chem. Commun.*, 2003, 190.
191. D.M. Guldi, H. Taieb, G.M.A. Rahman, N. Tagmatarchis and M. Prato, *Adv. Mater.*, 2005, **17**, 871.
192. A. Satake, Y. Miyajima and Y. Kobuke, *Chem. Mater.*, 2005, **17**, 716.
193. J.H. Rouse, P.T. Lillehei, J. Sanderson and E.J. Iochi, *Chem. Mater.*, 2004, **16**, 3904.
194. (a) B. Kim, H. Park and W.M. Sigmund, *Langmuir*, 2003, **19**, 2525; (b) B. Kim, H. Park and W.M. Sigmund, *Langmuir*, 2003, **19**, 4848.
195. D.-Q. Yang, J.-F. Rochette and E. Sacher, *J. Phys. Chem. B.*, 2005, **109**, 4481.
196. (a) H. Paloniemi, T. Ääritalo, T. Laiho, H. Luike, N. Kocharova, K. Haapakka, F. Terzi, R. Seeber and J. Lukkari, *J. Phys. Chem. B.*, 2005, **109**, 8634; (b) H. Paloniemi, M. Lukkarinen, T. Ääritalo, S. Areva, J. Leiro, M. Heinonen, K. Haapakka and J. Lukkari, *Langmuir*, 2006, **22**, 74; (c) D.M. Guldi, I. Zilbermann, G. Anderson, N.A. Kotov, N. Tagmatarchis and M. Prato, *J. Am. Chem. Soc.*, 2004, **126**, 14340; (d) D.M. Guldi, I. Zilbermann, G. Anderson, N.A. Kotov, N. Tagmatarchis and M. Prato, *J. Mater. Chem.*, 2005, **15**, 114; (e) D.M. Guldi and M. Prato,

*Chem. Commun.*, 2004, 2517; (*f*) A.A. Mamedov, N.A. Kotov, M. Prato, D.M. Guldi, J.P. Wicksted and A. Hirsch, *Nat. Mat.*, 2002, **1**, 190.
197. F. Tournus, S. Latil, M.I. Heggie and J.-C. Charlier, *Phys. Rev. B.: Cond. Matter and Mat. Phys.*, 2005, **72**, 075431/1.
198. R.J. Chen, Y. Zhang, D. Wang and H. Dai, *J. Am. Chem. Soc.*, 2001, **123**, 3838.
199. (*a*) D.M. Guldi, G.M.A. Rahman, N. Jux, N. Tagmatarchis and M. Prato, *Angew. Chem. Int. Ed.*, 2004, **43**, 5526; (*b*) D.M. Guldi, G.M.A. Rahman, N. Jux, D. Balbinot, U. Hartnagel, N. Tagmatarchis and M. Prato, *J. Am. Chem. Soc.*, 2005, **127**, 9830; (*c*) D.M. Guldi, G.M.A. Rahman, V. Sgobba, S. Campidelli, M. Prato and N.A. Kotov, *J. Am. Chem. Soc.*, 2006, **128**, 2315.
200. (*a*) L. Liu, T. Wang, J. Li, Z.-X. Guo, L. Dai, D. Zhang and D. Zhu, *Chem. Phys. Lett.*, 2003, **367**, 747; (*b*) V. Georgakilas, V. Tzitzios, D. Gournis and D. Petridis, *Chem. Mater.*, 2005, **17**, 1613; (*c*) Y.-Y. Ou and M.H. Huang, *J. Phys. Chem. B.*, 2006, **110**, 2031.
201. Y. Yan, M. Zhang, K. Gong, L. Su, Z. Guo and L. Mao, *Chem. Mater.*, 2005, **17**, 3457.
202. H. Murakami, T. Nomura and N. Nakashima, *Chem. Phys. Lett.*, 2003, **378**, 481.
203. G.M.A. Rahman, D.M. Guldi, S. Campidelli and M. Prato, *J. Mater. Chem.*, 2006, **16**, 62.
204. J. Chen and C.P. Collier, *J. Phys. Chem. B.*, 2005, **109**, 7605.
205. N. Nakashima, Y. Tomonari and H. Murakami, *Chem. Lett.*, 2002, **6**, 638.
206. K.A.S. Fernando, Y. Lin, W. Wang, S. Kumar, B. Zhou, S.-Y. Xie, L. T. Cureton and Y.-P. Sun, *J. Am. Chem. Soc.*, 2004, **126**, 10234.
207. (*a*) M. Zheng, A. Jagota, M.S. Strano, A.P. Santos, P. Barone, S.G. Chou, B.A. Diner, M.S. Dresselhaus, R.S. Mclean, G.B. Onoa, G.G. Samsonidze, E.D. Semke, M. Usrey and D.J. Walls, *Science*, 2003, **302**, 1545; (*b*) M. Zheng, A. Jagota, E.D. Semke, B.A. Diner, R.S. Mclean, S.R. Lustig, R.E. Richardson and N.G. Tassi, *Nature Mater.*, 2003, **2**, 338; (*c*) M.S. Strano, M. Zheng, A. Jagota, G.B. Onoa, D.A. Heller, P.W. Barone and M.L. Usrey, *Nano Lett.*, 2004, **4**, 543.
208. (*a*) S.M. Bachilo, M.S. Strano, C. Kittrell, R.H. Hauge, R.E. Smalley and R.B. Weisman, *Science*, 2003, **301**, 2361; (*b*) M.S. Strano, V.C. Moore, M.K. Miller, M.J. Allen, E.H. Haroz, C. Kittrell, R.H. Hauge and R.E. Smalley, *J. Nanosci. Nanotechnol.*, 2003, **3**, 81.
209. R. Krupke, F. Hennrich, H. Löhneysen and M.M. Kappes, *Science*, 2003, **301**, 344.
210. H. Li, B. Zhou, Y. Lin, L. Gu, W. Wang, K.A.S. Fernando, S. Kumar, L.F. Allard and Y.-P. Sun, *J. Am. Chem. Soc.*, 2004, **126**, 1014.
211. (*a*) E.V. Basiuk, V.A. Basiuk, J.-G. Banuelos, J.-M. Saniger-Blesa, V. A. Pokrovskiy, T.Y. Gromovoy, A.V. Mischanchuk and B.G. Mischanchuk, *J. Phys. Chem. B.*, 2002, **106**, 1588; (*b*) J. Kong and H. Dai, *J. Phys. Chem. B*, 2001, **105**, 2890.

212. Y. Maeda, S. Kimura, M. Kanda, Y. Hirashima, T. Hasegawa, T. Wakahara, Y. Lian, T. Nakahodo, T. Tsuchiya, T. Akasaka, J. Lu, X. Zhang, Z. Gao, Y. Yu, S. Nagase, S. Kazaoui, N. Minami, T. Shimizu, H. Tokumoto and R. Saito, *J. Am. Chem. Soc.*, 2005, **127**, 10287.
213. (*a*) P.G. Collions, M.S. Arnold and P. Avouris, *Science*, 2001, **292**, 706; (*b*) K. Balasubramanian, R. Sordan, M. Burghard and K. Kern, *Nano Lett.*, 2004, **4**, 827; (*c*) L. An, Q. Fu, C. Lu and J. Liu, *J. Am. Chem. Soc.*, 2004, **126**, 10520.

# Subject Index

absorption spectroscopy. *See also* transient absorption spectroscopy
   CNT-porphyrin polymer interactions, 355
   [60]fullerene and its derivatives, 80–1, 95–6
   fullerene inside dendrimers, 90–4, 96–8
   of fullerene-porphyrin arrays, 111–12
   polymer-fullerenes and solar emission, 234
   UV-Vis of LB films, 130, 133, 139
acetonitrile
   $C_{60}$ cluster formation, 283–7
   electrode deposition from, 289–92, 294
   fullerodendrimer absorption spectrum in, 90, 94, 98
   oxygen trapping, 99
acridinium, 9-mesityl-10-carboxymethyl-ion, 285
acylated glucose-based fullerodendrimers, 131
adamantyl fullerene derivatives, 319
AFM. *See* atomic force microscopy
aggregation
   $C_{60}$-$C_{60}$ interactions, 128, 130–4, 137
   carbon nanotubes, 331
   J-aggregates, 361
   polymer-fullerene bulk heterojunctions, 234
   toxicity of $C_{60}$ aggregates, 305
air-water interface, monolayers, 128, 136–7, 205–6, 252–3
β–alanine fullerene derivatives, 311
alkali metals, 347

alkanethiols, 272, 283
alkylation, 54–5, 339, 347–9
all-polymer type solar cells, 238
aluminosilicate minerals, 208
AM1.5 spectrum, defined, 226
amine derivatives of CNTs, 337–40
amine physisorption, 362–3
amino acid derivatives, 21, 301–2, 343
ammonium-crown ether complexes, 157–60, 165, 175, 196–7
amphiphilic derivatives, 72–3, 128, 132, 135–8
   as potential optical limiters, 205–11
androstane, 67–8
aniline, N,N-dialkyl derivatives, 249, 285
annealing, 232–3
antenna structures
   boron dipyrrin-alkanethiols, 272
   fulleropyrrolidine/OPV dyads, 84, 86, 88
   MDMO-PPV with $C_{60}$-phthalocyanines, 242
   perylenediimide-methanofullerenes, 239
   porphyrin containing complexes, 25, 274
   suitability for artificial photosynthesis, 153
anthracene derivative cycloadditions, 21–2, 24, 28–9
anthraquinone derivatives, 24, 62, 285
anti-cancer drugs, 308, 311
antibacterial activity, 317–19
antigenic properties, 29, 319–22
antioxidant activities, 52, 310–12
antiviral activity, 51, 317–19

## Subject Index

APFO Green, 235, 238
apoptosis, 314, 316
applications of CNTs, 114–15, 179, 330, 338
  biomedical applications, 308, 342–3, 357
  electronic devices, 259, 361–2
applications of fullerenes, 114–15
  biomedical applications, 51, 80, 172, 301–24
  materials science, 79, 149, 191–211
  molecular devices, 52, 295
  nanotechnology, 65–6, 134, 149
  nonlinear optics, 200–5
arc-discharge production, 3–4, 6, 331, 361
arginine derivatives, 302
aromaticity, 7, 53, 329–30
Arrhenius A-factors, 6
artificial photosynthesis, 63, 153, 191, 221
  candidate systems, 25, 28, 109
aryl substitution, 53, 55–9
arylation of CNTs, 348–50
association constants, 153, 157–8, 167
atomic force microscopy (AFM), 141, 176, 209, 315
  CNT investigations, 178, 331, 352, 354–5, 357–61
  CNTs on gold nanoparticles, 337, 345
  thin film characterization, 207, 293
autoimmune diseases, 320
axial coordination, 154–5, 193–5, 206
axle length, rotaxanes, 155
azacrowns, 128
azafullerene-based linear complexes, 194
aziridino-CNT derivatives, 345–6
aziridinofullerenes, 293–4, 315
azomethine ylides, 26–30, 36, 39, 340–2
azothiophene-fullerene dyads, 249–50

ball-milling. *See* mechanochemical functionalization
BAM (Brewster Angle Microscopy), 129–30, 206
barbituric acid derivatives, 162–4, 254, 257, 293–4
Bathocuproine, 244–6
Baytron-P, 228

BcP (2,9-dimethyl-4,7-diphenyl-1, 10-phenanthroline), 244–6
benzene combustion method, 4
benzocyclobutenes, 25
benzonitrile, 101, 103
bilayers
  heterojunction devices, 227–30, 243
  LBL deposition, 169
  vesicle formation, 283, 303
Bingel adduct, 60–1
Bingel reactions (cyclopropanation), 15–21
  carbene addition to CNTs, 344–5
  electron donor links, 192–3
  higher fullerenes, 42-43
  iodinated malonic acid, 323
  mesomorphic methanofullerenes, 65
  multiaddition reactions, 36–7
  porphyrin dyads, 192
  retro-Bingel reactions, 18–21, 42, 60–2, 69
bio-electrochemical sensors, 71–2
biomedical applications of CNTs, 342–3, 357
biomedical applications of fullerenes, 51, 80, 172, 301–24
bipyridine, 66, 254–5
Birch reductions, 346–7
birefringence, 140
*bis*-adducts
  [60]fullerene, 32, 36–7
  fulleropyrrolidines, 54
  with cholesterol groups, 129–30
*bis*-fulleropyrrolidinium salts, 317–19
1,5-*bis*-(hydroxyethoxyethoxy) naphthalene, 175
*bis*-(imidazolylporphyrinatozinc), 274
*bis*-malonates, 36–7
*bis*-methanofullerenes
  dendrimers, 91, 94–9, 106–8
  multilayer formation by, 134–7
  with acylated glucose units, 131
  with cholesterol groups, 129–30
*bis*-*o*-quinodimethanes, 40
bond-isomerization reaction, 58
bond length and reactivity, 41
bond strength, C-Cl in $C_{60}Ph_5Cl$, 57–9
boron-dipyrrin alkanethiols, 272
boron subphthalocyanines, 195

Brewster Angle Microscopy (BAM), 129–30, 206
bromination of higher fullerenes, 43
BTPF [3-(3,5-bis-trifluoromethyl-phenyl)-1-(4-nitrophenyl) pyrazolino[60/70]fullerenes], 238
bulk heterojunction devices, 224, 230–42, 257, 259
  alternatives to, 242–3
  dye-sensitized, 286–7, 289
  LBL assemblies as, 169
  vacuum deposition of, 281–3
butadiene sulfone, 197

$C_{60}$. See [60]fullerene under 'f'
$C_{120}$ dimer, 7
cadmium selenide, 222, 337
cadmium sulfide, 222
calixarenes, 128, 170, 172–3, 278–9
calmodulin, 319
carbazoles, 168
carbenes, 16, 344–5
  carbodiimides, 333
carbohydrate-based dendrimers, 131
carbon
  all-carbon cations, 72–3
  allotropes, 74, 329–30
  electronegativity and hybridization, 53
  in arc-discharge reactors, 3
  in chromatographic columns, 5
carbon monoxide (HiPco) process, 331, 347, 362
carbon nanotubes (CNT). See also applications of CNTs
  chemical doping, 349–50, 361
  conductivity, 329, 361
  electrocatalytic properties, 71
  electrostatic interactions, 178–81
  encapsulation by dendritic polymers, 354
  end tip/defect functionalization, 331–8
  fullerene interactions with, 74
  larger nanostructures from, 335
  non-covalent functionalization, 172–81, 350–61
  polymer composites with, 354
  polymer wrapping, 174–6
  production processes, 331, 361
  purification, 329, 333–5, 351
  separation of metallic and semiconducting, 361–3
  sidewall chemistry, 338–50
carboranes, 203
carboxylic acids
  CNT functionalization, 333–8
  fulleropyrrolidine carboxylic acids, 168–9, 267–8
  methanofullerene carboxylic acids, 292–3, 308–10
catalytic decomposition process for CNTs, 331, 361
catalyticaly active vanadyl CNT complexes, 335–6
catenanes, 17, 195
cationic behaviour of $C_{60}$, 72–3
C61–butyric acid, [6,6]-phenyl, methyl ester (PCBM), 229–30
CD. See cyclodextrines
cell permeability to CNTs, 335
charge separation
  bis-methanofullerene/OPV systems, 104–5
  donor-acceptor polyads, 198, 247
  fullerene-zinc porphyrin rotaxanes, 155–7
  fulleropyrrolidine/OPV systems, 83–4
  hydrogen bonded donor-acceptor systems, 154
charge transfer
  CT absorption bands, $C_{60}$-porphyrin arrays, 109–10
  metal-to-ligand-charge-transfer (MLCT) states, 66, 68, 105–6, 114
chemical adsorption onto electrodes, 292–5
chemically sensitive imaging, 337
chemotherapeutic potential, 324
chiral tethers, 37
chirality, 27, 142, 192
  amino acid derivatives, 301
  carbon nanotubes, 329, 335, 361–2
  helical copolymer arrays, 171
  higher fullerenes, 41, 43, 236
  tether controlled syntheses, 36–8
cholesterol, 129, 142
chromatography
  characterization of [60] and [70]fullerenes, v
  separation of carbon nanotubes, 335

Subject Index

separation of fullerenes, 5, 36, 41
separation of *incar*fullerenes, 7–8
clays, 208–9
CNT. See carbon nanotubes
$C_{120}O$ as a contaminant, 5
colours in solution, 1, 80
combustion of benzene, 4
concanavalin A, 303
conducting polymers, 71, 167, 176, 351–2
conductivity, 167, 350–1
conjugated oligomers, 28, 242, 245
  as potential electron donors, 71, 193
  compared to polymers, 247–8
  hydrogen bonding, 158–63
conjugated polymers, 168, 221, 293
  compared to oligomers, 247–8
  in heterojunctions with $C_{60}$, 228, 230, 242–3
  in photovoltaics, 86, 170, 254
controlled-potential electrolysis (CPE), 19–21, 62
coordination compounds, 67, 79, 105
copolymers, interpenetrating, 168
copper-*bis*-phenanthroline complexes, 106–9
copper phthalocyanines, 224, 244
corrannulene, 4
counter-ions, 308, 350
covalent functionalization of CNTs, 361
CPE (controlled-potential electrolysis), 19, 62
Criegee mechanism, 340, 342
cross-sectional area. See molecular area
18-crown-6 aldehydes, dibenzo-, 39
18-crown-6 ether, 20–1, 36–7
crown ethers
  $C_{60}$ inclusion polymers with, 170
  1,10-dinaphtho[38]crown-10 ether, 17
  fullerene-ammonium complexes, 160, 175, 196–7
  fullerene-porphyrin conjugates with, 157–8
  molecular devices containing, 69–70
  *syn*- and *anti*-conformations, 161
cup-and-ball conjugates, 157
Curtius rearrangement, 337
CV (cyclic voltammetry) experiments, 52, 54–6, 58–9, 69, 73
CVT (cyclotriveratrylene) inclusion complexes, 140–1

cyanobiphenyl groups, 144–6
cyclic voltammetry (CV) experiments, 52, 54–6, 58–9, 69, 73
  digital simluation and, 59, 61–2, 67
cycloaddition reactions. See also Bingel reactions
  1,3-dipolar cycloadditions of [60]fullerenes, 17, 34, 144, 148, 193
  1,3-dipolar cycloadditions of [70]fullerenes, 42–3, 238
  1,3-dipolar cycloadditions to CNTs, 340–3
  [4 + 2] cycloadditions (*See* Diels-Alder reactions)
  [2 + 1] cycloadditions of carbon nanotubes, 344–6
  [2 + 2] cycloadditions of carbon nanotubes, 346
  [3 + 2] cycloadditions of [60]fullerenes, 26–35
  liquid crystalline fullerene derivatives from, 144–8
  regioselectivity of, 36
cyclobis(paraquat-*p*-phenylene) cyclophane
  1,5-bis-[(2-hydroxyethoxy)ethoxy] naphthalene complex, 273–4
cyclodextrines (CD)
  $C_{60}$ inclusion polymers with, 170–1
  self-assembled multilayers, 279
  soluble $C_{60}$ conjugates with, 303
cyclodextrin–fullerene complexes, 311, 313–14
cyclopentadiene, 21–2
cyclophanes, 273–4
cyclopropanation. See Bingel reactions
cyclotriveratrylene (CVT) inclusion complexes, 140–1
cytochrome C, 281
cytotoxicity of fullerenes, 305, 307
cytotoxicity of singlet oxygen, 312

DABCO (1,4-diazabicyclo[2.2.2]octane), 277
dangling bonds, 5–6
DCC (1,3-dicyclohexylcarbodiimide), 333
debundling of CNTs, 331, 347, 359
defect functionalization of CNTs, 331–2

degradation in air, 5
dehydro(halogen)genation of PAHs, 4
dehydronaphthalenes, 4
dendrimers. *See also* fullerodendrimers
  *bis*-adducts to $C_{60}$, 36–7
  dendritic polymers and CNTs, 354
  dendritic porphyrins, 141, 279–89
  fulleropyrazolines combined with, 33
  fulleropyrrolidines combined with, 28
  PAMAM-$C_{60}$ multilayers, 279–80
  potential applications, 88–9
derivatization. *See* functionalization
1,4-diazabicyclo[2.2.2]octane (DABCO), 277
diazo compounds
  CNT arylation using, 349
  [60]fullerene cyclopropanation using, 16, 229
  metallic CNT separation with, 363
diblock fullerodendrimers, 131–2, 136, 138–9
diblock polymers, 168–9
dichloromethane
  fluoresecence spectra of $C_{60}$ derivatives in, 82
  fullerodendrimer absorption spectra in, 90, 93–4, 97
  fullerodendrimer fluorescence spectra in, 100–2
  spectra of $C_{60}$ derivatives in LB films and, 133
Diels-Alder reactions
  amino acid derivatives, 302
  avoiding, 17
  fullerene-porphyrin dyads, 192, 196
  fullerene-TTF dyads, 17, 202–4
  [60]fullerenes, 21–6, 39, 193, 302
  higher fullerenes, 42–3
  retro-Diels-Alder reactions, 22–3
  single-walled CNTs, 340, 344
diethyl ether, 111
dioxetane-CNT derivatives, 346
1, 3-dipolar cycloadditions, 148, 340–3
  with azomethine ylides, 17, 42–3
  CNTs, 340–3
  DPM-12, 238
  fulleropyrazolines, 34
  fulleropyrrolidines, 42–3, 144
  TTF-fullerenes, 17, 193

4,4′-dipyridinium, 1,1′-dihexyl-, diperchlorate, 275
DNA (deoxyribonucleic acid)
  complexing, 309
  photocleavage, 172, 312–16
  wrapping of CNTs, 362
donor-acceptor systems, 191–200
  as potential optical limiters, 202–5
  covalently linked, 192–3, 200
  heterojunctions, 225, 230–42
  polyads, 198–200
  self-assembly in, 210–11, 275, 277
  supramolecular interactions for, 193–8
donor-spacer-acceptor (DSA) systems, 52, 63, 67
  spacer flexibility effects, 67, 164, 192
"double-cable" polymers, 86, 242–3
doxorubicin, 311
DPM-12 [1,1-bis(4,4′-dodecyloxyphenyl)-(5,6)methanofullerene], 235–7
DPPH (2,2-diphenyl-1-picryhydrazyl) radical, 310
drug delivery applications, 80, 307–10, 323
DSA. *See* donor-spacer-acceptor systems
dumbbell-like molecules, 28, 33, 84–5, 87, 172
dye-sensitized solar cells, 223, 283, 286–7, 289

EDC (1-ethyl-3-(3-dimethylaminopropyl) carbodiimide hydrochloride), 333
EDOT (3,4-ethylenedioxythiophene), 169–70
EDX (energy-dispersive X-ray spectroscopy), 333
electrochemistry
  *incar*fullerenes, 9
  induced reactivity, 58–62
  of $C_{60}$ and $C_{70}$, 52–8
  oxidative, of $C_{60}$, 72–3
  spectroelectrochemistry, 67
electrodes, CNT applications as, 355
electrodes, fullerene fabrication onto
  chemical adsorption and spin coating, 292–5
  electrochemical deposition, 283–92

Langmuir-Blodgett films, 267–8
layer-by-layer deposition, 280–1
LiF/Al electrodes, 231
self-assembled monolayers, 269–80
vacuum deposition, 281–3
electroluminescence, 351
electrolysis, controlled potential, 19, 62
electron acceptors
  alkylfulleropyrrolidines as, 54
  aryl[70]fullerenes as, 56
  [60]fullerene derivatives as, 63
  tranullene derivatives as, 69, 71
electron donor candidates, 79, 83, 193
electron paramagnetic/spin resonance (EPR/ESR), 6, 310–11
electron transfer
  bilayer heterojunction devices, 227
  bis-methanofullerene/OPV systems, 100, 102–5
  copper-bis-phenanthroline complexes, 106–8
  fullerene-porphyrin systems, 109, 113, 172, 174, 277
  fulleropyrazoline/OPV systems, 85–7
  fulleropyrrolidine/OPV systems, 83
  intermolecular, 197
  intramolecular, 196
  metal complexes with $C_{60}$, 105
  non-photoactive electron donors, 193–4
  oligothienylenevinylene/$C_{60}$ systems, 162–3
  triphenylamine (TPA) -$C_{60}$ complexes, 165–6
  TTF-containing dyads, 70, 164
electronegativity increases, 53
electrophoresis, 335, 362
electropolymerization, 71
encapsulation. See DNA wrapping; endohedrals; fullerodendrimers; polymer wrapping
end-capping, 156, 168, 171
end tip/defect functionalization, CNTs, 331–8
endohedrals, 6–9, 115, 322–3
energy-dispersive X-ray spectroscopy (EDX), 333
energy-level diagrams, 84–5

energy transfer
  bis-methanofullerene/OPV systems, 100
  copper-bis-phenanthroline complexes, 106, 107
  fulleropyrrolidine/OPV dyads, 84
environmental impacts of fullerenes, 304–7
enzyme inhibition, 51–2, 311, 317–19
EPR. See electron paramagnetic resonance
erythropoietin delivery, 308
ESR. See electron spin resonance
ethanesulfonates, 278, 335
ether (diethyl), 111
ex-TTFs (π-extended tetrathiafulvalenes), 193, 200, 247
EXAFS (X-ray adsorption fine structure) spectroscopy, 331
exciton diffusion length, 244

ferrocene derivatives, 142, 203
  electrode deposition and, 282, 285
  electron donation by, 28, 193, 198–9
  porphyrin-$C_{60}$ triads in SAMs, 252, 271–2, 277–8
  SWNT sidewall functionalization, 343
ferrocene-formanilide-anthraquinone complexes, 285
field effect transistors, 363
field gating effects, 361
fill factors, photovoltaics, 225, 233, 257, 286, 289, 291
flexibility
  bis-methanofullerene/OPV systems, 103–4
  dendrimers, 148
  fullerene-[Ru(bpy)$_3$]$^{2+}$ systems, 67
  of linkages, 37, 67, 161, 164, 239
  prophyrin complexes, 25, 280
fluoranthenes, 4
fluorene copolymers, 235, 238
9-fluorenylmethyloxycarbonyl (Fmoc), 302, 320, 343
fluorescence spectra
  of $C_{60}$ and its derivatives, 80, 82
  of fullerodendrimers in various solvents, 92–4, 101–3
  of methano- and bis-methanofullerenes, 108

fluorescence switches, 23
fluorination of CNTs, 338–9
fluorofullerene $C_{60}F_{18}$, 53
Fmoc (9-fluorenylmethyloxycarbonyl), 302, 320, 343
formanilide–anthraquinone dyads, 285
Förster energy transfer, 83–6
FTIR (Fourier transform infrared spectrometry), 347, 356
fullerenes. *See also* applications of fullerenes; methanofullerenes *and individual fullerenes below*
  discovery of, v
  electrochemistry of, 53–8
  fabrication onto electrodes, 266–95
  higher order fullerenes, 1, 40–3, 72, 236
  incarcerated elements, 6–9, 115, 322–4
  interactions with porphyrins, 17, 25
  in molecular devices, 63–72
  oxidative electrochemistry of, 72–3
  reactivity compared to graphite and CNTs, 330
  stability, 5–6, 8
fullerene aqueous solutions (C60FAS), 313–14
fullerene-carborane dyads, 203
fullerene-oligonucletide conjugates, 313
fullerene-zinc porphyrin dyads, 155
[60]fullerene
  aggregation, 128, 132–4
  arrays with porphyrins, 109–13
  benzene combustion method, 4
  chromatographic elution, 5
  encapsulation in inclusion complexes, 140–1
  encapsulation in supramolecular polymers, 174
  fluorination, 53
  health and environmental impact, 304–7
  heterojunctions with conjugated polymers, 228
  incarceration of gases, 6–7
  photophysical effects, 80–1
  polyadducts, 39–40
  purification, 4–5
  pyrolytic condensation method, 4
  reduction potentials, 30–1, 53
  structure of [60-$I_h$]fullerene, 2
  sulfur-containing derivatives, linked to Au electrodes, 270
[60]fullerene, hexa(sulfo-*n*-butyl)-, 253, 314
[60]fullerene, 1,2-(4'-oxocyclohexano), 302
[60]fullerene-based polymers, 166–72, 238–9
[60]fullerene chemistry, 15–40
  *bis*-adducts, 32
  [3 + 2]cycloaddition reactions, 26–35
  cyclopropanation, 15–21
  Diels-Alder reactions, 21–6
  electrochemically induced reactivity, 58–62
  halogenation, 54
  multiaddition reactions, 35–40
  six-stage reduction, 52
  *tris*-adducts, 38–9
[60]fullerene clusters, 283
[60]fullerene dimer ($C_{120}$), 7
[70]fullerenes
  chirality, 236
  inert gas incarceration, 6–7
  porphyrin complexes, 291
  preparation, 4–5
  structure of [70-$D_{5h}$]fullerene, 2
[70]fullerene chemistry
  Diels-Alder reactions, 42–3
  electrochemical oxidation, 72
  halogenation, 54
  methanofullerene DPM-12, 237–8
  PCBM ester, 233, 236
  polyphenylated derivatives, 55
  reactivity, 40–3
  six-stage reduction, 52
[74]fullerenes, 7
[76]fullerenes, 2, 40–3, 72
[78]fullerenes, 2, 5, 72
[82]fullerenes, 7, 8, 322–3
[84]fullerenes, 2, 8–9, 40–3
fullerenols, 315, 318
  endohedral $C_{82}$-based, 322–3
  polyfunctional, 167, 305, 310–11
fulleroaziridines, 293–4, 315
fullerodendrimers, 22–3, 86
  as amphiphilic derivatives, 130, 132
  as optical limiters, 202, 204
  core protection in, 97, 131, 133–4, 146, 202

*Subject Index* 387

with acylated glucose, 131
diblock globular dendrimers, 131–2, 136, 138–9
fluorescence spectra of, 92–4, 101–3
fullerene inside approach, 89–99, 114, 129–32
fullerene outside approach, 99–105, 114, 134–7, 203
Langmuir-Blodgett films from, 128–39
liquid crystalline fullerene-ferrocene, 66, 142–8
methanofullerene-based, 129–39
photophysical properties, 88–105
phototoxicity, 306
polar headgroups and spreading behaviour, 128, 132–4, 137–8
fullerohelicates, 107
fulleroid conversion to methanofullerenes, 58–60, 229
[60]fulleropyrazolines, 32–5, 238
OPV dyads, 84–5, 87
[70]fulleropyrazolines, 42, 238
fulleropyrrolidines
*bis*-adducts, 38–9, 71, 253, 317–19
in fullerene-oligonucletide conjugates, 313
in fulleropeptides, 321–2
in Langmuir films, 205–6, 252–3, 267–8
mono- and *bis*-adducts, 54, 56–7
oligomer-tetrafullerene derivative, 242
porphyrin-containing triads, 271–4
preparation by cycloaddition, 28–30
ruthenium tris-bipyridine complexes, 67, 255
self-assembled monolayers, 269
self-assembled nanostructures, 208–11
smectite clay incorporation, 208
1,2,5-triphenyl-, composites with donors, 285
fulleropyrrolidine carboxylic acids, 168–9, 267–8
fulleropyrrolidine dendrimers
core protection in, 97, 131–3, 146, 202
fullerene-inside dendrimers, 90–4, 144–8, 202
mesomorphic, 66
fulleropyrrolidine dyads
as optical limiters, 201–2, 204
electrodeposition of aniline-linked dyads, 283

glutamic acid derivatives, 302–3, 320
hydrogen bonded dyads, 196–8
oligophenylenevinylene (OPV) dyads, 81–9, 148–9, 248–50
oligothiophene-$C_{60}$ systems, 249–50
phthalocyanine dyads, 192–3
porphyrin dyads, 192–5, 267–70, 275–6, 280–1, 284–5
pyrene-fulleropyrrolidine geometry, 74
TTF and ex-TTF dyads, 193
fulleropyrrolidine polymers, 170, 243
[70]fulleropyrrolidines, 42
fulleropyrrolidinium salts, 30–1, 54–5, 317–19
fulleroquinazolines, 302
fulleroquinoxalines, 24
functionalization
*bis*-functionalized fullerenes, 33
by fullerenes (*See entity functionalized*)
carbon nanotubes, 172–81, 331, 333–40, 349–61
electrochemical properties and, 52
electronic absorption effects, 80
fluorofullerenes and, 53
fulleropyrrolidines with 3-pyridyl groups, 194–5
for improved solubility, 127, 191, 301
liquid crystal malonates and, 65, 142
mechanochemical, 24, 28, 34, 331, 350
multisubstituted fullerene-polymer interactions, 167
non-covalent functionalization, 172–82, 350–61
porphyrin-[60]fullerene dyads, 63–4
temporary substituents, 40
tether-directed, 36, 38
via pyrazolinofullerenes, 33
with X-ray contrast agents, 323

genetic diagrams, 67
glucosamine, 335
glutamic acid derivatives, 302–3, 320
glutathione, 71, 305, 311
gold
adsorption of CNT-thiol derivatives, 337
$C_{60}$-porphyrin linking to, 252, 289–92
oligothiophene/$C_{60}$ linking to, 251

pyrene-CNT nanoparticles, 359
self-assembled monolayers on, 269–75
graphite. *See also* carbon
CNTs compared to, 329–30, 338, 357
fullerene production from, 3, 5
grazing incidence X-ray diffraction, 135, 139
griding. *See* mechanochemical functionalization
Grubbs catalyst, 274
guanidinium-carboxylate ion pairs, 164, 196
guanosine, 157–8, 197, 312–16

haemolysis, 307
halogenation of $C_{60}$ and $C_{70}$, 54
α-halomalonates, 16
health effects, 304–7
helical copolymer arrays, 171
helium, 3
heptapeptide-based antigens, 29
heterodimers, 159
heterojunction photovoltaics, 224, 244–9. *See also* bulk heterojunction devices
high speed vibration milling (HSVM), 24, 28, 34
higher fullerenes, 1, 40–3, 72, 236
highest occupied molecular orbital (HOMO), 224–5, 232, 244, 257
HOMO-LUMO gap, 248
HiPco (high presure carbon monoxide) process, 331, 347, 362
HIV (human immunodeficiency virus), 318–19
Hofmann rearrangement, 337
holography, 203
HOMO. *See* highest occupied molecular orbital
HSVM (high speed vibration milling), 24, 28, 34
Hufmann-Krätschmer method. *See* arc-discharge production
hybrid solar cells, 223
hydrazine derivatives (DPPH radical), 310
hydrazones, 32–3, 229
hydrogen, atomic, 346
hydrogen bonding, 153–4, 254, 257
$C_{60}$-based polymers, 166–72

$C_{60}$ donor assemblies, 154–63
multilayer LB films, 135–6
porphyrin-fullerene mixed films, 292–4
supramolecular donor-acceptor systems, 181, 195–7, 254, 257
TTF-fullerene derivatives, 163–6
hydrogen incarceration, 7
hydrogenation of CNTs, 346–8
hydrophilic head groups. *See* polar headgroups
hydroxyl radicals, 310–11
hysteresis curves, 130

I-V curve, photovoltaics, 224, 226
imaging ellipsometry, 357
imidazole, 1-vinyl-, 348
imidazolyl groups, fullerene-porphyrin complexes, 195
immunology, 29, 319–22
impurities and CNT defects, 332
incarcerated atoms. *See* endohedrals
*incar*fullerenes, 6–9
incident photon-current efficiency (IPCE) defined, 226
inclusion complexes, 140–1, 170–1
inert gases, 3, 6–7
infrared reflection absorption (IRRA) spectra, 293
infrared spectrometry, FTIR, 347, 356
interdigitation, 146
intermolecular interactions, Langmuir films, 130
internal quantum yield, 268–72, 274–9, 283, 292
intramolecular Bingel reactions, 17
iodinated malonic acids, 323
IPCE (incident photon-current efficiency), defined, 226
IRRA (infrared reflection absorption) spectra, 293
Isc (short-circuit photocurrent), 224, 228
isolation and purification, CNTs, 329, 333–5, 351
isolation and purification, fullerenes, 4–5
isomers
alkylated fulleropyrrolidines, 54–5
higher fullerenes, 41–2
*incar*fullerenes, 8

regioisomers, *bis*-functionalized
  fullerenes, 35
retro-Bingel reactions, 18–21
separation by chromatography, 1,
  7–8, 36, 41
isostearic acid fullerene derivatives, 311
IUPAC nomenclature, 35

J-aggregates, 361
Jurkat cells, 306, 316

Kiessig fringes, 135, 139

lactose conjugates, 304
Langmuir-Blodgett technique, 130,
  133–7
Langmuir films, 128–39
  amphiphilic fullerenes in, 205–8
  deposition on electrodes, 266–8
  in photovoltaic devices, 252–4
Langmuir-Schäfer techniques, 205, 207
laser ablation process, 331, 361–2
layer-by-layer (LBL) deposition
  carbon nanotubes, 179, 357
  fullerene-porphyrin bilayers, 208,
    253–4
  fullerene-PPE bilayers, 163
  fullerenes onto electrodes, 280–1
LiF/Al electrodes, 231
light-harvesting efficiency, 239, 283
light-induced effects, 80–1, 113–14.
  *See also* photophysical properties
liposomes, 314–15
liquid crystals
  covalent approach, 142–9
  methanofullerenes, 17, 65–6, 142–5
  non-covalent approach, 140–1
local density functional theory, 363
low bandgap polymers, 235, 238–9, 241
lowest unoccupied molecular orbital
  (LUMO), 224–5, 230, 232, 248, 257

macrocycles
  covalently-linked macrocycles, 192–3
  hydrogen-bonded
    metallomacrocycles, 154–8
  inclusion polymers with $C_{60}$, 170
  self-assembled nanostructures, 211
magnetic resonance imaging (MRI), 322
magnetic shielding, 7

malonates
  Bingel reaction with, 16, 60–1, 144
  biological activity of *tris*-adducts, 306,
    319
  *bis*-malonates, 36–7
  crown ethers containing, 69
  iodinated, 323
  liquid crystal, 65, 142, 144
mannopyranosyl-$C_{60}$ conjugates, 303–4
Marcus theory, 64, 84, 104, 112–13, 295
mass spectrometry, 8, 41
MDMO-PPV (poly[2-methoxy-5-(3′,7′-
    dimethyloctyloxy)-1,4-
    phenylenevinylene]), 228–30, 232–4,
    236, 238, 257
  phthalocyanine-$C_{60}$ blends, 242
mechanochemical functionalization, 24,
    28, 34, 331, 350
MEH-OPV5, 245–6
MEH-PPV [poly(2-methoxy-5-(2′-
    ethylhexyloxy)-1,4-
    phenylenevinylene], 228, 230, 239,
    245–6
melamine (1,3,5-triazine-2,4,6-triamine)/
    barbituric acid complexes, 162–4,
    257, 293–4
meso-linked porphyrin complexes,
    110–13
mesomorphic behaviour, 65–6, 140–2,
    144, 146–8
metal clusters, 277
metal complexes. *See also* ferrocene;
    phthalocyanines; porphyrins;
    ruthenium; zinc
  photophysical properties of arrays
    with $C_{60}$, 105–9, 114
metal nitrides, 9
metal-to-ligand-charge-transfer (MLCT)
    states, 5–106, 66, 68, 114
metallic carbon nanotubes, 361–3
metallocenes. *See* ferrocene; ruthenocene
metallomacrocycles, hydrogen-bonded,
    154–8
metalloporphyrins, 63, 65. *See also* zinc
    porphyrins
metals. *See also* copper, gold, ruthenium,
    zinc
  hydroxyfullerene reactions with, 305
  incarceration, 7–9, 322–3
  osmium, 277, 344

rhenium methanofullerene complexes, 105
silver nanocomposites, 204–5
methano adduct removal. See retro-Bingel reactions
(5,6)methanofullerene, 1,1-bis(4,4′-dodecyloxyphenyl)-, (DPM-12), 235–7
(6,6)methanofullerene, 1-(3-methoxycarbonyl)propyl-1-phenyl, 229
methanofullerene-cyclodextrin conjugates, 303–4
methanofullerene-OPV dyads, 88
methanofullerene-phthalocyanine dyads, 154
methanofullerene-porphyrin dyads and triads, 109, 276–7, 282
methanofullerene-TTF dyads, 17
methanofullerenecarboxylic acids, 292–3, 308–10
methanofullerenes. See also bis-methanofullerenes; PCBM
　alkanethiol linkages to gold, 269
　amphiphilic, as optical limiters, 201–3
　barbituric acid derivatives, 162–4, 254, 257, 293–4
　biological activity of malonic tris-adducts, 306, 319
　conjugated polyelectrolyte, 254, 256
　crown ether derivatives, 69–70, 157–9
　dendrimers, 129–39
　double cable polymers with, 243
　drug delivery using, 308
　fulleroid conversion, 58–60
　liquid crystals, 17, 65–6, 142–5
　mesomorphic, 65–6, 142–8
　metal complex arrays, 105–9
　mono- and bis- quenching, 107–8
　self-assembling polymer with calix[5]arene, 172–3
　supramolecular polymers with UP, 166–7
　synthesis, 15–16
　tris-adducts, 38–9, 306, 310
　with acylated glucose units, 131
methano[70]fullerenes, 233, 236–8
Methylene Blue (MB), 259
methylviologen, 251–2, 270–2, 292

micelle formation, 169, 202, 204–405.
　See also vesicles
microscopy. See AFM; BAM; EDX; POM; SEM; TEM
microwave irradiation
　azomethine ylide cycloaddition, 28, 30
　Diels-Alder reactions, 21, 23, 26, 344
　regioselective cycloaddition, 42
milling. See mechanochemical functionalization
MLCT (metal-to-ligand-charge-transfer) excited states, 66, 68, 105–6, 114
molecular area and monomolecular films, 129, 131–2, 135, 144, 205
molecular area and multilayer films, 146
molecular devices
　bio-electrochemical sensors from [60]fullerene, 71–2
　electropolymers from [60]fullerene, 71
　fullerenes and crown ethers, 69
　photoactive dyads, 63–6
　ruthenium bis-bipyridine complexes, 66–9
　trannulenes, 69–71
molecular dynamics simulation, 351
molecular π-donors, 245–9, 259
molecular ribbons, 110
molecular switches, 84, 281, 352
monolayer formation, 128–9, 131, 135, 205–6, 252–3
MPP (1-methyl-4-phenylpyridinium), 310
MPTP (1-methyl-4-phenyl-1,2,3,6-tetrahydropyridine), 310
MRI (magnetic resonance imaging), 322
multiaddition reactions, 35–40
multilayer LB films, 135–6, 139, 146
multilayers at ITO electrodes, 278–80
MWNTs (multi-walled nanotubes), 176, 330, 332–3. See also carbon nanotubes

nanocomposites, 204, 337, 351–2
nanorods, 208–11
nanotechnology
　molecular level data storage, 134
　photoactive liquid crystals, 65–6, 149
　photovoltaic cells, 249–54
nanotubes. See carbon nanotubes
naphthalene

1,5-*bis*-[(2-hydroxyethoxy)ethoxy], 175, 273–4
  dehydro, 4
1,10-dinaphtho[38]crown-10 ether, 17
2,3-naphthalocyanine, *bis*(di-isobutyloctadecylsiloxy)silicon (isoBOSiNc), 315
nematic phase, 142, 144, 203
network field effect transistors, 363
neuroprotective effects, 310–12
nevirapine, 319
nitrenes, 344
nitrile imines, 30–5, 340, 343
nitrogen incarceration, 6
NO synthase, 319
noble gases, 3, 6–7
nomenclature, bis-functionalized fullerenes, 35
non-conducting polymers, 353
non-covalent functionalization, 172–82, 350–61
non-covalent interactions. See also hydrogen bonding; π-π stacking
  in photosynthetic structures, 152
nonlinear optics, 9, 200–5
nucleic acids, 309, 312–16, 335
nucleophilic additions, 16, 41
nucleophilic alkylation of SWNTs, 347
nucleophilic substitutions, 30, 339, 357
numbering system, fullerene structures, 1

oligophenyleneethynylene (OPE)/$C_{60}$ systems, 88, 193, 200, 203, 249
oligophenylenevinylene. *See* OPV
oligosaccharides, 303
oligothienylenevinylene/$C_{60}$ systems, 162–3
oligothiophene/$C_{60}$ systems, 88, 163, 249–51
  anchored to gold electrodes, 271
  complexes with melamine, 164, 254, 257, 294
  complexes with porphyrins, 200
  radical anions from, 193
OPE (oligophenyleneethynylene)/$C_{60}$ systems, 88, 193, 200, 203, 249
optical limiting (OL), 200–5
optoelectronic properties, 152, 158, 182, 245
  carbon nanotubes, 330, 342, 352

OPV (oligophenylenevinylene)
  with *bis*-methanofullerene terminals, 99–105
  light harvesting dendrimers, 239, 241
  MEH-OPV5, 245–6
  molecular π-donors, 245
  OPV-fulleropyrazoline dyads, 84
  OPV-fulleropyrrolidine dyads, 81–4, 86, 89, 148–9, 160
  OPV-methanofullerene dyads, 84, 159–60
  OPV-methanofullerene triads, 161
  TTF-fullerene triads, 200
osmium, 277, 344
OTS (octadecyltrichlorosilane), 135
oxidation, 72–3, 332–3, 346
oxidation potentials, 83
oxidation states, 7
oxidative damage and toxicity, 305–6
oxygen. *See also* singlet oxygen
  benzene combustion method, 4
  fullerene degradation and, 5
  quenching in dissolved fullerodendrimers, 93, 96, 114
  reactive species, 305–6, 311, 315
ozone cycloaddition, 340

π- extended tetrathiafulvalenes (*ex*-TTFs), 193, 200, 247
π- orbital misalignment in CNTs, 330–1
π-π interactions
  CNT polymer wrapping, 351–7
  CNTs with molecular building blocks, 357–61
  porphyrin-fullerene, 286, 289
  SWNT/porphyrin, 360
  SWNT/pyrene, 180–1, 358–9
π-π stacking
  click complexation, 161–2
  porphyrin complexes, 157, 172, 197–8, 211
  SWNTs, 174–5, 354, 359
*p*-anisidine, N,N-dimethyl-, 285
PA-FTIR (photoacoustic Fourier transform infrared spectrometry), 356
paclitaxel, 308
PAHs (polycyclic aromatic hydrocarbons), 4, 16–17, 357–61

PAMAM (polyamidoamine dendrimer)-$C_{60}$ multilayers, 279–80
PAmPV [poly(5-alkoxy-*m*-phenylenevinylene)
-*co*-(2,5-dioctoxy-*p*-phenylene)vinylene], 352
 tethered derivatives as pseudorotaxanes, 174–5
PAT [poly(alkylthiophenes)], 228
PCBM [[6,6]-phenyl–C61–butyric acid methyl ester], 229–36, 238–9, 242, 257–9
 competitors, 239–40
 [70]fullerene version, 233
PCE (power conversion efficiency) defined, 225–6
PDDA [poly(diallyldimethylammonium)], 280–2, 354, 356–7
PDT (photodynamic therapy), 314–16
'peapods,' 74, 197–8
PEDOT-PSS [poly(3,4-ethylenedioxythiophene) poly(styrenesulfonate)], 228, 231, 236, 238, 244, 249
pentacene, 246
pentagon-heptagon pairs in CNTs, 332
pentathienylmelamine/$C_{60}$ systems, 163–4
peptides
 amino acid-containing fullerenes and, 301–2
 CNT linking, 335
 heptapeptide-based antigens, 29
 immunological properties, 319–22
 porphyrin-peptide oligomers, 289
periconjugation, 59
peripheral fullerenes. *See* fullerodendrimers, fullerene outside
perylenediimides, 239, 294
3,4,9,10-perylenetetracarboxylic acid *bis*-benzimidazole (PTCBI), 224, 244–5, 257
phase segregation
 carbon nanotubes, 179
 "double-cable" polymers and, 242
 molecular π-donors and, 245–9
 polymer-fullerene bulk heterojunctions, 234

1,10-phenanthroline, 2,9-dimethyl-4,7-diphenyl- (BcP), 244–6
phenanthroline complexes with $C_{60}$, 105–6, 195
phenothiazines, 68–9, 285
phenyl groups, 53, 55–9
phenylalanine derivatives, 302–3
phenyleneethynylene derivatives. *See* OPE, PPE
phenylenevinylene derivatives. *See* poly(phenylenelvinylenes)
photo-induced charge separation, 64–5, 191–2
photo-induced oxygen production, 51
photodynamic therapy (PDT), 314–16
photophysical properties
 [60]fullerene and its derivatives, 80–1, 221
 fullerene-porphyrin assemblies, 109–15, 157
 fullerodendrimers, 88–105
 metal complex arrays with [60]fullerenes, 105–9
 oligophenylenevinylene dyads, 81–9
photosensitization, 81, 312–13, 315–16
photosynthesis, 152. *See also* artificial photosynthesis
phototoxicity, 306
photovoltaic devices
 alternating donor and acceptor layers, 256
 carbon nanotubes in, 172, 181
 degradation mechanisms, 258
 double cable and supramolecular approaches, 86
 efficiency, 181, 222–3, 257–8
 fullerene-based, 114–15, 224–6
 fullerene modified electrodes, 266–95
 fulleropyrrolidine-OPV dyads, 81–2, 86, 149
 heterojunction approaches, 226–43, 246
 higher fullerenes, 42
 history of, 222–4, 258
 hydrogen bonded metallomacrocycles, 163
 LBL deposition, 163, 256
 parametric characterization of, 224–6
 predicted market for, 223

Subject Index

supramolecular nanostructures, 249–54
trannulenes and, 69
phthalocyanines
  bis(di-isobutyloctadecylsiloxy)silicon 2,3-naphthalocyanine (isoBOSiNc), 315
  diphenoxy-substituted fullerene complexes, 195
  fulleropyrrolidine dyads, 192, 242
  hydrogen bonded, 154
  linked to CNTs, 335
  optical limiting dispersions, 203, 206
  organic solar cells from, 224, 244
  subphthalocyanines, 28, 195
  vacuum deposition, 283
phytochlorin–fullerene dyads, 267–8
1-picryhydrazyl radical, 2,2-diphenyl-1- (DPPH), 310
plastic solar cells. See photovoltaic devices
P3MeT [poly(3-methylthiophene)], 353
PmPV [poly($m$-phenylenevinylene)] [-co-DOctOPV]-co-[2,5-dioctoxy-$p$-phenylene)-vinylene], 351–3
  CNT wrapping by, 174
polar headgroups
  amphiphilic fullerenes, 205
  carbon nanotubes, 180
  fullerodendrimer thin films, 128, 132–4, 137–9
polarized optical microscopy (POM), 140–1
polyadducts, 39–40, 43
poly(5-alkoxy-$m$-phenylenevinylene) (PAmPV)
  -co-(2,5-dioctoxy-$p$-phenylene)vinylene, 352
poly(alkylthiophenes) (PAT), 228, 230
poly(allylamine), 356
polyaniline (PANI), 167, 176–7, 352–3
polyarginine uptake, 308–9
poly(aryleneethynylene) polymers, 354
polycyclic aromatic hydrocarbons, 4, 16–17, 357–61
poly(diallyldimethylammonium) (PDDA), 280–2, 354, 356–7
poly[N-dodecyl-2-thienyl-5-(2'-(5-bis(2'-(2,1,3-benzothiadiazole)thienyl) pyrrole](PTPTB), 235

polyelectrolytes, 354, 356–7
poly(3,4-ethylenedioxythiophene) poly(styrenesulfonate) (PEDOT-PSS), 228
poly(3-hexylthiophene) (P3HT), 228, 230–4, 238, 242, 257–8
polymer wrapping of CNTs, 174–5, 351–7. See also DNA wrapping; [60]fullerene, encapsulation; fullerodendrimers
polymers
  carbon nanotubes and, 174–6, 340, 342, 351–7
  conducting, 71, 167, 176, 351–2
  double-cable, 242–3
  hydrogen bonded $C_{60}$-based, 166–72
  macrocycle-$C_{60}$ inclusion polymers, 170
  non-conducting, 353
  polymer-$C_{60}$ heterojunction devices, 226–43
  pyrene-linked, 355
polymers, conjugated. See conjugated polymers
poly[2-methoxy-5-(3',7'-dimethyloctyloxy)-1,4-phenylenevinylene] (MDMO-PPV), 229–30
poly(2-methoxy-5-(2'-ethylhexyloxy)-1,4-phenylenevinylene (MEH-PPV), 228, 230, 239, 245–6
polymethylmethacrylate (PMMA), 176, 178, 355
poly(3-methylthiophene) (P3MeT), 353
poly[3-(4-octylphenyl)thiophene] (POPT), 229
poly(3-octylthiophene) (P3OT), 232, 259
poly(phenylacetylene) (PPA), 352
polyphenylated [60]fullerene $C_{60}Ph_5Cl$, 57–8
poly($p$-phenyleneethynylene) (PPE), 169, 243, 350–1, 353
poly($m$-phenylenevinylene) (PmPV) [-co-DOctOPV]-co-[2,5-dioctoxy-$p$-phenylene)-vinylene], 351–3
  CNT wrapping by, 174
poly($p$-phenylenevinylene) (PPV)
  composite with CNTs, 354–5
  fullerene blends, 81
  hybrids with poly$p$-phenyleneethynylene, 243
  uracil-PPV, 168

poly-*p*-phenylenevinylenecarbazole, 168
poly(pyrenebutyric acid) (PPBA), 355
poly(2,6-pyridinylenevinylene) (PPyPV)
  -*co*-(2,5-dioctoxy-*p*-
  phenylene)vinylene, 352–3
polypyrrole (PPY), 352–3
polystyrene attachment to CNTs, 347
polystyrenesulfonate (PSS), 175, 178,
  280, 354, 356–7
polythienylenevinylene (PTV), 235
poly(vinylidene fluoride) (PVDF), 355,
  361
poly(1-vinylimidazole), 168
poly(4-vinylpyridine) (P4VPy)
  complexes, 168–9, 356
polyvinylpyrrolidine (PVP), 175, 354
POM (polarized optical microscopy),
  140–1
porphines. *See* porphyrins
porphyrin dimers, 110–14, 270–1
porphyrin-[60]fullerene arrays, 109–13,
  286
porphyrin-[60]fullerene dyads
  deposition of, on electrodes, 252–3,
    275, 280–1, 284–5
  donor-acceptor complexes containing,
    63–4, 192
  hydrogen bonded, 155–6
  Langmuir-Blodgett films, 267–8
  phototoxicity, 316
  self-assembled monolayers, 270–1,
    276–7
  self-organized clusters on gold
    nanoparticles, 289–90
porphyrin-[60]fullerene-porphyrin
  triads, 25, 253
"porphyrin jaws," 197–9
porphyrin-peptide oligomers, 289
porphyrins, 17, 25. *See also* zinc
  porphyrins
  $C_{60}$ inclusion polymers with, 170
  chromophoric, metalloporphyrins
    and, 63
  donor-acceptor polyads, 198
  fullerene inclusion complexes with,
    140–1
  fullerene mutual attraction with, 109,
    156, 197
  fulleropyrrolidine axial complexation,
    193–5, 206–7

hydrogen bonding *bis*-porphyrins,
  172, 174, 198
linked to CNTs, 335–6, 360
multiporphyrin photoactive arrays,
  274, 287–92
polymerized, self-assembled
  multilayers, 278–9
polymerized, SWNT association with,
  176, 355–6
self-assembled monolayers with, 271–
  4, 277–8
self-assembled nanostructures with,
  210–11
separation of CNTs using, 362
surface immobilized on PMMA, 176,
  178
tetraformyl-, 355
potential uses. *See* applications
power conversion efficiency (PCE)
  defined, 225–6
PPA [poly(phenylacetylene)], 352
PPBA [poly(pyrenebutyric acid)], 355
PPE (poly*p*-phenyleneethynylene), 169,
  243, 350–1, 353
PPV (poly*p*-phenylenevinylene), 81, 168,
  354–5
PPY (polypyrrole), 352–3
PPyPV [poly(2,6-pyridinylenevinylene)-
  *co*-(2,5-dioctoxy-*p*-
  phenylene)vinylene], 352–3
Prato's reaction, 26–8, 38, 341
pressure-area relationships, 129, 131–3,
  135–6
production methods, 3–4, 331, 361
protein immobilization on CNTs, 357–8
pseudorotaxanes
  ammonium-crown ether complexes,
    154, 196
  cyclophane-napthalene derivatives,
    273–4
  polymeric, 174–5
  TTF derivatives, 165–6
PSS (polystyrenesulfonate), 175, 178,
  280, 354, 356–7
PTCBI [3,4,9,10-perylenetetracarboxylic
  acid *bis*-benzimidazole], 224, 244–5,
  257
P3HT [poly(3-hexylthiophene)], 230–1,
  242, 257, 259
PTV (poly-thienylenevinylene), 235

purification, CNTs, 329, 342, 351
  oxidative, 332–3
  separation of metallic and
    semiconducting, 361–3
purification of fullerenes, 4–5. See also
  chromatography
PVDF [poly(vinylidene fluoride)], 355,
  361
PVP (polyvinylpyrrolidine), 175, 354
PyMMP [(1-pyrene)methyl 2-methyl-2-
  propenoate], 355
pyramidalization strain, CNTs, 330–1
pyrazoline-functionalized CNTs, 343
pyrazolinofullerenes. See
  fulleropyrazolines
pyrene derivatives
  diamino pyrenes, 355
  protein immobilization, 357–8
  soluble nanotubes and, 179–81, 358–9,
    361
pyrene-fulleropyrrolidines, 74
pyrene-linked polymers, 355
(1-pyrene)methyl 2-methyl-2-propenoate
  (PyMMP), 355
pyridazinofullerenes, 26
pyridine derivatives
  bipyridine, 66, 254–255
  2,6-diacylamino, 168
  4,4′-dipyridinium, 1,1′-dihexyl-,
    diperchlorate, 275
  1-methyl-4-phenyl-1,2,3,6-tetrahydro
    (MPTP), 310
  pyridinium, 1-methyl-4-phenyl-
    (MPP), 310
  pyridyl functionalization of fullerenes,
    194–5
pyrimidine ortho-quinodimethanes, 302
4[1H]pyrimidinone, 2-ureido- (UP), 88,
  153
  OPV heterodimers, 159–60
  supramolecular polymers of, 166–7
pyrolytic condensation of PAHs, 4
pyropheophorbide a, 316
pyrrole, 2,5-dithienyl, $C_{60}$ tethered, 274
pyrrolidine-functionalized CNTs, 342–3
pyrrolidinofullerenes. See
  fulleropyrrolidines

Q-bands, 110, 112, 176, 356, 358
quantum dots, 337

quenching
  electron *versus* energy transfer, 83, 107
  gold and ITO electrodes, 275, 279
  oligothienylenevinylene/$C_{60}$ systems,
    162
  photo-induced charge separation,
    64–5, 159
  triplet state quenching by, 94, 97–8
o-quinodimethanes, 24–6, 40, 42, 302,
  344
quinones, 63–4
quinoxalines, 24

radical additions, 41, 348–9
radical cations, 72, 110–11, 162, 165, 274
radical DPPH, 310
radical ion pairs, 154–5, 157, 164, 178–9,
  193, 358
radical scavengers, 310–12
radio-labelled fullerenes, 322–4
Raman spectroscopy, 208, 331, 351–2,
  362–3
reaction kinetics, retro-
  cyclopropanation, 62
reactions, [60]fullerene. See [60]fullerene
  chemistry
reactive oxygen species (ROS), 305–6,
  311, 315
redox process localization, 67
redox properties of fullerenes, 52, 57, 73
reduction potentials
  aryl-fullerenes, 55–7
  fluoro[60]fullerenes and $C_{60}$, 53
  [60]fullerene, 30–1
  fulleropyrrolidine/OPV dyads, 84
regioisomers, *bis*-functionalized
  fullerenes, 35
regioselective synthesis, 36–8, 40, 42, 60
  retro-Bingel reactions, 18–21, 43,
    60–2, 69
retro-cycloaddition, 30
retro-cyclopropanation. See retro-Bingel
  reactions
retro-Diels-Alder reactions, 22–3
reverse saturable absorption (RSA), 200
rhenium methanofullerene complexes,
  105
rigid anchors, 271
rigid linkages, 36–7, 39, 103–4, 193–3,
  252–3

rigid polymers, 354
rimantadine, 318
ring-closing methathesis, 274
ring-opening methathesis, 355
ROS (reactive oxygen species), 305–6, 311, 315
rotaxanes. *See also* pseudorotaxanes
   crown ether-ammonium complexes, 165–6, 196
   cyclodextrine-based polyrotaxanes, 171
   1,10-dinaphtho[38]crown-10 ether containing, 17
   phenanthroline based, 106, 195–6
   porphyrin-based, 155
   ruthenium bis-bipyridine complexes, 66–9, 105, 154, 254–5
ruthenium dyes, 281, 283
ruthenium tetraphenylporphyrin (RuCOTPP), 194
ruthenocene derivatives, 28, 193

SAMs. *See* self-assembled monolayers
scandium incarceration, 8
scanning electron microscopy (SEM), 172, 209, 331, 363
scanning tunneling microscopy (STM), 331
selenium, in photovoltaic devices, 222
self-assembled monolayers (SAMs), 29, 249–52
   on gold electrodes, 269–74
   on ITO electrodes, 275–80
   SWNTs on, 335–6
self-assembled nanostructures, 208–11, 303, 354
self-assembly copolymers, 172, 173
self-organization, 140–1, 208–11, 249, 259, 287, 289
SEM (scanning electron microscopy), 192, 209, 331, 363
semiconducting carbon nanotubes, 350, 361–3
semiconductors in solar cells, 222–3
sensitized singlet oxygen luminescence, 92–4, 96–8, 101–4
sexithiophene (6T), 245
short-cicuit photocurrent (Isc), 224, 228
silane, octadecyltrichloro- (OTS), 135

silicon 2,3-naphthalocyanine, *bis*(di-isobutyloctadecylsiloxy)- (isoBOSiNc), 315
silicon-based solar cells, 222–3
siloxane substituents, 201
silver nanocomposites, 204–5
single-walled nanotubes. *See* SWNT
singlet excited state, fullerenes, 80–1, 108
singlet oxygen
   cycloaddition to CNTs, 346
   photophysical effects, 80, 93, 98–9
singlet oxygen sensitization, 80–1, 90, 102, 312
6T (sexithiophene), 245
smectic phase, 142, 144, 146, 148
smectite clays, 208–9
sol-gel glasses, 201
solar cells. *See* photovoltaic devices
solubility
   carbon nanotubes, 179–80, 331, 335, 342, 354
   fluorinated CNTs, 338
   [60]fullerene, 15
   fullerodendrimers, 22–3
   functionalization to improve, 52, 127, 172–4, 191, 205, 301
   water-solubility, 22–3, 29–30, 205, 357
solvent effects
   organic photovoltaics, 230
   triplet state quenching by, 94, 97–8
solvent polarity effects
   energy and electron transfer processes, 113–14
   fullerene inside dendrimers, 92–9
   fullerene outside dendrimers, 99, 104–5
   OPV/fulleropyrazoline dyads, 84, 86
   polymer-wrapped nanotubes, 176
   porphyrin-methanofullerene arrays, 110, 112
solvent removal, 4
sonication, CNTs, 331, 333, 335
Soret bands, 110, 112, 176, 356, 358
spacer flexibility effects, 67, 164, 192
spectroelectrochemistry, 67
spectroscopy. *See* absorption; EXAFS; infrared; mass spectrometry; Raman; surface photovoltage; UV-Vis; X-ray
spin coating, 254, 292–5

Subject Index

stereochemistry. See chirality; isomers
STM (scanning tunnelling microscopy), 331
Stone–Wales defects, 332
structure-activity relationships
   $C_{60}$-tethered alkanethiols, 269
   fulleropyrrolidine/OPV photovoltaics, 83, 114
   π-conjugated oligomers, 248–9
   PCBM analogues, 239
subphthalocyanines, 28, 195
substitution reactions, CNTs, 349–350
sugars, 131, 303–4, 315–16, 354
sulfur contamination, 4
superoxide scavenging, 310
supramolecular chemistry, 79, 105, 110, 153, 191
   assembly of donor-acceptor systems, 193–8
supramolecular materials, 140–9, 287
supramolecular organization
   fullerene-porphyrins on gold, 289–90
   hydrogen bonding in, 181, 195–7, 254, 257
   photosynthesis as, 152
   photovoltaic cells, 86, 159, 249–54
   sugar-fullerene conjugates, 303
supramolecular polymers, 154, 166–7, 172
surface photovoltage spectroscopy, 268
surfactants, CNT-polymer matrices, 355
SWNTs (single-walled nanotubes), 74
   oxidation of MWNTs and, 332–3
   in photovoltaic devices, 172, 259
syn- and anti-conformations, 161

TA. See transient absorption spectroscopy
TEM (transmission electron micrography)
   carbon nanotubes, 331, 355, 357–9
   dispersed nanorods, 209–10
   inclusion polymers, 171–2
   polymer/SWNT associations, 176, 178
TEMPO fulleroid, 60
tether-controlled reactions, 36–9
tethered conjugated oligomer-C60 dyads, 158–63
tethered PAmPV pseudorotaxanes, 174–5

tetracene-[60]fullerene dyads, 17
tetraphenylmethane, 271
tetrathiafulvalene. See TTF
TGA-FTIR (thermogravimetric analysis - Fourier transform infrared spectrometry), 347
thermotropic liquid crystals, 140–9
thin films. See also Langmuir films
   organic heterojunction solar cells, 226
thiolation of CNTs, 337–9
thiophenes
   electropolymerizable, 71–2
   3,4-ethylenedioxy-, (EDOT), 169–70
   functionalized SWNTs, 339
   poly[3,4-ethylenedioxy- (See PEDOT)
   poly(3-hexylthiophene) (P3HT), 230
   poly(alkylthiophenes) (PAT), 228
   sexithiophene (6T), 245
"three-point" bonding, 168
titanium dioxide, 223, 280, 283–5, 287
toluene
   fullerene-porphyin absorption and luminescence spectra in, 111–12
   fullerodendrimer absorption spectra in, 90, 92, 94, 96, 100–1
   fullerodendrimer fluorescence spectra in, 100–1
p-toluidine, N,N-dimethyl-, 285
toxicity of carbon nanoparticles, 304–7
trannulenes, 69–71
transfer ratio (TR), films to solids, 135, 139
transient absorption (TA) spectroscopy
   [60]fullerene, 80
   fullerene inside dendrimers, 92, 96
   oligothienylenevinylene/$C_{60}$ systems, 162
   porphyrin-fullerene rotaxanes, 156–7
   porphyrin-fulleropyrrolidine films, 206–7
   porphyrin-methanofullerene arrays, 110–11
1,3,5-triazine-2,4,6-triamine (melamine), 162–4, 257, 293–4
triphenylamine (TPA)-$C_{60}$ complexes, 165–6
triphenylamine, 4,4',4''-tris(3-methylphenylphenylamino)-(m-MTDATA), 245–6
triplet excited state, 80–1

triplet lifetimes, 92–8, 102
triply fused porphyrin dimers, 110–13
*tris*-methanofullerene adducts, 38–9, 306, 310
TTF (tetrathiafulvalene), 245
  [60]fullerene derivatives, 17, 22–3, 193, 196, 202–5, 244
  fullerene-porphyrin triads with, 272–3
  fused with PTV polymers, 235
  hydrogen bonded fullerene derivatives, 163–6
  multiple donor-acceptor systems, 199–200
  trannulene dyads, 70
"two-point" binding, 156–7

ultrasoft pseudopotential plane-wave method, 363
ultrasonication, CNTs, 331, 333, 335
UP (2-ureido-4[1H]pyrimidinone), 88, 153
  OPV heterodimers, 159–60
  supramolecular polymers of, 166–7
uracil-PPV, 168
UV fullerene degradation, 3
UV-Visible spectra. *See also* absorption spectra
  of LB films, 130, 133, 139, 207

vacuum deposition, 281–3, 361
vacuum evaporation, 254
van der Waals forces, 170, 174, 280–1
  CNT aggregation, 172, 352
  complexing oppositely charged electron donors, 179–81, 358

vanadyl CNT complexes, 335–6
vesicles, 208, 283, 303. *See also* micelle formation
vibration milling. *See* mechanochemical functionalization
virus blocking, 51, 317–19

water-air interface, monolayers, 128, 136–7, 205–6, 252–3
water-solubility, 22–3, 29–30, 205, 357
Watson-Crick hydrogen bonded pairs, 157–8, 197
wrapping. *See* DNA wrapping; [60]fullerene, encapsulation; fullerodendrimers; polymer wrapping

X-ray adsorption fine structure (EXAFS), 331
X-ray contrast agents, 323
X-ray diffraction studies, 208, 331
  dendrimers, 140–1, 144
  grazing incidence, LB films, 135, 139
X-ray photoelectron spectroscopy (XPS), 331, 352, 356
X-ray spectroscopy, energy-dispersive (EDX), 333

zinc phthalocyanine, 154, 210–11
zinc porphyrins, 180–1, 293, 316–17
  charge separation in, 155, 198
  CNT complexes with, 358
  electrode deposition, 271, 275
  zinc tetraphenylporphyrin (ZnTPP), 194, 359